ANNALS OF
THE NEW YORK ACADEMY
OF SCIENCES

Volume 407

EDITORIAL STAFF
Executive Editor
BILL BOLAND
Managing Editor
JOYCE HITCHCOCK
Associate Editor
SHEILA KATES

The New York Academy of Sciences
2 East 63rd Street
New York, New York 10021

CELLULAR SYSTEMS FOR TOXICITY TESTING

ANNALS OF THE NEW YORK ACADEMY OF SCIENCES
Volume 407

CELLULAR SYSTEMS FOR TOXICITY TESTING

Edited by G. M. Williams, V. C. Dunkel, and V. A. Ray

The New York Academy of Sciences
New York, New York
1983

Library of Congress Cataloging in Publication Data

Main entry under title:

Cellular systems for toxicity testing.

   (Annals of the New York Academy of Sciences ; v. 407)
   Bibliography: p.
   Includes index.
   1. Toxicity testing—Congresses. 2. Cell culture—
Congresses. I. Williams, G. M. (Gary Murray), 1940–
II. Dunkel, V. C. (Virginia C.), 1934–
III. Ray, V. A., 1929–   IV. Series. [DNLM:
1. Mutagenicity tests—Congresses. 2. Cytological
techniques—Congresses. W1 AN626YL v.407 / QY 95 C393 1982]
Q11.N5   vol. 407   [RA1221]   500s   [615.9'07]   83-8346
ISBN 0-89766-206-7
ISBN 0-89766-207-5 (pbk.)

SP
Printed in the United States of America
ISBN 0-89766-206-7 (Cloth)
ISBN 0-89766-207-5 (Paper)

ANNALS OF THE NEW YORK ACADEMY OF SCIENCES
VOLUME 407
June 16, 1983

CELLULAR SYSTEMS FOR TOXICITY TESTING*

Editors and Conference Organizers
G. M. WILLIAMS, V. C. DUNKEL, AND V. A. RAY

———————◆———————

CONTENTS

*This volume is the result of a conference entitled Cellular Systems for Toxicity Testing, held
on October 5–7, 1982, by The New York Academy of Sciences.

Financial assistance was received from:

- AB ASTRA
- AIR FORCE OFFICE OF SCIENTIFIC RESEARCH
- AMERICAN CYANAMID COMPANY
- BURROUGHS WELLCOME CO.
- CIBA-GEIGY
- THE COCA-COLA COMPANY
- THE DOW CHEMICAL COMPANY
- E. I. DU PONT DE NEMOURS & COMPANY
- HOFFMANN-LA ROCHE INC.
- ICI AMERICAS INC.
- INTERNATIONAL BUSINESS MACHINES CORPORATION
- INTERNATIONAL MINERALS & CHEMICAL CORPORATION
- JOHNSON & JOHNSON PRODUCTS, INC.
- LILLY RESEARCH LABORATORIES
- MCNEIL PHARMACEUTICAL
- THE MILLENNIUM GUILD, INC.
- MOBIL OIL CORPORATION
- MONSANTO FUND
- ORTHO PHARMACEUTICAL CORPORATION
- PFIZER PHARMACEUTICALS / PFIZER LABORATORIES, ROERIG AND PFIPHARMECS DIVISIONS
- PHARMAKON RESEARCH INTERNATIONAL, INC.
- THE PROCTER & GAMBLE COMPANY
- REVLON HEALTH CARE GROUP / RESEARCH & DEVELOPMENT DIVISION
- R. J. REYNOLDS TOBACCO COMPANY
- SEARLE RESEARCH AND DEVELOPMENT / DIVISION OF G. D. SEARLE & CO.
- SHELL OIL COMPANY
- TEXACO INC.
- VELSICOL CHEMICAL CORPORATION

INTRODUCTION

Gary M. Williams

Naylor Dana Institute for Disease Prevention
American Health Foundation
Valhalla, New York 10595

The science of toxicology is the understanding of the mechanisms by which exogenous agents produce deleterious effects in biological systems. The actions of toxins are ultimately exerted at the cellular level. Isolated cell systems, therefore, represent important models for toxicologic studies.

A variety of prokaryotic and eukaryotic cell systems are now available for the study of different toxic effects, including general cytotoxicity, genotoxicity, mutagenesis, and carcinogenesis. This conference was an attempt to examine in detail the most advanced systems for the study of these effects.

An important component of toxicology is the identification of hazardous substances prior to their introduction into the environment. Thus, cell systems, because of their convenience and the fact that testing can be precisely controlled according to defined protocols, play an important role in the assessment of potential toxins. Moreover, at a time when chemical testing has reached gigantic proportions, cell systems offer one means of reducing the use of experimental animals.

It was clear from the broad support for this conference and the large participation in it that cell toxicology has arrived as a discipline and as a useful methodology. The hope of the organizers is that the conference and its proceedings will further contribute to advancement in these areas.

NOTES ON XENOBIOTIC METABOLISM

Silvio Garattini

Mario Negri Institute for Pharmacological Research
20157 Milan, Italy

The various types of chemicals—drugs, contaminants, food additives, natural products—to which we are exposed throughout our life span are frequently absorbed, distributed, and metabolized before they are eliminated. Therefore any single chemical may generate a large number of other chemical species in the body. Some of these may be irrelevant from the toxicological point of view, and others may be less or more toxic than the parent compound. Under certain conditions a chemical ingested may never even reach the bloodstream because it is rapidly and completely biotransformed by the intestinal flora. This makes *in vitro* toxicological studies extremely complicated, particularly when the findings are required for extrapolation to man in terms of toxicological risk.

It is impossible to summarize the enormous amount of information available on xenobiotic metabolism, and therefore these notes will only touch on some aspects of the matter, taking most of the examples from data generated by work done at this institute.

First, I shall show why the metabolism of any given chemical cannot be described in general terms because important qualitative and quantitative differences arise in relation to the conditions under which the chemical is studied. The numerous factors modulating xenobiotic metabolism always require continuous attention. Second, I shall discuss the roles of several barriers in cellular metabolic events; in some cases these barriers divide the metabolism into "compartments." Third, I shall look at the quantitative difficulties of deciding the significance of the concentration utilized *in vitro* in relation to the *in vivo* situation. Finally, we will touch on today's thorniest toxicological research problem, i.e., the fact that studies are necessarily carried out in controlled conditions, using only one chemical at a time, but man is always exposed at the same time or in succession to a mixture of chemicals, with all the opportunities for interaction between them.

FACTORS AFFECTING THE KINETICS AND METABOLISM OF XENOBIOTICS

The aim of animal studies in the field of toxicology is always to collect data that may be useful in predicting whether a chemical is likely to be harmful when applied to man. Usually several animal species are used and the chemical is given at increasing doses for different lengths of time until toxic effects are observed. The minimal dose inducing an appreciable toxic effect in the most sensitive animal species is then compared with the maximal dose likely to be given to man. The ratio thus obtained is considered a safety margin: the higher the ratio, the lower the risk the chemical is likely to induce in man.

This straightforward "classic" approach is unfortunately no longer tenable because comparing doses means comparing the amounts of a chemical outside the body. What appears more logical is to compare the amounts of the chemical present inside the body, i.e., in the blood or target organs of animals and man. This would involve comparing the blood or tissue concentration of a chemical inducing a noteworthy toxic

1

0077–8923/83/0407–0001 $01.75/0 © 1983, NYAS

effect in the most sensitive animal species with the concentration reached at the same site in man when the chemical is given at the maximal recommended dose.

This approach is dictated by the increasing knowledge that there may be no direct relation across animal species between the administered dose and the concentrations reached in various tissues. In fact chemicals are known to be absorbed, distributed, metabolized, and excreted in a manner that is frequently characteristic for each animal species. Therefore the same dose when given to different animal species may result in different blood and tissue concentrations. The question is further complicated by the fact that sometimes the toxic effects do not derive from the administered compound but from other chemical species (metabolites) formed by the living organism, as shown for carcinogenesis by the pioneer work of Miller and Miller.[1] Today we can in fact distinguish chemicals that may act directly as carcinogens, such as diepoxybutane, N-acetylimidazole, and sarcolysine, and chemicals that become carcinogens only after biotransformation, such as benzo[a]pyrene, vinyl chloride, and dimethylnitrosamine.[2]

All the processes that are important for the disposal of a chemical, including metabolism, may be drastically affected or modulated by a large number of endogenous and exogenous factors, summarized in TABLE 1. It is impossible here to discuss all these factors in detail. They have been covered in several reviews (see for instance

TABLE 1

FACTORS MODULATING XENOBIOTIC METABOLISM

The metabolism of a chemical can be defined only under a given set of conditions, which include among others:

Animal species	
Animal strain	
Sex	endogenously controlled
Age	
Pathology	
Route of administration	
Dose	
Diet	environmental influence
Presence of other chemicals	

References 3–6). Examples have been selected with particular reference to their influence on xenobiotic metabolism. It is essential to appreciate the complexity of this question in any attempt to extrapolate toxic effects from cellular systems to intact organisms and even more from animals to man.[7]

The *animal species* may be a critical factor in the metabolism of a given chemical because certain metabolic routes exist in some species but not in others.[8,9] In most cases the difference is quantitative rather than qualitative, because it is very difficult to exclude the formation of a metabolite without extensive work and sensitive enough analytical methods. An example of species differences is presented in TABLE 2, which shows the amount of dimethylxanthines found in blood after administration of caffeine. Caffeine can be N-demethylated in position 1, 3, or 7, giving rise respectively to theobromine (3,7 dimethylxanthine), paraxanthine (1,7 dimethylxanthine), or theophylline (1,3 dimethylxanthine). The dimethylxanthines found in the blood depend not only on the N-demethylation but also on the further metabolism of the dimethylxanthines and on the presence of other competing metabolic routes. The table shows that each animal species has a typical profile of dimethylxanthines in the blood, and it is impossible to find a species with a profile similar to man's.

TABLE 2

SPECIES DIFFERENCES IN THE METABOLISM OF CAFFEINE, ASSESSED BY MEASURING
PLASMA LEVELS OF DIMETHYLXANTHINES AFTER AN ORAL DOSE (10 MG/KG)
OF CAFFEINE SOLUTION*

Animal Species	Caffeine	Theophylline	Theobromine	Paraxanthine
Mouse	1022	36	479	205
Rat	2098	192	272	ND†
Rabbit	1000	242	116	1737
M. cynomolgus	4645	6487	747	417
Man‡	4706	301	424	1757

*Figures represent plasma AUC (μg/ml per minute).
†ND means not determined.
‡Caffeine given at 5 mg/kg.

Another metabolic route in the metabolism of caffeine is the opening of the purine ring with the formation of a uracil [1,3,7 trimethyl, 5 formylamino, 6 aminouracil (1,3,7 DAU)], a metabolite that is rapidly excreted in the urine. As indicated in TABLE 3, in rats this metabolite is found as a relatively large percentage of the administered dose, whereas only traces are present in mice and low levels in other animal species including man.[10] 1,3,7 DAU displays a mild cytotoxic activity in L 1210 leukemia-bearing mice,[11] so species differences in the formation and excretion of 1,3,7 DAU could be relevant in explaining the specific toxicological effects of caffeine.

Although species differences in xenobiotic metabolism are frequent, no generalizations can be made. There are for instance compounds, such as saccharine, that are not metabolized in several animal species including man.[12]

Within the same animal species, different *strains* have different capacity to metabolize xenobiotics. Most of the data available refer to mouse strains, but it is well known that genetic characters influence xenobiotic metabolism in man too. The classic

TABLE 3

1,3,7 DAU IN 48-HOUR URINE AS A PERCENTAGE OF THE ADMINISTERED
DOSE OF CAFFEINE (10 MG/KG ORALLY)*

CAFFEINE → 1,3,7 DAU

Animal Species	Percent of Administered Dose in Urine
Mouse	Traces
Rat	17.3
Guinea pig	4.9
Rabbit	2.1
Monkey	0.7
Man†	2.3

*Data from Reference 10.
†Caffeine given at 5 mg/kg orally.

experiments of Vesell have in fact shown that homozygotic twins are identical in metabolizing reference drugs while heterozygotic twins are quite different.[13] Individual differences in how patients react to drugs are a well-known situation that requires no additional comment.

Sex is another important factor in xenobiotic metabolism; major differences in xenobiotic toxicity have been observed depending on the sex of the species used in experimental studies. For instance hexachlorobenzene (HCB)–induced porphyria in rats is much slower in males than in females.[14,15] As reported in TABLE 4, this toxicological difference is associated with an increased formation of an HCB metabolite, pentachlorothiophenol, which is found in higher amounts in the liver and urine of female than of male rats.[16,17] Pentachlorothiophenol is the expression of the hydrolysis of pentachlorophenyl-*N*-acetylcysteine,[18] which in turn is derived *in vivo* from dechlorination of HCB and subsequent conjugation with glutathione. It is important to recall that HCB produces a high incidence of liver tumors in female rats.[19]

The xenobiotic metabolism matures, and therefore *age* is a variable that must be taken into account. Usually during the fetal and neonatal period, the xenobiotic

TABLE 4

SEX DIFFERENCES IN PORPHYRIA INDUCED BY HEXACHLOROBENZENE (HCB) AND IN THE METABOLISM OF HCB*

	Male	Female
Liver porphyrins†	0.7 ± 0.1	214 ± 38‡
Hexachlorobenzene	335 ± 20	339 ± 10
Pentachlorophenol	7 ± 0.4	8 ± 0.8
Pentachlorothiophenol	0.5 ± 0.1	2.0 ± 0.3‡
Bile pentachlorothiophenol§	2.8 ± 0.2	1.8 ± 0.1‡
Urine pentachlorothiophenol§	20	300

*Modified from Reference 16. HCB given at 50 μmol/kg orally.
†Nanomoles per gram.
‡p < 0.01 compared to males.
§Nanomoles per 24 hours per kilogram.

metabolism is slower than in adulthood.[20,21] Similarly, aged animals metabolize xenobiotics at a lower rate than adults do.[22] In some cases the difference can be striking; for instance the half-life of caffeine is about two hours in 3-month-old rats but it becomes about five hours in 24-month-old rats.[23]

The presence of *pathology* alters the xenobiotic metabolism in different ways. Liver or kidney impairment is known to affect the metabolism and excretion of xenobiotics. The amount of adipose tissue may be critical, particularly for lipophilic chemicals, in determining the concentration of xenobiotics available for metabolism.[24] The presence of a tumor may impair monooxygenase activity.[25] As depicted in FIGURE 1, a reduction in the number of erythrocytes can markedly influence the concentration of a chemical in plasma: in fact the fraction of adriamycin present in plasma increases proportionally to the degree of anemia.[26]

When chemicals are utilized for toxicological studies in animals, they may be administered through various *routes* including oral, subcutaneous, intravenous, dermal, and pulmonary. It should not be assumed that all these routes are equivalent in terms of metabolism. Before it reaches the various organs, a xenobiotic introduced

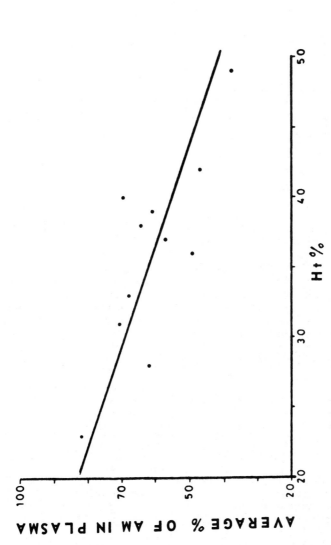

FIGURE 1. Plasma levels of adriamycin (ordinate) in patients with different hematocrit values (abscissa). (Modified from Reference 26.)

TABLE 5

DIFFERENCES IN THE KINETIC PARAMETERS OF BRAIN BUSPIRONE IN RELATION
TO THE ROUTE OF ADMINISTRATION (INTRAVENOUS OR ORAL)

BUSPIRONE ⟶ 1-PP

	Buspirone			1-PP		
Dose and Route	Peak*	$t_{1/2}$†	AUC‡	Peak*	$t_{1/2}$†	AUC‡
10 mg/kg intravenously	51 ± 4	25	909	8.1 ± 1	99	1433
10 mg/kg orally	<1	—	—	8.5 ± 1	124	1525

*Nanomoles per gram.
†Minutes.
‡Nanomoles per gram per minute.

orally must pass through the liver, where it may be actively metabolized (first-pass metabolism); this does not occur when a chemical is administered through other routes.

Buspirone, a drug with anxiolytic activity, is extensively metabolized; one metabolic pathway involves hydrolysis with the formation of 1-piperazinepyrimidine (1-PP).[27] As reported in TABLE 5, buspirone given intravenously is found in the brain at a concentration more than 50 times that reached when same dose of buspirone is given orally. This is not a problem of absorption because brain levels of the metabolite 1-PP are similar independent of the route of administration of buspirone.

During toxicological studies it is customary to test several *doses* including at least one that induces appreciable toxic effects. It must be understood that raising the dose does not always produce a proportional increase in metabolism, as the enzymes that metabolize xenobiotics may become saturated by a high concentration of the substrate. The substrate may therefore accumulate in blood or in tissue beyond what would be expected from the increase in dose. This phenomenon, known as dose-dependent kinetics, makes any extrapolation of toxic effect based solely on the administered dose difficult.

FIGURE 2 shows three examples of relationships between dose and blood or brain concentration of xenobiotics expressed as AUC, i.e., the area under the curve of xenobiotic concentrations in relation to time. For theobromine there is good agreement between doses and concentrations within a 100-fold range;[28] and for the congener caffeine, dose-dependent kinetics are already detectable at relatively low doses.[29] Even more striking is the case of *l*-fenfluramine, where an eightfold increase in the dose results in a brain level 172 times higher.[30]

Dose-dependent kinetics may in turn be modulated by a number of factors such as animal species, strain, sex, and age. For instance a 10-fold increase in the dose of caffeine gives an increase in the plasma caffeine AUC that is 12 times in mice, 16 times in rabbits, 12 times in *Macaca cynomolgus,* and 13 times in man, as opposed to 46 times for rats. These findings dictate a very cautious attitude in any extrapolation from *in vitro* to *in vivo*. They also suggest that any attempt to interpret toxicological data must be based on knowledge of the xenobiotic concentrations actually present *in vitro* or *in vivo*, independent of the doses given.

Diet is a frequently neglected factor, which may have a profound influence on xenobiotic metabolism. Several reviews are available concerning the influence on xenobiotic metabolism of specific nutrients,[31,32] micronutrients,[33,34] or nonnutritive substances present in the diet.[35] Low protein intake reduces[36,37] and high protein intake increases[38] the metabolic rate. Lipid intake, in terms of quantity and type, has been shown to affect xenobiotic metabolism, probably by altering the structure of the endoplasmic reticulum.[39,40] Even more important may be the influence of diet on cellular components such as glutathione, ascorbic acid, and tocopherol, which may be scavengers for electrophilic chemical species formed from the xenobiotic metabolism or which may prevent lipid peroxidation. Indeed peroxidation of lipids is known to alter rates of mixed-function oxygenase since it damages the membrane structure of the endoplasmic reticulum.[41]

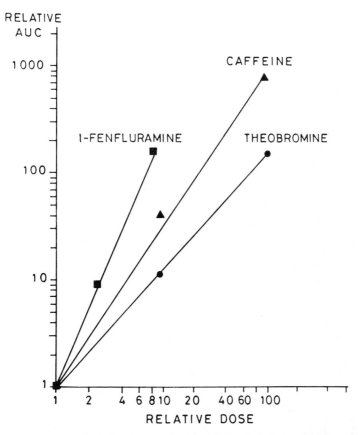

FIGURE 2. Examples of relations between increase of dose (abscissa) and increase of AUC (ordinate) for theobromine (circles), caffeine (triangles), and *l*-fenfluramine (squares). All the data of AUC refer to plasma with the exception of those of *l*-fenfluramine, which refer to brain.

The importance of microelements has also to be stressed. Animals made deficient in selenium show a reduction of selenium-dependent glutathione peroxidase,[42] which may be important in counteracting the deleterious consequences of radical reactions. Although no clear-cut correlation has been found, it is significant that selenium-deficient animals become more susceptible to the toxic action of certain xenobiotics such as adriamycin[43] and paraquat.[44]

Finally, xenobiotic metabolism is modulated by *interaction with other chemicals,* as will be described in the last section of this review.

Up to now the discussion has been centered mostly on metabolites stable enough to be measured in biological fluids or in tissues. However some metabolites are not directly detectable because they are so reactive that they immediately bind covalently to macromolecules. The toxicity of xenobiotics may therefore be correlated more to the labile than to the stable metabolites, although measurement of the latter is useful to suggest the presence of the former. Covalent binding of xenobiotics to DNA *in vivo* has been associated with cancer incidence,[45] while covalent binding to proteins is not always predictive of toxic effects.[46,47] Nevertheless protein-binding studies give the

TABLE 6

RANK ORDER OF HEXAMETHYLENMELAMINE (+ METABOLITES) LEVELS IN VARIOUS TISSUES AND AMOUNT OF COVALENT BINDING*

No.	Tissue	Level (nmol/g)	No.	Tissue	Covalent Binding (nmol/g)
1	Kidney	95 ± 5	1	liver	4.2 ± 1.9
2	Liver	80 ± 5	2	lung	1.5 ± 0.2
3	Lung	70 ± 5	3	spleen	1.4 ± 0.2
4	Tumor	63 ± 3	4	kidney	1.3 ± 0.2
5	Spleen	60 ± 2	5	tumor	0.8 ± 0.1
6	Brain	33 ± 4	6	heart	0.5 ± 0.1
7	Heart	19 ± 1	7	brain	0.4 ± 0.1

*Mouse bearing 20-day-old M 5076/73 A ovarian tumor.

signal that chemically reactive species are produced, and they suggest investigating for possible binding to DNA. Again it should be underlined that covalent binding of xenobiotics to macromolecules is modulated by all the factors previously discussed.

COMPARTMENTALIZATION OF XENOBIOTIC METABOLISM

Kinetic parameters and determination of metabolites in blood or urine represent the result of the overall metabolism. Therefore they cannot serve as a basis for prediction of the roles of different tissues and cells in any particular xenobiotic metabolism. Furthermore they cannot be used in extrapolation to the relation between metabolism and localized toxic effects such as carcinogenesis in specific organs. Compartmentalization of xenobiotic metabolism can be viewed from two angles: within the body and within the cell.

In vivo distribution of chemicals in the various tissues is frequently uneven depending on the chemical-physical characteristics of the agent (e.g., pK_a and lipophilicity) and the characteristics of the organ (e.g., blood flow, lipid content, and special barriers). TABLE 6 shows for instance the relative distribution of hexamethyl-melamine (plus metabolites) in different tissues of mice bearing an ovarian tumor (M

TABLE 7

COVALENT BINDING OF HEXAMETHYLMELAMINE (+ METABOLITES) TO VARIOUS TISSUES OF
MICE BEARING 20-DAY-OLD M 5076/73A OVARIAN TUMOR*

| Tissue | Percent of ^{14}C-HMM Metabolites Covalently Bound after | |
	2 Hours	40 Hours
Liver	5.2	22.7 (4.3)†
Heart	2.5	17.2 (6.8)
Spleen	2.3	18.7 (8.1)
Lung	2.1	20.9 (9.9)
Blood	1.6	11.2 (7.0)
Small intestine	1.4	2.7 (1.9)
Kidney	1.3	10.6 (7.6)
Brain	1.3	0.7 (0.5)
Tumor	1.2	8.8 (7.3)

*Modified from Reference 48.
†The increase over the amount covalently bound at 2 hours is shown in parenthesis.

5076/73A) two hours after drug administration. Distribution is uneven, and there is no clear-cut relation between the amount of the chemical in any one tissue and the amount of covalent binding, indicating that hexamethylmelamine is metabolized at different rates in different tissues to form the reactive chemical species.[48]

As time progresses, the drug and its metabolites are cleared from tissues at a rate higher than the clearance of the covalently bound drug (cell killing and DNA repair). Therefore the amount covalently bound tends to increase as a percentage of the total radioactivity present. TABLE 7 sets out these data, indicating that the changes range from 0.5 for the brain to 9.9 times for the lung.[48]

More precise indications about the potential metabolism of different tissues and cells can be obtained *in vitro*. However it is always difficult to be sure that the *in vivo* conditions are adequately mimicked. TABLE 8 shows that rat lymphocytes metabolize caffeine[49] differently than does the isolated, perfused rat liver;[50] the lymphocytes' principal metabolite is trimethylurine acid, the liver's is theobromine.

Species differences play a considerable role when certain cells are utilized for xenobiotic metabolic studies. As indicated in TABLE 9, rat, rabbit, and human lymphocytes do not metabolize caffeine the same way.[49] For chemical agents that become carcinogens after biotransformation, this regional metabolism may be more important than overall metabolic studies.

Distribution of xenobiotics is uneven, not only in tissues, but also at the subcellular

TABLE 8

METABOLISM OF CAFFEINE IN LYMPHOCYTES AND IN LIVER*

Metabolite of Caffeine	Rat Lymphocytes	Rat Perfused Liver
Theophylline	4.2	27
Theobromine	1.5	52
Paraxanthine	1	43
Trimethyluric acid	5.0	1

*Relative values taking 1 as the lowest concentration.

TABLE 9

METABOLITES OBTAINED AFTER INCUBATION OF CAFFEINE WITH LYMPHOCYTES
FROM DIFFERENT ANIMAL SPECIES*

	Lymphocytes from		
Metabolite of Caffeine	Rat	Rabbit	Man
Theophylline	1.96	—	—
Theobromine	0.66	1.80†	—
Paraxanthine	0.46	0.56	0.50
Trimethyluric acid	2.28†	1.02	0.54†

*Modified from Reference 49. Figures are pmol/hour per 10^6 cells.
†The most important metabolite for each species.

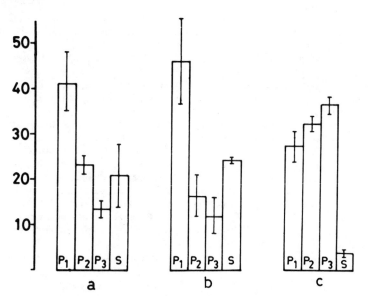

FIGURE 3. Distribution of radioactivity from ^{3}H-TCDD in mouse liver subcellular fractions. Male C57B1/6J mice were injected with 4 mCi, 25 mg/kg ^{3}H-TCDD ip, and killed either 7 (b) or 14 (c) days later. ^{3}H-TCDD, 88 ng/g, was added to livers from mice not given TCDD (a). P_1 (nuclear), P_2 (mitochondrial), and P_3 (microsomal) fractions were the sediments of centrifugations at respectively 670, 10,000, and 105,000 × g, and S (soluble fraction) was the high-speed supernatant. Apparent TCDD whole liver concentrations (ng/g ± SD) were 103.9 ± 14.0 and 95.0 ± 25.4 for b and c, respectively. (Reproduced from Reference 52 by courtesy of Marcel Dekker, Inc.)

level. When cells are homogenized and centrifuged to yield nuclear (P_1), mitochondrial (P_2), microsomal (P_3), and soluble (S) fractions, the resulting distribution patterns may differ considerably depending on the xenobiotic being investigated. For instance more than 50% of adriamycin accumulates in the liver nuclei as well as in cancer cells.[51] As time progresses, however, the subcellular distribution of a chemical may change. A typical example is shown in FIGURE 3, which reports the subcellular localization of 2,3,7,8 tetrachlorodibenzodioxine (TCDD) in $C_{57}Bl/6J$ mouse liver after administration of TCDD *in vivo* or addition *in vitro*. Fourteen days after treatment there is a decrease in the P_1 and S fractions and a marked increase in the P_2 and P_3 fractions. On account of the long half-life of TCDD, the absolute levels in the liver did not change.[52] It is also worth noting that the toxicity of TCDD only appears several weeks after administration of this agent.

The enzymatic systems relevant for xenobiotic metabolism are compartmentalized within the cell. Recent studies have indicated that cytochrome P_{450}, monooxygenases, epoxide hydrolase, and uridine diphosphate–glucuronyl transferase are present not only in the rat liver endoplasmic reticulum but also, to a lesser extent, in the nuclear membranes, inside the nuclei, Golgi apparatus, plasma membranes, and mitochondria.[53] Of particular importance appears to be the presence of xenobiotic metabolizing enzymes in the nuclear membrane.[54,55] The proximity of these membranes to the genomic material makes it possible for even very short-lived, highly reactive metabolites to reach the DNA and bind there. As represented in FIGURE 4, a xenobiotic entering a cell can reach the DNA directly or it may be metabolized by the endoplasmic reticulum system (ERS) (rough and smooth). In some cases the nuclear membranes may activate compounds already biotransformed by the ERS. It is interesting to note that the nuclear membranes are equipped with enzymes, such as epoxide hydrolase, that are frequently associated with detoxification of xenobiotics. Quantitative data are now available on enzymatic activities in the nuclei as opposed to the microsomes, with particular reference to the liver.

TABLE 10 shows that the ratio between rat liver microsomes and nuclei is about 15 for styrene monooxygenase and 19 for styrene epoxide hydroxylase; the ratio between styrene monooxygenase and epoxide hydrolase is relatively constant, respectively 3.0 and 3.6 for nuclei and microsomes.[56] However these ratios may change in relation to various types of induction; the nuclei are more inducible than microsomes by phenobarbital for styrene monooxygenase but less inducible than microsomes for styrene epoxide hydrolase.[56]

RELEVANCE OF XENOBIOTIC METABOLISM *IN VIVO* FOR CELLULAR SYSTEMS

In recent years there has been a remarkable development of cell systems for toxicity testing *in vitro*, with particular emphasis on investigating such parameters as transformation, mutations, or DNA damage, which may predict carcinogenic potential of xenobiotics.

The metabolism of xenobiotics in bacteria used for detecting mutagenic properties (Ames test) may be quite different from what happens in mammalian systems. The "Ames indirect test" was therefore developed, where the xenobiotic metabolism is achieved by adding to the bacteria different preparations of liver microsomal enzymes. However, these preparations cannot be assumed to retain all the metabolic activities of the liver. It has been shown for instance that liver microsomes of mice and rats actively *N*-demethylate aminopyrine but are inactive in *N*-demethylating caffeine, a metabolic pathway that occurs however in the isolated, perfused liver of rats.[50] This difficulty has

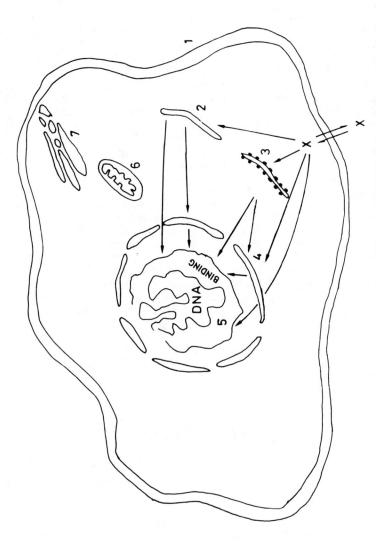

FIGURE 4. Schematic representation of a cell. X indicates the xenobiotic entering the cell. (1) Cell membrane; (2) smooth endoplasmic reticulum; (3) rough endoplasmic reticulum; (4) nuclear membrane; (5) genome; (6) mitochondria; (7) Golgi apparatus. The arrows indicate various possibilities of biotransformation of a chemical before it binds to DNA.

TABLE 10

STYRENE EPOXIDE HYDROLASE TO STYRENE MONOOXYGENASE RATIOS IN NUCLEI
AND MICROSOMES AFTER TREATMENT WITH PHENOBARBITAL (PB)
AND β-NAPHTHOFLAVONE (βNF)

	Control	PB	βNF
Styrene epoxide hydrolase/styrene monooxygenase			
Nuclei	3.0 (367/120)*	2.3 (900/388)	1.6 (462/270)
Microsomes	3.6 (7,125/1,950)	7.5 (23,649/3,159)	1.8 (9,799/5,493)
Microsomes/nuclei			
Styrene epoxide hydrolase	19 (7,125/367)	14 (23,649/900)	21 (9,799/462)
Styrene monooxygenase	15 (1,950/120)	8 (3,159/388)	20 (5,493/270)

*The number in parenthesis represents the mean enzymatic activity expressed as pmol/minute per mg protein.

been overcome to some extent by utilizing liver microsomal preparations obtained from rats pretreated with different inducers in order to increase selectively one of the multiple forms of cytochrome P_{450}.[57,58] Here again, however, there may be problems, particularly when a xenobiotic has multiple competing metabolic pathways. As shown in TABLE 11, styrene may be activated to form an epoxide or an arene oxide. Phenobarbital enhances the formation of the epoxide but not of the arene oxide, while the opposite happens with 3-methylcholanthrene as an inducer.[59] It is worth noting that the epoxide hydrolase that transforms the styrene epoxide into phenylethyleneglycol[60] makes possible a further activation via an arene oxide. Covalent binding can in fact be obtained by activating either styrene or its metabolite phenylethyleneglycol.[59]

Intact hepatocytes have been used instead of liver microsomal preparations in an effort to combine the metabolizing system and the target of toxic effect. However even these isolated cells can pose problems in maintaining their metabolic activity without

TABLE 11

COVALENT BINDING TO PROTEINS AFTER INCUBATION OF RAT LIVER MICROSOMES
WITH STYRENE OR PHENYLETHYLENEGLYCOL*

Styrene		Phenylethyleneglycol	
C	2.56 ± 0.39	C	0.06 ± 0.03
PB	4.12 ± 0.16†	PB	0.07 ± 0.04
3 MC	2.60 ± 0.21	3 MC	0.11 ± 0.08†

*Figures are nmol/mg per 2 hours. C = controls. PB = phenobarbital pretreatment. 3 MC = 3 methylcholanthrene pretreatment.
†p < 0.05 compared to control.

continuous removal of the metabolites formed. The metabolic activity of hepatocytes is regulated by a number of factors, which are not yet completely understood.[60] In this respect it may be of interest to recall that the cellular supply of scavengers for electrophilic metabolites, such as glutathione, is depleted in a short time when hepatocytes are incubated *in vitro*;[61] this contrasts with the stable levels existing *in vivo* and may make the isolated hepatocytes more sensitive to the action of xenobiotic metabolites that react with macromolecules.

In studies with cellular systems, it is important to verify what actually happens to the added xenobiotic. An example is given in FIGURE 5. VP 16 [4'-demethylepipodo-phyllotoxin-5-(4,6-*O*-)ethylidene-D glycopyranoside] was added to rat liver micro-somes and rat hepatocytes at two different concentrations (10 and 100 μg/ml) and incubated with appropriate cofactors for three hours. VP 16 is not appreciably metabolized, as shown by the drug assays performed by high-pressure liquid chroma-tography (HPLC) at one, two, and three hours. In contrast over 90% of the added VP 16 disappeared from the perfusing medium when the drug was added to the isolated, perfused rat liver at a concentration of 100 μg/ml. The disappearance of VP 16 from rat plasma compared well with the isolated, perfused liver but not with the rat liver microsomes or hepatocytes (M. D'Incalci *et al.*, unpublished data). Obviously any toxic effect on microsomes and hepatocytes, or lack of toxicity, observed in these conditions for VP 16 has very little to do with the kinetics of VP 16 *in vivo*.

This kind of observation raises another question, i.e., On what basis should concentrations of xenobiotics for *in vitro* use be selected when the kinetics of the xenobiotic are known *in vivo*? VP 16 again illustrates this point. The kinetics of VP 16 in 3LL-bearing mice have been investigated,[62] and the sensitivity to VP 16 of the same cancer cells has been investigated *in vitro*.[63] The conclusion is that for VP 16 cytotoxicity, the time of exposure is more important than the absolute concentration of the drug. In fact as shown in FIGURE 6, 0.1 μg/ml of VP 16 incubated for 72 hours had the same effect as 1 μg/ml incubated for 24 hours. This suggests that for VP 16, the AUC could be a more reliable parameter than the peak level (C_{max}) as a basis for extrapolations from *in vitro* to *in vivo* cytotoxicity.

These findings have been recently extended by studies on murine neuroblastoma cells; it was established that there are chemicals, such as melphalan and bleomycin, that have similar cytotoxic effect when incubated for 1 or 24 hours while other chemicals, such as VM 26 and vincristine, are active only after a long, but not a short, incubation.[64] It is suggested that for the first class of drugs the cytotoxicity is related to the *in vivo* C_{max} while for the second class the AUC is more important.

How well cellular systems can predict DNA damage *in vivo* is far from established on a quantitative basis. A series of experiments were recently made to establish the covalent binding of cyclophosphamide—a drug that has to be metabolized to become cytotoxic[65]—to different *in vitro* systems. Preliminary results, reported in TABLE 12, concern the covalent binding of cyclophosphamide to proteins and to DNA when incubated with rat liver microsomes (plus DNA), nuclei, hepatocytes, or isolated, perfused liver or when injected *in vivo*.[59] In each test the starting concentration of the drug in the incubating medium or in plasma was the same. The amount of covalent binding depends on the test employed, and only the isolated, perfused liver predicts with enough accuracy the amount of covalent binding, particularly the ratio between binding to proteins and to DNA. The other systems are less predictive, nuclei tending to magnify the binding to DNA, and hepatocytes the binding to proteins, compared to what was observed *in vivo*. The reasons for this discrepancy probably lie in the rate of metabolism of cyclophosphamide and in substrate availability and removal, but it would certainly be inaccurate to make quantitative predictions from the *in vitro* (cellular systems) to the *in vivo* data in this case.

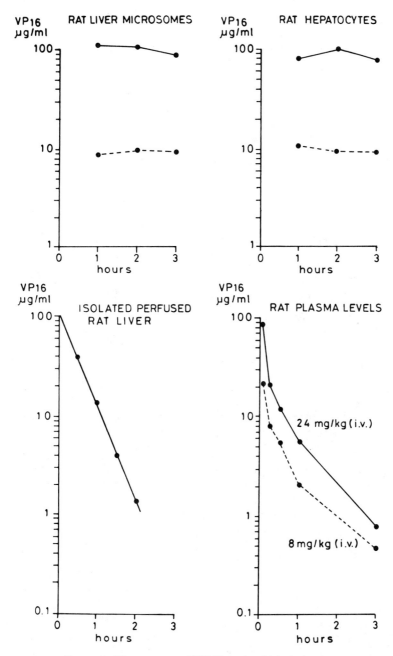

FIGURE 5. Disappearance of VP16 in various biological systems.

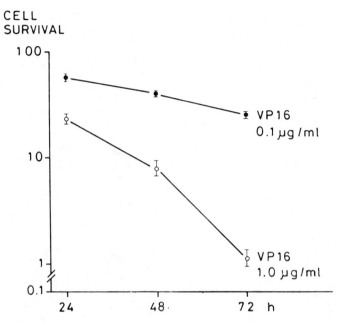

FIGURE 6. Effect of the time of incubation *in vitro* on the cytotoxicity of VP 16 for Lewis lung carcinoma (3LL) cells.

Kinetic and metabolic considerations and extrapolations should never be dissociated from knowledge concerning the mechanisms of toxic effects, since in fact this knowledge ultimately determines which chemical species and concentrations are meaningful.

A xenobiotic may be cytotoxic in relation to more than one mechanism, depending on the target cells and the experimental conditions utilized. For instance it is generally believed that adriamycin, a powerful antitumoral agent, exerts its cytotoxic activity because it forms strong complexes with double-stranded DNA, which can be understood in terms of the known intercalation model.[66,67] According to this hypothesis, which fits a number of experimental results, adriamycin would act per se without

TABLE 12

COVALENT BINDING OF CYCLOPHOSPHAMIDE METABOLITES TO LIVER PROTEINS
AND DNA IN VARIOUS EXPERIMENTAL SYSTEMS

| | Covalent Binding of Cyclophosphamide Metabolites (pmol/mg protein or DNA per 2 hours) | |
Experimental Model (rat liver)	Proteins	DNA
Microsomes + DNA	109 ± 2 (100)	9 ± 2 (100)
Nuclei	8 ± 1 (7)	3 ± 1 (33)
Hepatocytes	80 ± 10 (70)	1 ± 0.1 (11)
Isolated, perfused organ	22 ± 5 (20)	1 ± 0.2 (11)
In vivo	12 ± 2 (11)	0.6 ± 0.1 (6)

requiring metabolic transformation. Other hypotheses suggest that adriamycin may act by stimulating DNase I[68] or by inhibiting DNA polymerase.[69,70] More recently however it was found that adriamycin may also bind covalently to proteins[71,72] and to DNA *in vitro*[73] and *in vivo*;[74] this mechanism requires metabolic activation—which however does not appear, at least for rat isolated nuclei,[73] to depend on cytochrome P_{450} mixed monooxygenase but perhaps on the formation of highly reactive radicals.[75-77] The extent of covalent binding of activated adriamycin is reported in TABLE 13 where this irreversible reaction is obtained using rat liver microsomes or nuclei in the presence of a reduced nicotinamide adenine dinucleotide phosphate (NADPH)–generating system.

All the mechanisms described to explain the cytotoxic effect of adriamycin require penetration and/or metabolism of this agent in the cells. It is therefore surprising to learn that *in vitro* adriamycin may exert a cytotoxic effect even when it cannot become available to the cells because it has been immobilized on an insoluble agarose support.[78]

TABLE 13

IRREVERSIBLE INTERACTION OF ^{14}C-ADRIAMYCIN AND METABOLITES WITH NUCLEAR PROTEINS AND DNA IN THE PRESENCE OF SKF 525A*

Experimental Conditions		Proteins (1 mg)	DNA (1 mg)
Control	− NADPH	22.7 ± 0.7	17.7 ± 1.0
	+ NADPH	36.1 ± 1.2†	90.1 ± 5.9†
SKF 525A	− NADPH	25.9 ± 1.6	18.3 ± 1.9
	+ NADPH	37.0 ± 1.8†	81.7 ± 2.8†

*[14]C-adriamycin was added to the incubation mixture at a final concentration of 2.5 μM. SKF 525A, an inhibitor of liver monooxygenases, was added to the incubation mixture at a final concentration of 1 mM. Each value is the mean ± standard error of four determinations.

†p < 0.01 compared to nuclei − NADPH.

This finding suggests a further mechanism for adriamycin, involving an interaction on the cell surface. This is an example of how the interpretation of kinetic and metabolic data on adriamycin is considerably affected by the mechanism of toxicity envisaged.

INTERACTION WITH OTHER XENOBIOTICS

Xenobiotic toxicity and metabolism are studied under well-defined, controlled conditions so as to minimize interference from unknown factors. In real life however this seldom happens, and most often man is exposed to a given xenobiotic in connection with a mixture of many other chemicals. For example, hundreds of compounds are present together with benzopyrene during cigarette smoking and hundreds of compounds accompany TCDD during incineration of wastes. The toxicologist therefore has the formidable task of interpreting his data in the framework of the "natural" conditions of man's exposure to xenobiotics. There is no way around this problem because the interaction of only two chemicals given concomitantly or in succession is already very complex.

The question that has been most widely studied concerns the interaction on the cytochrome P_{450} mixed monooxygenase system, where chemicals can be classified as inducers or as inhibitors when they enhance or depress the activity of these enzymes.

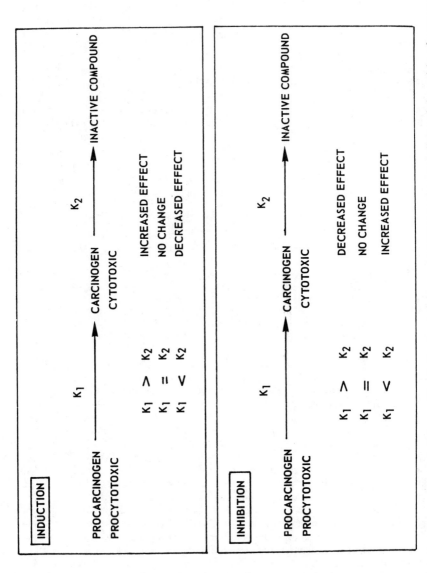

FIGURE 7. Schematic representation of the possible effects of an inducer or an inhibitor on the metabolism and the activity of a procarcinogen or a procytotoxic agent.

The consequence of these effects in relation to another chemical given during activation or inhibition of mixed oxygenases is relatively simple if the chemical exerts toxic effects per se; activation increases the detoxification, while inhibition results in higher concentrations of the toxic compound. The problem becomes more complicated when a chemical exerts its toxic effects per se and/or through the action of a metabolite. FIGURE 7 shows the general case of an agent that is activated to form a reactive chemical species, which is then detoxified. Different effects can be obtained depending on how the inducer or the inhibitor influences the rates of biotransformation of the chemical.

The situation however is frequently more complicated because of the presence of more than one toxic metabolite or because an inactive metabolite may be subsequently activated. The theoretical number of different possibilities makes any prediction almost impossible unless specific, time-consuming studies are made for the compound under investigation.

Even considering an interaction between only two chemicals, a number of factors may codetermine the final effect. For instance an inducer in one animal species may be inactive in another species. This is the case of the oral contraceptive agents, which stimulate liver microsomal enzymes in rats and mice but have no such effect in guinea

TABLE 14

EFFECT OF LYNESTRENOL (L) + MESTRANOL (M) ON LIVER MONOOXYGENASES OF DIFFERENT ANIMAL SPECIES*

Treatment (mg/kg orally)	Species	Aminopyrine N-Demethylation (control = 100)	Styrene Epoxide Synthetase (control = 100)
LM (1.25 + 0.075)	mouse	155†	—
LM (1.25 + 0.075)	rat	164†	158†
LM (5.0 + 0.3)	guinea pig	101	98

*Modified from Reference 79.
†$p < 0.001$.

pigs,[79] as reported in TABLE 14. This finding may be related to the already mentioned multiplicity of cytochrome P_{450} species, with different sensitivity to various inducing agents.

The efficacy of the same inducer may vary considerably in relation to the strain utilized within the same animal species. For instance certain strains of mice are relatively resistant to induction of their liver microsomal enzyme activity by certain agents. A genetic basis for this effect has been described recently.[80] Whatever the mechanism involved, the AKR and DBA$_2$ strains are relatively insensitive to the induction of arylhydrocarbonhydroxylase (AHH) by TCDD or TCDF (tetrachlorodibenzofuran) when compared to other strains such as C$_3$H and C$_{57}$Bl$_6$.[81,82] It is noteworthy that strains that are poorly inducible are also much less sensitive to the thymic atrophy induced by TCDD[83] and TCDF,[84] as shown in TABLE 15. Similar results have been obtained in regard to humoral immunosuppressant activity.[83,84] Similar studies have been reported by other authors. For instance, mouse liver nuclear AHH is stimulated only twofold by 3-methylcholanthrene pretreatment in AKR and DBA$_2$ mice as opposed to the sevenfold increase obtainable in C$_3$H and C$_{57}$Bl$_6$ strains.[85] Awareness of these differences is crucial in determining cell systems suitable for use *in vitro* and in extrapolating data.

TABLE 15

DIFFERENT EFFECTS OF TETRACHLORODIBENZODIOXIN (TCDD) AND
TETRACHLORODIBENZOFURANE (TCDF) IN TWO STRAINS OF MICE

Mouse Strain	TCDD (μg/kg ip)*	TCDF (μg/kg ip)	Thymus Weight (controls = 100)	PFC ($\times 10^6$ splenocytes) (controls = 100)
C57Bl6	6	—	70†	20†
C57Bl6	—	180	60†	19†
DBA2	6	—	84	70†
DBA2	—	180	85	65†

*ip is intraperitoneally.
†p < 0.01.

The effect of an inducer is not predictable for all substrates. As shown in TABLE 11, two pathways of a given substrate may have different sensitivity to an inducer. In this respect the formation of an areneoxide (styrene-3,4-oxide) represents a secondary metabolic pathway in the *in vitro* biotransformation of styrene because it occurs at the picomolar level in terms of covalent binding, whereas styrene-7,8-oxide is bound at nanomolar concentrations. The fact that these two metabolic pathways can be activated by phenobarbital and 3-methylcholanthrene respectively is not without significance, considering that styrene-3,4-oxide was far more mutagenic in *in vitro* tests than was styrene-7,8-oxide.[86]

As already alluded to, the fact is that in man exposure to a toxic agent usually occurs in the presence of many chemicals that may compete for absorption, metabolism, and excretion at the same time as being inhibitors or inducers for the toxic agent. This confounds any attempt at reasonable prediction. In addition competition may occur for the same receptor. As reported in TABLE 16, a strongly immunodepressant dose of TCDD may be antagonized by the simultaneous administration of an inactive dose of TCDF. This effect should be remembered for instance in interpreting the lack of immunodepressant effect of a mixture of polyCDD and polyCDF in the effluent of an incinerator.[87,88]

In closing these notes, it is important to recall that kinetic and metabolic information on xenobiotics represents only part of the problem of extrapolating data across species. A toxic response depends not only on the concentration of the xenobiotic (and/or metabolites) present at a given location but also on the "sensitivity" of the system. This means that equal concentrations are not a guarantee of equal toxic responses because all the factors (animal species, strain, sex, age, diet, etc.) that modulate metabolism are just as important in modulating the "receptors" capable of

TABLE 16

IMMUNODEPRESSANT EFFECT OF TCDD IN COMBINATION WITH TCDF
OR VARIOUS TETRA CDD AND TETRA CDF

TCDD (μg/kg ip)	TCDF (μg/kg ip)	PFC/Spleen (controls = 100)
—	10	80
1.2	—	19
1.2	10	53*
Mixture of tetra CDD and tetra CDF		83

*p < 0.01 compared to TCDD.

toxic responses. Therefore no generalization can be made but individual cases must be considered. For instance saccharin given at high doses in the diet (1–5%) induces marked immunodepression in rats.[89] Similarly saccharin *in vitro* inhibits the mitogenic activity of phytohemoagglutinin (PHA) in lymphocytes of mice and rats.[90] However the concentrations of saccharin effective in rats do not have the same effect on human lymphocytes; for instance 0.5 mg/ml significantly inhibits the PHA response of rat lymphocytes, but 2 mg/ml of saccharin are ineffective on human lymphocytes. This is a clear indication that equal concentrations do not always mean equal toxic effects across various animal species.

CONCLUSIONS

There is no doubt that the advent of sensitive analytical techniques that have made possible quantitative studies of xenobiotic kinetics and metabolism has opened new perspectives in interpreting the biological and toxicological effects of xenobiotics. However reliance on concentrations of the xenobiotic under investigation and on precise knowledge of xenobiotic metabolic pathways rather than on doses is not yet widespread. Similarly precise information about concentrations and biotransformation of xenobiotics is necessary when working with cellular systems. Only an integrated approach taking into consideration *in vitro* and *in vivo* differences as well as differences between animal species offers a rational basis for extrapolating data from animals to man. Despite limited success in interpreting the toxic effects of some chemicals, we still lack the understanding of toxicological mechanisms essential for any extrapolation. The prediction of toxicological effects of xenobiotics must therefore be considered an open field of investigation where cross-fertilization among various approaches and disciplines is the best hope for obtaining sounder results.

REFERENCES

1. MILLER, E. C. & J. A. MILLER. 1981. Mechanisms of chemical carcinogenesis. Cancer **47:** 1055–1064.
2. WEISBURGER, J. H. & G. M. WILLIAMS. 1982. Metabolism of chemical carcinogens. *In* Cancer. A Comprehensive Treatise. F. F. Becker, Ed. 2nd edit. **1:** 241–333. Plenum Press. New York, N.Y.
3. PARKE, D. V. 1982. Significance of the metabolism of xenobiotics for toxicological evaluation. *In* Animals in Toxicological Research. I. Bartosek, A. Guaitani & E. Pacei, Eds.: 127–145. Raven Press. New York, N.Y.
4. DANIEL, J. W. 1982. Safety evaluation: the role of pharmacokinetic and comparative metabolic studies. *In* Animals in Toxicological Research. I. Bartosek, A. Guaitani & E. Pacei, Eds.: 115–126. Raven Press. New York, N.Y.
5. BRODIE, B. B. & J. R. GILLETTE, Eds. 1971–1975. Handbook of Experimental Pharmacology. **28** (three parts). Springer-Verlag. Berlin, Federal Republic of Germany.
6. COON, M. J., A. H. CONNEY, R. W. ESTABROOK, H. V. GELBOIN, J. R. GILLETTE & P. J. O'BRIEN, Eds. 1980. Microsomes, Drug Oxidation, and Chemical Carcinogenesis. Academic Press. New York, N.Y.
7. GARATTINI, S. Difficulties in extrapolating toxicological data from animals to man. *In* International Conference on Safety Evaluation and Regulation. Karger. Basel, Switzerland. (In press.)
8. WILLIAMS, T. 1959. Detoxication Mechanisms. Chapman and Hall. London, England.
9. SMITH, R. L. & L. G. DRING. 1970. Patterns of metabolism of β-phenylisopropylamines in man and other species. *In* Amphetamines and Related Compounds. E. Costa & S. Garattini, Eds.: 121–139. Raven Press. New York, N.Y.

10. LATINI, R., M. BONATI, E. MARZI & S. GARATTINI. 1981. Urinary excretion of an uracilic metabolite from caffeine by rat, monkey and man. Toxicol. Lett. **7:** 267–272.
11. GARATTINI, S., E. ERBA, L. MORASCA, G. PERI, A. MANTOVANI, S. FILIPPESCHI, F. SPREAFICO & M. ARNAUD. 1982. In vitro and in vivo cytotoxicity of 6-amino-5-formylmethylamino-1,3-dimethyl uracil, a uracilic metabolite of caffeine. Toxicol. Lett. **10:** 313–319.
12. BALL, L. M., A. G. RENWICK & R. T. WILLIAMS. 1977. The fate of [^{14}C]saccharin in man, rat and rabbit and of 2-sulphamoyl[^{14}C]benzoic acid in the rat. Xenobiotica **7:** 189–203.
13. VESELL, E. S. 1975. Genetically determined variations in drug disposition and response in man. In Handbook of Experimental Pharmacology. J. R. Gillette & J. R. Mitchell, Eds. **28**(Part 3): 169–212. Springer-Verlag. Berlin, Federal Republic of Germany.
14. SAN MARTIN DE VIALE, L. C., A. A. VIALE, S. NACHT & M. GRINSTEIN. 1970. Experimental porphyria induced in rats by hexachloro-benzene. A study of the porphyrins excreted by urine. Clin. Chim. Acta **28:** 13–23.
15. GRANT, D. L., F. IVERSON & G. V. HATINA. 1974. Effects of hexachlorobenzene on liver porphyrin levels and microsomal enzymes in the rat. Environ. Physiol. Biochem. **4:** 159–165.
16. RIZZARDINI, M. & A. G. SMITH. 1982. Sex differences in the metabolism of hexachlorobenzene by rats and the development of porphyria in females. Biochem. Pharmacol. **31:** 3543–3548.
17. RICHTER, E., G. RENNER, J. BAYERL & M. WICK. 1981. Differences in the biotransformation of hexachlorobenzene (HCB) in male and female rats. Chemosphere **10:** 779–785.
18. RENNER, G., E. RICHTER & K. P. SCHUSTER. 1978. N-Acetyl-S-(pentachlorophenyl)cysteine, a new urinary metabolite of hexachlorobenzene. Chemosphere **7:** 663–668.
19. SMITH, A. G. & J. R. CABRAL. 1980. Liver-cell tumours in rats sed hexachlorobenzene. Cancer Lett. **11:** 169–172.
20. SOYKA, L. F. & G. P. REDMOND, Eds. 1981. Drug Metabolism in the Immature Human. Raven Press. New York, N.Y.
21. MORSELLI, P. L., Ed. 1977. Drug Disposition during Development. Spectrum. New York, N.Y.
22. SCHMUCKER, D. L. 1979. Age-related changes in drug disposition. Pharmacol. Rev. **30:** 445–456.
23. LATINI, R., M. BONATI, E. MARZI, M. T. TACCONI, B. SADURSKA & A. BIZZI. 1980. Caffeine disposition and effects in young and one year old rats. J. Pharm. Pharmacol. **32:** 596–599.
24. BIZZI, A., M. T. TACCONI, G. TOGNONI, P. L. MORSELLI & S. GARATTINI. 1978. Distribution of fenfluramine in normal and obese mice. Int. J. Obesity **2:** 1–5.
25. VILLA, P., A. GUAITANI & I. BARTOSEK. 1978. Differences in pentobarbital disappearance rate in rats bearing two lines of Walker carcinosarcoma 256. Biochem. Pharmacol. **27:** 811–812.
26. PIAZZA, E., M. BROGGINI, A. TRABATTONI, N. NATALE, A. LIBRETTI & M. G. DONELLI. 1981. Adriamycin distribution in plasma and blood cells of cancer patients with altered hematocrit. Eur. J. Cancer Clin. Oncol. **17:** 1089–1094.
27. CACCIA, S., S. GARATTINI, A. MANCINELLI & M. MUGLIA. 1982. Identification and quantitation of 1-(2-pyrimidinyl)-piperazine an active metabolite of the anxiolytic agent buspirone, in rat plasma and brain. J. Chromatogr. **252:** 310–314.
28. BONATI, M., R. LATINI, B. SADURSKA, E. RIVA, F. GALLETTI, J. E. BORZELLECA, S. TARKA, M. J. ARNAUD & S. GARATTINI. Kinetics and metabolism of theobromine in male rats. I. Toxicol. Appl. Pharmacol. (In press.)
29. LATINI, R., M. BONATI, D. CASTELLI & S. GARATTINI. 1978. Dose-dependent kinetics of caffeine in rats. Toxicol. Lett. **2:** 267–270.
30. CACCIA, S., G. DAGNINO, S. GARATTINI, G. GUISO, R. MADONNA & M. G. ZANINI. 1981. Kinetics of fenfluramine isomers in the rat. Eur. J. Drug Metab. Pharmacokinet. **6:** 297–301.
31. CAMPBELL, T. C. 1979. Influence of nutrition on metabolism of carcinogens. Adv. Nutr. Res. **2:** 29–55.
32. WADE, A. E. & W. P. NORRED. 1976. Effect of dietary lipid on drug-metabolizing enzymes. Fed. Proc. **35:** 2475–2479.

33. BECKING, G. C. 1976. Hepatic drug metabolism in iron-magnesium- and potassium-deficient rats. Fed. Proc. **35:** 2480–2485.
34. ZANNONI, V. G. & P. H. SATO. 1976. The effect of certain vitamin deficiencies on hepatic drug metabolism. Fed. Proc. **35:** 2464–2469.
35. WATTENBERG, L. W., W. D. LOUB, L. K. LAM & J. L. SPEIER. 1976. Dietary constituents altering the responses to chemical carcinogens. Fed. Proc. **35:** 1327–1331.
36. ALCANTARA, E. N. & E. W. SPECKMANN. 1976. Diet, nutrition, and cancer. Am. J. Clin. Nutr. **29:** 1035–1047.
37. JUNGE, O. & K. BRAND. 1975. Mixed function oxidation of hexobarbital and generation of NADPH by the hexose monophosphate shunt in isolated rat liver cells. Arch. Biochem. Biophys. **171:** 398–406.
38. ANDERSON, K. E., A. H. CONNEY & A. KAPPAS. 1976. Nutritional influence of cytochrome P-450 mediated hepatic drug oxidation in normal males. Gastroenterology **71:** A1/894.
39. CENTURY, B. 1973. A role of the dietary lipid in the ability of phenobarbital to stimulate drug detoxification. J. Pharmacol. Exp. Ther. **185:** 185–194.
40. MARSHALL, W. J. & A. E. M. MCLEAN. 1971. A requirement for dietary lipids for induction of cytochrome P-450 by phenobarbitone in rat liver microsomal fraction. Biochem. J. **122:** 569–573.
41. TAPPEL, A. L. 1973. Lipid peroxidation damage to cell components. Fed. Proc. **32:** 1870–1874.
42. ALBERTS, D. S., Y-M. PENG & T. E. MOON. 1978. α-Tocopherol pretreatment increases adriamycin bone marrow toxicity. Biomedicine **29:** 189–191.
43. FACCHINETTI, T., F. DELAINI, M. SALMONA, M. B. DONATI, S. FEUERSTEIN & A. WENDEL. The influence of selenium intake on chronic adriamycin toxicity and lipid peroxidation in rats. Toxicol. Lett. (In press.)
44. BURK, R. F., R. A. LAWRENCE & J. M. LANE. 1980. Liver necrosis and lipid peroxidation in the rat as the result of paraquat and diquat administration. J. Clin. Invest. **65:** 1024–1031.
45. LUTZ, W. K. 1979. In vivo covalent binding of organic chemicals to DNA as a quantitative indicator in the process of chemical carcinogenesis. Mutat. Res. **65:** 289–356.
46. GILLETTE, J. R. 1974. A perspective on the role of chemically reactive metabolites of foreign compounds in toxicity. I. Correlation of changes in covalent binding of reactivity metabolites with changes in the incidence and severity of toxicity. Biochem. Pharmacol. **23:** 2785–2794.
47. TYNDALL, J. 1979. Knowledge once gained casts a light beyond its own immediate boundaries. Drug Metab. Dispos. **7:** 451–453.
48. GARATTINI, E., T. COLOMBO, M. G. DONELLI & C. PANTAROTTO. Distribution, metabolism and irreversible binding of hexamethylmelamine in mice bearing ovarian carcinoma. Cancer Chemother. Pharmacol. (In press.)
49. BONATI, M., L. JIRITANO, F. GALLETTI, F. TURSI & G. BELVEDERE. Caffeine metabolism in vivo and in peripheral blood lymphocytes in different species. Toxicol. Lett. (In press.)
50. SZCZAWINSKA, K., E. GINELLI, I. BARTOSEK, C. GAMBAZZA & C. PANTAROTTO. 1981. Caffeine does not bind covalently to liver microsomes from different animal species and to proteins and DNA from perfused rat liver. Chem. Biol. Interact. **34:** 345–354.
51. BROGGINI, M., P. GHERSA & M. G. DONELLI. Subcellular distribution of adriamycin in the liver and tumor of 3LL bearing mice. Eur. J. Cancer Clin. Oncol. (In press.)
52. MANARA, L., P. COCCIA & T. CROCI. 1982. Persistent tissue levels of TCDD in the mouse and their reduction as related to prevention of toxicity. Drug Metab. Rev. **13:** 423–446.
53. STASIECKI, P., F. OESCH, G. BRUNDER, E. D. JARASCH & W. W. FRANKE. 1980. Distribution of enzymes involved in metabolism of polycyclic aromatic hydrocarbons among rat liver endomembranes and plasma membranes. Eur. J. Cell Biol. **21:** 79–92.
54. BRESNICK, E., B. HASSUK, P. LIBERATOR, W. LEVIN & P. E. THOMAS. 1980. Nucleolar cytochrome P-450. Mol. Pharmacol. **18:** 550–552.
55. GARATTINI, E., G. GAZZOTTI & M. SALMONA. 1980. Is nuclear styrene monooxygenase activity a microsomal artifact? Chem. Biol. Interact. **31:** 341–346.
56. GAZZOTTI, G., E. GARATTINI & M. SALMONA. 1981. Nuclear metabolism II. Further studies on epoxide hydrolase activity. Chem. Biol. Interact. **35:** 311–318.

57. LU, A. Y. H. & S. B. WEST. 1980. Multiplicity of mammalian microsomal cytochromes P-450. Pharmacol. Rev. **31:** 277–295.
58. LU, A. Y. H. 1979. Multiplicity of liver drug metabolizing enzymes. Drug Metab. Rev. **10:** 187–208.
59. PANTAROTTO, C., J. BAGGOTT, A. GUAITANI, M. G. DONELLI & E. GARATTINI. Secondary metabolism and covalent binding to cellular macromolecules. *In* Short-Term Tests for Carcinogenesis: Quo Vadis? Excerpta Medica. Amsterdam, the Netherlands. (In press.)
60. VAINIO, H., K. HEMMINKI & E. ELOVAARA. 1977. Toxicity of styrene and styrene oxide on chick embryos. Toxicology **8:** 319–325.
61. THURMAN, R. G. & F. C. KAUFFMAN. 1980. Factors regulating drug metabolism in intact hepatocytes. Pharmacol. Rev. **31:** 229–251.
62. COLOMBO, T., M. BROGGINI, L. TORTI, E. ERBA & M. D'INCALCI. 1982. Pharmacokinetics of VP16-213 in Lewis lung carcinoma bearing mice. Cancer Chemother. Pharmacol. **7:** 127–131.
63. D'INCALCI, M., E. ERBA, M. VAGHI & L. MORASCA. 1982. *In vitro* cytotoxicity of VP 16 on primary tumor and metastasis of Lewis lung carcinoma. Eur. J. Cancer Clin. Oncol. **18:** 377–380.
64. HILL, B. T., R. D. H. WHELAN, H. T. PUPNIAK, L. Y. DENNIS & M. A. ROSHOLT. 1981. A comparative assessment of the in vitro effects of drugs on cells by means of colony assays or flow microfluorimetry. Cancer Chemother. Pharmacol. **7:** 21–26.
65. TORKELSON, A. R., J. A. LaBUDDE & J. H. WEIKEL, JR. 1974. The metabolic fate of cyclophosphamide. Drug Metab. Rev. **3:** 131–165.
66. DI MARCO, A. & F. ARCAMONE. 1975. DNA complexing antibiotics: daunomycin, adriamycin and their derivatives. Arzneim.-Forsch. **25:** 368–375.
67. SCHWARTZ, H. S. 1976. Mechanisms and selectivity of anthracycline aminoglycosides and other intercalating agents. Biomedicine **24:** 317–323.
68. FACCHINETTI, T., A. MANTOVANI, R. CANTONI, L. CANTONI, C. PANTAROTTO & M. SALMONA. 1977. Effect of daunomycin, adriamycin and its congener AD32 on the activity of DNase I from bovine pancreas. Biochem. Pharmacol. **26:** 1953–1954.
69. DI MARCO, A. 1975. Adriamycin (NSC-123127): mode and mechanism of action. Cancer Chemother. Pharmacol. **6** (part 3): 91–106.
70. PIGRAM, W. J., W. FULLER & L. D. HAMILTON. 1972. Stereochemistry of intercalation: interaction of daunomycin with DNA. Nature London New Biol. **235:** 17–19.
71. GHEZZI, P., M. G. DONELLI, C. PANTAROTTO, T. FACCHINETTI & S. GARATTINI. 1981. Evidence for covalent binding of adriamycin to rat liver microsomal proteins. Biochem. Pharmacol. **30:** 175–177.
72. SINHA, B. K. & J. L. GREGORY. 1981. Role of one-electron and two-electron reduction products of adriamycin and daunomycin in deoxyribonucleic acid binding. Biochem. Pharmacol. **30:** 2626–2629.
73. GARATTINI, E., M. G. DONELLI, P. CATALANI & C. PANTAROTTO. Intact rat liver nuclei catalyze adriamycin irreversible interactions with DNA and nuclear proteins. Toxicol. Lett. (In press.)
74. SINHA, B. K. & R. H. SIK. 1980. Binding of [^{14}C]-adriamycin to cellular macromolecules *in vivo.* Biochem. Pharmacol. **29:** 1867–1868.
75. LOWN, J. W., S.-K. SIM, K. C. MAJUMDAR & R.-Y. CHANG. 1977. Strand scission of DNA by bound adriamycin and daunorubicin in the presence of reducing agents. Biochem. Biophys. Res. Commun. **76:** 705–710.
76. NAKANISHI, Y. & E. L. SCHNEIDER. 1979. In vivo sister-chromatid exchange: a sensitive measure of DNA damage. Mutat. Res. **60:** 329–337.
77. BACHUR, N. R., S. L. GORDON & M. V. GEE. 1978. A general mechanism for microsomal activator of quinone anticancer agents to free radicals. Cancer Res. **38:** 1745–1750.
78. TRITTON, T. R. & G. YEE. 1982. The anticancer agent adriamycin can be actively cytotoxic without entering cells. Science **217:** 248–250.
79. JORI, A., M. SALMONA, L. CANTONI & G. GUISO. 1977. Effect of contraceptive drugs on liver monooxygenases in several animal species. *In* Pharmacology of Steroid Contraceptive Drugs. S. Garattini & H. W. Berendes, Eds.: 313–325. Raven Press. New York, N.Y.
80. KAHL, G. F., D. E. FRIEDERICI, S. W. BIGELOW, A. B. OKEY & D. W. NEBERT. 1980.

Ontogenetic expression of regulatory and structural gene products associated with the Ah locus. Comparison of rat, mouse, rabbit and sigmoden hispedis. Dev. Pharmacol. Ther. **1:** 137–162.

81. POLAND, A. & E. GLOVER. 1975. Genetic expression of aryl hydrocarbon hydroxylase by 2,3,7,8-tetrachlorodibenzo-*p*-dioxin: evidence for a receptor mutation in genetically non-responsive mice. Mol. Pharmacol. **11:** 389–398.

82. GREENLEE, W. F. & A. POLAND. 1979. Nuclear uptake of 2,3,7,8-tetrachlorodibenzo-*p*-dioxin in C57BL/6J and DBA/2J mice. J. Biol. Chem. **254:** 9814–9821.

83. GARATTINI, S., A. VECCHI, M. SIRONI & A. MANTOVANI. 1982. Immunosuppressant activity of TCDD in mice. *In* Chlorinated Dioxin and Related Compounds. O. Hutzinger, R. W. Frei, E. Merian & F. Pocchiari, Eds.: 403–409. Pergamon Press. Oxford, England.

84. VECCHI, A., M. SIRONI, M. A. CANEGRATI & S. GARATTINI. Comparison of the immunosuppressive effect of 2,3,7,8-tetrachlorodibenzo-*p*-dioxin (TCDD) and 2,3,7,8-tetrachlorodibenzofuran (TCDF) in mice. *In* Symposium on Chlorinated Dioxins and Dibenzofurans in the Total Environment. L. H. Keith, G. Choudhary & C. Rappe, Eds. Ann Arbor Science. Ann Arbor, Mich. (In press.)

85. BRESNICK, E. 1979. Nuclear activation of polycyclic hydrocarbons. Drug Metab. Rev. **10:** 209–223.

86. WATABE, T., M. ISOBE, T. SAWAHATA, K. YOSHIKAWA, S. YAMADA & E. TAKABATAKE. 1978. Metabolism and mutagenicity of styrene. Scand. J. Work Environ. Health 4(suppl. 2): 142–155.

87. RAPPE, C., H. R. BUSER & H.-P. BOSSHARDT. 1979. Dioxins, dibenzofurans and other polyhalogenated aromatics: production, use, formation, and destruction. Ann. N.Y. Acad. Sci. **320:** 1–18.

88. GIZZI, F., R. REGINATO, F. BENFENATI & R. FANELLI. 1982. Polychlorinated dibenzo-*p*-dioxins (PCDD) and polychlorinated dibenzofurans (PCDF) in emissions from an urban incinerator. I. Average and peak values. Chemosphere **6:** 577–583.

89. LUINI, W., A. MANTOVANI & S. GARATTINI. 1981. Effects of saccharin on primary humoral antibody production in rats. Toxicol. Lett. **8:** 1–6.

90. MANTOVANI, A., W. LUINI, G. P. CANDIANI, M. SALMONA, F. SPREAFICO & S. GARATTINI. 1980. In vitro effects of saccharin on cell-mediated host defence mechanisms. Toxicol. Lett. **5:** 287–295.

METABOLIC PROPERTIES OF *IN VITRO* SYSTEMS*

Anthony Dipple and C. Anita H. Bigger

Chemical Carcinogenesis Program
Basic Research Program–LBI
Frederick Cancer Research Facility
Frederick, Maryland 21701

INTRODUCTION

The toxic properties of many chemicals do not reside in the chemicals themselves but in one or more of their metabolites. For such chemicals, the biological system involved actually contributes to its own damage by providing the enzymes and cofactors necessary to generate these toxic metabolites. For example, Andrianov *et al.* showed that of various cultured cells examined, only those that extensively metabolized benzo[a]pyrene were subject to its toxic effects.[1] Gelboin *et al.* reached essentially the same conclusion, but more importantly, they also showed that 3-hydroxybenzo[a]pyrene, a metabolite of benzo[a]pyrene, was cytotoxic to several cell lines irrespective of their capacity to metabolize, and their sensitivities to, the parent compound.[2] The importance of metabolism in toxicity *in vivo* is well illustrated by the metabolic activation of chemical carcinogens, where it has been shown that certain metabolites of chemical carcinogens are more potent inducers of tumors in experimental animals than is the parent carcinogen.[3]

In vitro systems for toxicity testing require then not only an appropriate end point but also an appropriate metabolizing system if they are to be sensitive to chemicals that exert their toxic effects via a metabolite. At present, it is not possible to create an ideal metabolic activation system for use *in vitro* that would be suitable for all chemicals and for all toxic effects. This would require foreknowledge of the mechanisms through which various toxicities are expressed and the identities of all the toxic metabolites involved, and a much greater understanding of and ability to manipulate the enzyme systems involved in metabolism than presently exist. Because of these deficiencies in our knowledge, we presently utilize, for *in vitro* testing purposes, metabolic activation systems that are present in cellular systems or subcellular fractions, but it is possible to begin to evaluate how suitable these systems are for various purposes.

Perhaps the most obvious approach to comparing the metabolism of chemicals in various systems would be to monitor the metabolites formed. Much valuable work of this nature has been undertaken, but it is not an easy task when dealing with an *in vivo* system because metabolites are transported and excreted. Urinary and fecal metabolites give an overall view of the ultimate fate of most of a given chemical, and experimental procedures involving perfusion of various organs or cannulation of various ducts or vessels can be used to obtain information on metabolism in specific organs. It is easier to monitor metabolism *in vitro* because the system is contained. In either case, however, because metabolism involves the sequential attack of several enzyme systems on a substrate molecule, the distribution of metabolites present

*Research sponsored by the National Cancer Institute, Department of Health and Human Services, under contract no. NO1-CO-23909 with Litton Bionetics, Inc. The contents of this publication do not necessarily reflect the views or policies of the Department of Health and Human Services, nor does mention of trade names, commercial products, or organizations imply endorsement by the United States government.

0077–8923/83/0407–0026 $01.75/0 © 1983, NYAS

changes in a complex fashion with time, such that this approach is somewhat analogous to looking at a single frame in a motion picture. In intersystem comparison, it is not too obvious which individual frames from each system should be compared.

For these reasons, many laboratories, including our own, have chosen to monitor metabolism in various systems by analysis of the DNA-bound metabolites formed in these systems. This is a noninvasive procedure in that the metabolizing system, *in vivo* or *in vitro,* is not disturbed experimentally during the time interval being investigated. It also has the advantage that DNA-reactive metabolites are accumulated over the experimental period, such that a cumulative history of metabolic events is recorded. The accuracy of this record can be compromised in some systems, however, by the intervention of DNA-repair processes. Another disadvantage to this approach is that only a small fraction of the total metabolites, i.e., the DNA-reactive metabolites, are examined and that not all toxic effects are mediated by reaction with DNA. For example, 3-hydroxybenzo[a]pyrene, discussed above, is not known to be involved in DNA binding, and Iype et al. have demonstrated differential cytotoxic effects for 7,12-dimethylbenz[a]anthracene in normal and malignant rat liver epithelial cells despite similar levels of DNA binding.[4] On the other hand, such toxic effects as mutation and carcinogenesis, for example, do seem to be associated with DNA-reactive metabolites so that analysis of carcinogen/mutagen-DNA products formed in various metabolic activation systems does seem an appropriate initial approach to the comparison of various systems with respect to these toxicities.

Analyses of carcinogen-DNA interactions have mostly involved the administration of a radioactive carcinogen, isolation of DNA at appropriate times, enzymic hydrolysis of the DNA to deoxyribonucleosides, and chromatographic separation of the resultant radioactive carcinogen-deoxyribonucleoside adducts. This yields some form of "fingerprint" of the product formed, which can be compared with analogous "fingerprints" obtained in other metabolic activation systems even in cases where the exact structure of each specific product is unknown. Theoretically this allows fairly wide application of this approach (though herein we will limit our discussion to just a few examples), but in fact, unless very sophisticated separations are achieved and synthetic, characterized markers are available, it is very difficult to demonstrate that the products formed in two different systems are the same. In contrast, without markers and with limited chromatographic resolution, it is often easy to demonstrate differences between different systems. While most of the information discussed here has been obtained as summarized above, the recent development of the enzymic ^{32}P-postlabeling method by Randerath et al.[5] eliminates the need for a radioactive carcinogen/mutagen and will considerably broaden the applicability of this general approach.

IN VIVO SYSTEMS

The primary concern in most toxicity testing is the protection of man, but there is little information of course on carcinogen-DNA adducts formed in man *in vivo.* Development of the use of antibodies to detect the products of carcinogen-DNA interactions, recently reviewed by Poirier,[6] has permitted the tentative detection of DNA modified by the dihydrodiol epoxide of benzo[a]pyrene in white blood cells of some foundry workers and roofers[7] and in lung tissue from some hospital patients.[8] Another relevant observation is that Herron and Shank were able to detect both 7-methylguanine and O^6-methylguanine in DNA isolated from the liver of a male victim of probable dimethylnitrosamine poisoning.[9] These findings suggest that these two carcinogens are converted to the same reactive metabolites in both man and experimental animals.

Because of the paucity of information in man, experimental animals are the primary system for evaluating carcinogenic effects in man. The relationship between man and the animal systems is not always clear, however. For example, coal tar was recognized to be a skin carcinogen in man[10] and subsequently shown to be a skin carcinogen in mice,[11] while 2-naphthylamine, a human bladder carcinogen, produces liver tumors rather than bladder tumors in mice.[12]

In the literature on carcinogen-DNA adducts formed in experimental animals, there are large quantitative differences in extents of binding in different tissues or different species. However, subject to the difficulties mentioned earlier about absolutely proving adducts to be identical, it appears that the same pathways of metabolic activation to DNA-reactive metabolites largely prevail irrespective of the species or tissue investigated. For example, it is generally agreed that the majority of DNA-bound benzo[a]pyrene in mouse skin arises from reaction of the *anti*-7,8-dihydrodiol-9,10-epoxide with DNA[13-15] even when mice of different strains are used.[16] The same major product has been reported in DNA from rat skin,[15] mouse lung and liver,[17,18] and several different rat tissues.[19] One report, using a very short exposure to this carcinogen, concludes that the DNA adducts formed in rat lung and liver arise primarily from a metabolite other than this dihydrodiol epoxide,[20] and there are some differences in the other reports with respect to the minor adducts present. These will probably be clarified by further study.

Clearly defined differences with respect to a minor, but not insubstantial, DNA adduct have been documented for N-hydroxy-2-acetylaminofluorene in the Sprague-Dawley rat. In male or female liver or kidney, N-(deoxyguanosin-8-yl)-2-aminofluorene is found to be the major adduct, but N-(deoxyguanosin-8-yl)-2-acetylaminofluorene, attributed to activation by a sulfotransferase, is also found only in the male rat liver, not in kidney or in the female liver.[21] Furthermore, during the course of feeding 2-acetylaminofluorene to rats, the ratio of acetylated to deacetylated adduct changes from an initial value of 20:80 to 3:97 after 15 days.[22] Thus, while there are examples of the metabolic activation of a chemical carcinogen being largely similar irrespective of tissue or species, there are also examples where there are differences, which could be important toxicologically, between male and female organs of the same strain or even in the same tissue after exposure to the carcinogen.

In Vitro Systems

Tissue Explants

Tissue explants are closer to the integrity of *in vivo* animal systems than are most *in vitro* systems. Human tissues have been studied extensively with this approach, and this work, recently reviewed by Autrup and by Harris *et al.*,[23,24] shows that several carcinogens are extensively metabolized and bound to DNA in various cultured human tissues, with interindividual quantitative differences being in the range of 10- to 150-fold. So far, DNA adduct analyses suggest that the same routes of activation of chemical carcinogens found in experimental animals are operative in man. Thus, in analyses of benzo[a]pyrene-DNA adducts formed in tracheobronchial tissues from mice, rats, hamsters, bovines, and humans, Autrup *et al.* found that the (+) *anti*-dihydrodiol epoxide was responsible for the majority of the adducts in all cases except the rat, where the (−) enantiomer made an equal or greater contribution to the total products.[25] In explants of other human tissue, such as the colon[26] or bladder,[27] the major adduct corresponds to that in the respiratory system tissue mentioned above.

Cultured Cells

In studies of DNA-bound adducts in cultured cells, some greater variability with respect to metabolic activation becomes apparent. This is perhaps partially because these systems are easier to work with than whole animals and, therefore, more studies with them have been undertaken, but some of this variability probably reflects on the systems themselves. In a recent review, Pelkonen and Nebert have discussed many factors that can influence the metabolic capacities of cultured cells,[28] and recent work from Williams' laboratory clearly points out that different culture conditions may be optimal for primary cultures of hepatocytes from different animal species.[29] It has long been known that the cytochrome P_{450} content of liver-derived cell cultures can be quite different from that of liver *in vivo*, leading Owens and Nebert in 1975 to counsel caution with respect to extrapolations between such systems.[30]

This caution is further justified by subsequent DNA-binding investigations. For example, in contrast to the findings discussed earlier for rat liver *in vivo*,[21,22] the major DNA adduct found in hepatocyte cultures exposed to either 2-acetylaminofluorene[31] or to its *N*-hydroxy derivative[32] is the acetylated adduct *N*-(deoxyguanosin-8-yl)-2-acetylaminofluorene. The perversity of this difference between *in vivo* and *in vitro* is compounded by the findings that rat or mouse epidermal or fibroblast cell cultures generate the same deacetylated major adduct formed in rat liver *in vivo*.[31] In this same study, Poirier *et al*. noted that in response to removal of serum from culture medium, quantitative and qualitative changes in DNA-bound adducts could result.[31] In studies with benzo[a]pyrene, there have been differing reports on DNA binding in hepatocyte preparations (as opposed to cultures). Jernström *et al*. found binding to DNA largely involved an activated species derived from 9-hydroxybenzo[a]pyrene,[33] while Shen *et al*. found this route to be responsible for minor adducts only, with the dihydrodiol epoxide accounting for most of the binding.[34] In cell-culture systems generally, benzo[a]pyrene largely seems to be activated for DNA binding through the bay region dihydrodiol epoxide route, but differences in product distributions are observed depending on the cell system being used. So far, it is not altogether clear as to what extent these differences may be attributable to selective DNA-repair systems in different cells or to differences in metabolic activation systems. It is not possible to review all these studies comprehensively, but findings seem to range between Lo and Kakunaga's report of only one benzo[a]pyrene-DNA adduct in BALB/3T3 mouse cells[35] to the Eastman *et al*. report of 15 different adducts in hamster tracheal epithelial cells.[36] While the differences were not so extreme, Weinstein and his collaborators have noted a less complex distribution of adducts in mouse C3H10T 1/2 cells[37] than in hamster embryo cells.[38] Several years ago, Baird and Diamond showed that in hamster embryo cells, the isomeric *syn*-dihydrodiol epoxide as well as the *anti*-dihydrodiol epoxide makes a large contribution to DNA binding, depending on the length of time for which the cells are exposed to benzo[a]pyrene.[39] This accounts then for some of the differences between hamster cells and mouse cells with respect to the number of different adducts formed. It also raises the issue that the stereoselectivity of metabolism of benzo[a]pyrene may be different in different systems, and this is an important issue with respect to the activation of chemicals for toxicity testing. In general, less sophisticated chromatographic separations and no synthetic markers have been available for similar studies with 7,12-dimethylbenz[a]anthracene so that it would be difficult to determine, at present, contributions from different isomeric bay region dihydrodiol epoxides. However, Bigger *et al*. have been able to distinguish adducts formed through the bay region dihydrodiol epoxide route from those formed through other routes,[40-42] and in studies of binding of this carcinogen to DNA in cultures of human skin cells, rat liver cells, hamster embryo cells, or mouse embryo cells, the DNA adducts all exhibit similar chromatographic properties[43,44] and correspond to the bay region dihydrodiol epoxide adducts formed in mouse skin *in vivo*.[40,45]

Perhaps the cell-culture systems resemble overall the *in vivo* systems in that one can find many examples where the metabolic activation system in different cell-culture systems seems to be qualitatively the same despite large quantitative differences. If the subtleties of these systems are examined, however, differences from system to system in terms of whether an adduct is acetylated or not or in terms of the relative stereochemistry of the reactive metabolite can be revealed.

Subcellular Systems

Perhaps not surprisingly, some large changes in metabolic activation are observed when making the transition to a subcellular system. In some studies, crude tissue homogenates are used, but it is more common to use the supernatant from centrifugation of the crude homogenate at 9,000 $\times$ g (S-9 fraction)[46] or to use a microsomal preparation, which is usually the washed pellet resulting from centrifugation of the S-9 fraction at 100,000 $\times$ g for one hour.[47,48] Since in each of these steps some components of the original tissue are removed, the metabolic properties of these three subcellular systems will not be identical. In particular, the microsomal preparation will result in the loss of soluble glutathione transferases, sulfotransferases, and cofactors such as glutathione and uridine diphosphate (UDP)–glucuronic acid. For activity, these systems require the addition of reduced nicotinamide adenine dinucleotide phosphate (NADPH), and an NADPH-generating system consisting of glucose-6-phosphate, glucose-6-phosphate dehydrogenase, and Mg^{2+} may also be added.[47,48]

For some chemical carcinogens, these subcellular systems seem to mimic satisfactorily the metabolic activation that occurs *in vivo*. For example, either microsomes[49] or an S-9 fraction[50] from livers of rats induced with phenobarbital converts aflatoxin B_1 to the same DNA-reactive epoxide formed in rat liver *in vivo*. Similarly, S-9 fraction from an Aroclor 1254–pretreated male rat catalyzed the binding of N-hydroxy-2-acetylaminofluorene to the DNA of *Salmonella typhimurium* to yield only the deacetylated N-(deoxyguanosin-8-yl)-2-aminofluorene adduct.[51] While this is the major adduct formed in rat liver *in vivo*, it is usually accompanied by some acetylated adduct in the male rat liver. For benzo[*a*]pyrene with either 3-methylcholanthrene- or Aroclor-induced mouse liver S-9 fraction, the major product found in DNA is reported to be the same as the major *in vivo* product, but quite different products were obtained if phenobarbital induction was used.[52] In a similar study using liver microsomes from 3-methylcholanthrene-induced rats, the benzo[*a*]pyrene dihydrodiol epoxide again was found to be the major reactive metabolite produced,[53] but its adducts were accompanied by small amounts of adducts believed to result from a reactive 4,5-epoxide generated from the 9-phenol.[54] These reports for benzo[*a*]pyrene are not entirely typical of the whole field. For example, King *et al.* in an early study of microsome-catalyzed binding of benzo[*a*]pyrene to DNA found that the major adduct resulted from the generation of some reactive product—later identified as the 9-phenol-4,5-epoxide[54]—other than the dihydrodiol epoxide.[55] The latter could account for only a small fraction of the total adducts in these studies. Similarly, in our studies with 7,12-dimethylbenz[*a*]anthracene it was clear that there was little resemblance between the DNA adducts formed *in vivo* and those formed in the presence of rat liver microsomes or S-9 fraction.[40–42] Moreover, it was clear that in either of these subcellular systems, the qualitative nature of the DNA adducts formed was dependent on the hydrocarbon concentration,[41,42] though this is not the case for cell cultures or mouse skin *in vivo*.[44] For this carcinogen, the K-region epoxide accounted for most of the binding in the subcellular systems at high substrate concentrations while as yet unidentified other reactive metabolites were preponderant at lower substrate concentrations. Our overall conclusion is that for 7,12-dimethylbenz[a]anthracene, the metabolic activation observed *in vivo* is largely reproduced *in vitro* provided the integrity of a cellular system is maintained.[43] Similar conclusions have recently been reported for benzo[a]pyrene also.[56]

While all of the reasons for the different metabolic activation of hydrocarbons in subcellular systems *in vitro* are not totally clear, recent studies on the effects of various additions to microsomal systems are beginning to clarify this situation substantially in the case of benzo[a]pyrene. With 3-methylcholanthrene-induced rat liver microsomes, most authors have found that benzo[a]pyrene primarily gives the phenol-epoxide adducts with DNA along with some dihydrodiol epoxide adducts.[54-61] If UDP–glucuronic acid is added to this system however, the ratio of adducts formed from these two reactive derivatives is shifted somewhat in favor of the dihydrodiol epoxide.[57,58] Guenthner *et al.*[59] also found that the products arising from the phenol-epoxide route could be selectively substantially reduced in the microsome-catalyzed system by the addition of albumin,[59] which they presumed selectively trapped the phenol-epoxide and not the dihydrodiol epoxide. These and other observations led these authors to suggest that the reactive phenol-epoxide may well be produced in intact cellular systems also but that in this environment, it is trapped more easily before it can modify DNA than in the microsomal systems. Addition of purified epoxide hydrolase to the microsomal incubations also has the dramatic effect of almost totally eliminating phenol-epoxide DNA adducts and making the microsomal binding picture look very much like the *in vivo* situation.[60,61] This same effect can be achieved if prior to preparation of microsomes, the rats are injected with *trans*-stilbene oxide, which leads to an approximately threefold induction of epoxide hydrolase levels.[61,62] These studies provide a good rationalization for the differences observed in subcellular metabolism systems compared with the intact cell. Further, they suggest that by appropriate manipulation, the subcellular systems can be modified to reflect the intact cell situation.

CONCLUSIONS

Analysis of DNA-bound carcinogen in different metabolic activation systems suggests that where activation occurs it is likely to be through the same metabolic pathway irrespective of the animal species, organ, or cell-culture system considered. Subtle differences in the route of metabolic activation in different organs or cells have been documented, however, and these differences may be important toxicologically. Subcellular systems can fairly reflect metabolic activation *in vivo* for some chemicals, but for others there can be drastic differences between the events in these two systems. In cases such as the latter, it may be possible to correct these differences by making modifications to the subcellular system to make it more like a cellular system. This may be appropriate for some purposes. For others it may be better not to destroy the cellular integrity in the first place.

ACKNOWLEDGMENTS

The authors wish to acknowledge helpful discussions with Mimi Poirier, Willie Lijinsky, Curtis Harris, and Fred Beland.

REFERENCES

1. ANDRIANOV, L. N., G. A. BELITSKY, O. J. IVANOVA, A. Y. KHESINA, S. S. KHITROVO, L. M. SHABAD & J. M. VASILIEV. 1967. Br. J. Cancer **21:** 566–575.
2. GELBOIN, H. V., E. HUBERMAN & L. SACHS. 1969. Proc. Nat. Acad. Sci. USA **64:** 1188–1194.
3. MILLER, E. C., J. A. MILLER & H. A. HARTMANN. 1961. Cancer Res. **21:** 815–824.
4. IYPE, P. T., J. E. TOMASZEWSKI & A. DIPPLE. 1979. Cancer Res. **39:** 4925–4929.
5. RANDERATH, K., M. V. REDDY & R. C. GUPTA. 1981. Proc. Nat. Acad. Sci. USA **78:** 6126–6129.
6. POIRIER, M. C. 1981. J. Nat. Cancer Inst. **67:** 515–519.
7. HARRIS, C. C. Personal communication.
8. PERERA, F. P., M. C. POIRIER, S. H. YUSPA, J. NAKAYAMA, A. JARETZKI, M. M. CURNEN, D. M. KNOWLES & I. B. WEINSTEIN. 1982. Carcinogenesis **3:** 1405–1410.
9. HERRON, D. C. & R. C. SHANK. 1980. Cancer Res. **40:** 3116–3117.
10. BUTLIN, H. T. 1892. Br. Med. J. (2): 66–71.
11. TSUTSUI, H. 1918. Gann **12:** 17–21.
12. BONSER, G. M., D. B. CLAYSON, J. W. JULL & L. N. PYRAH. 1952. Br. J. Cancer **6:** 412–424.
13. KOREEDA, M., P. D. MOORE, P. G. WISLOCKI, W. LEVIN, A. H. CONNEY, H. YAGI & D. M. JERINA. 1978. Science **199:** 778–781.
14. VIGNY, P., Y. M. GINOT, M. KINDTS, C. S. COOPER, P. L. GROVER & P. SIMS. 1980. Carcinogenesis **1:** 945–950.
15. BAER-DUBOWSKA, W. & K. ALEXANDROV. 1981. Cancer Lett. **13:** 47–52.
16. PHILLIPS, D. H., P. L. GROVER & P. SIMS. 1978. Int. J. Cancer **22:** 487–494.
17. WILSON, A. G. E., H.-C. KUNG, M. BOROUJERDI & M. W. ANDERSON. 1981. Cancer Res. **41:** 3453–3460.
18. EASTMAN, A., J. SWEETENHAM & E. BRESNICK. 1978. Chem. Biol. Interact. **23:** 345–353.
19. KLEIHUES, P., G. DOERJER, M. EHRET & J. GUZMAN. 1980. Arch. Toxicol. Suppl. **3:** 237–246.
20. BOROUJERDI, M., H.-C. KUNG, A. G. E. WILSON & M. W. ANDERSON. 1981. Cancer Res. **41:** 951–957.
21. BELAND, F. A., K. L. DOOLEY & C. D. JACKSON. 1982. Cancer Res. **42:** 1348–1354.
22. POIRIER, M. C., B. TRUE & B. A. LAISHES. 1982. Cancer Res. **42:** 1317–1321.
23. AUTRUP, H. 1982. Drug Metab. Rev. **13**(4): 603–646.
24. HARRIS, C. C., B. F. TRUMP, R. GRAFSTROM & H. AUTRUP. 1982. J. Cell. Biochem. **18:** 285–294.
25. AUTRUP, H., F. C. WEFALD, A. M. JEFFREY, H. TATE, R. D. SCHWARTZ, B. F. TRUMP & C. C. HARRIS. 1980. Int. J. Cancer **25:** 293–300.
26. AUTRUP, H., C. C. HARRIS, B. F. TRUMP & A. M. JEFFREY. 1978. Cancer Res. **38:** 3689–3696.
27. STONER, G. D., F. B. DANIEL, K. M. SCHENK, H. A. J. SCHUT, P. J. GOLDBLATT & D. W. SANDWISCH. 1982. Carcinogenesis **3:** 195–201.
28. PELKONEN, O. & D. W. NEBERT. 1982. Pharmacol. Rev. **34:** 189–222.
29. MASLANSKY, C. J. & G. M. WILLIAMS. 1982. In Vitro **18:** 683–692.
30. OWENS, I. S. & D. W. NEBERT. 1975. Mol. Pharmacol. **11:** 94–104.
31. POIRIER, M. C., G. M. WILLIAMS & S. H. YUSPA. 1980. Mol. Pharmacol. **18:** 581–587.
32. HOWARD, P. C., D. A. CASCIANO, F. A. BELAND & J. G. SHADDOCK, JR. 1981. Carcinogenesis **2:** 97–102.
33. JERNSTRÖM, B., S. ORRENIUS, O. UNDEMAN, A. GRASLUND & A. EHRENBERG. 1978. Cancer Res. **38:** 2600–2607.
34. SHEN, A. L., W. E. FAHL & C. R. JEFCOATE. 1980. Arch. Biochem. Biophys. **204**(2): 511–523.
35. LO, K.-Y. & T. KAKUNAGA. 1981. Biochem. Biophys. Res. Commun. **99:** 820–829.
36. EASTMAN, A., B. T. MOSSMAN & E. BRESNICK. 1981. Cancer Res. **41:** 2605–2610.
37. BROWN, H. S., A. M. JEFFREY & I. B. WEINSTEIN. 1979. Cancer Res. **39:** 1673–1677.

38. IVANOVIC, V., N. E. GEACINTOV, H. YAMASAKI & I. B. WEINSTEIN. 1978. Biochemistry **17:** 1597–1603.
39. BAIRD, W. M. & L. DIAMOND. 1977. Biochem. Biophys. Res. Commun. **77:** 162–167.
40. BIGGER, C. A. H., J. E. TOMASZEWSKI & A. DIPPLE. 1978. Biochem. Biophys. Res. Commun. **80:** 229–235.
41. BIGGER, C. A. H., J. E. TOMASZEWSKI & A. DIPPLE. 1980. Carcinogenesis **1:** 15–20.
42. BIGGER, C. A. H., J. E. TOMASZEWSKI, A. W. ANDREWS & A. DIPPLE. 1980. Cancer Res. **40:** 655–661.
43. BIGGER, C. A. H., J. E. TOMASZEWSKI, A. DIPPLE & R. S. LAKE. 1980. Science **209:** 503–505.
44. BIGGER, C. A. H. & A. DIPPLE. 1983. *In* Organ and Species Specificity in Chemical Carcinogenesis. R. Langenbach, S. Nesnow & J. M. Rice, Eds.: 587–603. Plenum Press. New York, N.Y.
45. MOSCHEL, R. C., W. M. BAIRD & A. DIPPLE. 1977. Biochem. Biophys. Res. Commun. **76:** 1092–1098.
46. GARNER, R. C., E. C. MILLER & J. A. MILLER. 1972. Cancer Res. **32:** 2058–2066.
47. GELBOIN, H. V. 1969. Cancer Res. **29:** 1272–1276.
48. GROVER, P. L. & P. SIMS. 1968. Biochem. J. **110:** 159–160.
49. ESSIGMANN, J. M., R. G. CROY, A. M. NADZAN, W. F. BUSBY, JR., V. N. REINHOLD, G. BUCHI & G. M. WOGAN. 1977. Proc. Nat. Acad. Sci. USA **74:** 1870–1874.
50. STARK, A. A., J. M. ESSIGMANN, A. L. DEMAIN, T. R. SKOPEK & G. N. WOGAN. 1979. Proc. Nat. Acad. Sci. USA **76:** 1343–1347.
51. BERANEK, D. T., G. L. WHITE, R. H. HEFLICH & F. A. BELAND. 1982. Proc. Nat. Acad. Sci. USA **79:** 5175–5178.
52. SANTELLA, R. M., D. GRUNBERGER & I. B. WEINSTEIN. 1979. Mutat. Res. **61:** 181–189.
53. FAHL, W. E., D. G. SCARPELLI & K. GILL. 1981. Cancer Res. **41:** 3400–3406.
54. KING, H. W., M. H. THOMPSON & P. BROOKES. 1976. Int. J. Cancer **18:** 339–344.
55. KING, H. W., M. H. THOMPSON & P. BROOKES. 1975. Cancer Res. **34:** 1263–1269.
56. ASHURST, S. W. & G. M. COHEN. 1982. Carcinogenesis **3:** 267–273.
57. FAHL, W. E., A. L. SHEN & C. R. JEFCOATE. 1978. Biochem. Biophys. Res. Commun. **85:** 891–899.
58. NEMOTO, N., T. HIRAKAWA & S. TAKAYAMA. 1980. Carcinogenesis **1:** 115–120.
59. GUENTHNER, T. M., B. JERNSTRÖM & S. ORRENIUS. 1980. Carcinogenesis **1:** 407–418.
60. ALEXANDROV, K., P. M. DANSETTE & C. FRAYSSINET. 1980. Biochem. Biophys. Res. Commun. **93:** 611–616.
61. GUENTHNER, T. M. & F. OESCH. 1981. Cancer Lett. **11:** 175–183.
62. LESCA, P., T. M. GUENTHNER & F. OESCH. 1981. Carcinogenesis **2:** 1049–1056.

BIOLOGICAL SIGNIFICANCE OF END POINTS

Virginia C. Dunkel

Bureau of Foods
Food and Drug Administration
Washington, D.C. 20204

Traditionally, long-term animal studies have been accepted as the only experimental laboratory method suitable for providing conclusive evidence for the carcinogenicity of a chemical. However, scientific advances in carcinogenesis, genetics, and molecular biology as well as recognition that alternative methods are needed to aid in the evaluation of the numerous chemicals that may pose a hazard have led to a better recognition of the utility of short-term *in vitro* test methods for identifying carcinogens. Many different systems have been proposed for such testing purposes, and in a review article Hollstein *et al.* described over 100 short-term tests that have been used to various degrees for testing chemicals.[1] However, only a relatively small portion of this array of test methods has received significant, widespread use. Several factors contribute to the selective process, such as the ease with which the assay can be performed, the data base that provides information on predictive value, and the nature of the end point measured in the test system. The biological end points of *in vitro* test systems that are available for assessing the carcinogenic potential of chemicals are both genetic and nongenetic and can be measured in both prokaryotic and eukaryotic cells.

The end points used to assess the interaction between various microorganisms and chemical carcinogens include mutation induction, DNA damage, and chromosomal effects. A large number of assays using bacteria and fungi as test organisms measure mutation induction. The mutations are detected as phenotypic changes and can result from alterations in the structure of DNA such as base substitutions, frameshifts, large deletions, insertions, and translocations. The most extensively studied and widely used microbial gene mutation assay is the Ames *Salmonella*/mammalian microsome test.[2] Other microorganisms that have been used for testing but to a much lesser degree include other strains of bacteria such as *Escherichia coli*,[3] the yeast *Saccharomyces cerevisiae*,[4] and the mold *Neurospora crassa*.[5] A general problem encountered in using both yeasts and molds, however, is the permeability of the cell wall, which can restrict entry of certain chemicals into the cell. In addition, the eukaryotic microorganisms usually require higher concentrations of chemicals for induction of mutations than do bacteria, probably because yeast and molds exhibit more powerful detoxification mechanisms and have more DNA-repair pathways.

In the microbial assays measuring DNA damage, two isogenic strains of bacteria are used, which are identical in all respects except for their ability to repair DNA lesions. One of the strains is DNA-repair deficient and the other is competent in repairing DNA lesions. The effects of a compound are therefore scored as the preferential killing of the DNA-repair-deficient strains. Several different kinds of bacteria including *E. coli*,[6] *Bacillus subtilis*,[7] and *Salmonella typhimurium*[2] have been employed in these tests.

Chromosomal effects such as multilocus chromosome deletions and chromosomal nondisjunction can be measured in fungi. However, such test methods have not been generally used for determining the genotoxic effects of chemicals.

The end points used in mammalian cell test systems are similar to those of bacteria and fungi and include gene mutation, chromosomal effects, and DNA repair and

34

0077–8923/83/0407–0034 $01.75/0 ©1983, NYAS

damage. Importantly, neoplastic transformation, an end point of significant relevance and value in assessing chemicals for carcinogenic potential, is also measured in mammalian cells.

Gene mutations can be conveniently measured in several mammalian cell systems, and such events are detected as phenotypic changes, as is the case with bacteria and yeast. The cells most frequently utilized in these assays include V79[8-10] and CHO[11] Chinese hamster cells and L5178Y mouse lymphoma cells.[12,13] The markers commonly used in V79 and CHO cells to detect the mutagenic events are resistance to 8-azaguanine or 6-thioguanine, which arises from a loss of hypoxanthineguanine phosphoribosyl transferase (HGPRT) activity, or resistance to ouabain, which is due to alteration of membrane receptor sites. In L5178Y mouse lymphoma cells, the marker generally used is resistance to bromodeoxyuridine (BUdR) or trifluorothymidine (TFT), which results from a loss of thymidine kinase activity. Since these cell lines have limited capacity to metabolize all chemicals to their ultimate reactive form, exogenous metabolic activation is provided either by liver enzyme S-9 preparations or intact cells. Both primary fibroblasts[14] and liver epithelial cells[15] have been used. Cell-mediated mutagenesis appears to be a better indicator of *in vivo* metabolic pathways[16-19] and may also reflect the organ specificity of the chemicals tested.

Chromosomal aberrations observed in mammalian cells are characteristic of damage sustained in G_1 cells, which are translated into breakage/exchange figures prior to chromosome replication. The aberrations, such as breaks, terminal and interstitial deletions, rings, translocations, and dicentrics, can be detected in a wide range of mammalian cells. As a group, Chinese hamster cells are probably the most well suited for such determinations, since they have short cell cycles and small numbers of large chromosomes.[20,21] In addition to direct testing *in vitro*, an *in vivo* approach can also be used to assess chromosomal effects. In the *in vivo* studies, cells, generally peripheral blood lymphocytes[22] or bone marrow cells,[23,24] are cultured after an *in vivo* exposure to the chemical under test. The subsequent techniques used for processing and analysis of the cells are then similar to those used in the direct *in vitro* method. In addition to the gross morphological effects on chromosomes, assessments can also be made for sister chromatid exchanges (SCEs).[25] SCEs involve equal symmetrical exchanges between sister chromatids and do not result in any gross alteration in chromosome morphology. The exchange events between sister chromatids are visualized after they have been grown for two cell cycles in BUdR and then stained.

A variety of techniques can be used to measure DNA damage and consequent DNA repair. Strand breakage can be assessed by either alkaline sucrose gradient centrifugation[26,27] or by alkaline elution from membrane filters.[28,29] However, great care is required in interpreting the results of such assays, since dead or dying cells will display DNA fragmentation as a result of the action of degradative nucleases. Repair synthesis or unscheduled DNA synthesis can be measured either by autoradiography[30,31] or by liquid scintillation counting.[32,33] Only with autoradiography is it possible to unequivocally distinguish repair synthesis from replicative DNA synthesis, and this technique is preferred for screening.[34] Among the cell types that have been used for repair studies are continuous cell lines such as WI-38[35] and HeLa.[36] Cell lines require inhibition of replicative DNA synthesis, however, and such techniques as hydroxyurea suppression or arginine deprivation are employed. Since the metabolic capabilities of the cells are limited, they also require addition of exogenous systems for metabolic activation. Both of these problems have been overcome with the development of a DNA-repair assay in primary rat hepatocyte cultures.[37,38] The freshly isolated, nondividing liver cells that are used have the capacity to metabolize carcinogens and respond with DNA repair. Studies to date have shown that this assay has substantial sensitivity and reliability with activation-dependent chemicals.[34,39,40]

The last end point that can be used in mammalian cells is morphological transformation. Among all the *in vitro* end points, it is the most relevant to the *in vivo* situation and can be considered as carcinogenesis *in vitro*. Both epithelial and fibroblastic cells can be used for transformation, but the test systems now of major importance all employ fibroblasts. The morphological change observed, whether as altered colonies or foci in a monolayer of normal cells, is generally characterized by piling up of the cells in an irregular, crisscross pattern representing a loss of growth inhibition and cell-cell orientation. This is in contrast to the regular and orderly growth pattern of the nontransformed cells. With cell passage, the transformed cells acquire other characteristics such as the ability to grow in semisolid medium and the capacity to produce tumors after transplantation into syngeneic or immunosuppressed animals. Probably the system that has been most extensively used in transformation studies is the Syrian hamster embryo clonal assay,[41–44] in which primary or early passage cells are used as the target population. The cells are diploid and appear to have enzyme systems that allow metabolism of a fairly wide spectrum of chemicals to their active forms.

Continuous lines that can be used for transformation include C3H/10T1/2[45] and BALB/c 3T3[46–48] mouse cells. In assays with these cells, foci of morphologically transformed cells appear on a monolayer of normal cells. In contrast to hamster embryo cells, the mouse cell lines are aneuploid and have a more limited capacity to metabolize chemicals.

Another system that has been used for testing chemicals utilizes the BHK-21 baby hamster kidney cell line.[49,50] The end point in this assay is not morphological transformation as in the other assays but the ability of the treated cells to form colonies in soft agar. BHK-21 cells can be considered premalignant because they can form colonies in semisolid medium and, when grown under the proper conditions, can produce tumors after transplantation into the hamster.[51] Although the assay may be useful for testing chemicals for biological activity, it is not in character with the other transformation assays in which the cell undergoes definite morphological changes.

Considering the end points available and the number of cell systems that can be used for measuring them, we are concerned about the biological relationship of the various *in vitro* methods for testing chemicals for carcinogenic potential. *In vitro* transformation, or carcinogenesis *in vitro* as it has been referred to, can probably be considered as the most relevant end point because it corresponds to an observed *in vivo* effect. Although the cells when initially transformed do not have the capacity to induce tumors, on transplantation into the appropriate host they do acquire this characteristic after a number of cell divisions. In certain respects the *in vitro* progression of morphologically transformed "preneoplastic" cells to a malignant population is a counterpart to tumor development *in vivo*.

Historically, the somatic mutation theory of cancer was the basis for the development of many of the aforementioned genetic tests. Some of the evidence to support this concept and thus in turn the use of genetic test methods includes the following facts:

1. The conversion of a normal cell to a neoplastic cell results in a heritable change with the neoplastic characteristic transferred to daughter cells on division.
2. A number of neoplastic diseases such as neuroblastoma, retinoblastoma, and Wilm's tumor are inherited as autosomal dominant or autosomal recessive conditions.[52]
3. Chromosomal alterations are observed in tumors, and such changes appear to occur nonrandomly.[53]
4. There is a higher incidence of cancer in individuals with certain genetic

disorders such as Bloom's syndrome, Fanconi's anemia, and xeroderma pigmentosa. Individuals with xeroderma pigmentosa are predisposed to an increased incidence of skin tumors, and studies have shown that an enzyme defect in the cell results in inadequate repair of DNA damage induced by ultraviolet radiation. In Fanconi's anemia and Bloom's syndrome a number of different tumors have been observed, and both of these disorders are associated with chromosomal fragility.[52]

5. Carcinogens are metabolized to electrophilic species,[54] which react with macromolecules including DNA, RNA, and protein.

A number of aspects of carcinogenesis are inconsistent with the somatic mutation theory, however. Some agents such as asbestos[55,56] and other foreign bodies[57] produce tumors without evidence of direct interaction with DNA. Moreover, the incidence of tumors is higher in immunologically impaired individuals,[58] and it has been proposed that the development of lymphoid tumors results from impaired regulation of lymphoid tissue.[59] In addition, experimental studies with mouse teratocarcinoma cells have shown that after these malignant cells are transplanted into mouse blastocysts, they give rise to a variety of normal tissues in mosaic mice.[60] The implication is that the neoplastic conversion did not involve a structural change in the DNA but rather a change in gene expression. Since these other aspects of the carcinogenic process require consideration, justification for *in vitro* tests cannot at present be based on mechanistic events.

In view of this correspondence between an *in vivo* effect and an *in vitro* end point as a reason for using transformation, and the somatic mutation theory of cancer as a reason for using genetic end points, one might be tempted to conclude that any single end point could be construed as sufficient for making a determination about the potential carcinogenicity of a chemical. However, in the actual application of *in vitro* test methods, the significance of end points resides mainly in the correlations between the results obtained in well-conducted and definitive long-term carcinogenicity bioassays.

We all recognize the importance of correlation studies in establishing a foundation for the use of *in vitro* assays. Their purpose is to gain the necessary perspective for reasonable application of the methodology; in addition to having a sufficient base of test information in terms of numbers of chemicals tested, it is important to know what types of responses will be obtained with compounds in the different chemical classes. Absolute precision in percent correlations cannot be expected, since both biological and practical considerations have an impact on both the *in vitro* and the *in vivo* evaluation of a chemical.

Essentially an *in vivo* bioassay, as conducted according to the National Cancer Institute (NCI)–National Toxicology Program protocol,[61] is a battery of four tests consisting of an experiment on the male mouse, the female mouse, the male rat, and the female rat. Although 50 animals are used per group, the bioassay is relatively insensitive, since it is designed to detect levels of tumor incidence of 5–10% as compared to none in the control.[62] In addition there is well-documented variability of response and assays for which no definitive conclusion can be reached. Of 203 NCI carcinogenicity bioassays (data summarized in Reference 56), 67 gave no significant evidence for carcinogenicity and 15 were identified as inadequate for evaluation. The remaining 121 assays then fell into a number of different categories. In 20 (16.5%) assays there was positive evidence for carcinogenicity in both sexes of both species; in 23 (19.0%) assays there was positive evidence in at least one sex of both species; in 53 assays (43.8%) there was positive evidence in only one species; and finally, in 25 assays (20.6%) there was only suggestive evidence of a carcinogenic effect. Evaluation of *in vivo* bioassay data is difficult and requires a mix of toxicological, pathological, and statistical interpretations. This, then, is the standard (denominator) against which *in vitro* tests are measured—a standard with its own set of problems in data evaluation.

The numerator of the correlation equation contains the results obtained by using the various *in vitro* end points. An important factor here is the potential of any *in vitro* test to generate nonconcordant results, which unfortunately have been referred to as false negatives or false positives. A false-negative result is one in which a short-term test gives a negative result for a known carcinogen or, more importantly, for carcinogens within a specific class of chemicals, and a false-positive result is one in which the test substance or related chemicals induce measurable responses in the *in vitro* test systems but have not induced tumors in long-term tests. It would be unrealistic to expect complete agreement between results obtained *in vitro* and those obtained *in vivo,* and there are some obvious explanations for this situation. Nonconcordant responses could be related to the method used for biotransformation of particular substances to their ultimate reactive forms. Activating systems, either whole cells or the liver homogenates that are added to *in vitro* test systems, cannot duplicate the entire metabolic capability of the intact animal, and it can be expected that some procarcinogens may fail to show activity in certain short-term test systems. On the other hand, the innate metabolic capabilities of the target cells must also be considered. The processing of certain classes of chemicals could possibly lead to the production of end products that would not be produced *in vivo* but that would have *in vitro* activity. Different substances may act through different mechanisms to induce tumors in the whole animal, and *in vitro* tests may not be capable of mimicking such interactions. For example, some agents may act as promoters, simply accelerating the expression of cells that have already been initiated; with other substances, such as asbestos, the physical interaction with the affected tissue is important. It can also be expected that sensitive *in vitro* tests may respond to the presence of impurities at levels too low to be detected in whole-animal bioassays.

These factors do not lessen the utility of *in vitro* test systems but should be taken into consideration in the design and use of a battery of tests so that meaningful information can be obtained on carcinogenic potential and a consistent basis can be established for control of chemical carcinogens in our environment.

REFERENCES

1. HOLLSTEIN, M., J. MCCANN, F. A. ANGELOSANTE & W. W. NICHOLS. 1979. Short-term tests for carcinogens and mutagens. Mutat. Res. **65:** 133–226.
2. AMES, B. N., J. MCCANN & E. YAMASAKI. 1975. Methods for detecting carcinogens with the Salmonella/mammalian-microsome mutagenicity test. Mutat. Res. **31:** 347–364.
3. MOHN, G. R., J. ELLENBERGER & P. J. VAN BLADEREN. 1980. Evaluation and relevance of *Escherichia coli* test systems for detecting and for characterizing chemical carcinogens and mutagens. *In* Applied Methods in Oncology. G. M. Williams *et al.,* Eds. **3:** 27–41. Elsevier/North-Holland Biomedical Press. Amsterdam, the Netherlands.
4. ZIMMERMAN, F. K. 1973. Detection of genetically active chemicals using various yeast systems. *In* Chemical Mutagens: Principles and Methods for Their Detection. A. Hollaender, Ed. **3:** 209–258. Plenum Press. New York, N.Y.
5. DE SERRES, F. J. & H. V. MALLING. 1971. Measurement of recessive lethal damage over the entire genome and at two specific loci in the ad-3 region of a two-component heterokaryon of *Neurospora crassa. In* Chemical Mutagens: Principles and Methods for Their Detection. A. Hollaender, Ed. **2:** 311–342. Plenum Press. New York, N.Y.
6. ROSENKRANZ, H. S. & Z. LEIFER. 1980. Determining the DNA-modifying activity of chemicals using DNA-polymerase-deficient *Escherichia coli. In* Chemical Mutagens:

Principles and Methods for Their Detection. F. J. de Serres & A. Hollaender, Eds. **6:** 109–147. Plenum Press. New York, N.Y.

7. KADA, T., K. TUTIKAWA & Y. SADAIE. 1972. In vivo and host-mediated rec-assay for screening chemical mutagens and phloxine, a mutagenic red dye detected. Mutat. Res. **16:** 165–174.

8. CHU, E. H. Y. & H. V. MALLING. 1968. Mammalian cell genetics. II. Chemical induction of specific locus mutations in Chinese hamster cells in vitro. Proc. Nat. Acad. Sci. USA **61:** 1306–1313.

9. HUBERMAN, E. & L. SACHS. 1976. Mutability of different genetic loci in mammalian cells by metabolically activated carcinogenic polycyclic hydrocarbons. Proc. Nat. Acad. Sci. USA **73:** 188–192.

10. KUROKI, T. C., C. DREVON & R. MONTESANO. 1977. Microsome-mediated mutagenesis in V79 Chinese hamster cells by various nitrosamines. Cancer Res. **37:** 1044–1050.

11. HSIE, A. W., D. B. COUCH, J. P. O'NEILL, P. A. SAN SEBASTIAN, R. BREMER, J. C. MACHANOFF, A. P. RIDDLE, J. C. LI, N. FUSCOE, N. FORBES & M. H. HSIE. 1979. Utilization of a quantitative mammalian cell mutation system, CHO/HGPRT, in experimental mutagenesis and genetic toxicology. In Strategies for Short-Term Testing for Mutagens/Carcinogens. B. E. Butterworth, Ed.: 39–54. CRC Press. West Palm Beach, Fla.

12. CLIVE, D., K. O. JOHNSON, J. F. S. SPECTOR, A. G. BATSON & M. M. M. BROWN. 1979. Validation and characterization of the L5178Y/TK$^{+/-}$ mouse lymphoma mutagen assay system. Mutat. Res. **59:** 61–108.

13. AMACHER, D. E., S. PAILLET & V. A. RAY. 1979. Point mutations at the thymidine kinase locus in L5178Y mouse lymphoma cells. I. Applications to genetic toxicological testing. Mutat. Res. **64:** 391–406.

14. HUBERMAN, E. & L. SACHS. 1974. Cell mediated mutagenesis of mammalian cells with chemical carcinogens. Int. J. Cancer **13:** 326–333.

15. LANGENBACH, R., H. J. FREED & E. HUBERMAN. 1978. Liver cell–mediated mutagenesis of mammalian cells by liver carcinogens. Proc. Nat. Acad. Sci. USA **75:** 2863–2867.

16. BIGGER, C. A. H., J. E. TOMASZEWSKI, A. DIPPLE & R. S. LAKE. 1980. Limitations of metabolic activation systems used with in vitro tests for carcinogens. Science **209:** 503–505.

17. DYBING, E., E. SØDERLUND, L. TIMM HAUG & S. S. THORGEIRSSON. 1979. Metabolism and activation of 2-acetylaminofluorene in isolated rat hepatocytes. Cancer Res. **39:** 3268–3275.

18. SCHMELTZ, I., J. TOSK & G. M. WILLIAMS. 1978. Comparison of the metabolic profiles of benzo[a]pyrene obtained from primary cell cultures and subcellular fractions derived from normal and methycholanthrene-induced rat liver. Cancer Lett. **5:** 81–89.

19. SELKIRK, J. K. 1977. Divergence of metabolic activation systems for short-term mutagenesis assays. Nature **270:** 604–605.

20. ISHIDATE, M. & S. ODASHIMA. 1977. Chromosome tests with 134 compounds on Chinese hamster cells in vitro—a screening for chemical carcinogens. Mutat. Res. **48:** 337–354.

21. MUTSUOKA, A., M. HAYASHI & M. ISHIDATE, JR. 1979. Chromosomal aberration tests on 29 chemicals combined with S9 mix in vitro. Mutat. Res. **66:** 277–290.

22. EVANS, H. J. 1976. Cytological methods for detecting chemical mutagens. In Chemical Mutagens: Principles and Methods for Their Detection. A. Hollaender, Ed. **4:** 1–29. Plenum Press. New York, N.Y.

23. COHEN, M. M. & K. HIRSCHHORN. 1972. Cytogenetic studies in animals. In Chemical Mutagens: Principles and Methods for Their Detection. A. Hollaender, Ed. **2:** 515–534. Plenum Press. New York, N.Y.

24. LEGATOR, M. S., K. A. PALMER & I. A. ADLER. 1973 A collaborative study of in vivo cytogenetic analysis. I. Interpretation of slide preparations. Toxicol. Appl. Pharmacol. **24:** 337–350.

25. PERRY, P. E. 1980. Chemical mutagens and sister chromatid exchange. In Chemical Mutagens: Principles and Methods for Their Detection. F. J. de Serres & A. Hollaender, Eds. **6:** 1–39. Plenum Press. New York, N.Y.

26. LETT, J. T., I. R. CALDWELL, C. J. DEAN & P. ALEXANDER. 1967. Rejoining of x-ray induced breaks in the DNA of leukemia cells. Nature **214:** 790–792.

27. MCGRATH, R. A. & R. W. WILLIAMS. 1966. Reconstruction in vivo of irradiated *Escherichia coli* deoxyribonucleic acid, the rejoining of broken pieces. Nature **212**: 534–535.
28. KOHN, K. W. 1979. DNA as a target in cancer chemotherapy: measurement of macromolecular DNA damage produced in mammalian cells by anticancer agents and carcinogens. Methods Cancer Res. **16**: 291–345.
29. SWENBERG, J. A., G. L. PETZOLD & P. R. HARBACK. 1976. In vitro DNA damage/alkaline elution assay for predicting carcinogenic potential. Biochem. Biophys. Res. Commun. **72**: 732–738.
30. PAINTER, R. B. & J. E. CLEAVER. 1969. Repair replication, unscheduled DNA synthesis and the repair of mammalian DNA. Radiat. Res. **37**: 4151–4166.
31. RASMUSSEN, R. E. & R. B. PAINTER. 1964. Evidence for repair of ultraviolet damaged deoxyribonucleic acid in cultured mammalian cells. Nature **203**: 1360–1362.
32. EVANS, R. G. & A. NORMA. 1968. Radiation stimulated incorporation of thymidine into the DNA of human lymphocytes. Nature **216**: 455–456.
33. STICH, H. F. & R. H. C. SAN. 1970. DNA repair and chromatid anomalies in mammalian cells exposed to 4-nitroquinoline 1-oxide. Mutat. Res. **10**: 389–404.
34. International Agency for Research on Cancer. 1980. DNA damage and repair in mammalian cells. *In* Long-term and Short-term Screening Assays for Carcinogens: A Critical Appraisal. Report 7 (Suppl. 2): 201–226. IARC. Lyon, France.
35. SAN, R. H. C. & H. F. STICH. 1975. DNA repair synthesis of cultured human cells as a rapid bioassay for chemical carcinogens. Int. J. Cancer **16**: 284–291.
36. MARTIN, D. N., A. C. MCDERMID & R. C. GARNER. 1978. Testing of unknown carcinogens and non-carcinogens for their ability to induce unscheduled DNA synthesis in the HeLa cells. Cancer Res. **38**: 2621–2627.
37. WILLIAMS, G. M. 1976. Carcinogen-induced DNA repair in primary rat liver cell cultures; a possible screen for chemical carcinogens. Cancer Lett. **1**: 231–236.
38. WILLIAMS, G. M. 1977. The detection of chemical carcinogens by unscheduled DNA synthesis in rat liver primary cell cultures. Cancer Res. **37**: 1845–1851.
39. WILLIAMS, G. M. 1980. The predictive value of DNA damage and repair assays for carcinogenicity. *In* Applied Methods in Oncology. G. M. Williams *et al.*, Eds. **3**: 213–230. Elsevier/North-Holland Biomedical Press. Amsterdam, the Netherlands.
40. PROBST, G. S., L. E. HILL & B. BEWSEY. 1980. Comparison of three in vitro assays for carcinogen-induced DNA damage. J. Toxicol. Environ. Health **6**: 333–349.
41. BERWALD, Y. & L. SACHS. 1963. In vitro cell transformation with chemical carcinogens. Nature **200**: 1182–1184.
42. DIPAOLO, J. A., R. L. NELSON & P. DONOVAN. 1971. In vitro transformation of Syrian hamster embryo cells by diverse chemical carcinogens. Nature **235**: 278–280.
43. PIENTA, R. J., J. A. POILEY & W. B. LEBHERZ III. 1977. Morphological transformation of early passage golden Syrian hamster embryo cells derived from cryopreserved primary cultures as a reliable in vitro bioassay for identifying diverse carcinogens. Int. J. Cancer **1**: 642–655.
44. PIENTA, R. J. 1980. A transformation bioassay employing cryopreserved hamster embryo cells. *In* Mammalian Cell Transformation by Chemical Carcinogens. Advances in Environmental Toxicology. N. Mishra, V. C. Dunkel & M. Mehlman, Eds. **1**: 47–83. Senate Press, Inc. Princeton Junction, N.J.
45. REZNIKOFF, C. A., J. S. BERTRAM, D. N. BRANKOW & C. HEIDELBERGER. 1973. Quantitative and qualitative studies of chemical transformation of cloned C3H mouse embryo cells sensitive to postconfluence inhibition of cell division. Cancer Res. **33**: 3239–3249.
46. DIPAOLO, J. A., K. TAKANO & N. C. POPESCU. 1972. Quantitation of chemically induced neoplastic transformation of BALB/3T3 cloned cell lines. Cancer Res. **32**: 2686–2695.
47. KAKUNAGA, T. 1973. A quantitative system for assay of malignant transformation by chemical carcinogens using a clone derived from BALB/3T3. Int. J. Cancer **12**: 463–473.
48. SIVAK, A., M. C. CHAREST, L. RUDENKO, D. M. SILVEIRA, I. SIMONS & A. M. WOOD. 1980. BALB/c-3T3 cells as target cells for chemically induced neoplastic transformation. *In* Mammalian Cell Transformation by Chemical Carcinogens. Advances in Environ-

mental Toxicology. M. Mishra, V. C. Dunkel & M. Mehlman, Eds. **1:** 133–180. Senate Press, Inc. Princeton Junction, N.J.

49. DiMAYORCA, G., M. GREENBLATT, T. TRAUTHEN, A. SOLLER & R. GIORDANO. 1973. Malignant transformation of BHK 21 clone 13 in vitro by nitrosamines, a conditional state. Proc. Nat. Acad. Sci. USA **70:** 46–99.

50. PURCHASE, I. F. H., E. LONGSTAFF, J. ASHBY, J. A. STYLES, D. ANDERSON, P. A. LEFEVRE & F. R. WESTWOOD. 1978. An evaluation of six short-term tests for detecting organic chemical carcinogens. Br. J. Cancer **37:** 873–959.

51. ISHII, Y., J. A. ELIOTT, N. K. MISHRA & M. W. LIEBERMAN. 1977. Quantitative studies of transformation by chemical carcinogens and UV-radiation using a subclone of BHK21 clone 13 Syrian hamster cells. Cancer Res. **37:** 2023–2029.

52. MULVIHILL, J. J. 1975. Congenital and genetic diseases. *In* Persons at High Risk of Cancer: An Approach to Cancer Etiology and Control. J. F. Fraumeni, Jr., Ed.: 3–35. Academic Press, Inc. New York, N.Y.

53. LEVAN, G. & F. MITELMAN. 1977. Chromosomes and the etiology of cancer. *In* Chromosomes Today. A. de la Chapelle & M. Sorsa, Eds.: 363–371. Elsevier/North-Holland Biomedical Press. Amsterdam, the Netherlands.

54. MILLER, E. C. & J. A. MILLER. 1966. Mechanisms of chemical carcinogenesis: nature of proximate carcinogens and interactions with macromolecules. Pharmacol. Rev. **18:** 805–838.

55. STANTON, M. F. 1973. Tumors of the pleura induced with asbestos and fibrous glass. J. Nat. Cancer Inst. **51:** 317–319.

56. STANTON, M. F. & C. WRENCH. 1972. Mechanisms of mesothelioma induction with asbestos and fibrous glass. J. Nat. Cancer Inst. **48:** 797–821.

57. BRAND, K. G. 1982. Cancer associated with asbestosis, schistosomiasis, foreign bodies, and scars. *In* Cancer a Comprehensive Treatise. Etiology: Chemical and Physical Carcinogenesis. F. F. Becker, Ed.: 661–692. Plenum Press. New York, N.Y.

58. MELIEF, C. J. & R. S. SCHWARTZ. 1982. Immunocompetence and malignancy. *In* Cancer a Comprehensive Treatise. Etiology: Chemical and Physical Carcinogenesis. F. F. Becker, Ed.: 161–199. Plenum Press. New York, N.Y.

59. SCHWARTZ, R. S. 1972. Immunoregulation, oncogenic viruses and malignant lymphomas. Lancet **1:** 1266.

60. MINTZ, B. & K. ILLMENSEE. 1975. Normal genetically mosaic mice produced from malignant terotocarcinoma cells. Proc. Nat. Acad. Sci. USA **72:** 3585–3589.

61. SONTAG, J. A., N. P. PAGE & U. SAFFIOTTI. 1976. Guidelines for Carcinogen Bioassay in Small Rodents. Carcinogenesis Technical Report Series 1. Department of Health, Education and Welfare Publication No. (NIH) 76-801. U.S. Government Printing Office. Washington, D.C.

62. CHU, K. C., C. CUETO, JR. & J. M. WARD. 1981. Factors in the evaluation of 200 National Cancer Institute bioassays. J. Toxicol. Environ. Health **8:** 251–280.

MECHANISMS OF TOXIC INJURY

James W. Bridges,*† Diane J. Benford,* and Susan A. Hubbard*

*Robens Institute of Industrial and Environmental Health and Safety
†Department of Biochemistry
University of Surrey
Guildford, Surrey, GU2 5XH, United Kingdom

INTRODUCTION

Chemically induced cell injury (cytotoxicity) can, for convenience, be considered as occurring in a sequence of phases, i.e.:

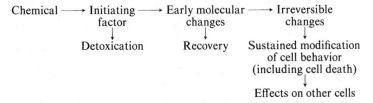

There would appear to be no intrinsic differences between the forms of cell injury caused by endogenous chemicals and those produced by natural or synthetic "foreign" chemicals (xenobiotics). The nature and magnitude of cell injury produced by a chemical will depend on a number of factors including:

1. The dose and persistence of the chemical at cellular sites producing the initiating factor.
2. The rate of production of initiating factor(s) and its rate of detoxication.
3. The accessibility of the initiating factor to the crucial "target" cellular molecules and the extent and persistence of the reaction between them.
4. The biological role of these "target" molecules (which may vary with cell type, physiological state of the cell, etc.).
5. The ability of cellular defense mechanisms to repair, replace, or compensate for these altered "target" molecules.
6. The nature and extent of release of products from the injured cell that may cause cytotoxic or stimulatory effects on other cells of the same or different cell types.

Initiating Factor(s)

Cell injury may be initiated by the formation of a stable (noncovalent) complex with an enzyme, receptor site, cofactor, etc., or via the formation of highly chemically reactive species or by provoking physicochemical changes (e.g., pH, redox, ionic composition, solubility, solvation or pO_2 changes) within the cell or in its immediate environment (see TABLE 1).

Although the direct effects of chemicals on the inhibition of enzymes involved in energy production and protein synthesis have been extensively investigated by biochemists, in recent years attention in toxicological research has been largely

42

0077-8923/83/0407-0042 $01.75/0 © 1983, NYAS

TABLE 1
TYPES OF CHEMICALLY INDUCED TOXICITY

	Chemical	Initiating Factor	Early Molecular Target	Reference
Enzyme inhibition	parathion	paraxon	cholinesterase	54
	fluoroacetate	fluorocitrate	aconitase	55
	dinitrophenol	dinitrophenol	uncouples oxidative phosphorylation	56
	rotenone	rotenone	inhibits oxidative phosphorylation	57
Cofactor depletion	galactosamine	UDP galactosamine*	UTP depletion	58
	ethionine	S-adenosylethionine	adenine/ATP	59
	menaquinone	semiquinone	NAD(P)H depletion	60
Receptor interactions	TCDD	TCDD	binding to cytosolic receptor	10
Physicochemical changes				
(redox)	aniline	phenylhydroxylamine	hemoglobin to methemoglobin	61
(redox)	paracetamol	quinoneimine	oxidative damage	62
(solubility)	sulfapyridine	acetylsulfapyridine	precipitation in tubules	63
(solubility)	harmol	harmolglucuronide	biliary cholestasis	64
(physical damage)	silica	silica	lysosomal membranes	65
(solvation)	hexane	hexane	membranes	66

*UDP and UTP are uridine diphosphate and triphosphate, respectively.

focused, probably unduly so, on the initiation of toxicity by chemically unstable species that are produced within the cell[1] and react rapidly with cellular constituents.

Among the chemically unstable species that may be formed by cells are:

1. Electrophiles, commonly formed from C-, N-, or S-containing compounds that are metabolized to species that are deficient, or potentially deficient, in an electron pair.

2. Free radicals, i.e., species that contain an odd number of electrons and may possess a positive, a negative, or no charge.

3. Other neutral reactive species, e.g., carbenes and nitrenes in which two electrons are lost. There is as yet very little information on the contribution of the formation of this type of species to cell injury.[1]

Electrophiles

The hypothesis that many carcinogens mediate their effect via a metabolic conversion to electrophilic species that subsequently covalently bind to DNA[2] provided the stimulus for research both on the enzyme systems responsible for generating such electrophiles and the range of toxicological consequences that could follow the production of such highly chemically reactive species. Cytochrome P_{450} has come under

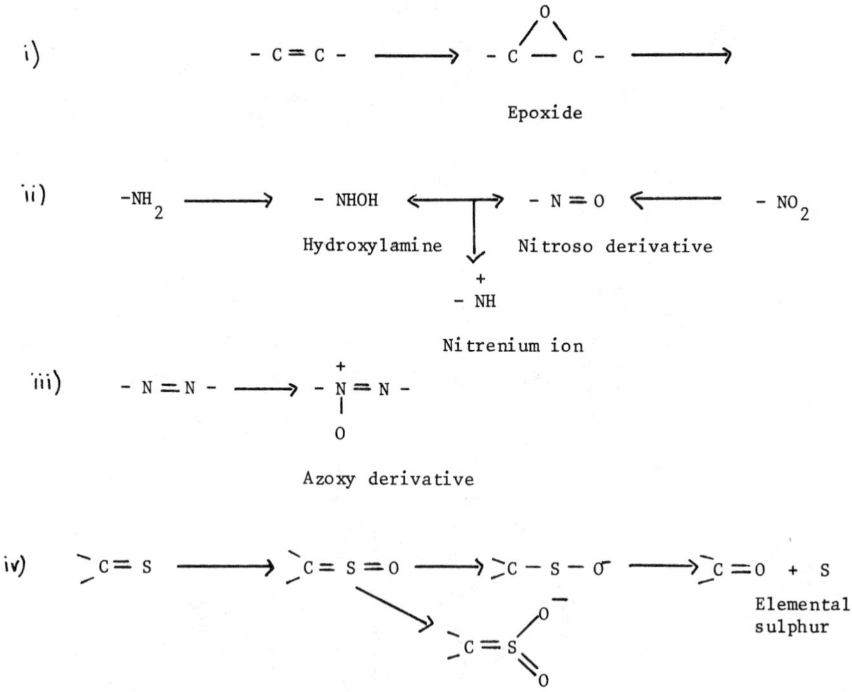

FIGURE 1. Some of the activation reactions that may be catalyzed by one or more forms of cytochrome P_{450}.

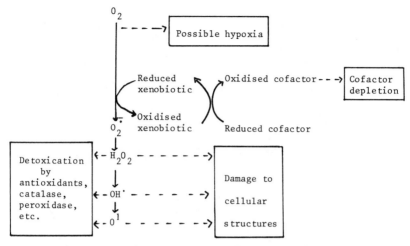

FIGURE 2. Oxygen-initiated toxicity.

particular scrutiny as a primary source of such electrophilic species because of its ability to metabolize a very wide range of organic structures. It is undoubtedly the most versatile enzyme system both for producing toxic metabolites and for detoxifying biologically active xenobiotics.[2]

Some of the activation reactions that may be catalyzed by one or more forms of cytochrome P_{450} are shown in FIGURE 1. A number of other enzymes have also been claimed to generate reactive electrophiles, albeit their individual substrate specificity is probably much narrower than that of the major forms of cytochrome P_{450}. These enzymes include prostaglandin synthetase, nitro reductase, and xanthine oxidase.

Free Radicals

Toxicity may occur directly as a result of chemicals being converted to free radicals or via the formation of superoxide as a by-product of their metabolism. Despite its potential importance, organic free radical–induced toxicity has been much less extensively investigated than has that mediated by electrophiles. Likely organic precursors for radical formation include quinones, nitroaromatics, carbon tetrachloride, bipyridyls, and aromatic amines.[3,4] Although cytochrome P_{450} is undoubtedly involved in the reduction of carbon tetrachloride to form free radicals ($CCl_3\cdot$ and $CCl_3O_2\cdot$), more commonly reduced nicotinamide adenine dinucleotide phosphate (NADPH) cytochrome c (P_{450}) reductases or other flavin-containing reductases are responsible for free radical generation.[5]

Univalent or bivalent reduction of oxygen leading to the formation of O_2^- (superoxide) and H_2O_2 (hydrogen peroxide) is now well established to be a common metabolic process. Most endogenous and exogenous chemicals that are readily reduced by intracellular enzymes are in turn able to reduce oxygen by electron transfer. These chemicals include catechols (e.g., adrenaline and dopa), quinones (e.g., adriamycin, mitomycin C, alloxan), bipyridyls (e.g., paraquat), and nitroaromatics (e.g., metronidazole).[5] Superoxide is not considered itself to be a particularly toxic species because it is relatively stable chemically; indeed it can diffuse to sites remote from that of its formation. However, it may be converted chemically and/or enzymically to a number of more toxic forms of active oxygen including $HO_2\cdot$ (protonated forms of superoxide), O' (singlet oxygen), $OH\cdot$ (hydroxyl radical), and H_2O_2 (see FIGURE 2). $OH\cdot$ is a very highly reactive species and may cause direct damage to many endogenous molecules.[6]

Two aspects of free radical formation are of particular potential importance from a toxicological viewpoint. Firstly, many free radicals can readily participate in recycling reactions, thus producing a sustained level of free radicals in the cell, i.e.,

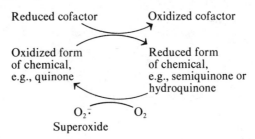

Secondly, free radicals may initiate chain reactions, e.g., lipid peroxidation (see below). Other possible consequences of redox cycling are a depletion of reduced cofactors and hypoxia. It follows from these properties of free radicals that whereas the highest achievable stoichiometry of the reaction between electrophiles and their molecular targets is normally 1:1, for free radical reactions a "multiplier" effect can occur, greatly enhancing the potential toxicity contribution of each organic radical molecule.

In vivo and Cellular Aspects of the Formation of Reactive Metabolites of Xenobiotics and Oxygen

Cells typically have a number of defense systems that limit or prevent the achievement of high intracellular concentrations of initiating factors. These include the drug-metabolizing enzymes, antioxidants, binding proteins, and active oxygen–metabolizing enzymes. The levels of initiating factor available for reaction with "target" cellular molecules are dependent on the balance of the rate of formation and the rate of inactivation reactions. This balance may vary between cell types[7,8] and may be changed markedly from the in vivo situation, when cells are cultured. For example, "phenobarbitone-type" P_{450}, the form of this enzyme most commonly concerned with xenobiotic metabolism, tends to fall markedly and somewhat selectively when rat hepatocytes are cultured, while most cell lines contain very low or nondetectable P_{450}-dependent activities.[9] In addition to being affected by culture conditions, levels of antioxidants may also vary in vivo depending on diet. Oxygen tension can also influence the formation of reactive metabolites; redox cycling of free radicals (see above) is likely to be particularly susceptible to oxygen-tension changes. Very high or gross hypoxic conditions may reduce redox cycling, while mild hypoxia may exacerbate this process.

Early Molecular Targets

Early targets of initiating factors may be proteins (e.g., enzymes, structural proteins) nucleic acids, lipids (unsaturated fatty acids), cofactors, etc. Some initiating factors appear to have rather specific early "targets," e.g., dinitrophenol and 2,3,7,8-tetrachlorodibenzodioxin (TCDD),[10] while others bind to a very wide range of targets, e.g., benzpyrene.

The majority of the carcinogens that have been studied would appear to yield, directly or via the formation of metabolites, very stable complexes with DNA. However, the role of free radicals, which can produce nucleic acid damage without leaving bound adducts, may also be important in the initiation of cancer. Changes invoked in membrane transport or in the ratios of adenosine triphosphate (ATP) to adenosine diphosphate (ADP) or reduced cofactor to oxidized cofactor are other potentially important changes that may result from the early attentions of initiating factors in cells. (NB, the early molecular targets that may lead to cellular necrosis are given more detailed consideration below.)

The toxicological importance of these early changes is dependent on the extent to which the biological functioning of the "target" molecules is modified, the duration of the effect, and how crucial the biological function is for the cell's short- and long-term survival. The effect(s) may be tempered by the qualitative and quantitative effectiveness of cellular mechanisms for the replacement or repair of compromised molecules and by the extent to which other cellular processes can be brought into play to compensate for the modification of the "target" molecules. These compensation processes may themselves result in toxicity. For example, the plasticizer diethylhexylphthalate is metabolized to the chemically stable products monoethylhexylphthalate and 2-ethylhexanol, which inhibit hepatic mitochondrial β-oxidation of fatty acids. To compensate for the buildup of fatty acids consequent to this inhibition in rodents, peroxisome proliferation occurs. However, the peroxisomes so formed differ from normal peroxisomes in having a reduced catalase content. As a result, the hydrogen peroxide generated by the peroxisomes during their oxidation of fatty acids is not completely detoxicated in the peroxisomes and, consequently, some hepatocellular damage may occur.[11]

Toxic Consequences for the Cell

Because most cells incorporate a variety of very active defense and repair systems, commonly they recover completely from early chemically induced cellular changes. However at high or prolonged doses of a chemical, these cellular defense and repair mechanisms may be overwhelmed and sustained changes may occur. Imposed changes may be of a quantitative or qualitative nature. In cell culture, effects on growth rate, state of differentiation, and degree of aging are commonly observed. These changes may in turn influence the growth and other characteristics of the surrounding cells (see TABLE 2).

In the following sections, attention is confined to the processes involved in cell necrosis.

CELL NECROSIS

From a biochemical viewpoint a cell may be considered to be dying when the dynamic state of the plasma membrane is sufficiently disturbed that a gross irreversible reduction occurs in the maintenance of the differential composition of major constituents between the cell and its surrounding medium.

TABLE 2

SOME CHANGES OTHER THAN CELL DEATH TO INDICATE TOXICITY

End Point	System		Example	References
Genotoxicity (mutagenicity, DNA repair, cytogenetic changes)	bacteria, fungi, mammalian cells	*in vitro*	most chemical carcinogens, e.g., aromatic amines, polycyclic aromatic hydrocarbons	67,68
	whole animals			
Carcinogenicity	rodents		tumor production by a variety of agents	69
Presence of N-substituted porphyrins (green pigments)	liver	*in vivo*	2-allyl-2-isopropylocatamide	70
Inflammatory response	skin	*in vivo*	sodium lauryl sulfate	71
Allergic response	skin, blood	*in vivo*, *in vitro*	aluminum hydroxide	72
Centrilobular lipid accumulation	liver	*in vivo*	CCl_4	74
Peroxisome proliferation	liver	*in vivo*, *in vitro*	hypolipidemic drugs, e.g., clofibrate	11,73,75,76,77,78
Lipid peroxidation	liver	*in vivo*		
Lipofuscin accumulation	liver	*in vivo*		
Growth-rate changes	liver enlargement	*in vivo*	ponceaux MX	75,76

Common Morphological Changes in Cell Necrosis

Chemically mediated injury may be so severe that cells are killed almost instantaneously, e.g., high concentrations of osmium tetroxide or formaldehyde, or the effects may be rather more subtle, e.g., safrole, so that a progressive series of degenerative changes are observed prior to death. For all chemicals other than very rapidly acting necrotic agents, the events in cell necrosis may be considered as occurring in two phases, the early reversible changes and the late irreversible changes.

The early reversible changes usually include mild cytoplasmic edema, dilation of the endoplasmic reticulum, slight mitochondrial swelling, disaggregation of polysomes, and occurrence of small aggregates of chromatin around the nucleus.[12,13] The late irreversible changes are commonly extensive mitochondrial swelling with cristae disruption, gross cytoplasmic swelling with dissolution of organelles, plasma membrane rupture, and nuclear dissolution.[12,13] Because it is very difficult to discern the point of cell death, it remains to be established whether all of these late changes occur prior to death or whether some are consequences of cell death.

Because gross loss of plasma membrane function is probably incompatible with cell survival, many workers have used measurement of leakage of cellular constituents or failure to exclude polar dyes as a marker of necrosis.

Biochemical Events in Cell Necrosis

It remains to be established whether biochemical processes involved at the later stages in cell necrosis are common regardless of the initiating processes and the cell type involved. Based on the premise that irreversible breakdown of the plasma membrane is crucial to cell necrosis, a number of mechanisms may be postulated by which necrosis might be triggered (see Figure 3), including reduction in ATP levels, alterations in membrane transport or containment, depressed RNA or protein synthesis, or cellular pH changes. Changes in various cell organelles have been proposed as crucial to cell death. Although dissolution of the nucleus is often observed in necrotic cells, it is unlikely that alterations in DNA are a common cause of cell death (except perhaps in the case of rapidly dividing cells), for nucleic acid synthesis inhibitors do not generally precipitate cell death. Rupture of lysosomes with the consequent release of hydrolytic enzymes can result in cell damage and even cell death in cell types that take up damaging particles such as silica or uric acid, but chemically induced lysosomal damage would not appear to be an important early common factor in cell death.

Direct Effects on Membranes

A number of chemicals including heavy metals, ions, organic solvents, aliphatic aldehydes, acids, and bases may exert a direct action on the plasma membrane causing either modification of specific transport sites or nonspecific changes in membrane permeability (such as alterations in lipid-lipid and lipid-protein relationships). Gross membrane changes may result in rapid cell death. (More subtle early membrane changes that may result in cell death are considered below.)

Particular caution must be taken in interpreting the results of investigations of the direct or indirect effects of chemicals on the plasma membranes of isolated cells, particularly those that are freshly isolated, because many membrane features may be modified considerably by the isolation process and by the lack of all-round contact with other cells in the cell culture.

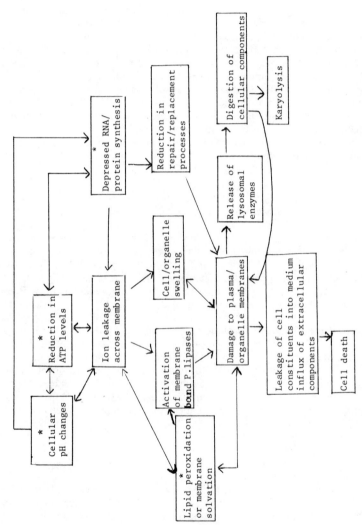

FIGURE 3. Biochemical changes that may result in cellular necrosis. (Asterisks indicate probable initiation points.)

Membrane Effects and Lipid Peroxidation

Unsaturated fatty acids appear to be essential constituents of cellular membranes and are believed to be important for normal membrane functioning. However, unsaturated fatty acids are vulnerable to oxidative attack initiated by free radicals (either organic or superoxide radicals), leading to a chain reaction (see FIGURE 4) in which lipid free radicals ($L\cdot$) are formed, which in the presence of oxygen result in lipid peroxy free radicals, which in turn form lipid free radicals. As a result, the chain length of the fatty acid may be altered or the structure changed in other ways, e.g., formation of alcohols with consequent effects on membrane fluidity. Other reactive species, e.g., aldehydes and ketones such as malondialdehyde and 4-hydroxynonenal, are also liberated and may react with amino, sulfhydryl, and hydroxy groups of proteins, polysaccharides, and cofactors, etc., causing direct or indirect effects on membrane permeability and transport mechanisms. The extent of lipid peroxidation that occurs will depend on a number of factors including the qualitative nature and amount of the initiating factor reaching the membrane, the membrane content of unsaturated fatty acids and their accessibility, the oxygen tension, the presence of iron ions,[5] the cellular content of antioxidants, e.g., β-carotene and α-tocopherol, and the activity of enzymes that can terminate the chain reaction, e.g., glutathione peroxidase.

Chemicals that are widely held to cause cell necrosis through the triggering of extensive lipid peroxidation include carbon tetrachloride,[14] paraquat,[15] ozone, and nitrogen oxides.[16] However, in each case it remains to be established whether lipid peroxidation of the plasma membrane or of other cellular membranes is the most important event in the early stages of the lesion and whether lipid peroxidation is the sole cause of the cellular necrosis.

Carbon tetrachloride has been claimed to initiate lipid peroxidation via its metabolic conversion to $CCl_3\cdot$ or more probably $CCl_3O_2\cdot$. Whichever is the critical species, it is apparent that the initial metabolic step in forming the reactive metabolite is reductive metabolism by cytochrome P_{450}. The need for reducing conditions may be the explanation of the observation that hypoxia tends to enhance carbon tetrachloride–invoked necrosis. On this reasoning the initial, largely centrilobular location of carbon tetrachloride–induced hepatotoxicity is probably due to two factors, namely, the lower oxygen tensions and the high cytochrome P_{450} concentrations in the centrilobular region of the liver compared with those in the midzonal and periportal areas. If the oxygen tension falls too far, lipid peroxidation will of course become rate limited by the availability of oxygen (see FIGURE 4).

It is apparent from the foregoing that in designing *in vitro* studies to investigate the role of chemically induced lipid peroxidation in cellular necrosis, due regard must be paid to the choice of oxygen tension. Use of pure carbogen (95% O_2/5% CO_2) would not appear to be generally appropriate if it is intended to extrapolate the results of the findings to the *in vivo* situation.

Many other chemicals undoubtedly stimulate lipid peroxidation, including alloxan, tri- and diethyltin dichloride, iron salts, ethyl alcohol, and paracetamol.[5,14,17]

ATP Levels and Ion Transport

Adequate levels of ATP are essential for a range of transport, repair, and synthetic processes. It is, therefore, not surprising that chemically induced, prolonged depletion of ATP either as a consequence of its impaired synthesis or enhanced destruction can result in toxicity. Inhibition of ATP synthesis may arise in a number of ways, e.g., by

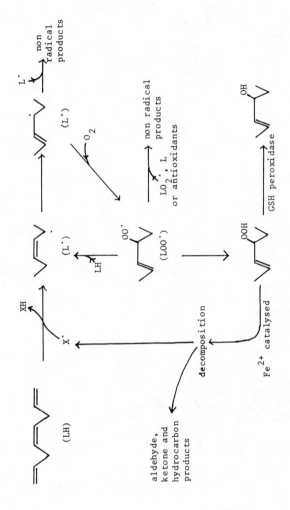

FIGURE 4. Lipid peroxidation.

inhibition of one or more processes in pathways forming reduced cofactors, by depletion of reduced cofactor levels by futile cycling (see above), or by inhibition or uncoupling of oxidative phosphorylation. Interference with ATP levels may occur either because of chemical specific affinity for a particular enzyme concerned in maintaining ATP levels (e.g., cyanide, carbon monoxide, dinitrophenol) or through the damaging effects of reactive metabolites (see Covalent Binding).

Changes in the volume of intracellular components and altered dispersal of their contents are characteristic of cells undergoing necrosis. Such changes may be ascribed to alterations in ion and water transport as a consequence of effects either on the cellular membranes themselves or on the provision of energy for active transport.

It is now well established that a number of inorganic ions play a very crucial role in a variety of cellular processes. Control of the concentration of ions such as Na^+, K^+, Ca^{2+}, Mg^{2+}, and Mn^{2+} appears to be essential for cellular transport, energy production, protein and nucleic acid synthesis, cell shape, control of the cell cycle, cell-cell communication, and cell functions. In assessing whether changes in a metal ion concentration may be associated with toxicity, it is important to distinguish between metal ion–dependent and metal ion–regulating processes.

The role of calcium changes during the course of cell death has attracted particular interest. In part this has arisen from the observation that calcification is a very frequent occurrence in necrosed cells. A number of workers have suggested that a buildup in intracellular free calcium may be a common phenomenon in all cells prior to death and could initiate the later stages leading to cellular demise. Cellular concentrations of calcium are typically $\sim 10^{-7}$ M while those in extracellular fluids are $\sim 10^{-3}$ M. Much of the cellular Ca^{2+} however is not in the free form but is bound to proteins in the mitochondria, endoplasmic reticulum, and cytosol, e.g., calmodulin. Following chemically induced alterations in membrane function, the membrane translocation of Ca^{2+} may be modified causing an inhibition of Ca^{2+} uptake into mitochondria or an increased Ca^{2+} uptake from the medium. As a consequence, increased cytoplasmic concentrations of Ca^{2+} may arise. Laiho and Trump have observed a direct correlation between cellular Ca^{2+} concentrations and cell death in Ehrlich's ascites tumor cells exposed to *p*-chloromercuribenzene sulfonate, antimycin C, iodoacetate, and 2,4-dinitrophenol.[19] They have also shown that in this system, cell death correlates inversely with ATP concentration. It has also been reported that primary maintenance cultures of hepatocytes are killed more rapidly by chemicals in Ca^{2+}-rich media than in control media.[20] However other workers using freshly isolated hepatocytes have not observed such an effect except for A23187.[21,22] It is not clear whether these differences can be ascribed solely to the differences in the physiological state of the hepatocytes or whether other factors are also involved.

There is insufficient evidence as to whether the plasma membranes of the freshly isolated hepatocytes are more or less physiological than those of primary maintenance cultures of hepatocytes; however these data do cast doubt on the concept that elevated Ca^{2+} levels are an obligatory requirement for cell death. In order to clarify the status of Ca^{2+} in cell death, investigations will need to be made of free cytosolic Ca^{2+} with time in cells of various types exposed to a range of toxic agents and doses.

A buildup of free calcium ions may have a number of potentially adverse consequences on the cell. Of particular interest are the effects of Ca^{2+} on membrane-bound phospholipase activation and microtubule organization.[13] Phospholipases may be activated by Ca^{2+} in various membranes including the plasma, mitochondrial, and endoplasmic reticulum membranes. Changes in membrane phospholipids may result in alterations in membrane permeability, which may be exacerbated by the disruptive action of the released fatty acids and lysophosphatids on membrane organization. As a consequence, Ca^{2+} transport and transport of other ions and small molecules may be further disrupted. Loss of specialized phospholipids such as cardiolipin may also bring about changes in particular enzyme activities, e.g., cytochrome oxidase. Modification of microtubules may influence cell shape, causing effects such as blebbing of the plasma membrane, which in turn may affect permeability. Chemicals such as vinblastine and cytochalasin may cause similar effects via a direct interaction with microtubular elements. Ca^{2+} may also cause other effects, e.g., cross-linking of carboxy groups of glutamic acid and ϵ-amino groups of lysine, resulting in high molecular weight protein aggregates.[18] Progressive destruction of membrane phospholipids and the increased stretching of the membranes due to increased fluid uptake and damage to the cytoskeleton may eventually give rise to gross failure to retain cellular constituents and to exclude medium components, resulting in cell death. It remains to be established under what circumstances changes in free Ca^{2+} concentrations are a common pathway in cell death.

It is well established that one ion may modify the toxicological properties of another, e.g., zinc can reduce the toxic action of cadmium.[23] It is possible that changes in the balance of concentrations between a number of ions may turn out to be the crucial factor in cell death, rather than changes in Ca^{2+} concentrations per se.

Covalent Binding

Reactive metabolites may bind covalently to cellular macromolecules such as nucleic acids, proteins, cofactors, lipids, and polysaccharides, thereby modifying their biological properties. As a consequence, a wide variety of effects may occur including loss of energy production, change in membrane permeability, reduced synthesis of macromolecules, etc. The relationship between binding to nucleic acids and carcinogenicity and mutagenicity is considered elsewhere. Little is known of the extent or significance of covalent binding to cofactors, lipids, or polysaccharides, therefore only the correlation between covalent binding to proteins and cell necrosis will be considered here.

Evidence to support the important role of covalent binding in cell death is largely circumstantial and based on the correlation between the extent of binding and the severity of the accompanying lesion *in vivo*. Binding is often highest in the tissue most

TABLE 3

EXAMPLES OF CHEMICALS SHOWING COVALENT BINDING APPARENTLY
RELATED TO TOXICITY

Chemical	Proposed Active Metabolite	Site of Toxicity and Binding	References
Bromobenzene	epoxide	liver	26
Paracetamol	imidoquinone	liver	27
CCl_4	$CCl_3 \cdot$, $CCl_3O_2^-$	liver	28
Allyl alcohol	acrolein	liver	29
Cocaine	N-oxidation product	liver	30
Methimazole	S	liver	31
p-Aminophenol	quinoneimine	kidney	32,33
2-Furamide	furan epoxide	kidney	34
Nitrofurazone	hydroxylamine	kidney/liver	35
4-Ipomeanol	furan epoxide	lung	36
Thiourea	S	lung	31

TABLE 4
INHIBITION OF PROTEIN SYNTHESIS IN ISOLATED RAT HEPATOCYTES*

Chemical	Dosage Causing 50% Inhibition after 1 Hour Exposure (M)
Cycloheximide	5.0×10^{-7}
Paracetamol	9.0×10^{-3}
Phenacetin	2.5×10^{-3}
Acetanilide	7.5×10^{-3}
Aniline	1.1×10^{-2}
Phenobarbitone	3.75×10^{-3}
Safrole	1.4×10^{-3}
Diethylmaleate	2.0×10^{-3}

*Freshly isolated rat hepatocytes were incubated for 1 hour with various concentrations of test compound, and the approximate concentration causing 50% inhibition of incorporation of ^{14}C-leucine into protein was determined. (Data from Reference 39.)

susceptible to toxicity. One of the earliest studies to implicate covalent binding in cytotoxicity was the demonstration by Brodie *et al.* that the hepatotoxic agent bromobenzene covalently bound to sites of necrosis and that the extent of both binding and necrosis was related to the rate of metabolism of bromobenzene.[24] Paracetamol is another example of a hepatotoxic agent that exhibits dose-related covalent binding to liver protein, which is mirrored by a decrease in glutathione content.[25] Reactive metabolites capable of binding covalently to proteins include epoxides, free radicals, hydroxylamines, aldehydes, and quinones (TABLE 3).

Although covalent binding often shows general correlation with tissue and species specificity in toxicity, there are often anomalies. For example, Wiley *et al.* found no direct relationship between the relative covalent binding and toxicities of a series of bromobenzene analogues.[37] Also covalent binding is often observed in tissues that are not affected by a particular toxic substance and in target tissues under conditions and doses where toxicity is not seen.[38] These data may be interpreted in terms of the occurrence of a threshold level of binding below which covalent binding is not a cause of toxicity. Because binding of radiolabel to protein is the usual criterion of covalent binding that is measured, rather than the effect on particular proteins concerned with energy production, membrane function, etc., it is not surprising that correlation between covalent binding and cell death is frequently not straightforward. It is likely that the majority of the cellular proteins that are associated with covalently bound residues are not essential to the short-term survival of the cell. Moreover, the concentrations of these proteins, e.g., ligandin, may vary considerably according to the physiological state of the cell. For highly lipophilic chemicals in particular, technical problems may also be experienced in distinguishing between tightly bound and covalently bound material.

Inhibition of Protein Synthesis

Inhibition may occur at several stages of protein synthesis:

1. Formation of the aminoacyl t-RNA may be blocked by amino acid analogues (e.g., ethionine).
2. Binding of aminoacyl t-RNA to ribosomes is blocked by various antibiotics (e.g., tetracyclines).
3. Formation of peptide bonds, catalyzed by peptidyl transferase, is inhibited by

chloramphenicol in mitochondrial protein synthesis and by cycloheximide in extramitochondrial protein synthesis.

    4. Inhibition occurs as a result of inhibition of RNA synthesis.

    Inhibition of protein synthesis is unlikely to be a common important contributor to cell death because the immediate impact of inhibition on cell membrane permeability and energy production will probably be relatively slight. Only if the inhibitory effects are sufficiently extensive and prolonged that replacement and repair processes become key factors in the cell's survival is protein synthesis inhibition likely to be a cause of cell death. In support of this view, investigations on a range of chemicals using isolated rat hepatocytes have failed to demonstrate a relationship between protein synthesis inhibition and either hepatocyte viability *in vitro* or hepatic necrosis *in vivo*[39] (TABLE 4). Also cycloheximide, a known inhibitor of protein synthesis, was found to protect against the cytotoxic effects of certain hepatotoxic agents.[40]

Reduced Glutathione Depletion

    Glutathione has several functions in the maintenance of cellular integrity, including:

    1. Protection of SH groups, particularly those at the active sites of enzymes, against inactivation by heavy metals or oxidation by drugs such as organic mercurials.

    2. Reactivation of SH enzymes that have been inactivated by oxidation.

    3. Conjugation with toxic chemicals or metabolites.

    4. Detoxication of endogenous peroxides and reactive oxygen species.

    Because of its high concentration (~5 mM), glutathione has a particularly important protective role in the liver. It has been argued that a reduction in hepatic glutathione concentrations is an essential prerequisite for reactive metabolites to reach "target" molecules in sufficient concentrations to initiate cell damage. Decreased levels of glutathione by any of the routes shown in FIGURE 5 may be either directly or indirectly involved in cellular damage. It is notable that hepatocytes within 100 μm of the central vein contain less glutathione than do those in other regions,[41] which may be a contributory factor to the prevalence of centrilobular necrosis with some hepatotoxic agents.

    We have investigated the effects of some benzodioxole compounds on reduced glutathione content and viability of freshly isolated rat hepatocytes (TABLE 5). Isosafrole was found to be more toxic, and caused a greater depletion of glutathione, than safrole or dihydrosafrole, thus correlating well with the relative hepatotoxicities *in vivo*. Diethylmaleate also caused a marked depletion of reduced glutathione but had little effect on cell viability, suggesting that the depletion of glutathione per se was not the direct cause of cell death. It is probable that isosafrole and safrole deplete glutathione by forming glutathione conjugates or by a redox cycling reaction perhaps involving their catechol metabolites. Depletion of glutathione also increases the hepatotoxicity of paracetamol and bromobenzene, possibly by permitting reactive metabolites to bind covalently to cellular macromolecules in increasing amounts.[25,26]

    Hogberg and colleagues have shown that a short depletion of glutathione by, for example, diethylmaleate is apparently harmless to isolated rat hepatocytes and that glutathione can be rapidly resynthesized.[42,43] However, prolonged depletion of glutathione for periods greater than one hour may lead to increased leakage of cellular constituents. In contrast, Babson *et al.* found that the toxicity of adriamycin to isolated hepatocytes was increased by bischloroethylnitrosourea (BCNU), which specifically

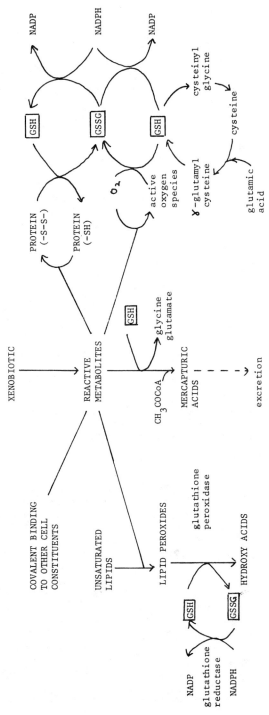

FIGURE 5. Glutathione metabolism and reactive metabolite formation.

inactivates glutathione reductase, but not by diethylmaleate.[44] Adriamycin toxicity is
thought to be mediated via the formation of a semiquinone capable of redox cycling to
generate reactive oxygen species that would be removed by reduced glutathione. Thus,
if adriamycin toxicity occurred as a direct result of glutathione depletion, it should
have been potentiated by diethylmaleate treatment. It was deduced that even at low
reduced glutathione levels (25% of control), glutathione re-formation was sufficient to
protect against adriamycin toxicity provided that glutathione reductase activity was
not impaired. (NB, other parameters of toxic change in hepatocytes have been the
subject of several recent reviews.)[45,46]

Correlation between the Early Molecular Targets and Cell Death

It is generally agreed that cell death can be recognized by the breakdown of
membrane integrity whether measured by uptake of vital dyes or leakage of enzymes
and cofactors. There is little difference between these parameters, although leakage of

TABLE 5

COMPARISON OF THE EFFECTS OF BENZODIOXOLE COMPOUNDS ON GLUTATHIONE CONTENT
AND VIABILITY OF ISOLATED RAT HEPATOCYTES*

Concentration (mM)	Isosafrole	Safrole	Dihydro-safrole	Benzo-dioxole	Diethylmaleate
Glutathione content (% control)					
0.1	88	96	84	87	68†
0.5	68†	84†	81†	84	53†
1.0	44†	48†	65†	87	50†
Cell viability (% control)					
0.1	96	95	89	100	95
0.5	89	91	91	95	91
1.0	42†	80†	84†	95	92

*Freshly isolated rat hepatocytes were incubated with test compounds for 1 hour before
determination of trypan blue exclusion and glutathione content.
†Significantly different from control ($p < 0.05$).

the enzyme arginosuccinate lyase has been suggested to occur before the other events
in isolated hepatocytes.[47] However, there is little agreement on the relative importance
of the events that precede membrane leakage. This may possibly be due to the use of
different techniques and conditions in different laboratories. For example, Stacey et al.
found little difference between the use of thiobarbituric acid–reacting substances and
ethane production as indices of lipid peroxidation in isolated hepatocytes treated with
carbon tetrachloride.[48] In contrast, Smith et al. found ethane was the more sensitive
index of lipid peroxidation.[49] One possible explanation for the discrepancy is that
Smith and colleagues used an atmosphere of 93.5% O_2/6.5% CO_2 for incubations
whereas the former group carried out incubations in air. Increased oxygen levels would
tend to increase the basal level of lipid peroxidation but decrease the reductive
bioactivation of carbon tetrachloride (see above), the overall effect being a much
decreased induction of lipid peroxidation.[48] Similarly, pretreatment of animals or
incubation of hepatocytes in Fe^{2+}-containing buffers may result in increased basal
levels of lipid peroxidation, thereby decreasing the apparent effect of toxic chemicals.
　The importance of lipid peroxidation in cell death is still poorly understood. Stacey
and Klaassen found that inhibition of lipid peroxidation with antioxidants did not

TABLE 6

TOXICITY OF ETHYLTIN DERIVATIVES TO ISOLATED RAT HEPATOCYTES*

Treatment	Trypan Blue Exclusion (%)	LDH Retention (%)	Thiobarbituric Acid Reactant Production (nmol/hour per 2×10^6 cells)	ATP Content (nmol/10^6 cells)	O_2 Uptake (nmol/minute per 10^6 cells)
None	86 ± 5	93 ± 8	1.4 ± 0.3	18.0 ± 2.1	16.7 ± 2.0
Triethyltin-Br					
100 μM	30 ± 13†	35 ± 14†	3.5 ± 0.2†	4.2 ± 1.2†	11.8 ± 1.1†
10 μM	74 ± 12	93 ± 5	1.7 ± 0.4	5.5 ± 1.0†	11.2 ± 1.4†
Diethyltin-Cl					
100 μM	48 ± 14†	72 ± 9†	4.8 ± 0.2†	9.0 ± 1.3†	10.6 ± 0.6†
10 μM	75 ± 8	—	2.0 ± 0.3†	13.0 ± 3.2†	10.8 ± 0.8†

*Data from Reference 17.
†Significantly different from the control.

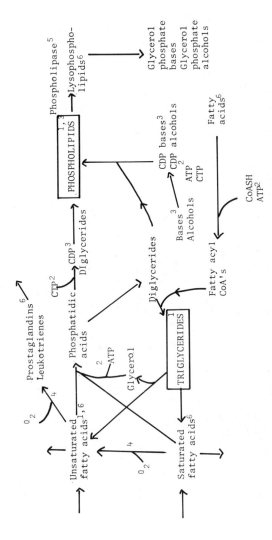

FIGURE 6. Sites of action of chemicals on membrane lipid metabolism that may lead to an increase in membrane permeability. (1) Major sites of free radical attack causing lipid peroxidation. (NB, free radicals and lipid peroxidation products may also cause damage akin to that caused by covalent binding.) (2) Pathways directly affected by ATP depletion. (3) Likely covalent binding sites. (NB, membrane proteins, glycoproteins, cofactors, and individual enzymes in the above pathways or the processes that synthesize them may also be subjected to covalent binding.) (4) Reactions affected by hypoxia. (NB, free radical and electrophile formation as well as lipid peroxidation may also be affected.) (5) Sites affected by changes in free Ca^{2+} concentration. (6) May also modify membrane permeability adversely. Bases = choline, ethanolamine, and serine, for example. Alcohols = inositol, for example. CDP and CTP = cytidine diphosphate and triphosphate, respectively.

reduce membrane damage or glutathione depletion caused by sodium iodoacetamide or diethylmaleate,[50] indicating that lipid peroxidation was not the cause of cell death. However, Smith *et al.* found that carbon tetrachloride–induced lipid peroxidation occurred prior to cell death, while in the case of bromobenzene, lipid peroxidation occurred later, possibly as a result of cell necrosis.[49] In contrast, Casini *et al.* found that cell death due to bromobenzene was reduced by antioxidants and that this reduction was proportional to the reduction in lipid peroxidation.[51] Antioxidants did not decrease covalent binding by bromobenzene, suggesting that the active metabolite responsible for cell death was different from that involved in binding covalently to cellular proteins.

Hogberg and coworkers have suggested that severe glutathione depletion inevitably results in lipid peroxidation,[42,43] but Babson *et al.* found no correlation between glutathione depletion and adriamycin-related lipid peroxidation and cell death.[44] Reiter and Wendel also concluded that it is not the absence of glutathione per se that causes lipid peroxidation, but the decreased removal of toxic metabolites that interact with the microsomal monooxygenase system to produce reactive oxygen species.[52] Another view is that glutathione depletion leads to an increase in intracellular iron available for activation of lipid peroxidation.[53]

The other early changes discussed above have been much less extensively studied. Wiebkin *et al.* compared ATP content and oxygen uptake with thiobarbituric acid reactants and membrane leakage of isolated hepatocytes treated with organotin compounds[17] (TABLE 6). There was a good correlation between lipid peroxidation and membrane leakage, but ATP content and oxygen uptake were apparently more sensitive indices of toxicity and were affected by a lower concentration of toxicant than were the other parameters. A decrease in ATP content of isolated hepatocytes has been found to be the most rapid change occurring as a result of exposure to toxic substances.[39] This could lead to a decrease in energy-requiring processes such as ion transport and protein synthesis, but this relationship has yet to be fully investigated.

The lack of agreement between different laboratories may be ascribed in part to the variety of experimental conditions employed. In many studies inadequate dose-response investigations are reported or there is little information provided on the development of particular changes with time. Some standardization of methodology on widely used cell types, e.g., isolated hepatocytes, is desirable in order that findings from different laboratories can be compared more readily. From the published literature and our own findings, one may deduce however that the relative importance of particular early molecular events varies from chemical to chemical. The observation that the glutathione redox cycle can function on relatively low levels of glutathione[44] may be an indicator that the cell is well protected against single toxic events and that it is only when the protective mechanisms are overwhelmed by several toxic changes that cell death ensues.

CONCLUSIONS

Present evidence would indicate that chemically mediated toxicity may involve a number of molecular events. No single early process would appear to predominate in response to all chemicals, even in a defined cell preparation (FIGURE 6). Whether a common final obligatory sequence of changes precludes cell death remains controversial. Selection of incubation conditions, e.g., oxygen tension, incubation medium, cell type, etc., would appear to have a very important influence on the observed processes of toxic change induced by an individual chemical in isolated cells. It is probably inappropriate in our present state of knowledge to employ a single biochemical parameter as an indicator of serious cytotoxic change.

REFERENCES

1. SNYDER, R., D. V. PARKE, J. J. KOCSIS, D. J. JOLLOW, G. G. GIBSON & C. M. WITMER, Eds. 1982. Biological Reactive Intermediates. **1** & **2**. Plenum Press. New York, N.Y.
2. MILLER, E. C. & J. A. MILLER. 1966. Pharmacol. Rev. **18:** 805.
3. COON, M. J., A. H. CONNEY, R. W. ESTABROOK, H. V. GELBOIN, J. R. GILLETTE & P. J. O'BRIEN, Eds. 1980. Microsomes and Drug Oxidations and Chemical Carcinogenesis. **2**. Academic Press, Inc. New York, N.Y.
4. MASON, R. P. & C. F. CHIGNELL. 1982. Pharmacol. Rev. **33:** 189.
5. KAPPUS, H. & H. SIES. 1981. Experientia **37:** 1233.
6. HALLIWELL, B. 1982. Trends Biochem. Sci. **7:** 270.
7. CONNOLLY, J. & J. W. BRIDGES. 1980. *In* Progress in Drug Metabolism. J. W. Bridges & L. F. Chasseaud, Eds. **5:** 1. John Wiley, Ltd. Chichester, England.
8. GRAM, T. E., Ed. 1980. Extrahepatic Metabolism of Drugs and Other Foreign Compounds. MTP Press, Ltd. Lancaster, England.
9. BRIDGES, J. W. & J. R. FRY. 1978. *In* The Induction of Drug Metabolism. R. W. Estabrook & E. Lindenlaub, Eds. **14:** 343. Symposia Medica Hoechst, F. K. Schattauer Verlag. Stuttgart & New York.
10. POLAND, A. & J. C. KNUTSON. 1982. Annu. Rev. Pharmacol. Toxicol. **22:** 517.
11. MANN, A. H., S. C. PRICE, F. E. MITCHELL, R. H. HINTON & J. W. BRIDGES. (Submitted for publication.)
12. WYLLIE, A. H. 1981. *In* Cell Death in Biology and Pathology. I. D. Bowen & R. A. Lockshin, Eds.: 9–34. Chapman and Hall. London & New York.
13. TRUMP, B. F., I. K. BEREZESKY & A. R. OSORNIO-VARGAS. 1981. *In* Cell Death in Biology and Pathology. I. D. Bowen & R. A. Lockshin, Eds.: 209–242. Chapman and Hall. London & New York.
14. SLATER, T. F. 1972. Free Radical Mechanisms in Tissue Injury. Pion, Ltd. London, England.
15. BUS, J. S. & J. E. GIBSON. 1979. Rev. Biochem. Toxicol. **1:** 125.
16. MENZEL, D. B. 1979. Environ. Health Perspect. **29:** 105.
17. WEIBKIN, P., R. A. PROUGH & J. W. BRIDGES. 1982. Toxicol. Appl. Pharmacol. **62:** 409.
18. RICE-EVANS, C. A. & M. J. DUNN. 1982. Trends Biochem. Sci. **7:** 282.
19. LAIHO, K. U. & B. F. TRUMP. 1974. Virchows Arch. B **15:** 267.
20. SCHANNE, F. A. X., A. B. KANE, E. E. YOUNG & J. L. FARBER. 1979. Science **206:** 700.
21. SMITH, M. T., H. THOR & S. ORRENIUS. 1981. Science **1257.**
22. STACEY, N. H. & C. D. KLAASSEN. 1982. J. Toxicol. Environ. Health **9:** 267.
23. BONNER, F. W., L. J. KING & D. V. PARKE. 1979. Chem. Biol. Interact. **27:** 343.
24. BRODIE, B. B., W. D. REID, A. K. CHO, G. SIPES, G. KRISHNA & J. R. GILLETTE. 1971. Proc. Nat. Acad. Sci. USA **68:** 160.
25. POTTER, W. Z., S. S. THORGEIRSSON, D. J. JOLLOW & J. R. MITCHELL. 1974. Pharmacology **12:** 29.
26. JOLLOW, D. J., J. R. MITCHELL, N. ZAMPAGLIONE & J. R. GILLETTE. 1974. Pharmacology **11:** 151.
27. HINSON, J. A., L. R. POHL, T. J. MARKS & J. R. GILLETTE. 1981. Life Sci. **29:** 107.
28. SIEGERS, C. P., J. G. FILSER & H. M. BOLT. 1978. Toxicol. Appl. Pharmacol. **46:** 709.
29. REID, W. D. 1972. Experientia **28:** 1058.
30. EVANS, M. A. & R. D. HARBISON. 1978. Toxicol. Appl. Pharmacol. **45:** 739.
31. NEAL, R. A. 1980. Rev. Biochem. Toxicol. **2:** 131.
32. CALDER, I. C., A. C. YOUNG, R. A. WOODS, C. A. CROWE, K. N. HAM & J. TANGE. 1979. Chem. Biol. Interact. **27:** 245.
33. CROWE, C. A., A. C. YOUNG, I. C. CALDER, K. N. HAM & J. D. TANGE. 1979. Chem. Biol. Interact. **27:** 235.
34. MCMURTY, R. J. & J. R. MITCHELL. 1977. Toxicol. Appl. Pharmacol. **42:** 285.
35. MCCALLA, D. R., A. REUVERS & C. KAISER. 1971. Biochem. Pharmacol. **20:** 3532.
36. BOYD, M. R. 1980. Rev. Biochem. Toxicol. **2:** 71.
37. WILEY, R. A., R. P. HANZLIK & T. GILLESSE. 1979. Toxicol. Appl. Pharmacol. **49:** 249.
38. GORROD, J. W. 1981. *In* Testing for Toxicity. J. W. Gorrod, Ed.: 77. Taylor-Francis, Ltd. London, England.

39. GWYNN, J. 1981. PhD Thesis. University of Surrey. Guildford, England.
40. POPP, J. A., H. SHINIZUKA & E. FARBER. 1974. Am. J. Pathol. **74:** 58A.
41. SMITH, M. T., N. LOVERIDGE, E. D. WILLS & J. CHAYEN. 1979. Biochem. J. **182:** 103.
42. HOGBERG, J. & A. KRISTOFERSON. 1978. Acta Pharmacol. Toxicol. **42:** 271.
43. ANUNDI, I., J. HOGBERG & A. H. STEAD. 1979. Acta Pharmacol. Toxicol. **45:** 45.
44. BABSON, J. R., N. S. ABELL & D. J. REED. 1981. Biochem. Pharmacol. **30:** 2299.
45. BRIDGES, J. W. 1981. *In* Testing for Toxicity. J. W. Gorrod, Ed.: 125. Taylor-Francis, Ltd. London, England.
46. FRY, J. R. & J. W. BRIDGES. 1979. Rev. Biochem. Toxicol. **1:** 201.
47. ACOSTA, D., D. C. ANUFORO & R. V. SMITH. 1980. Toxicol. Appl. Pharmacol. **53:** 306.
48. STACEY, N. H., H. OTTENWALDER & H. KAPPUS. 1982. Toxicol. Appl. Pharmacol. **62:** 421.
49. SMITH, M. T., H. THOR, P. HARTZELL & S. ORRENIUS. 1982. Biochem. Pharmacol. **31:** 19.
50. STACEY, N. H. & C. D. KLAASSEN. 1981. Toxicol. Appl. Pharmacol. **58:** 8.
51. CASINI, A., M. GIATI, R. J. HYLAND, A. SERRONI, D. GILFAR & J. L. FARBER. 1982. J. Biol. Chem. **257:** 6721.
52. REITER, R. & A. WENDEL. 1982. Chem. Biol. Interact. **40:** 365.
53. LEVINE, W. G. 1982. Life Sci. **31:** 779.
54. CAGE, J. C. 1953. Biochem. J. **54:** 426.
55. PETERS, R. A. 1963. Biochemical Lesions and Lethal Synthesis. Macmillan Publishing Co., Inc. New York, N.Y.
56. LEHNINGER, A. L. 1982. Principles of Biochemistry. Worth Publishers, Inc. New York, N.Y.
57. O'BRIEN, R. D. 1967. Insecticides, Action & Metabolism. Academic Press, Inc. New York, N.Y.
58. FARBER, J. L. & E. I. MOLTY. 1975. Am. J. Pathol. **81:** 237.
59. FARBER, E. 1967. Adv. Lipid Res. **5:** 119.
60. IYANAGI, T. & I. YAMAZAKI. 1969. Biochim. Biophys. Acta **172:** 370.
61. KIESE, M. 1959. Naunyn-Schmeidebergs Arch. Exp. Pathol. Pharmakol. **235:** 354.
62. CORCORAN, G. B. & J. R. MITCHELL. 1982. *In* Biological Reactive Intermediates II. R. Snyder, D. V. Parke, J. J. Kocsis, D. J. Jollow, G. G. Gibson & C. M. Witmer, Eds. (Part B): 1085–1098. Plenum Press. New York, N.Y.
63. WILLIAMS, R. T. 1959. Detoxication Mechanisms. Chapman and Hall. London, England.
64. MULDER, G. 1982. Personal communication.
65. ALLISON, A. C., J. S. HARRINGTON & M. BIRBECK. 1966. J. Exp. Med. **124:** 141.
66. DOULL, J., C. D. KLAASSEN & M. O. AMDUR. 1982. Casarett & Doull's Toxicology. The Basic Science of Poisons. 2nd edit. Macmillan Publishing Co., Inc. New York, N.Y.
67. STICH, H. F. & R. H. C. SAN, Eds. 1981. Short Term Tests for Chemical Carcinogens. Plenum Press. New York & London.
68. DE SERRES, F. & A. HOLLAENDER, Eds. 1982. Chemical Mutagens, Principles & Methods for Their Detection. 1–7. Springer-Verlag. New York, N.Y.
69. ROE, F. J. C. 1981. *In* Testing for Toxicity. J. W. Gorrod, Ed.: 29–43. Taylor-Francis, Ltd. London, England.
70. DE MATTEIS, F., A. H. GIBBS, & A. P. UNSELD. 1982. *In* Biological Reactive Intermediates II. R. Snyder, D. V. Parke, J. J. Kocsis, D. J. Jollow, G. G. Gibson & C. M. Witmer, Eds. (Part B): 1319. Plenum Press. New York & London.
71. MIDDLETON, M. C. 1981. *In* Testing for Toxicity. J. W. Gorrod, Ed.: 275. Taylor-Francis, Ltd. London, England.
72. PARRISH, W. E. 1981. *In* Testing for Toxicity. J. W. Gorrod, Ed.: 297. Taylor-Francis, Ltd. London, England.
73. LOMBARDI, B. & G. UGAZIO. 1965. J. Lipid Res. **6:** 498.
74. GRASSO, P. & T. J. B. GRAY. 1977. Toxicology **7:** 327.
75. REDDY, J. K., D. L. AZARNOFF & C. E. HIGNITE. 1980. Nature **283:** 397.
76. COHEN, A. J. & P. GRASSO. 1981. Food Cosmet. Toxicol. **19:** 585.
77. REDDY, J. K., N. D. LALWANI, M. REDDY & S. A. QURESHI. 1982. Cancer Res. **42:** 259.
78. MITCHELL, A., C. ELCOMBE, R. H. HINTON & J. W. BRIDGES. (Submitted for publication.)

SCREENING OF TOXIC COMPOUNDS
IN MAMMALIAN CELL CULTURES

Björn Ekwall

Toxicology Laboratory
National Food Administration
S-751 26 Uppsala, Sweden

INTRODUCTION

This communication will deal with the possibility of detecting high acute toxicity of chemicals by the use of simple and inexpensive cell-culture tests. The research area has been reviewed recently,[1-5] so I will just outline some general features.

In the last decade only 50 or so laboratories throughout the world have been testing general toxicity using tissue cultures—the field thus is rather small. Some of these laboratories have used differentiated cells, such as liver,[6-8] nerve, blood, and lung cells, as reviewed by Paganuzzi-Stammati and colleagues.[2] Some problems arise with organ-specific cultures for screening purposes.[3] Therefore, most screening has been done with less differentiated cells, which are easy to cultivate and which do not change from test to test. Such cells have been used to screen for local toxicity of plastics,[9] dental materials,[9,10] minerals,[11] and tobacco chemicals[12,13] and also for the systemic toxicity in man of reuse water,[14] minerals,[15] metals,[16] solvents,[17] drugs,[18,19] and compounds to be marketed by chemical companies.[20,21] Different cells and methods have been used in all laboratories, which makes validation procedures difficult. However, some results from different methods have been compared,[9,13,20] as have results obtained *in vitro* and *in vivo*.[9,14-16,21] Often these correlations have been good, suggesting a relevant screening. Unfortunately, very few large-scale validation studies have been made to date.[3,22] Hence, *in vitro* methods to screen for general toxicity, as opposed to genetic toxicity, cannot yet be considered valid and have also not been generally accepted.

Lately, some new tendencies have been seen. First, the success of mutagenicity screening *in vitro* has created expectations that general toxicity would be similarly evaluated. Second, antivivisectionist campaigns against animal experiments have resulted in a drive to develop so-called alternative test methods, of which cell culture is one. These expectations have stimulated investigators to develop new strategies.

One such strategy is to refine the choice of cells and end points of one method. Thus, human corneal cells are now used to screen for local eye toxicity of chemicals, with a method employing sophisticated end points.[23] Another strategy is to use batteries of tests with different cell types, to cover most aspects of injury. The cell types could either represent various organ-specific functions[24] or various aspects of basic cell functions.[48] A third strategy is to do more basic research into fundamental mechanisms of toxicity. When such mechanisms have been clarified, rational *in vitro* models could be set up.[25] By contrast, a fourth approach ignores whether the toxic mechanism screened for is known or not. As long as the end point of the test correlates well with *in vivo* toxicity, the test may be used.[26] In the following, a fifth strategy or rationale for using cell cultures to screen for general toxicity will be described.

64

0077-8923/83/0407-0064 $01.75/0 © 1983, NYAS

The Possibility of Testing Basal Cytotoxicity *In Vitro*

Most human cells have two aspects (FIGURE 1). One aspect is the structures and functions common to all of them, which may be called the basal cell functions. Another aspect consists of structures and functions typical to each cell, tissue, or organ, which may be called the organ-specific functions. The basal cell functions always support the

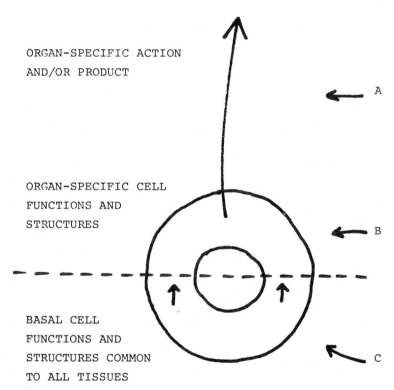

ORGAN-SPECIFIC ACTION AND/OR PRODUCT

A

ORGAN-SPECIFIC CELL FUNCTIONS AND STRUCTURES

B

BASAL CELL FUNCTIONS AND STRUCTURES COMMON TO ALL TISSUES

C

FIGURE 1. The two aspects of every human cell or organ: (1) the organ-specific functions and structures typical to each cell; and (2) the basal functions and structures common to all cells, which support the organ-specific cellular functions. Note that the organ-specific action and/or product of the cells may be interfered with by toxic action of chemicals at three levels of organization: (A) by extracellular interference at an organizational level; (B) by interference with specific cell functions in the cell (organ-specific cytotoxicity); and (C) by interference with basal cell functions, which in turn influence the specific functions (basal cytotoxicity).

specific cell functions. Hence, toxicity of chemicals to the basal functions, which may be called basal cytotoxicity, may disturb organ-specific action of the cells.

In 1967, Krasovsky and Ilnjitsky proposed toxicity tests in tissue cultures for agents that interfere with general cell functions, but not for agents with an organ-specific toxicity.[27] Somewhat later, Christian and Nelson also concluded that toxicity of agents to basic cellular processes, as opposed to specialized functions, is best studied *in vitro*.[28]

Many years ago my colleagues and I asked ourselves if the cultured cell could be used in toxicity tests as an indicator of basal cellular functions, and thus measure basal cytotoxicity of chemicals.[29] In that case the actual basal cytotoxicity of an agent in man and animals might be appreciated by a comparison between the toxicity of the agent *in vitro* and *in vivo*. If agents with an actual basal cytotoxicity at the lethal and/or therapeutic dosage in man could be shown to be common, studies of the toxicity to man of such agents would be feasible *in vitro*.

To prove our hypothesis we had to show (1) that a cell culture may indicate basal cytotoxicity and (2) that a substantial number of chemicals in fact are basally cytotoxic at lethal dosage or at exposure for man. We thought that these issues both might be validated by comparisons of results from standard tissue-culture toxicity tests of varied agents with other *in vitro* results and with animal and human toxicity of the same agents. A similar reaction of less differentiated and highly differentiated cultured cells to toxicity would support the idea that less differentiated cells may measure basal cytotoxicity. A correlation between similar toxic dosage of agents *in vitro* and *in vivo* and a known cytotoxic action of the same agents *in vivo* would also indicate valid cell tests of basal cytotoxic action. Finally, the number of agents with a similar *in vitro* and *in vivo* toxicity, both quantitatively and qualitatively, would determine how common basal cytotoxicity is. We therefore tested agents for their toxicity to HeLa cells in the same standard system. Later, we used the accumulated toxicity data for validation studies to prove the hypothesis about a common basal cytotoxic action of chemicals in man that is possible to study in tissue culture.

TOXICITY TESTS OF CHEMICALS ON HeLa CELLS IN THE MIT-24 SYSTEM

The metabolic inhibition test supplemented by a 24-hour toxicity determination, called the MIT-24 system, has been used by us to test the toxicity of various chemicals to an established cell line, the HeLa cells.[30,32] The same test system has now been employed for many years to produce comparable results. Drugs have been selected for testing for validation purposes because of their well-known toxicity in man.[22]

Two chemicals were usually tested in one microtiter plate (FIGURE 2). The agents were diluted in steps of 1:5 in five identical channels, with reference cultures between the test areas. HeLa cells, made round in a temporary suspension culture, were distributed to all cups, and the cups were sealed with liquid paraffin and a plastic film and then incubated at 37°C for 7 days. At the microscopic, 24-hour viability estimation, cultures with 100% round cells were considered totally inhibited while spindle-shaped cells indicated viability. After 7 days, viability was determined with use of the color change of the pH indicator phenol red included in the medium. Violet (pH 8.5) cultures were considered totally inhibited, while orange (pH 6.5) and red (pH 7.5) cultures were considered uninhibited or partially inhibited, respectively. Microphotographs of the cell reaction were taken routinely after 24 hours, which for some agents revealed specific cell injuries. For comparison with such injuries FIGURE 3 shows normal spindle-shaped cells of a reference culture. A very common injury of the cells was the cytoplasmic blebs,[38] thought to consist of cytosol that has leaked through a damaged cell membrane (FIGURE 4). Another very common cell reaction was vacuolization (FIGURE 5). Vacuoles were often induced by agents at noninhibitory concentrations (FIGURE 6).

Advantages of the MIT-24 system are the simple procedure, the monitoring of drug-induced pH changes and precipitates, the two objectively determined end points

MICROSCOPY, 24 HOURS:

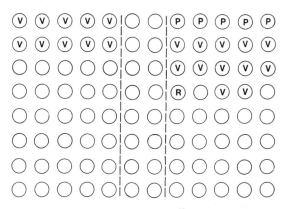

INDICATOR CHANGE, 7 DAYS:

FIGURE 2. Tray record from an MIT-24 test of two drugs.[37] Each drug was diluted (from above downward in the protocol) in a 5- by 8-cup area of a microtiter tray, separated from the other test area by 2 × 8 reference cultures. The pH values of the 24-hour record indicate pH of the medium induced by high concentrations of drugs at the start of incubation. T = total absence of fusiform cells, implying 100% round cells. D = deficient outgrowth of fusiform cells compared with outgrowth of these cells in the reference cultures. P = precipitate. V = violet color, indicating metabolic inhibition. R = red color, indicating partial inhibition. Empty circles = normal outgrowth after 24 hours of incubation, and normal orange-yellow color of reference cups after 7 days of incubation.

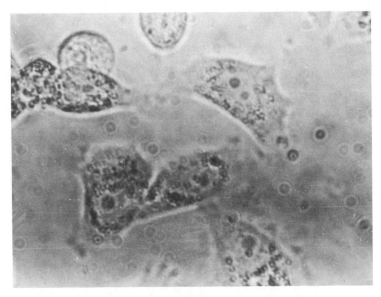

FIGURE 3. Normal fusiform HeLa cells in a reference culture of the MIT-24 test after 24 hours of incubation.

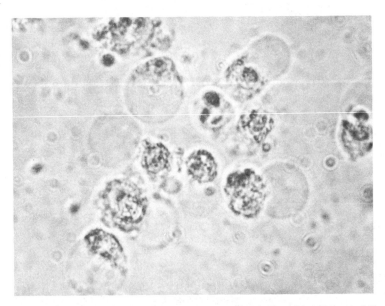

FIGURE 4. The reaction of HeLa cells after 24 hours of incubation with 0.013 μg/ml digoxin. Every cell has a cytoplasmic bleb, probably consisting of cytosol escaped from the rest of the cell, suggesting injury of the cell membrane.

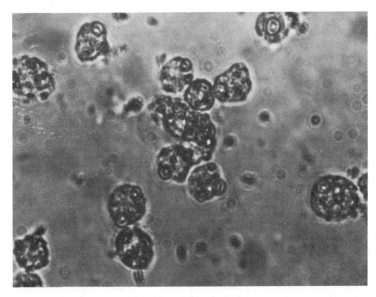

FIGURE 5. The reaction of HeLa cells after 24 hours of incubation with 1,000 μg/ml amphetamine sulfate. Every cell has several large vacuoles.

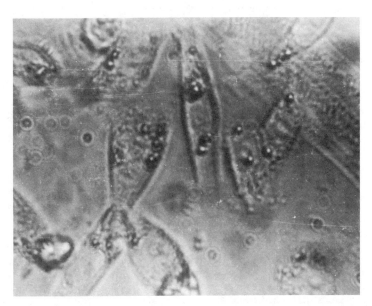

FIGURE 6. The reaction of HeLa cells after 24 hours of incubation with 8.0 μg/ml propranolol HCl. Fusiform cells with vacuoles.

of toxicity at two incubation times, and finally the possibility of screening for toxic injuries by microscopy.

VALIDATION STUDIES TO PROVE THE HYPOTHESIS ON BASAL CYTOTOXICITY

The first type of validation study compared results from HeLa cells in the MIT-24 system with results of toxicity tests of chemicals on other cells in other systems. One study compared the toxicity *in vitro* of 27 drugs to several cell types, including established cells and primary cultures of differentiated cells.[33] About the same toxicity was found in all systems. Another study compared the cytotoxicity of 14 agents to cell lines and rat hepatocytes.[31] Four specifically hepatotoxic agents had a high differential toxicity to the liver cells, proving the metabolic capacity of the hepatocytes, while most agents had the same toxicity to both cell types, which may be interpreted as basal cytotoxicity for the differentiated liver cells.

The second type of validation study compared the HeLa cytotoxicity with the dosage and concentrations of a great variety of chemicals *in vivo*. The similarity between *in vitro* and *in vivo* toxic dosage was then compared with known organ-specific or cytotoxic action *in vivo* of the chemicals.

The toxicity in vitro was compared with toxic dosage and concentrations in animals and man in the form of ratios (TABLE 1). Thus, the 50% inhibitory concentration (IC_{50}) after 7 days of incubation for HeLa cells was divided by the intravenously administered (iv) 50% lethal dosage (LD_{50}) for mice, to make up the specific index for the lethal dosage (SI_{ld}).[22,29] Also, similar SI_{ld} based on approximate iv lethal doses in man could be calculated. Later comparisons of concentrations led to so-called cytotoxic quotients (CQ_L and CQ_T).[35,36] Either the human blood concentration for a lethal dose of an agent could be divided by the 24-hour IC_{50} for HeLa cells to make up a cytotoxic quotient for lethal dosage (CQ_L), or a human blood concentration from a therapeutic dose of a drug (exposure of a chemical) could be divided by the IC_{50} for HeLa cells to make up a cytotoxic quotient for therapeutic dosage (CQ_T). A similar toxicity *in vitro* and *in vivo* is indicated by low SI_{ld} and high CQ_L and CQ_T values near unity, while a comparatively lower dosage *in vivo* suggesting organ-specific action is indicated by high SI_{ld} and low CQ_L and CQ_T values.

The first validation study of this type calculated SI_{ld} in the mouse and man for 41 mainly neurospecific drugs, using literature data.[22,29] The tissue-culture toxicity data came from fetal chicken nerve cells in primary cultures.[39] The variation of SI_{ld} was well correlated with organ-specific and basal cell toxicity, respectively.

A second study calculated mouse and human SI_{ld} based on results with HeLa cells for about 50 drugs.[22,34] The variation of SI_{ld} was likewise well correlated to organ specificity or basal cytotoxicity of the chemicals. Moreover, most of the randomly selected drugs (85%) had an SI_{ld} of 50 or less, suggesting basal cytotoxicity for most agents.

A third study of SI_{ld} has been done with 20 plasticizers.[32] All compounds had an SI_{ld} around 4, suggesting basal cytotoxicity.

However, it could be shown that dosage comparisons in the form of SI_{ld} may be misleading for some compounds.[22] Therefore later validations were done with CQ_L and CQ_T values, calculated for the same compounds as in earlier studies.[22,34]

One such study compared CQ_L values for 46 drugs with lethal effects in man.[35] CQ_L was well correlated with known cytotoxic action of the agents. Depressants of the central nervous system had a CQ_L around 0.1, suggesting a basal cytotoxicity to extra vulnerable nerve cells, while drugs with a cytotoxic action to organs outside the central nervous system, because they could not penetrate the blood-brain barrier, had a CQ_L

TABLE 1

QUOTIENTS USED TO COMPARE *IN VITRO* AND *IN VIVO* TOXICITY*

Quotient between	Name	Short Form	Range
7-Day IC_{50} for cells/mouse or human LD_{50}	specific index for lethal dosage	SI_{ld}	0.1–10,000
Human lethal blood concentration/24-hour IC_{50} for cells	cytotoxic quotient for lethal dosage	CQ_L	0.0001–5.0
Human therapeutic blood concentration/24-hour or 7-day cell IC_{50}	cytotoxic quotient for therapeutic dosage†	CQ_{T_d} CQ_{T_r}	0.00001–2.0

*Data from earlier publications.[22,29,35,36]
†CQ_{T_d} connected with directly measured single dosage and 24-hour cell IC. CQ_{T_r} connected with repeated dosage and 7-day IC_{50} for cells.

near unity. As before, about 80% of the drugs seemed to have a basal cytotoxicity at lethal dosage in man, now suggested by a CQ_L of 0.01 or higher.

Another study related CQ_T with known cytotoxic side effects of the same drugs.[36] There was a good correlation. Some drugs, such as the sedatives, seemed to have therapeutic effects due to basal cytotoxic mechanisms.

To summarize results from two types of validation studies with more than 100 agents, simple cell cultures probably can be used to test a basal cytotoxic action of chemicals. Furthermore, most chemicals seem to interfere lethally with man and animals by such a basal cytotoxic action. For some chemicals the basal cytotoxic concentration in man may be reached at therapeutic dosage (for drugs) or at industrial exposure. As a consequence of the possibility of testing basal cytotoxicity *in vitro,* comparisons of toxic dosage and, above all, concentrations *in vitro* and *in vivo* in the form of SI_{ld} and CQ_L and CQ_T values will be predictive.

The predictive value for SI_{ld} and CQ_L can be exemplified by methadone, phenobarbital, and quinidine (TABLE 2). Methadone has an organizational therapeutic and lethal action on specific receptors in the brain, predicted by the high human SI_{ld} and the low CQ_L and CQ_T values. Phenobarbital probably interferes cytotoxically with brain cells at lethal dosage, predicted by an SI_{ld} of 10 and a CQ_L of 0.1. Quinidine has lethal and side effects by a probable cytotoxic interference with many organs outside the brain, notably the heart, as predicted by the low SI_{ld} and high CQ_L near unity. Note that the mouse SI_{ld} can be a poor indicator of human toxicity due to species differences, as shown by methadone and phenobarbital.

THREE APPLICATIONS OF A BATTERY TO TEST BASAL CYTOTOXICITY

The findings from our preliminary validation studies open new possibilities of using cell cultures (1) to study basally cytotoxic mechanisms of agents and (2) to test chemicals for basal cytotoxicity. Since every test method has its blind spot, this should not be done with a single test, but with a standard battery of test methods, employing a few different cell types and a few balanced end points. An established cell line, human diploid fibroblasts, and rat hepatocytes have been proposed[3,22] to provide tests with little variation, normal basic functions, and metabolic capacity, respectively. The methods must be standardized to produce reproducible and comparable results.[3] It is probably important to perform IC_{50} tests of every compound, also for nontoxic agents and for screening purposes, to make results definite.[3] Finally, qualitative records of the cell reaction, incubation times easy to compare with *in vivo* situations, and equal sensitivity of test cells and tissues *in vivo* would be of value. Such a battery could be applied in three areas of toxicology.

The first use could be to study systematically the basal cellular reaction to toxic concentrations of compounds.[3,30,40] This is a sector of basic science that tries to explain cytotoxic mechanisms of chemicals, which later would be applied to increase knowledge of the toxicity in man of relevant agents. These studies may be performed stepwise:

1. Toxicity tests by the battery of a large number of compounds.
2. Studies on the type of injury (morphology, biochemistry).
3. Structure-activity relationship (SAR) and other analyses to explain the cell reactions in terms of physicochemical activity of the compounds.
4. Studies to determine the relevance of findings *in vitro:* (a) *in vitro/in vitro* comparisons; (b) dosage comparisons *in vitro/in vivo;* (c) comparisons of concentrations *in vitro/in vivo;* (d) comparisons of toxic effects *in vitro/in vivo.*

TABLE 2

CALCULATED AND ACTUAL BASAL CYTOTOXICITY FOR THREE DRUGS IN MAN*

Drug	SI_{ld} in Mice	SI_{ld}	CQ_L	CQ_{T_d}	CQ_{T_r}	Cytotoxicity at Lethal Dosage	Cytotoxicity at Therapeutic Dosage
Methadone	6	200	0.01	0.003	0.005	no	no
Phenobarbital	3	10	0.1	0.02	0.05	yes, to CNS	yes, CNS
Quinidine	0.2	0.7	1.0	0.05	0.3	yes	yes

*For definitions of SI_{ld}, CQ_L, CQ_{T_d}, and CQ_{T_r}, see TABLE 1. Table based on data from previous research.[22,35,36]

The second use of the battery would be for routine, acute toxicity testing, as a supplement to conventional tests. This would parallel the use of *in vitro* mutagenicity tests in described programs of carcinogenicity tests.[41,42] To stimulate discussion, I will now propose a program that stresses *in vitro* pretests of various kinds as well as a search for toxic mechanisms, called "the multistep *in vitro/in vivo* acute toxicity test."[22]

1. Determination of physical properties of the formula, such as electronic distribution, oxidation-reduction potentials, dissociation constants, partition coefficients, and surfactant activity. Simple *in vitro* standard tests of binding to protein, DNA, and subcellular fractions,[43] uptake into cultured cells, hemolytic and antihemolytic activity. The data may predict toxicity as well as pharmacokinetics.

2. Tests of the potential basal cytotoxicity of the agent in man by the battery.

3. Determination of iv and peroral (po) approximate LD_{50} for mice and rats, with use of a small number of animals and with a recording of symptoms, plasma and tissue concentrations, and autopsy.[44] Cytotoxicity test results may guide dosage.

4. Calculation of CQ_L to provisionally determine the basal cytotoxicity of the agent. While mechanisms of organ-specific agents must be studied by conventional techniques, toxic mechanisms of basally cytotoxic agents (probably a majority) could be preliminarily studied in the test battery, to be confirmed later by animal experiments. For some agents, the toxic mechanism could be relevant also for chronic, low-dose administration of the compound.

A third use of the test battery would be to screen basal cytotoxicity of compounds. The incentive would be the myriad of chemicals and natural toxicants that ought to be evaluated for their toxicity with inexpensive and ethical methods.[26] Compounds could either be screened for their potential basal cytotoxicity, that is, if their *in vitro* cytotoxicity is very high, the chance for a human toxicity of the agent also would be high. Or they could be screened for their actual basal cytotoxicity, which could be appreciated by a comparison of the toxicity *in vitro* with exposure (CQ_T value). Disadvantages of such screening are the inevitable false results. In screening for systemic toxicity, the lack of capacity of the battery to detect organ-specific toxicity will lead to false-negative results. Atypical pharmacokinetics of some agents *in vivo* and methodological difficulties in comparing *in vitro* and *in vivo* toxicity could lead to both false positives and false negatives.[22] Used to screen local toxicity, the local pharmacokinetics of some compounds may lead to difficulties. Common experience from *in vitro* screening for local toxicity is that high toxicity *in vitro* does not correspond to local irritation for 30% of the agents.[9,49] This could depend on a rapid elimination of the agent from the local site by the circulation. The lack of blood vessels in the cornea may lead to fewer such false positives in the screening for local eye toxicity, however.

In spite of the errors mentioned, the battery could be used to screen for both systemic and local toxicity. An agency forced to set priorities for conventional tests could use the battery to select compounds with a high cytotoxicity for further testing. A company developing chemicals could take away highly cytotoxic agents from further processing, by use of the battery. Pairs and mixtures of chemicals with a provisionally determined high actual cytotoxicity (SI_{ld}, CQ_L, and CQ_T values) could be screened for their combined systemic or local toxicity by modifications of the tests in the battery.[45,46] This is especially valuable since most real-life situations involve combined toxicity of agents and since animal tests are too expensive to screen for combined toxicity.

The use of the same standard tests for all described purposes would bring mechanism studies and screening closer together. Screening data could be used for mechanism studies, and vice versa. The performance of single experiments *in vitro* would be possible, the results of which could be compared with a wide frame of reference.

PRACTICAL MEASURES TO PROMOTE BASAL CYTOTOXICITY TESTS

Much work remains to be done to prove the value of the discussed ideas. The basic strategy as well as the features of an optimal test battery must first be discussed by *in vitro* cytotoxicologists.

As a contribution to this discussion, a few remarks on the relative merits of end points to test basal cytotoxicity will be made. There are many good end points in use already, such as growth determined by protein analysis[9,14,15,19] and cell counts,[12] membrane damage determined by enzyme release,[18,23] release of radioactive markers[13,23] or exclusion of colors,[17] cellular synthesis and metabolism measured by uptake of markers,[10,23] and finally plating efficiency of cells.[11,16,20] Of these end points growth has one disadvantage, i.e. it cannot be used for nondividing cells or be compared with injury to these cells. Interference of toxicants with active processes such as the MIT-24 end points, growth, and plating efficiency will give few false-negative results, while interference with passive structures in the cell such as the membrane could result in many false negatives, i.e., agents that kill cells by a fixative action that preserves the structure of the membrane. On the other hand, the former end points may result in many false positives, which is not the case with the latter. Active processes and structures would therefore supplement one another as indicators of injury. Also, sensitivity of the end points deserves discussion. While some of the tests in the battery clearly should measure lethal end points, other tests probably should measure subtle sublethal injury. Recently, a rapid screening test on hamster CHO-K1 cells was described, which measured labile energy metabolites, such as adenosine 5′-monophosphate, adenosine triphosphate, and reduced nicotinamide adenine dinucleotide, in cells by isotachophoresis.[21] Changes were noted at 1/1,000 of the lethal dose.

After discussions of the ideal test battery, candidates for tests to be included must be validated. The validation procedures must consist of the described comparisons between different *in vitro* results for many chemicals and comparisons on a large scale between *in vitro* and *in vivo* toxicity (SI_{ld}, CQ_L, and CQ_T values). SI_{ld} and CQ_L values ought to be related to known toxic action of compounds. Besides the series of compounds with known *in vivo* toxicity, agents with a quite unknown toxicity could be tested also, followed by animal toxicity tests of the most cytotoxic compounds. These procedures could be concerted, multilaboratory efforts[47] but could also be performed as literature studies or by an agreement between two laboratories to test the same substances.

It is hoped that chemical companies, governmental agencies, and other donators of funds will support the outlined studies to validate basal cytotoxicity tests *in vitro*. This rationale has a chance of providing results just now expected in many circles. It is possible that very few, if any, toxicological methods are able to produce such key information at such a low cost as the outlined battery to test basal cytotoxicity. It is also possible that the battery in the future will be considered as indispensable as the short-term mutagenicity and carcinogenicity tests.

SUMMARY

The cytotoxic concentrations of about 100 randomly selected drugs and chemicals, tested on HeLa cells in the MIT-24 system and/or in primary cultures of fetal chicken cells, were compared with the lethal doses and/or concentrations of the agents in the mouse and in man. Most agents (80%) had a similar toxicity *in vitro* and *in vivo*, suggesting a lethal interference in man with basal functions common to all specialized human tissues as well as cultured cells, i.e., basal cytotoxicity. This high frequency of basally cytotoxic agents opens possibilities for screening chemicals for toxicity and for studying cytotoxic mechanisms with a standard battery of a few appropriate cell tests. This battery may be used in three ways: (1) to study cytotoxic mechanisms of all chemicals, and apply the resulting knowledge to understanding toxicity in man of basally cytotoxic agents; (2) to supplement conventional animal tests in acute toxicity test programs; (3) to screen chemicals and extracts for their potential basal cytotoxicity. To validate these ideas and to select suitable tests for the battery, results from cytotoxicity tests on a wide variety of chemicals in several *in vitro* systems must be compared with one another and with the toxicity of the agents in animals and man.

REFERENCES

1. BERKY, J., Ed. 1978. In Vitro Toxicity Testing. Franklin Institute Press. Philadelphia, Pa.
2. PAGANUZZI-STAMMATI, A., V. SILANO & F. ZUCCO. 1981. Toxicology 26: 91–153.
3. EKWALL, B. 1980. Toxicology 17: 127–142.
4. REES, K. R. 1980. J. R. Soc. Med. 73(4): 261–264.
5. NARDONE, R. 1983. Cell culture techniques for the assessment of acute toxicity. Acta Pharmacol. Toxicol. (In press.)
6. GRISHAM, J. W., R. K. CHARLTON & D. G. KAUFMAN. 1978. Environ. Health Perspect. 25: 161–171.
7. INMON, J., A. STEAD, M. D. WATERS & J. LEWTAS. 1981. In Vitro 17: 1004.
8. WILLIAMS, G. 1981. *In* Organ-directed Toxicity/Chemical Indices and Mechanisms (IUPAC). S. S. Brown & D. S. Davies, Eds.: 131–145. Pergamon Press. Oxford, England.
9. AUTIAN, J. 1977. Artif. Organs 1: 53–60.
10. WENNBERG, A. 1980. Scand. J. Dent. Res. 88: 46–52.
11. REISS, B., J. R. MILLETTE & G. WILLIAMS. 1980. Environ. Res. 22: 315–321.
12. PILOTTI, Å., K. ANCKER, E. ARRHENIUS & C. ENZELL. 1975. Toxicology 5: 49–62.
13. THELESTAM, M., M. CURVALL & C. ENZELL. 1980. Toxicology 15: 203–217.
14. CHRISTIAN, R. T., T. E. CODY, R. S. CLARK, R. LINGG & E. J. CLEARY. 1973. AIChE Symp. Ser. 70: 15–21.
15. CHRISTIAN, R. T., J. B. NELSON, T. E. CODY, E. LARSON & E. BINGHAM. 1979. Environ. Res. 20: 358–365.
16. ANDREWS, T. K., JR., B. B. THOMPSON, M. A. WINCEK & M. E. FRAZIER. 1978. In Vitro 14: 381.
17. HOLMBERG, B. & T. MALMFORS. 1974. Environ. Res. 7: 183–192.
18. GOTO, Y., C. A. DUJOVNE, D. W. SHOEMAN & K. ARAKAWA. 1976. Toxicol. Appl. Pharmacol. 36: 121–130.
19. HORVATH, S. 1980. Toxicology 16: 59–66.
20. WININGER, M., F. A. KULIK & W. D. ROSS. 1978. In Vitro 14: 381.
21. WININGER, M. T., J. M. LAVOIE & W. D. ROSS. 1982. In Vitro 18: 290.
22. EKWALL, B. 1983. Correlation between cytotoxicity in vitro and LD50-values. Acta Pharmacol. Toxicol. (In press.)
23. DOUGLAS, W. H. J. 1981. Development of in vitro alternatives to the Draize test. Paper presented at the Third Symposium on Alternative Techniques Sponsored by Lord Dowding Fund, London, England, November 5.
24. NARDONE, R. M. 1982. In Vitro 18(3/2): 289.
25. GOLDBERG, A. Personal communication.
26. NARDONE, R. M. 1980. Toxicology 17: 105–111.
27. KRASOVSKY, G. N. & A. P. ILNJITSKY. 1967. Gig. Sanit. 32: 66–70.

28. CHRISTIAN, R. T. & J. B. NELSON. 1971. In Vitro 6(5): 377.
29. EKWALL, B. & B. SANDSTRÖM. 1971/1980. Specific Index—A Measure of Drug Specificity. Department of Anatomy, University of Uppsala. Uppsala, Sweden.
30. EKWALL, B. 1980. Toxicology 17: 273–295.
31. EKWALL, B. & D. ACOSTA. 1982. Drug Chem. Toxicol. 5: 219–231.
32. EKWALL, B., C. NORDENSTEN & L. ALBANUS. 1982. Toxicology 24: 199–210.
33. EKWALL, B. & A. JOHANSSON. 1980. Toxicol. Lett. 5: 299–307.
34. EKWALL, B. 1980. Toxicol. Lett. 5: 309–317.
35. EKWALL, B. 1980. Toxicol. Lett. 5: 319–321.
36. EKWALL, B. 1981. Toxicol. Lett. 7: 359–366.
37. EKWALL, B. 1980. Acta Univ. Ups. No. 353.
38. EKWALL, B. 1982. In Vitro 18(3/2): 289.
39. POMERAT, C. M. & C. D. LEAKE. 1954. Ann. N.Y. Acad. Sci. 58: 1110.
40. EKWALL, B. 1981. Principles and prospects of in vitro cytotoxicology. Paper presented at the Third Symposium on Alternative Techniques Sponsored by Lord Dowding Fund, London, England, November 5.
41. BRIDGES, B. A. 1973. Environ. Health Perspect. 6: 221–227.
42. WEISBURGER, J. H. & G. WILLIAMS. 1980. In Casarett and Doull's Toxicology. J. Doull, C. D. Klaassen & M. O. Amdur, Eds.: 84–138. Macmillan Publishing Co. New York, N.Y.
43. BICKEL, M. H. & J. W. STEELE. 1974. Chem. Biol. Interact. 8: 151–162.
44. ZBINDEN, G. & M. FLURY-ROVERSI. 1981. Arch. Toxicol. 47: 77–99.
45. EKWALL, B. & B. SANDSTRÖM. 1978. Toxicol. Lett. 2: 285–292.
46. EKWALL, B. & B. SANDSTRÖM. 1978. Toxicol. Lett. 2: 293–298.
47. BALLS, M. Personal communication.
48. PETTERSSON, B., M. CURVALL & C. R. ENZELL. 1980. Toxicology 18: 1–15.
49. DILLINGHAM, E. Personal communication.

ROLE OF CALCIUM IN CYTOTOXIC INJURY
OF CULTURED HEPATOCYTES*

Daniel Acosta and Elsie M. B. Sorensen

Department of Pharmacology and Toxicology
College of Pharmacy
University of Texas
Austin, Texas 78712

INTRODUCTION

The role of calcium in irreversible cell death has been the subject of much controversy recently. Documented *in vitro* studies from several laboratories have shown that calcium can have paradoxical beneficial or detrimental effects on liver cells exposed to hepatotoxic agents.[1-6] The present study reviews various types of *in vitro* liver models for the study of hepatoxins and provides data supporting the idea that calcium protects hepatocytes from cytotoxic injury by a well-known hepatotoxin.

Comparison of In Vitro Hepatic Models

Of the *in vitro* hepatic models available to investigate the metabolism and toxicity of xenobiotics, the following preparations have been utilized the most extensively: perfused liver,[7] liver cell lines,[8,9] isolated hepatocytes in suspension,[10-12] and primary cultures of hepatocytes.[13-16] TABLE 1 summarizes the advantages and disadvantages of the various liver models. On the basis of this comparison we developed a primary culture system of rat hepatocytes to evaluate the cytotoxicity and metabolism of xenobiotics.

Development of Primary Hepatocyte Culture for Cytotoxicity Studies

Our major goal has been to develop a primary culture system that retains differentiated functions and responses that are characteristic of the intact tissue *in vivo*. Major emphasis has been placed on the development of rat hepatocyte cultures because of the important role of the liver in the metabolism of xenobiotics to inactive, pharmacologically active, or toxic metabolites. To establish the functionality of the hepatocyte cultures and their similarity to the liver *in vivo*, we have examined a number of parameters that are considered characteristic of functional hepatocytes. These include sulfobromophthalein uptake,[17] glutathione conjugation,[17] L- and M-pyruvate kinase activity,[17] cytochrome P_{450} levels,[18] O-demethylation and conjugation activity,[19] metabolic activation of xenobiotics to toxic intermediates,[14] total urea,[20] maintenance of stable lactate-to-pyruvate ratios,[20] and cytotoxic injury after exposure to well-known hepatotoxic agents.[13,20,21] These activities can be monitored to insure the continued functionality and uniformity of cell populations and to reveal the injurious effects of toxic chemicals as well. Therefore, studies using this toxicity-testing system are more reliable in predicting the likelihood of liver injury *in vivo* than are those using

*This work was supported in part by BRSG S07RRO5849 awarded by the Biomedical Research Support Grant Program of the National Institutes of Health.

78

other models as outlined. By examining a variety of hepatotoxic agents (such as acetaminophen, cadmium, papaverine, tetracycline, ethanol, norethindrone, tricyclic antidepressants, and carbon tetrachloride),[13,20–22] we have demonstrated that our primary liver cell culture system has proven useful as an experimental model for hepatotoxicity studies.

Role of Calcium in Cytotoxic Injury of the Liver

Unrelated hepatotoxins produce liver injury that is accompanied by marked accumulation of intracellular calcium.[23–25] Because the extracellular concentration of

TABLE 1

IN VITRO LIVER SYSTEMS USED FOR CYTOTOXICITY AND METABOLISM STUDIES

System	Advantages	Disadvantages
Perfused liver	Retention of structural integrity; maintenance of cell-to-cell inter-relationships.	Short viability period of a few hours; complex and costly perfusion apparatus; reproducibility problems among laboratories; statistical sampling problems.
Liver cell lines	Increased viability period; easier to maintain than primary cultures.	Loss of differentiated liver functions; characteristics of cancer or transformed cells.
Freshly isolated hepatocytes	Ease of isolation; comparable drug-metabolizing activity as intact liver; ability to evaluate toxicity and metabolism of xenobiotic in same system.	Lack of cell-to-cell contact; short viability period (a few hours); damage to membranes by isolation procedures; leakage of cofactors and enzymes; impaired intermediary metabolism.
Primary hepatocyte culture	Increased longevity over perfused liver and isolated hepatocytes; recovery from trauma and damage of isolation procedure; retention of several differentiated liver functions; capability for use in conduction of acute and chronic toxicity and metabolism studies of xenobiotics.	Loss of cytochrome P_{450} with time in culture; loss of several differentiated functions upon subculturing the cells (functional liver lines difficult to establish).

calcium is very high ($\sim 10^{-3}$ M) and the intracellular concentration is very low (10^{-7} to 10^{-6} M), calcium influx into injured cells may simply be the result of passive equilibration of the cation, not the initiator of cell injury.[26] However, recent investigations suggest that the influx of calcium may indeed initiate cell injury and be the ultimate mediator of toxic cell death produced by a variety of toxic chemicals.[1–3]

Extracellular calcium has been shown to have paradoxical beneficial and detrimental effects on primary cultures or isolated suspensions of rat hepatocytes that have been exposed to cytotoxic agents. Schanne *et al.* showed that 10 different membrane-active toxins caused more cell damage in the presence than in the absence of 3.6 mM calcium chloride as assessed by viability, plating efficiency, and dye hydrolysis.[1] Carbon tetrachloride hepatotoxicity was reported to occur in the presence of 3.6 mM calcium chloride when the ionophore A23187 was used to disrupt intracellular calcium homeostasis; cultured hepatocytes showed marked swelling and blebbing, as well as increased LDH (lactate dehydrogenase) and GPT (glutamic-pyruvic transaminase) leakage.[3] Casini and Farber found that carbon tetrachloride decreased cellular viability with increasing levels of extracellular calcium from 0.3 to 3.6 mM.[2]

Although these three studies from two laboratories showed detrimental effects of calcium on hepatocyte cultures less than two hours old, other laboratories have reported either beneficial effects or no effects of calcium on cultures or isolated cells of a similar age. Smith et al. found that several cytotoxic agents caused less damage to cultured hepatocytes in the presence of 2.6 mM calcium than under calcium-free conditions.[4] Edmondson and Bang, furthermore, reported that freshly isolated hepatocytes incubated in calcium-free medium lacked microvilli and nuclear contents, rapidly lost the ability to exclude trypan blue and to retain LDH, and failed to accumulate α-aminoisobutyric acid.[5] Stacey and Klaassen found that the addition of 1 mM calcium chloride to the incubation medium did not change hepatocyte viability or leakage of potassium or AST (aspartate aminotransferase) from freshly isolated liver cells exposed to cadmium, copper, amphotericin B, or lysolecithin.[6] Omission of calcium from the medium, however, reversed the toxicity of the calcium ionophore A23187 in this same study.

The conflicting results of these reported studies have led to a continuing controversy in explaining the role of calcium in toxic cell death. Mechanistically, irreversible toxic cell death is thought to result from hepatotoxin-induced plasma membrane damage followed by a specific disturbance in the homeostasis of intracellular calcium[1] or other ions such as sodium or potassium.[4] Although other ions may exert different physiological effects, calcium influx is considered capable of considerable disruption of cellular integrity by affecting calcium-sensitive enzyme systems,[27] oxidative phosphorylation pathways,[28] or structural proteins such as microtubules.[29] Because the previously cited cytotoxicity studies utilized either freshly isolated hepatocytes or hepatocytes cultured for two hours or less, it is likely that even slight plasma membrane damage, caused by the rigorous isolation procedures usually employed, would allow altered hepatotoxin influx and subsequent damage more severe than would be observed if cells were allowed to recover following isolation.

In fact, freshly isolated parenchymal hepatocytes are reported to be severely altered cells (metabolically, morphologically, and functionally) primarily because of the use of collagenase and mechanical agitation for dissociation of the intact liver into individual cells or clusters of cells. Freshly isolated cells have reduced ornithine decarboxylase activity,[30] reduced protein synthesis with altered response to hormones, disaggregation of polysomes, lower albumin, transferrin, and fibrinogen levels,[31] and lower 5'-nucleotidase and alkaline phosphatase activities[32] when compared to intact liver cells or to cells cultured for one or more days. Although freshly isolated rat hepatocytes maintain their overall shape and dye-exclusion capacities, these cells generally lose about 10% of their metabolic capacity per hour[33] and thus are probably less suited for chemical challenge than are those cells allowed to recover during culture. Primary cultures of rat hepatocytes, moreover, have been documented to recover from the trauma of isolation[34] and to regain metabolic activities characteristic of the normal liver.[35] For this reason, we utilized primary rat hepatocyte cultures that were about 24 hours old to study the effects of extracellular calcium on several structural and functional parameters of cadmium-challenged hepatocytes.

METHODS

Primary cultures of parenchymal hepatocytes were prepared from 7- to 10-day-old Sprague-Dawley rats as previously described.[17,21] This procedure employs retrograde perfusion of the portal vein with 0.03 to 0.05% collagenase in calcium-free Hanks' balanced salt solution and dissociation of liver cells with subsequent collagenase

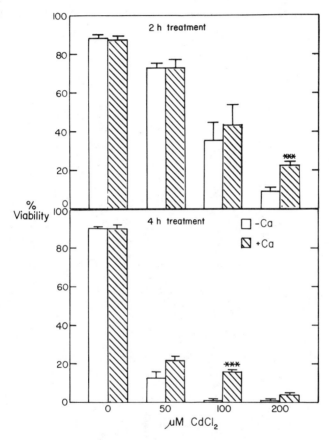

FIGURE 1. Percentage of viable hepatocytes (about 24 hours old) after a 2-hour or 4-hour exposure to 0–200 μM CdCl$_2$ · 2½H$_2$O, with or without 1.8 mM CaCl$_2$, as determined by *in situ* counts of about 300 to 500 cells per culture. Values are expressed as mean ± one standard error; two and five cultures (2 to 10 dishes/treatment) were assayed for the 2-hour and 4-hour exposures, respectively. Asterisks (***) indicate significant difference from the appropriate control group incubated in the absence of calcium: $p \leq 0.001$.

treatments. About 5×10^6 parenchymal hepatocytes were plated into plastic tissue-culture dishes and maintained approximately 24 hours in arginine-deficient, Dulbecco-Vogt's modification of Eagle's minimum essential medium at 37°C in a humidified environment of 5% carbon dioxide and 95% air. After about 24 hours, cultures were rinsed thoroughly and incubated for 2 or 4 hours in Earle's balanced salt solution containing from 0 to 200 μM cadmium chloride in the presence or absence of calcium chloride (1.8 or 3.6 mM).

Following incubation, the structural and functional status of the cells was determined using several procedures on approximately 30 cultures. For determination of the functional integrity of cultures, we assayed LDH leakage from cells into the medium,[36] total urea in media and cells,[37] and cellular viability using an *in situ* modification of the Tolnai trypan blue exclusion method.[38] Statistical differences in functional parameters were computed using Student's t-test. Structural features of

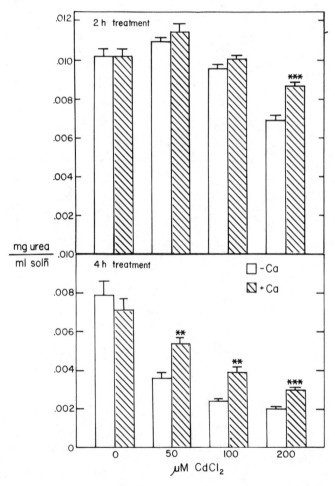

FIGURE 2. Total urea in cells and media following a 2-hour or 4-hour cadmium exposure (0–200 μM CdCl$_2$ · 2½H$_2$O) of approximately 24-hour-old hepatocytes in the presence or absence of 1.8 mM CaCl$_2$. Values are expressed as mean (± one standard error); three and two cultures (8 to 14 dishes/treatment) were assayed for the 2-hour and 4-hour exposures, respectively. Asterisks indicate significant difference from the appropriate control group incubated in the absence of calcium: p ≤ 0.01 (**) and p ≤ 0.001 (***).

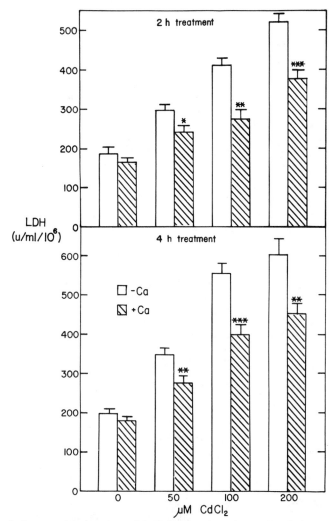

FIGURE 3. Lactate dehydrogenase (LDH) leakage into the media of about 24-hour-old hepatocyte cultures treated for 2 hours or 4 hours with $CdCl_2 \cdot 2\frac{1}{2}H_2O$ (0 to 200 μM) in the presence or absence of 1.8 mM $CaCl_2$. Enzyme activity is expressed as units/ml media per 10^6 cells. Mean values (± one standard error) are presented. Three and seven cultures (10 to 42 dishes/treatment) were assayed for the 2-hour and 4-hour exposures, respectively. Asterisks indicate significant difference from the appropriate control group incubated in the absence of calcium: $p \leq 0.05$ (*), $p \leq 0.01$ (**), and $p \leq 0.001$ (***).

parenchymal hepatocytes were assessed at the optical or electron microscopic level, using cells fixed in 3% glutaraldehyde in 0.1 M phosphate buffer followed by either Giemsa staining and examination using an inverted phase microscope or routine preservation procedures for transmission electron microscopy.[39] In either case, cells were processed entirely within the culture dishes in which they had been grown and treated; at no point were attached cells scraped from the culture dishes for morphological assessments.

RESULTS

Cadmium-exposed liver cells provided with 1.8 mM calcium chloride in the media survived longer than those cells incubated without calcium (FIGURE 1), although differences were not significant for every treatment group. The percentage of viable control cells did not vary with respect to the length of exposure or the presence of calcium chloride in the incubation medium. Similar trends were observed for total urea in cells and media; no statistical differences were observed in control cultures incubated with or without added extracellular calcium (FIGURE 2). After a two-hour incubation, however, 200 μM cadmium chloride–exposed cultures provided with 1.8 mM calcium chloride had higher total urea levels. These trends were repeated in the four-hour exposures; those cultures incubated with calcium showed higher total urea levels, although levels were considerably lower than those observed in the two-hour incubations.

As observed with cellular viability and total urea, the leakage of LDH from control cells into the media was similar regardless of exposure time or the presence of 1.8 mM calcium chloride in the incubation medium (FIGURE 3). In the cadmium-treatment groups, however, significantly less leakage of LDH from hepatocytes occurred when calcium was provided during incubation. Elevation of extracellular calcium levels from 1.8 to 3.6 mM resulted in significant reductions in the LDH leakage from parenchymal hepatocytes exposed to higher cadmium concentrations (FIGURE 4). No inactivation of

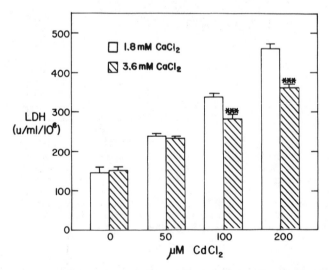

FIGURE 4. Lactate dehydrogenase (LDH) leakage into the media of approximately 24-hour-old hepatocyte cultures exposed for 4 hours to 0–200 μM CdCl$_2$·2$\frac{1}{2}$H$_2$O in the presence or absence of 1.8 mM CaCl$_2$. Enzyme activity is expressed as units/ml medium per 10^6 cells for the two cultures assayed for these data. Asterisks (***) indicate significant difference from the appropriate control group incubated in the presence of 1.8 mM CaCl$_2$: $p \le 0.001$.

+Ca   -Ca

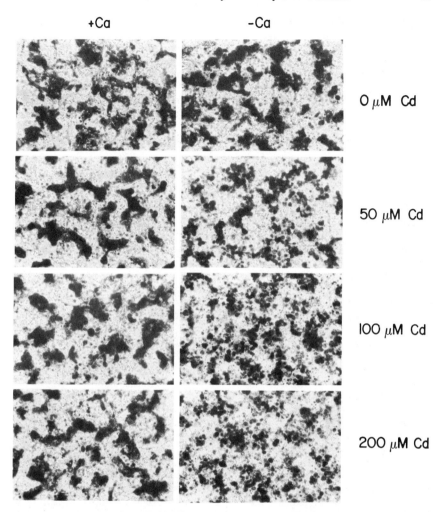

0 μM Cd

50 μM Cd

100 μM Cd

200 μM Cd

FIGURE 5. Representative fields from primary liver cell cultures showed that at higher cadmium levels, the absence of calcium in the incubation media resulted in greater dispersion of parenchymal hepatocytes. Giemsa stain, × 118.14.

LDH resulted from exposure of the enzyme to a maximum of 200 μM cadmium chloride.

Morphologically, the 24-hour-old control parenchymal hepatocyte cultures were composed of cords of about 5 to 40 cells which associated in bilaminar arrays akin to those observed in the intact rat liver (FIGURE 5). Generally, this condition was observed in all cultures provided with 1.8 mM extracellular calcium chloride, although more cellular debris was present in those cultures exposed to higher levels of cadmium. In cultures deprived of additional calcium in the incubation medium, especially those exposed to higher cadmium concentrations, greater dispersion of hepatocytes was evident with a considerable reduction in the number of hepatocyte cords remaining after treatment (FIGURE 5).

Ultrastructurally, control parenchymal hepatocytes resembled their *in vivo* counterparts in a number of respects. Numerous hepatocytes formed and/or maintained normal bile canaliculi, microvillar projections from areas of the plasma membrane, normal Golgi complexes, autophagic vacuoles, lysosomelike organelles, and rather typical mitochondria with dense matrices (FIGURE 6). Cadmium-exposed parenchymal hepatocytes, however, showed numerous abnormalities including necrosis, degenerative mitochondrial changes, deterioration of rough endoplasmic reticulum, proliferation of smooth endoplasmic reticulum, and loss of microvilli (FIGURES 6 and 7).

DISCUSSION

The utilization of primary cultures of rat hepatocytes for the evaluation of cytotoxicity or irreversible cell injury has been limited to a few laboratories around the world.[1-3,13-16,20-22,40] Primary cultures of liver cells have principally been exploited as activation or assay systems for the detection and study of carcinogens and mutagens, as highlighted by this conference and the previous conference.[41] Our laboratory has been particularly interested in developing primary cultured hepatocytes to investigate the mechanisms of cytotoxicity and cellular damage produced by toxic agents. In this study, we focused attention on the role of calcium in metal-induced toxicity.

The exact reason for the reported paradoxical effects of calcium on hepatotoxin-induced cell damage has not been elucidated at present; however, several possibilities exist in addition to the altered metabolic capacity of freshly isolated or younger (i.e., less than two-hour) cultures. Differences in the exposure media, for example, might explain such discrepancies. Complete tissue-culture media were used in several studies that showed that calcium augmented the detrimental effects of a number of hepatotoxins,[1,3] whereas several studies (including the present work) that showed a beneficial or nil effect utilized buffers or salt solutions as the exposure solution.[4,6] Other possibilities include differences in hepatocyte-isolation procedures and inherent mitochondrial calcium loads for various species of rats provided with water or nutrient supplies containing different levels of calcium.

Functionally, the addition of cadmium to the incubation media in our study caused time- and dose-dependent reductions in cellular viability and total cellular urea with increases in the leakage of lactate dehydrogenase from cultured hepatocytes. Reported cadmium-induced hepatotoxicity includes reduced intracellular potassium, increased AST leakage, and increased lipid peroxidation.[12,42] Cadmium also was reported to be a potent inhibitor of hepatic drug metabolism, reducing cytochrome P_{450} levels and inhibiting metabolism of hexobarbital, *p*-nitroanisole, ethylmorphine, aniline, and zoxazolamine *in vivo*,[43] although the exact mechanism (or mechanisms) of cadmium-induced cell death has not been established at present.

Isolated rat hepatocytes, maintained in medium 60 to 75 minutes prior to exposure to 3 μM cadmium chloride, accumulated twice as much cadmium (3.8 nmol/mg protein) as zinc (2 nmol/mg protein) during a 20-hour incubation period.[44] In the same study, it was observed that, once accumulated by parenchymal hepatocytes, cadmium was less mobile during exit-exchange processes than was its biologically related congener zinc; moreover, cadmium influx enhanced zinc efflux. Since zinc and cadmium share a common transport mechanism[45] and since zinc can effectively prevent a number of toxic manifestations of cadmium,[46] zinc and perhaps other divalent cations such as calcium could compete with intracellular enzymes or metallothionein-binding sites. Alternatively (and perhaps concurrently), divalent

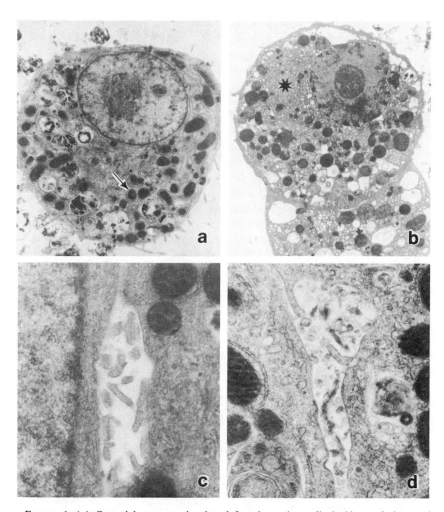

FIGURE 6. (a) Control hepatocyte incubated four hours in media lacking cadmium and calcium. Subcellular organelles were similar morphologically to *in vivo* counterparts; microvilli, autophagic vacuoles, mitochondria (arrow), rough endoplasmic reticulum, and the nucleus were normal, × 3792. (b) Cadmium-exposed parenchymal hepatocyte (four hours, 50 μM CdCl$_2$ · 2½H$_2$O, no calcium) showing peripheral vacuolation, rough endoplasmic reticulum reductions, smooth endoplasmic reticulum proliferation (asterisk), and a pleomorphic, pyknotic nucleus, × 3792. (c) Normal bile canaliculus from a culture exposed four hours to media lacking cadmium or calcium, × 8295. (d) Abnormal bile canaliculus from a culture exposed four hours to 50 μM CdCl$_2$ · 2½H$_2$O; irregular-shaped microvilli and precipitate are present, × 8295. Uranyl acetate and lead citrate.

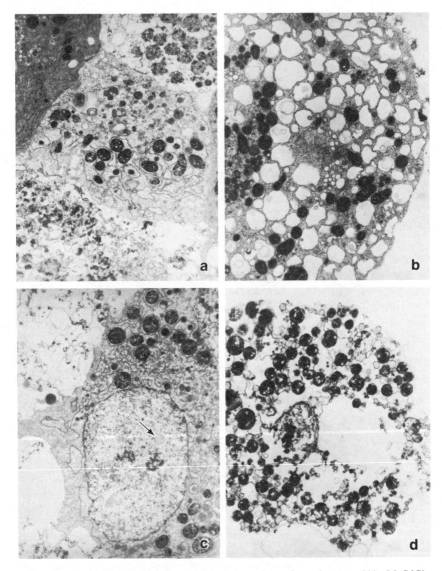

FIGURE 7. (a) Mitochondrial changes in a culture exposed two hours to 200 μM CdCl$_2$ ·
2½H$_2$O with calcium. Mitochondria are normal in the upper left hand cell, slightly swollen with
swollen cristae in the middle cell, more swollen in the upper right hand cell, and completely
disrupted in the lower left hand cell. (b) Cisternal swelling of smooth endoplasmic reticulum from
a cell exposed four hours to 50 μM CdCl$_2$ · 2½H$_2$O without calcium. (c) Nuclear inclusions
(arrow), mitochondrial cisternal swelling, peripheral blebbing, and reductions in microvilli of a
cell exposed two hours to 200 μM cadmium chloride in the presence of calcium. (d) Necrotic
parenchymal hepatocyte exposed four hours to 50 μM cadmium chloride without calcium.
Disruption of the nucleus (center), mitochondria, and plasma membrane are evident. Uranyl
acetate and lead citrate, × 4522.88.

cations such as calcium might decrease lipid fluidity of the plasma membrane[47] by binding directly to anionic sites of the lipid bilayer, probably to phospholipid polar head groups and to sialic acid residues,[48] and triggering phase separations[49–52] that circumscribe protein carrier molecules to effectively shut off carrier function.[50] Such a mechanism might explain why the presence of 1.8 or 3.6 mM calcium chloride in the incubation medium of primary cultures of young rat hepatocytes served a protective function in terms of reducing LDH leakage and cellular mortality as seen in our studies. Such calcium-induced functional changes could be expected to reduce the morphological damage observed in cultured hepatocytes challenged with cadmium.

The severity of cadmium-induced hepatocyte damage is well documented. Intravenous injection of 0.06 mg cadmium/kg in adult rats resulted in *in situ* congestion and occasional parenchymal hepatocyte necrosis at the optical microscopic level[53] and numerous profound changes at the ultrastructural level.[54] These included parenchymal cell necrosis, deterioration of rough endoplasmic reticulum, proliferation of smooth endoplasmic reticulum, autophagocytosis, and degeneration of mitochondria. These changes closely parallel those observed *in vitro* in our hepatocyte cultures (FIGURES 6 and 7), as previously reported.[55]

In the presence of calcium in the incubation medium, cadmium-induced morphological changes were less severe, paralleling the functional data accumulated in this study. This might be interpreted as a less severe manifestation of cadmium-induced parenchymal hepatocyte necrosis due to calcium-induced competition for carrier-binding sites and/or membrane phase separations that shut off carrier function of certain plasma membrane proteins, as previously mentioned.

In summary, our primary cultures of rat hepatocytes serve as useful experimental models to evaluate the cytotoxicity of xenobiotics. We have demonstrated that our system of cultured hepatocytes retains liver-specific functions and shows dose- and time-dependent responses to recognized hepatotoxins. As such, this *in vitro* system is credible and relevant to the *in vivo* condition. We have further utilized this culture system to investigate the role of calcium in cytotoxic injury of the liver by cadmium. Regardless of the length of cadmium exposure (two to four hours), hepatocytes incubated in media containing 1.8 mM calcium released less lactate dehydrogenase into the media, showed higher total urea levels, and had increased viability (as measured by trypan blue exclusion) when compared with cultures incubated in media to which calcium had not been added. These effects were both dose and time dependent. At higher cadmium concentrations, moreover, the elevation of the calcium levels from 1.8 mM to 3.6 mM resulted in significant reductions in the leakage of LDH into the media. Morphologically, those cells exposed to higher levels of cadmium in the absence of calcium did not adhere together to form cords of hepatocytes as readily as did those cells incubated in 1.8 mM extracellular calcium. At the ultrastructural level, those parenchymal hepatocytes exposed to higher levels of cadmium, especially in the absence of calcium in the medium, showed more severe necrosis with peripheral vacuolation, proliferation of smooth endoplasmic reticulum, reduction in rough endoplasmic reticulum, mitochondrial swelling, and formation of precipitate in bile canaliculi. The present study, therefore, showed that the presence of extracellular calcium lessened cadmium-induced cytotoxicity of cultured parenchymal hepatocytes in terms of less severe morphological abnormalities, reduced LDH leakage, increased viability, and increased total urea.

ACKNOWLEDGMENTS

Ultrastructural work was conducted in the Electron Microscopy Laboratory, which is under the direction of Dr. Dennis T. Brown. Ruben Ramirez-Mitchell provided valuable suggestions regarding the ultrastructural work.

REFERENCES

1. SCHANNE, F. A. X., A. B. KANE, E. E. YOUNG & J. L. FARBER. 1979. Calcium dependence of toxic cell death: a final common pathway. Science **206**: 700–702.
2. CASINI, A. F. & J. L. FARBER. 1981. Dependence of the carbon-tetrachloride-induced death of cultured hepatocytes on the extracellular calcium concentration. Am. J. Pathol. **105**: 138–148.
3. CHENERY, R., M. GEORGE & G. KRISHNA. 1981. The effect of ionophore A23187 and calcium on carbon tetrachloride–induced toxicity in cultured rat hepatocytes. Toxicol. Appl. Pharmacol. **60**: 241–252.
4. SMITH, M. T., H. THOR & S. ORRENIUS. 1981. Toxic injury to isolated hepatocytes is not dependent on extracellular calcium. Science **213**: 1257–1259.
5. EDMONDSON, J. W. & N. U. BANG. 1981. Deleterious effects of calcium deprivation on freshly isolated hepatocytes. Am. J. Physiol. **241**: C3–C8.
6. STACEY, N. H. & C. D. KLAASSEN. 1982. Lack of protection against chemically induced injury to isolated hepatocytes by omission of calcium from the incubation medium. J. Toxicol. Environ. Health **9**: 267–276.
7. THURMAN, R. G. & L. A. REINKE. 1979. The isolated perfused liver: a model to define biochemical mechanisms of chemical toxicity. *In* Reviews in Biochemical Toxicology. E. Hodgson, J. R. Bend & R. M. Philpot, Eds. **1**: 249–285. Elsevier. New York, N.Y.
8. DUJOVNE, C. A., D. SHOEMAN, J. BIANCHINE & L. LASAGNA. 1972. Experimental bases for the different hepatotoxicity of erythromycin preparations in man. J. Lab. Clin. Med. **79**: 832–844.
9. GOTO, Y., C. A. DUJOVNE, D. W. SHOEMAN & K. ARAKAWA. 1976. Hepatotoxicity of general anesthetics on rat hepatoma cells in culture. Proc. Soc. Exp. Biol. Med. **151**: 789–794.
10. ZIMMERMAN, H. J., J. KENDLER, S. LIBBER & L. LUKACS. 1974. Hepatocyte suspensions as a model for demonstration of drug hepatotoxicity. Biochem. Pharmacol. **23**: 2187–2189.
11. FRY, J. R. & J. W. BRIDGES. 1977. The metabolism of xenobiotics in cell suspensions and cell cultures. *In* Progress in Drug Metabolism. J. W. Bridges & L. F. Chasseaud, Eds. **2**: 71–118. John Wiley & Sons. London, England.
12. STACEY, N. H., L. R. CANTILENA, JR & C. D. KLAASSEN. 1980. Cadmium toxicity and lipid peroxidation in isolated rat hepatocytes. Toxicol. Appl. Pharmacol. **53**: 470–480.
13. ACOSTA, D., D. C. ANUFORO & R. V. SMITH. 1980. Cytotoxicity of acetaminophen and papaverine in primary cultures of rat hepatocytes. Toxicol. Appl. Pharmacol. **53**: 306–314.
14. ACOSTA, D. & D. B. MITCHELL. 1981. Metabolic activation and cytotoxicity of cyclophosphamide in primary cultures of postnatal rat hepatocytes. Biochem. Pharmacol. **30**: 3225–3230.
15. FRY, J. R. & J. W. BRIDGES. 1979. Use of primary hepatocyte cultures in biochemical toxicology. *In* Reviews in Biochemical Toxicology. E. Hodgson, J. R. Bend & R. M. Philpot, Eds. **1**: 201–247. Elsevier. New York, N.Y.
16. GRISHAM, J. W. 1979. Use of hepatic cell cultures to detect and evaluate the mechanisms of action of toxic chemicals. Int. Rev. Exp. Pathol. **20**: 123–210.
17. ACOSTA, D., D. ANUFORO & R. V. SMITH. 1978. Primary monolayer cultures of postnatal rat liver cells with extended differentiated functions. In Vitro **14**: 428–436.
18. NELSON, K. F., D. ACOSTA & J. V. BRUCKNER. 1982. Long-term maintenance and induction of cytochrome P-450 in primary cultures of rat hepatocytes. Biochem. Pharmacol. **31**: 2211–2214.
19. ACOSTA, D., D. ANUFORO, R. MCMILLIN, W. SOINE & R. V. SMITH. 1979. Cytochrome P-450 levels and O-demethylation activity in cultures of rat hepatocytes. Fed. Proc. **38**: 366.
20. SANTONE, K. S., D. ACOSTA & J. V. BRUCKNER. 1982. Cadmium toxicity in primary cultures of rat hepatocytes. J. Toxicol. Environ. Health **10**: 169–177.
21. ANUFORO, D. C., D. ACOSTA & R. V. SMITH. 1978. Hepatotoxicity studies with primary cultures of rat liver cells. In Vitro **14**: 981–988.
22. MITCHELL, D. B. & D. ACOSTA. 1981. Evaluation of the cytotoxicity of tricyclic antidepressants in primary cultures of rat hepatocytes. J. Toxicol. Environ. Health **7**: 83–92.

23. REYNOLDS, E. S. 1963. Liver parenchymal cell injury. I. Initial alterations of the cell following poisoning with carbon tetrachloride. J. Cell Biol. **19:** 139–157.
24. JUDAH, J. D., K. AHMED & A. E. M. McLEAN. 1964. Possible role of ion shifts in liver injury. In Ciba Foundation Symposium on Cellular Injury. A. V. S. de Reuck & J. Knight, Eds.: 187–205. Churchill. London, England.
25. FARBER, J. L. & S. K. EL-MOFTY. 1975. The biochemical pathology of liver cell necrosis. Am. J. Pathol. **81:** 237–250.
26. FARBER, J. L. 1981. The role of calcium in cell death. Life Sci. **29:** 1289–1295.
27. WONG, P. Y. K. & W. Y. CHEUNG. 1979. Calmodulin stimulates human platelet phospholipase A_2. Biochem. Biophys. Res. Commun. **90:** 473–480.
28. BYGRAVE, F. L. 1977. Mitochondrial calcium transport. Curr. Top. Bioenerg. **6:** 259–318.
29. MARCUM, J. M., J. R. DEDMA, B. R. BRINKLEY & A. R. MEAN. 1978. Control of microtubule assembly-disassembly by calcium-dependent regulator protein. Proc. Nat. Acad. Sci. USA **75:** 3771–3775.
30. JUST, W. W. & H. SCHIMASSEK. 1980. Lack of ornithine decarboxylase activity in isolated rat liver parenchymal cells. Eur. J. Cell. Biol. **22:** 649–653.
31. TANAKA, K., M. SATO, Y. TOMITA & A. ICHIHARA. 1978. Biochemical studies on liver functions in primary cultured hepatocytes of adult rats. I. Hormonal effects on cell viability and protein synthesis. J. Biochem. **84:** 937–946.
32. KATO, S., K. AOYAMA, T. NAKAMURA & A. ICHIHARA. 1979. Biochemical studies on liver functions in primary cultured hepatocytes of adult rats. III. Changes of enzyme activities on cell membranes during culture. J. Biochem. **86:** 1419–1425.
33. KREBS, H. A., P. LUND & M. EDWARDS. 1979. Features of isolated liver cells. In Cell Populations. E. Reid, Ed.: 1–6. Ellis Horwood Limited. Chichester, England.
34. BISSELL, D. M. & P. S. GUZELIAN. 1980. Phenotypic stability of adult rat hepatocytes in primary monolayer cultures. Ann. N.Y. Acad. Sci. **349:** 85–98.
35. ICHIHARA, A., T. NAKAMURA, K. TANAKA, Y. TORNITA, K. AOYAMA, S. KATO & H. SHINNO. 1980. Biochemical functions of adult rat hepatocytes in primary culture. Ann. N.Y. Acad. Sci. **349:** 77–84.
36. MITCHELL, D. B., K. S. SANTONE & D. ACOSTA. 1980. Evaluation of cytotoxicity in cultured cells by enzyme leakage. J. Tissue Culture Methods **6:** 113–116.
37. GUTMANN, L. & H. BERGMEYER. 1974. Determination of urea with glutamate dehydrogenase as indicator enzyme. In Methods of Enzymatic Analysis. H. Bergmeyer, Ed.: 1794–1798. John Wiley & Sons. London, England.
38. TOLNAI, S. 1975. A method for viable cell count. In Tissue Culture Association Manual. V. J. Evans & M. M. Vincent, Eds. **1:** 37–39. Tissue Culture Association. Rockville, Md.
39. SORENSEN, E. M. B. 1976. Ultrastructural changes in the hepatocytes of green sunfish, *Lepomis cyanellus* (R.), exposed to solutions of sodium arsenate. J. Fish Biol. **8:** 229–240.
40. McQUEEN, C. A. & G. M. WILLIAMS. 1982. Cytotoxicity of xenobiotics in adult rat hepatocytes in primary culture. Fundam. Appl. Toxicol. **2:** 139–144.
41. BOREK, C. & G. M. WILLIAMS, Eds. 1980. Differentiation and Carcinogenesis in Liver Cell Cultures. Ann. N.Y. Acad. Sci. **349**.
42. STACEY, N. H. & C. D. KLAASSEN. 1981. Interaction of metal ions with cadmium-induced cellular toxicity. J. Toxicol. Environ. Health **7:** 149–158.
43. SCHNELL, R. C. 1978. Cadmium-induced alteration of drug action. Fed. Proc. **37:** 28–34.
44. FAILLA, M. L., R. J. COUSINS & M. J. MASCENIK. 1979. Cadmium accumulation and metabolism by rat liver parenchymal cells in primary monolayer culture. Biochim. Biophys. Acta **583:** 63–72.
45. HUAN, P. C., B. SMITH, P. BOHDAN & A. CORRIGAN. 1980. Effect of zinc on cadmium influx and toxicity in cultured CHO cells. Biol. Trace Element Res. **2:** 211–220.
46. FLICK, D. F., H. F. KRAYBILL & J. M. DIMITROFF. 1971. Toxic effects of cadmium: a review. Environ. Res. **4:** 71–85.
47. GORDON, L. M., R. D. SAUERHEBER & J. A. ESGATE. 1978. Spin label studies on rat liver and heart plasma membranes: effects of temperature, calcium, and lanthanum on membrane fluidity. J. Supramol. Struct. **9:** 299–326.
48. SCHLATZ, L. & G. V. MARINETTI. 1972. Calcium binding to the rat liver plasma membrane. Biochim. Biophys. Acta **290:** 70–83.

49. Trauble, H. & H. Eibl. 1974. Electrostatic effects on lipid phase transitions: membrane structure and ionic environment. Proc. Nat. Acad. Sci. USA **71:** 214–219.

50. Ohnishi, S. & T. Ito. 1974. Calcium-induced phase separations in phosphatidylserine-phosphatidylcholine membranes. Biochemistry **13:** 881–887.

51. Jacobson, K. & D. Papahadjopoulos. 1975. Phase transitions and phase separations in phospholipid membranes induced by changes in temperature, pH, and concentration of bivalent cations. Biochemistry **14:** 152–161.

52. Hartmann, W., H. J. Galla & E. Sackman. 1977. Direct evidence of charge-induced lipid domain structure in model membranes. FEBS Lett. **78:** 169–172.

53. Cook, J. A., E. O. Hoffmann & N. R. Di Luzio. 1975. Influence of lead and cadmium on the susceptibility of rats to bacterial challenge. (39117). Proc. Soc. Exp. Biol. Med. **150:** 741–747.

54. Hoffmann, E. O., J. A. Cook, N. R. Di Luzio & J. A. Coover. 1975. The effects of acute cadmium administration in the liver and kidney of the rat: light and electron microscopic studies. Lab. Invest. **32:** 655–664.

55. Sorensen, E. M. B. & D. Acosta. 1982. Protective effect of calcium on cadium-induced cytotoxicity in cultured rat hepatocytes. In Vitro **18**(3): 288.

ADAPTATION OF THE DNA-REPAIR AND MICRONUCLEUS TESTS TO HUMAN CELL SUSPENSIONS AND EXFOLIATED CELLS*

Hans F. Stich, Richard H. C. San, and Miriam P. Rosin

Environmental Carcinogenesis Unit
British Columbia Cancer Research Center
Vancouver, British Columbia
Canada V5Z 1L3

INTRODUCTION

In vitro assays for genotoxicity have been applied to over 11,000 chemicals, many of which have been found to have the capacity to induce mutations, chromosome aberrations, sister chromatid exchanges, mitotic recombination, gene conversion, nondisjunction, DNA fragmentation, DNA repair, and a host of other nuclear anomalies. The results of these observations have profoundly changed our attitude toward environmental carcinogenesis. Only two decades ago, chemicals with carcinogenic and mutagenic properties were considered to be rare. The induction of a DNA alteration and mutation in somatic cells was thought to be an extremely rare event. Today, one has the impression that we are living in a brew of carcinogenic and genotoxic compounds and that hundreds, if not thousands, of mutagens and clastogens enter man daily through the regular diet. Previously, the development of cancer was thought to occur in individuals accidentally exposed to an excessive dose of one carcinogen. Now one is inclined to accept the idea that, since relatively large amounts of carcinogens enter man and carcinogen-DNA adducts are formed in many organs, cancer will develop primarily in individuals who ingest too small an amount of anticarcinogenic agents or who have a defective defense mechanism. If this assumption is correct, then *in vitro* tests, which lack the complete metabolic activation and inactivation systems of a whole organism, may predict a potential but not the actual carcinogenic hazard to man. The design, development, and validation of short-term *in vivo* tests that are applicable to man are needed to obtain a higher and more reliable predictive value.

VERSATILITY OF THE DNA-REPAIR ASSAY

A host of various techniques have been used to measure DNA repair in mammalian cells. They include unscheduled DNA synthesis (UDS), shifts in sucrose gradient sedimentation profiles, survival of DNA-bond carcinogens, and removal of DNA adducts. The choice among the available procedures will depend on one's objective. UDS on cultured mammalian cells appears to be best suited for screening a large series of compounds. On the other hand, the removal of DNA adducts would appear to be the method of choice for yielding information on organ-specific DNA-repair capacities

*These studies were supported by the Strategic Grants Program of the Natural Sciences and Engineering Research Council Canada and by the National Cancer Institute of Canada. Dr. H. F. Stich is a Research Associate of the National Cancer Institute of Canada. Dr. M. P. Rosin is a staff member of the British Columbia Cancer Foundation of Vancouver, B.C., Canada.

93

and the survival of particular alkylated bases after exposure to alkylating carcinogens. The great versatility of the DNA-repair tests makes them unique among the 50 or so genotoxic assays reported up to now. A few examples may suffice to show the multitude of issues that can be tackled by applying DNA-repair assays.

Detection of Hypersensitive Disorders

Based on studies on DNA-repair-deficient bacterial mutants, one would expect to see a correlation between increased sensitivity to the lethal, mutagenic, and transforming actions of carcinogens and defects in the DNA-repair systems. Actually, only patients afflicted with xeroderma pigmentosum (XP) and ataxia telangiectasia show definite abnormalities in excision repair. A host of other genetically based "hypersensitive disorders," including Fanconi's anemia, Cockayne's syndrome, Huntington's chorea, and bilateral retinoblastoma, exhibit abnormal replication of damaged DNA rather than a defect in the DNA-repair system.[1,2] A hypersensitivity of cells toward carcinogens, as is the case for tissue of familial polyposis patients to alkylating agents, may not implicate, per se, a reduced DNA-repair capacity.[3]

Modulators of DNA Repair

A reduced DNA repair can be caused by several factors which are not necessarily genetic in nature. For example, the postreplicative repair has been reduced in rat cells infected with Rauscher leukemia virus and exposed to ultraviolet (UV) or 4-nitroquinoline-1-oxide (4NQO).[4] However, no change in UDS levels was observed in Chinese hamster ovary (CHO) cells abortively infected with simian adenovirus type 7 (SA7) and treated with 4NQO (TABLE 1). A host of chemicals, including tumor promoters, reduce UDS initiated by UV irradiation.[5,6] The significance of the chemically induced DNA-repair inhibition is somewhat difficult to interpret, since all the examined repair inhibitors also suppress semiconservative replication. Thus the existence of specific DNA-repair inhibitors has been questioned.[7] The issue of specific or nonspecific repair inhibitors should not detract from the possible biological significance of a reduced DNA repair. For example, an elevation in temperature to 41.5–41.8°C for two hours temporarily suppresses DNA-repair synthesis.[8] Such a hyperthermia significantly delays the recovery of UV-irradiated adenovirus (AD12).[9] The response of a cell with a temporarily incapacitated repair appears to be comparable to that found in XP cells with a permanently reduced repair capacity (FIGURE 1).

Correlation between DNA Repair and Genotoxicity

Despite considerable efforts, no simple link has emerged as yet between DNA repair and induction of mutations, chromosome aberrations, or sister chromatid exchanges. The DNA-repair-deficient XP cells do not show an increased sensitivity to all clastogens, as can be seen from TABLE 2. Conversely, cells with an elevated response to clastogenic agents need not necessarily carry a defect in their repair capacity. On the other hand, the application of caffeine reduces postreplicative DNA repair, increases lethality, elevates mutations, and increases chromosome aberrations in carcinogen-treated cells, indicating that the same type of DNA damage is responsible for all the observed effects.[10]

TABLE 1

LACK OF CORRELATION BETWEEN DNA-REPAIR SYNTHESIS (UDS), CHROMOSOME ABERRATIONS, AND NEOPLASTIC TRANSFORMATION IN CULTURED EMBRYONAL SYRIAN HAMSTER CELLS

	UDS Grains/Nucleus	Metaphase Plates (%) with Chromosome Aberrations	Neoplastic Transformation Colonies/2×10^6 Survivors
Simian adenovirus 7 (SA7)	0	21.5	16
4-Nitroquinoline-1-oxide (4NQO)	68	27.8	0
SA7 + 4NQO	62	54.1	243

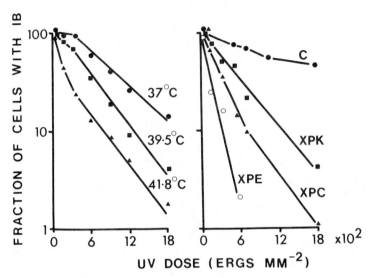

FIGURE 1. (Left) The capacity of UV-irradiated AD12 to form intranuclear inclusion bodies in cultured human fibroblasts following a 7-hour exposure to hyperthermia. Samples were taken 48 hours after infection. (Right) The capacity of UV-irradiated AD12 to form intranuclear inclusion bodies in cultured fibroblasts of three xeroderma pigmentosum patients (XPK, XPC, and XPE) and control. Samples were taken 48 hours after infection. C: control fibroblasts from normal person.

Correlation between DNA Repair and Neoplastic Transformation

The high frequency of skin tumors in sun-exposed areas of XP patients seems to indicate a link between DNA-repair deficiency and predisposition to cancer. Such an idea is supported by animal experiments using alkylating agents. A high frequency of cancers has been found in tissues having a slow removal rate of O^6-methylguanine adducts.[11] An involvement of DNA repair in transformation is also indicated by "confluent holding recovery" experiments.[12] If cell proliferation is blocked after carcinogen exposure, the transformation frequency is greatly reduced. However, not all observations support such a simple, direct relationship. Formation and removal of benzo[a]pyrene (BaP)–DNA adducts have been found to occur at the same rate in cultured mouse cells that are highly susceptible or resistant to transformation by BaP.[13] In addition, no differences in UDS levels were seen between 4NQO-treated, uninfected and SA7-infected hamster cells, although they differed greatly in the frequency of transformation (TABLE 1). The general validity of the link between slow removal of O^6-methylguanine and the development of cancers in such a tissue can be questioned, since various DNA adducts can survive for prolonged periods in organs that do not develop neoplasms.

Suitability of UDS Tests in Screening for Potential Carcinogens

We have repeatedly pointed out the need for using diverse screening assays that have different end points and that employ different organisms.[14,15] Exclusive reliance on the response of one gene in one particular cell type could distort our view of genotoxic hazards and could lead to an unrealistic priority list of chemicals that are of concern to regulatory agencies. The pros and cons of UDS have been repeatedly reviewed,[16,17] and the validity of the tests using autoradiographic detection or liquid scintillation counting was assessed on 136 and 147 chemicals, respectively.[18] If one compares chemicals within a molecular group, then a good quantitative relationship can be observed between the level of UDS and the extent of carcinogenicity[19] or clastogenicity.[20] The great advantage of UDS assays is their applicability to all cells of all organisms, including human ones.

IN SEARCH OF RELEVANT TEST SYSTEMS

The 50 or so proposed genotoxicity assays seem to belong to two main categories: tests using specific end points (e.g., his$^+$ revertants in *Salmonella typhimurium*) and tests that respond to a wide spectrum of mutagen- or carcinogen-induced lesions. The pros and cons of these two approaches have been argued for years by supporters and detractors. This controversy can be boiled down to a simple issue. By using, for example, a well-defined mutation to quantitate the genotoxic effect of a chemical, one may learn about its action on specific nucleotide sequences. However, one could grossly under- or overestimate mutagenicity, since different loci may respond differently.[21,22] On the other hand, a UDS assay may respond to a host of DNA alterations that are induced by a great variety of different chemical and physical carcinogens. The UDS results, however, reveal little, if anything, about the mechanism of action or the biological consequences of a DNA-carcinogen interaction. When discussing the usefulness of a test system, one should, in addition, consider the issue of relevance. How good is an assay in revealing the mutagenic, clastogenic, recombinogenic, etc., properties of a chemical, and how reliable is an assay in predicting that a compound will be genotoxic for man at doses to which human populations are exposed? It is becoming quite clear that the relevance of the test results is an unresolved question haunting regulatory agencies. To improve relevance, the UDS test has been modified

TABLE 2

DIFFERENT RESPONSES OF UDS AND CHROMOSOME ABERRATIONS TOWARD CARCINOGENS

Groups	Carcinogens/Mutagens*	UDS Control	UDS XP	Chromosome Aberrations Control	Chromosome Aberrations XP
Reduced UDS	4NQO, N-acetoxy-2-AAF, aflatoxin B₁ + S-9, DMN + S-9, K-region epoxides of BA, MCA	+ +	+	+	+ +
	ultraviolet light	+ +	+	+	+ +
Nonreduced UDS	MNNG, MMS, MNU	+ +	+ +	+ +	+ +
No UDS	daunomycin, human and simian adenoviruses, herpes simplex, and cytomegaly	−	−	+ +	+ +

*AAF is acetylaminofluorene; BA is benz[a]anthracene; MCA is 3-methylcholanthrene; MMS is methylmethanesulfonate; MNU is methylnitrosourea.

and adapted to different cells and organ cultures. A few cases exemplify the new developments.

The use of S-9 mixtures for activating precarcinogens and the interpretation of the results have always been a matter of concern.[23] The metabolites of various precarcinogens may differ depending on the animal species used for preparing the microsomal fraction or the compounds used to induce liver enzymes. The *in vitro* activation and deactivation by an S-9 mixture may not reflect the metabolism in intact liver cells of a whole organism. This issue was resolved by using short-term cultures of hepatocytes or suspensions of freshly isolated liver cells.[24–27] Recently, the hepatocyte primary culture/DNA-repair system (HPC/DNA-repair test) was validated on a series of coded carcinogens and noncarcinogens.[28] A wide spectrum of precarcinogens and complex environmental mixtures, including diesel, coke oven, and tar pot emissions,[29] gave unequivocal positive results, indicating the usefulness of this test in screening compounds requiring metabolic activation.

DNA-repair tests can also be used for detecting the organotropic effect of carcinogens and precarcinogens that may be activated only in particular tissues. The basic technique consists of exposing small tissue pieces or biopsies to carcinogens and tritiated thymidine (^{3}H-TdR), fixing, sectioning, and performing an autoradiographic analysis.[30] This procedure has been successfully applied for estimating the repair capacity of esophageal[30] and tracheal[31] epithelium to various carcinogens. The advantages of measuring UDS in individual cells of a tissue should not be underestimated. The level and length of UDS can be estimated in the different cell types of an organ, in areas of inflammation, and in regions of dysplasia. This method may bridge the gap between cytopathological analysis and strictly biochemical studies, including survival of DNA adducts or shifts in sucrose gradient sedimentation profiles, which can only provide average values for an organ.

The measurement of UDS in organ pieces can be improved by administering the carcinogen *in vivo* and estimating repair synthesis *in vitro*. In this way, an organ-specific distribution of carcinogens or tissue-specific activation of precarcinogens can be readily detected. The method includes administering a genotoxic compound to an animal, sacrificing the animal at certain time intervals, removing organ pieces, incubating these pieces with ^{3}H-TdR, and processing the specimens for autoradiographic analysis. The organ-specific UDS, following injection of 4NQO or dimethylnitrosamine (DMN) into mice, corresponds to the targets of these carcinogens.[32–34]

The use of primary rat hepatocytes for activating precarcinogens and detecting UDS induction should be generally applicable. Such studies should yield information on the activation as well as repair capacities of various human tissues that are obtainable through biopsies. The successful application of UDS to freshly isolated human as well as mouse intestinal mucosal cells, which have been exposed *in vitro* to the direct-acting carcinogen N-methyl-N'-nitro-N-nitrosoguanidine (MNNG) and the precarcinogen aflatoxin B_1, demonstrates the feasibility of this approach (TABLE 3).[35,36]

TOWARD GENOTOXICITY TESTS IN MAN

The main objective in screening thousands of compounds and mixtures for their genotoxicity has been to recognize a possible hazard to human populations. This laudable approach resulted, however, in an unforeseen dilemma. By improving the sensitivity of the bioassays, introducing a wide array of test organisms, and selecting various DNA-repair-deficient indicator cells, one is finding a rapidly increasing percentage of chemicals to be positive in at least a few of the bioassays that are currently in vogue. Today one can predict that all foods and beverages containing browning reaction or pyrolysis products will give a positive genotoxic result.[37,38] Similarly, vegetables, fruits, and fruit juices containing phenolics will induce a genotoxic effect.[39,40] This means that the overwhelming majority of food items must elicit a positive response in an *in vitro* test due to the presence in them of phenolic acids, flavonoids, browning reaction, and pyrolysis products.

Considering such observations, the question must be asked as to what to do with all the *in vitro* genotoxicity data, despite the fact that they are quantitative, reliable, and reproducible. It will be a herculean task to identify, among the vast number of genotoxic agents, those mutagens, clastogens, and recombinogens that actually pose a significant hazard to man. Should one focus on new man-made chemicals to which man and mammals have never been previously exposed and against which they may not have evolved the appropriate defense mechanisms? Or should we be more concerned with food chemicals that are ingested in milligram or even gram quantities?

TABLE 3

CARCINOGEN-INDUCED UNSCHEDULED DNA SYNTHESIS (UDS) IN FRESHLY ISOLATED MUCOSAL CELLS FROM HUMAN SMALL INTESTINE

Carcinogen	UDS (Grains/ Nucleus)*	Percent Cells with Lightly Labeled Nuclei	Percent Cells with Heavily Labeled Nuclei
N-Methyl-N'-nitro-N- nitrosoguanidine			
3×10^{-4} M	16.2	88	0
3×10^{-5} M	8.5	84	2
Aflatoxin B_1			
3×10^{-4} M	19.2	88	2
2×10^{-4} M	10.8	84	0
1×10^{-4} M	5.8	86	2
3×10^{-5} M	4.0	90	2
Solvent control (1% dimethylsulfoxide)	0.6	0	0

*Suspensions of mucosal cells isolated by pronase digestion were exposed concomitantly to carcinogen and ^{3}H-TdR (10 μCi/ml) for two hours and processed for autoradiography.

One could speculate that, over the millions of years of evolution, our ancestors will have developed the necessary detoxification systems for this class of substances, thus making us resistant to our unavoidable daily dose of ingested food mutagens, clastogens, and carcinogens.

The dilemma just posed could be resolved by measuring the genotoxic damage directly in the tissue that is the target for the carcinogens and from which tumors will later arise. However, only a few of the hitherto developed short-term assays can be applied to man. The more promising ones are reviewed in TABLE 4. Many of these techniques are restricted to particular cell types because of the difficulties in obtaining human tissue specimens. This limitation may again somewhat reduce their predictive value. Red or white blood cells and spermatocytes or sperms, which are utilized in some of the bioassays, do not represent tissues from which the majority of human cancers arise. In the search for an easily obtainable tissue, we have explored the feasibility of

TABLE 4

METHODS SUITABLE FOR DIRECT DETECTION OF GENOTOXICITY IN MAN

Cell Type	End Point	Reference
Peripheral lymphocytes	chromosome aberrations	44
	sister chromatid exchanges	45
	6-thioguanine resistance	46
	benzo[a]pyrene-DNA adducts	47
Erythrocytes	variant hemoglobins	48
Sperm	abnormal morphology	49
	number of fluorescent bodies	50
Exfoliated cells:		
Buccal mucosa	micronuclei	42,43
Urinary bladder	micronuclei	51
Sputum	micronuclei	unpublished data

TABLE 5

FREQUENCIES OF MICRONUCLEATED CELLS IN THE BUCCAL MUCOSA OF POPULATION GROUPS AT AN ELEVATED RISK FOR ORAL CANCER

	Micronucleated Mucosa Cells (%)	
Chewing Habit	Individual Cases	Average
Betel quid*		
<4 per day	2.6, 2.4, 2.1, 1.4, 1.4, 1.3, 1.0, 0.7, 0.6	1.50
>15 per day	11.7, 9.8, 9.0, 8.2, 6.7, 6.3, 5.4, 3.4, 3.0	7.05
Raw betel nut†	7.5, 7.4, 6.0, 5.5, 5.4, 5.3, 5.3, 4.8, 4.7, 4.6, 4.1, 4.1, 3.5, 3.2, 3.0, 2.7, 2.5	4.68
Khaini tobacco‡	4.9, 4.8, 4.1, 3.9, 3.9, 3.2, 2.9, 2.8, 2.6, 2.5, 2.3, 2.1, 1.8, 1.8, 1.5, 1.5, 1.3, 1.3, 1.3, 1.3, 1.3, 1.2, 1.1, 1.1, 0.9, 0.8, 0.8	2.18
Smokers and drinkers§	2.2, 2.1, 1.2, 1.1, 1.0, 0.6	1.36

*The betel quid used by the population group examined in Orissa (India) consisted of betel nut (*Areca catechu*), betel leaf (*Piper betle*), tobacco, lime, catechu, and spices.

†The "kwa" used by the examined Khasis, a hill tribe in Meghalaya (India), consist of one-quarter of a raw betel nut, betel leaf, and lime.

‡Powdered Khaini tobacco is mixed by many inhabitants of Bihar (India) with lime and placed into the gingivolabial groove.

§The examined individuals were Caucasian residents of British Columbia, Canada. They smoked one-and-a-half packs or more of cigarettes per day and consumed alcoholic beverages daily (wine and/or heavy liquor).

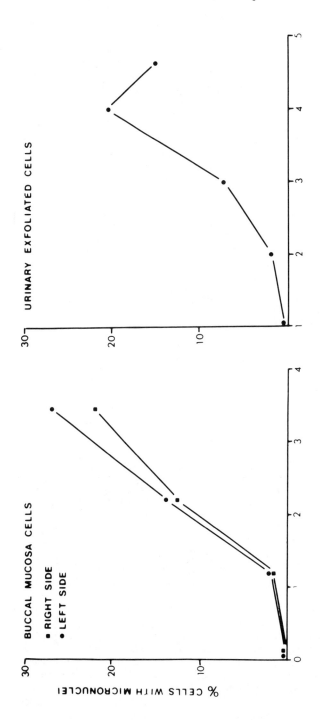

FIGURE 2. (Left) The frequency of micronucleated buccal mucosa cells of a patient receiving radiotherapy to the head region (total of 3,400 rads over a period of 3½ weeks). (Right) The frequency of micronucleated cells recovered from the urine of a patient who received radiotherapy to the pelvic region (20 doses over about 4½ weeks; total of 5,000 rads).

using exfoliated human cells. These can be readily collected by noninvasive procedures from buccal mucosa, urinary bladder (urine), bronchi (sputum), colon and cervix. With proper instrumentation, they can furthermore be obtained from the nasopharynx, esophagus, and stomach. However, they can by no means be considered as ideal cells for genetic or cytogenetic studies. They do not divide and are moribund. Nevertheless, they reflect cytogenetic anomalies that have occurred in dividing cell populations of the basal epithelial layers.[41] Micronuclei, which result from acentric chromatid or chromosome fragments or aberrant chromosomes, survive during the passage of cells from the basal layer to the surface of the mucosa of the oral cavity or urinary bladder. The frequency of micronucleated exfoliated cells provides a measure of the genotoxic damage in human mucosa of various sites. We recently applied this micronucleus test to exfoliated cells of the buccal mucosa of population groups at high risk for oral

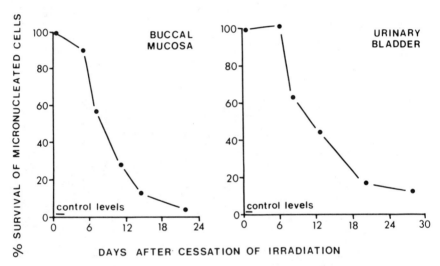

FIGURE 3. (Left) The disappearance of micronucleated buccal mucosa cells from seven patients after termination of radiotherapy (total estimated dose to the mucosa ranged from 5,000–6,000 rads). (Right) The disappearance of micronucleated exfoliated cells from the urine of a patient after termination of radiotherapy (total estimated dose to the urinary bladder was 4,000 rads).

cancer, including chewers of betel quid (pan), mixtures of raw betel nut, betel leaf, and lime, and a tobacco/lime mixture.[42,43] The results show an elevated frequency of micronucleated cells in all examined quid or tobacco chewers (TABLE 5). The frequency of micronucleated buccal mucosa cells differed in groups with different chewing habits. Chewers of only a few quids per day had a significantly lower frequency of micronucleated cells than did those using higher numbers.

The micronucleus test on exfoliated cells of the oral mucosa, bladder, and cervix can be validated on patients receiving radiotherapy to the head/neck or pelvic region, respectively. The increases in the frequency of micronucleated cells during the course of radiotherapy are shown in FIGURE 2. After cessation of radiation treatment, the micronuclei gradually disappear from the exfoliated cells (FIGURE 3). Their reduction is delayed by about five to seven days, which suggests that this is the time interval required for the nonaffected new cells to arrive at the tissue surface. The methodology and some of the possible pitfalls have already been reviewed.[41] However, it is possible to state that a careful application of the micronucleus test to exfoliated human cells can be used as an "internal dosimeter" to reveal the genotoxic effect of carcinogens in their target tissues. We are currently exploring the possibility of utilizing this procedure for estimating the success of various intervention programs. Whether exfoliated cells can also be used for other genotoxicity tests remains to be seen. The possibility of applying UDS to these cells as evidence of a carcinogenic action should be explored.

REFERENCES

1. CLEAVER, J. E., D. CHAR, W. C. CHARLES & N. RAND. 1982. Repair and replication of DNA in hereditary (bilateral) retinoblastoma cells after X-irradiation. Cancer Res. **42:** 1343–1347.
2. CLEAVER, J. E. 1980. DNA damage, repair systems and human hypersensitive diseases. J. Environ. Pathol. Toxicol. **3:** 53–68.
3. BARFKNECHT, T. & J. B. LITTLE. 1982. Abnormal sensitivity of skin fibroblasts from familial polyposis patients to DNA alkylating agents. Cancer Res. **42:** 1249–1254.
4. WATERS, R., N. MISHRA, N. BOUCK, G. DIMAYORCA & J. D. REGAN. 1977. Partial inhibition of postreplication repair and enhanced frequency of chemical transformation in rat cells infected with leukemia virus. Proc. Nat. Acad. Sci. USA **74:** 238–242.
5. GAUDIN, D., R. S. GREGG & K. L. YIELDING. 1971. DNA repair inhibition: a mechanism of action of cocarcinogens. Biochem. Biophys. Res. Commun. **45:** 630–636.
6. GAUDIN, D., R. S. GREGG & K. L. YIELDING. 1972. Inhibition of DNA repair by cocarcinogens. Biochem. Biophys. Res. Commun. **48:** 945–949.
7. CLEAVER, J. E. & R. B. PAINTER. 1975. Absence of specificity in inhibition of DNA repair replication by DNA-binding agents, cocarcinogens and steroids in human cells. Cancer Res. **35:** 1773–1778.
8. STICH, H. F., R. H. C. SAN, P. LAM, J. KOROPATNICK & L. LO. 1977. Unscheduled DNA synthesis of human cells as a short-term assay for chemical carcinogens. *In* Origins of Human Cancer. H. H. Hiatt, J. D. Watson & J. A. Winsten, Eds. (Book C): 1499–1512. Cold Spring Harbor Laboratory. Cold Spring Harbor, N.Y.
9. LAM, P. & H. F. STICH. 1978. Hyperthermia and host-cell reactivation of adenovirus 12. Can. J. Genet. Cytol. **20:** 35–40.
10. ROBERTS, J. J. 1977. DNA repair, mutations and carcinogenesis. Colloq. Int. CNRS **256:** 237–253.
11. RAJEWSKY, M. F., L. H. AUGENLICHT, H. BIESSMANN, R. GOTH, D. F. HULSER, O. D. LAERUM & L. Y. LOMAKINA. 1977. Nervous-system-specific carcinogenesis by ethylnitrosourea in the rat: molecular and cellular aspects. *In* Origins of Human Cancer. H. H. Hiatt, J. D. Watson & J. A. Winsten, Eds. (Book B): 709–726. Cold Spring Harbor Laboratory. Cold Spring Harbor, N.Y.
12. KAKUNAGA, T., K. LO, J. LEAVITT & M. IKENAGA. 1980. Relationship between transformation and mutation in mammalian cells. *In* Carcinogenesis: Fundamental Mechanisms and Environmental Factors. B. Pullman, P. O. P. Ts'o & H. V. Gelboin, Eds.: 527–541. D. Reidel Publishing Co. Dordrecht, the Netherlands.
13. LO, K-Y. & T. KAKUNAGA. 1982. Similarities in the formation and removal of covalent DNA adducts in benzo[*a*]pyrene-treated BALB/3T3 variant cells with different induced transformation frequencies. Cancer Res. **42:** 2644–2650.
14. STICH, H. F., L. WEI & P. LAM. 1978. The need for a mammalian test system for mutagens: action of some reducing agents. Cancer Lett. **5:** 199–204.
15. STICH, H. F., P. LAM, L. W. LO, D. J. KOROPATNICK & R. H. C. SAN. 1975. The search for relevant short term bioassays for chemical carcinogens: the tribulation of a modern Sisyphus. Can. J. Genet. Cytol. **17:** 471–492.
16. SAN, R. H. C. & H. F. STICH. 1975. DNA repair synthesis of cultured human cells as a rapid bioassay for chemical carcinogens. Int. J. Cancer **16:** 284–291.

17. STICH, H. F., R. H. C. SAN, P. P. S. LAM, D. J. KOROPATNICK, L. W. LO & B. A. LAISHES. 1976. DNA fragmentation and DNA repair as an *in vitro* and *in vivo* assay for chemical procarcinogens, carcinogens and carcinogenic nitrosation products. IARC Sci. Publ. **12:** 617-636.
18. MITCHELL, A. D., D. A. CASCIANO, M. L. MELTZ, D. E. ROBINSON, R. H. C. SAN, G. M. WILLIAMS & E. S. VON HALLE. 1983. Unscheduled DNA synthesis tests: a report of the "Gene-Tox" program. Mutat. Res. (In press.)
19. STICH, H. F., R. H. C. SAN & Y. KAWAZOE. 1971. DNA repair synthesis in mammalian cells exposed to a series of oncogenic and non-oncogenic derivatives of 4-nitroquinoline-1-oxide. Nature London **229:** 416–419.
20. SAN, R. H. C., W. STICH & H. F. STICH. 1977. Differential sensitivity of xeroderma pigmentosum cells of different repair capacities towards the chromosome breaking action of carcinogens and mutagens. Int. J. Cancer **20:** 181–187.
21. HANNAN, M. A. & A. NASIM. 1978. Genetic activity of bleomycin: differential effects on mitotic recombination and mutations in yeast. Mutat. Res. **53:** 309–316.
22. HANNAN, M. A., A. NASIM & T. BRYCHCY. 1978. Mutagenic and antimutagenic effects of bleomycin in *Saccharomyces cerevisiae*. Mutat. Res. **58:** 107–110.
23. BIGGER, C. A. H., J. E. TOMASZEWSKI, A. DIPPLE & R. S. LAKE. 1981. Limitations of metabolic activation systems used with *in vitro* tests for carcinogens. Science **209:** 503–505.
24. WILLIAMS, G. M. 1977. Detection of chemical carcinogens by unscheduled DNA synthesis in rat liver primary cultures. Cancer Res. **37:** 1845–1851.
25. WILLIAMS, G. M., E. BERMUDEZ & D. SCARAMUZZINO. 1977. Rat hepatocyte primary cell cultures. III. Improved dissociation and attachment techniques and the enhancement of survival by culture medium. In Vitro **13:** 809–817.
26. WILLIAMS, G. M. 1978. Further improvements in the hepatocyte primary culture DNA repair test for carcinogens: detection of carcinogenic biphenyl derivatives. Cancer Lett. **4:** 69–75.
27. BROUNS, R. E., M. POOT, R. DE VRIND, T. VAN HOEK-KON & P. T. HENDERSON. 1979. Measurement of DNA-excision repair in suspensions of freshly isolated rat hepatocytes after exposure to some carcinogenic compounds. Its possible use in carcinogenicity screening. Mutat. Res. **64:** 425–432.
28. WILLIAMS, G. M., M. F. LASPIA & V. C. DUNKEL. 1982. Reliability of the hepatocyte primary culture/DNA repair test in testing of coded carcinogens and noncarcinogens. Mutat. Res. **97:** 359–370.
29. VED BRAT, S., C. TONG & G. M. WILLIAMS. 1982. Detection of genotoxic airborne chemicals in rat liver culture systems. *In* Genotoxic Effects of Airborne Agents. R. R. Tice, D. L. Costa & K. M. Schaich, Eds.: 619–631. Plenum Publishing Corp. New York, N.Y.
30. STICH, H. F. & D. J. KOROPATNICK. 1977. The adaptation of short-term assays for carcinogens to the gastrointestinal system. *In* Pathophysiology of Carcinogenesis in Digestive Organs. E. Farber *et al.*, Eds.: 121–134. University of Tokyo Press. Tokyo, Japan. University Park Press. Baltimore, Md.
31. ISHIKAWA, T., S. TAKAYAMA & F. IDE. 1980. Autoradiographic demonstration of DNA repair synthesis in rat tracheal epithelium treated with chemical carcinogens *in vitro*. Cancer Res. **40:** 2898–2903.
32. STICH, H. F. & D. KIESER. 1974. Use of DNA repair synthesis in detecting organotropic actions of chemical carcinogens. Proc. Soc. Exp. Biol. Med. **145:** 1339–1342.
33. LAISHES, B. A., D. J. KOROPATNICK & H. F. STICH. 1975. Organ-specific DNA damage induced in mice by the organotropic carcinogens 4-nitroquinoline-1-oxide and dimethylnitrosamine. Proc. Soc. Exp. Biol. Med. **149:** 978–982.
34. STICH, H. F., D. KIESER, B. A. LAISHES, R. H. C. SAN & P. WARREN. 1975. DNA repair of human cells as a relevant, rapid, and economic assay for environmental carcinogens. Gann Monogr. Cancer Res. **17:** 3–15.
35. FREEMAN, H. J. & R. H. C. SAN. 1980. Use of unscheduled DNA synthesis in freshly isolated human intestinal mucosal cells for carcinogen detection. Cancer Res. **40:** 3155–3157.
36. STICH, H. F., R. H. C. SAN & H. J. FREEMAN. 1981. DNA repair synthesis (UDS) as an *in*

vitro and *in vivo* bioassay to detect precarcinogens, ultimate carcinogens and organotropic carcinogens. *In* Short-Term Tests for Chemical Carcinogens. H. F. Stich & R. H. C. San, Eds.: 65–82. Springer-Verlag. Berlin & New York.

37. STICH, H. F., M. P. ROSIN, C. H. WU & W. D. POWRIE. 1982. The use of mutagenicity testing to evaluate food products. *In* Mutagenicity: New Horizons in Genetic Toxicology. J. A. Heddle, Ed.: 117–142. Academic Press, Inc. New York, N.Y.

38. STICH, H. F., W. STICH, M. P. ROSIN & W. D. POWRIE. 1980. Mutagenic activity of pyrazine derivatives: a comparative study with *Salmonella typhimurium, Saccharomyces cerevisiae* and Chinese hamster ovary cells. Food Cosmet. Toxicol. **18**: 581–584.

39. STICH, H. F. & W. D. POWRIE. 1982. Plant phenolics as genotoxic agents and as modulators for the mutagenicity of other food components. *In* Carcinogens and Mutagens in the Environment. Food Products. H. F. Stich, Ed. **1**: 135–145. CRC Press. Boca Raton, Fla.

40. STICH, H. F., M. P. ROSIN, C. H. WU & W. D. POWRIE. 1981. The action of transition metals on the genotoxicity of simple phenols, phenolic acids and cinnamic acids. Cancer Lett. **14**: 251–260.

41. STICH, H. F. & M. P. ROSIN. 1983. Micronuclei in exfoliated human cells as an internal dosimeter for exposures to carcinogens. *In* Carcinogens and Mutagens in the Environment. Naturally Occurring Compounds: Endogenous Formation and Modulation. H. F. Stich, Ed. **2**. CRC Press. Boca Raton, Fla. (In press.)

42. STICH, H. F., W. STICH & B. B. PARIDA. 1982. Elevated frequency of micronucleated cells in the buccal mucosa of individuals at high risk for oral cancer: betel quid chewers. Cancer Lett. **17**: 125–134.

43. STICH, H. F., J. R. CURTIS & B. B. PARIDA. 1982. Application of the micronucleus test to exfoliated cells of high cancer risk groups: tobacco chewers. Int. J. Cancer **30**: 553–559.

44. FORNI, A., A. CAPELLINI, E. PACIFICO & E. C. VIGILIANI. 1971. Chromosome changes and their evolution in subjects with past exposure to benzene. Arch. Environ. Health **22**: 373–378.

45. MUSILOVA, J., K. MICHALOVA & J. URBAN. 1979. Sister-chromatid exchanges and chromosomal breakage in patients treated with cytostatics. Mutat. Res. **67**: 289–294.

46. ALBERTINI, R. J. & E. F. ALLEN. 1981. Direct mutagenicity testing in man. *In* Human Risk Analysis. P. J. Walsh, C. R. Richmond & E. D. Copenhaver, Eds. Franklin Institute Press. Philadelphia, Pa.

47. HARRIS, C., F. PEREIRA & M. C. POIRIER. 1982. As cited in Development and Possible Use of Immunological Techniques to Detect Individual Exposure to Carcinogens. IARC Internal Technical Report No. 82/001. International Agency for Research on Cancer. Lyon, France.

48. STAMATOYANNOPOULOS, G. 1979. Possibilities for demonstrating point mutations in somatic cells, as illustrated by studies of mutant hemoglobins. *In* Genetic Damage in Man Caused by Environmental Agents. K. Berg, Ed.: 49. Academic Press, Inc. New York, N.Y.

49. WYROBEK, A. J. 1981. Methods for human and murine sperm assays. *In* Short-Term Tests for Chemical Carcinogens. H. F. Stich & R. H. C. San, Eds.: 408–419. Springer-Verlag. Berlin & New York.

50. KAPP, R. W., D. PICCIANO & C. B. JACOBSON. 1979. Y-Chromosomal nondisjunction in dibromochloropropane-exposed workmen. Mutat. Res. **64**: 47–51.

51. ROSIN, M. P. & H. F. STICH. 1983. The identification of antigenotoxic/anticarcinogenic agents in food. *In* Diet and Cancer: From Basic Research to Policy Implications. Alan R. Liss. New York, N.Y. (In press.)

THE SIGNIFICANCE OF DNA-DAMAGE ASSAYS IN TOXICITY AND CARCINOGENICITY ASSESSMENT

Kurt W. Kohn

Laboratory of Molecular Pharmacology
National Cancer Institute
Bethesda, Maryland 20205

INTRODUCTION

Integrity of DNA, the genetic material, is obviously essential for the genetically intact survival of cells and organisms. Organisms are continuously subjected to a variety of physical and chemical hazards that may damage their DNA, and have evolved a complex set of biochemical repair mechanisms that can to various degrees cope with different types of damage. When assessing the significance of various DNA-damage assays, one must consider the specific DNA effects that are measured and their biological significance. For example, DNA strand breaks may be due to direct DNA damage by an exogenous chemical agent, or may be produced indirectly during DNA-repair processes or other physiologic responses in the cell. If a DNA-repair process is involved, then the question arises as to what its capacity is to deal with the type and magnitude of damage to which individuals may be exposed, and to what extent the repair is prone to error that could lead to genetic damage. Some individuals may be especially susceptible to certain types of DNA damage because of an inherited deficiency in a DNA-repair mechanism. Although a substantial amount of information has been amassed relating to questions such as these, it still only scratches the surface of this complex problem. This paper will survey the types of DNA damage that can now be measured by methods suitable for large-scale testing and their biological significance. The focus will be on filter methods for measuring DNA damage.

TYPES OF DNA DAMAGE THAT CAN BE MEASURED BY FILTER METHODS

Filter methods have been developed that can distinguish several types of DNA damage.[6,34,35,39] The essential methodology is as follows.

Mammalian cells are deposited on a membrane filter in an apparatus that permits a solution to be pumped at a controlled rate through the filter. The cells are lysed on the filter by means of a detergent, such as 2% sodium dodecyl sulfate in the presence of 0.02 M ethylenediaminetetraacetic acid (EDTA) at pH 10. Under these conditions, most of the cell substance is solubilized and flows out through the filter. The nuclear DNA however remains quantitatively on the filter. At this point, an optional enzymatic digestion step can be included in order to digest proteins that may be linked to the DNA, or to introduce strand breaks at the sites of specific lesions recognized by a specific endonuclease that acts in excision repair.[20] DNA strands are then dissociated by means of an alkaline solution of tetrapropyl- or tetraethyl-ammonium hydroxide plus EDTA at pH >12. This "eluting solution" is pumped through the filter at a steady rate, and eluted fractions are collected. Eluted DNA is measured either by scintillation counting, the cells having been prelabeled for one to two cell-cycle times with ^{3}H- or ^{14}C-thymidine, or by a fluorometric method when radioactive labeling of the cells is impractical.[19]

106

The rate or extent of elution of DNA from the filters is governed by two main factors. First, the average time required for a DNA single strand to pass through the filter is proportional to the length of the strand. Secondly, the elution of DNA is in some cases limited by DNA-protein cross-links, because of the adsorption of DNA-bound proteins to the filter material. This adsorption can be made very efficient or can be nearly eliminated by the choice of filter type and detergent. In addition, DNA-protein cross-links can be eliminated by the use of a suitable proteinase enzyme. In general, adsorptive conditions are used for the assay of DNA-protein cross-links and nonadsorptive conditions are used for the assay of all other types of DNA lesions, including strand breaks and interstrand cross-links.

BIOLOGICAL EFFECTS OF DNA LESIONS

Single-Strand Breaks

Single-strand breaks (SSBs) can be produced in cells by direct chemical damage of DNA or by enzymatic action (reviewed by Roberts).[48] Enzymatic action may be due to repair endonucleases, topoisomerases, or autolytic nucleases.

Chemically Induced SSBs

The chemical reactions that are known to generate SSBs proceed by free radical mechanisms; free radicals may be generated by ionizing radiation, oxidation-reduction reactions, or photochemical reactions. Cells have efficient mechanisms for the repair of at least some types of SSBs, so that these lesions may not always be deleterious. SSBs may differ in regard to the chemical nature of the strand termini, and this may affect the efficiency of enzymatic strand rejoining.

SSBs are the most prominent macromolecular DNA lesions produced by ionizing radiation. Most or all of these SSBs are repaired within a few hours in all mammalian cells that have been examined. In human embryo (WI38) cells, whether in early passage or senescent, 99% of the SSBs generated by 5,000 rad of x-ray were resealed by six hours after radiation.[3] In addition, x-ray produces DNA-protein cross-links (DPCs) at a frequency of approximately 6% of that of the SSBs.[38] The repair of x-ray-induced DPCs may be slower than that of the SSBs or may be incomplete.[24] In addition, x-ray produces double-strand breaks (DSBs) at a frequency of 2–10% of that of the SSBs, and these DSBs are repaired at a rate similar to the rate of SSB repair.[6] (X-ray-induced DSBs may actually be closely spaced SSBs on opposite strands and might be repaired by essentially the same mechanism that repairs SSBs.) Aside from the macromolecular DNA lesions that have been considered, ionizing radiation is known to produce a variety of chemical base damages, and these could be major causes of genetic and toxic effects.

SSBs are prominent macromolecular lesions in the action of the antitumor antibiotic bleomycin in mammalian cells.[30,54] This drug is cytotoxic and produces chromosome aberrations. The chemical mechanism of action involves weak binding of the drug to DNA and generation of free radicals by an interaction between bound ferrous ion and molecular oxygen.[10,28,29,53] The chemical effect on DNA involves release of free bases, especially thymine, and destruction of deoxyribose residues. Perhaps because of the presence of a fragment of deoxyribose residues at the strand termini, many of the SSBs remain unrepaired and may be responsible for the genetic and toxic damage.

Exposure of mammalian cells to light from fluorescent lamps generates large numbers of SSBs and produces mutations in V79 cells.[4,7,17] More than one type of DNA damage is involved in producing these effects, however. Essentially all of the SSBs are rejoined within two hours. Most of the SSBs are generated even by wavelengths greater than 345 nm and are associated neither with mutagenesis nor with cytotoxicity. Mutagenesis appears to be confined to wavelengths below 345 nm, and may or may not be related to a subclass of the observed SSBs. The free radical scavengers dimethylsulfoxide, glycerol, and KI reduced SSB production but did not reduce mutagenesis. SSBs were observed in cells incubated with photoproducts formed by illuminating solutions of tryptophan and riboflavin, but these photoproducts did not produce mutations. Hydrogen peroxide also was observed to produce large numbers of reparable SSBs, which were not associated with mutagenesis and produced very little cytotoxicity.[2]

Thus there are divers types of chemically induced SSBs, which differ greatly in biological significance. The extent or rate of repair may be an indicator of the possible genetic or toxic significance of SSBs produced by particular agents. Photochemical reactions, hydrogen peroxide, and x-ray produce much more genetic damage. The most damaging free radical species produced both by x-ray and by hydrogen peroxide are hydroxyl radicals. Why then is x-ray-induced damage so much more significant than hydrogen peroxide–induced damage? A possible reason may be that whereas hydrogen peroxide–induced radicals would be expected to be produced randomly so that the consequent DNA lesions are randomly distributed, x-ray-induced radicals are formed in clusters so that there is a much higher probability of production of multiple lesions close to each other. There is therefore a much higher probability of generation of DSBs. Furthermore, neighboring lesions may pose a more severe problem for repair than do isolated lesions.

Some other factors should be mentioned relating to the significance of SSBs arising by different mechanisms. (1) It is important to distinguish SSB repair from loss of degraded DNA from the system, which may occur in dying or disrupted cells. (2) The deleterious effects of otherwise relatively innocuous SSBs may be enhanced by agents that inhibit or retard SSB repair. Several such agents are known, including carbamo-ylating agents,[23,31–33] inhibitors of poly-(adenosine diphosphoribose) synthesis,[64] and inhibitors of DNA-repair synthesis.

SSBs Induced during Excision Repair

DNA strand breaks occurring as an intermediate in excision repair can be distinguished by the use of inhibitors of DNA-repair synthesis, such as arabinosylcyto-sine or hydroxyurea.[16] The number of strand breaks seen during excision repair is a steady-state level determined by the balance between the rate of SSB production by a repair endonuclease and the rate of SSB resealing by polynucleotide ligase. Between these two steps, DNA-repair synthesis is required in order to replace the excised strand segments. Inhibition of repair synthesis therefore prevents resealing and causes the endonuclease-induced SSBs to accumulate. In this way, for example, SSBs have been accumulated in the course of the repair of DPCs, thus providing evidence that DPCs are repaired by a nucleotide excision mechanism.[20]

The action of certain excision repair mechanisms can also be inferred by compari-son of normal with appropriate repair-deficient cells, such as particular types of xeroderma pigmentosum cells.[20,22]

SSBs Induced by DNA-Intercalating Agents

A variety of DNA-intercalating agents have been noted to produce SSBs that appear to be associated with covalently bound protein.[47,50,51,64,66,67] The frequency of the SSBs (measured using proteinase and nonadsorptive conditions) is similar to the frequency of DPCs, and the DPCs appear to be localized at or near the sites of the SSBs. The proteins may in fact be linked to one terminus of the strand breaks, as is known to occur in the action of topoisomerases.[60,61] It has been proposed that these intercalator-induced DNA "lesions" are due to the action of a topoisomeraselike enzyme in response to DNA intercalation.[51] In the presence of intercalator, the topoisomeraselike enzyme might be trapped in its DNA-bound configuration. Intercalators have been found to produce both single- and double-strand breaks, both associated with DNA-bound protein; and the ratio of the two types of breaks has been found to vary greatly for different intercalators.

The extent of production of either type of protein-associated strand break is not correlated with the cytotoxicity of various intercalating compounds.[67] The detection of protein-associated DNA breaks would nevertheless be indicative of DNA intercalation, which could produce mutagenic or carcinogenic effects by this or by another mechanism.

DNA Degradation Due to Cell Autolysis or Disruption

Dosages of cytotoxic compounds often cause autolysis of DNA in some proportion of cells that are in the process of dying.[15] Disruption of a proportion of the cell population can also be produced artifactually during sample preparation.[58] DNA degradation in dying or disrupted cells can be distinguished on the basis of the small size of such DNA, which causes it to elute as a fast component. When an appropriate series of treatment dosages is used, such fast-eluting components are easily distinguishable from the more gradual elution expected for genetically significant DNA damage.

Alkali-Labile Sites

The formation of alkali-labile sites (ALSs) may be an indicator of genetically significant DNA damage, especially in the case of alkylating agents. Alkylating agents react with DNA at multiple sites, which differ in nucleophilic reactivity and in sensitivity in regard to the production of genetic damage. The guanine-O6 position, which is probably a major site in the production of genetic damage, has a relatively weak nucleophilic reactivity and hence reacts extensively only with the more potent alkylating agents. O^6-Alkyl-guanine residues do not constitute macromolecular DNA damage such as would be detectable by the techniques under consideration. These lesions are repaired mainly by an alkyl transferase mechanism that does not involve strand breaks as an intermediate.[40,41] Alkylation of weak nucleophilic sites however would be indicated by the presence of ALSs. The degree of reactivity of simple alkylating agents at guanine-O6 correlates with their degree of reactivity at DNA phosphates, which also are weakly nucleophilic sites. Alkylated DNA phosphate groups tend to form strand breaks in alkali. Therefore, the formation of these ALSs would correlate with alkylation of weak nucleophilic sites, including some that may be important sites in the production of genetic damage.

Alkylations at guanine-N7, the major strong nucleophilic site of DNA, are mainly repaired by glycosylase and alkaline phosphate (AP)–endonuclease reactions, which involve the transient formation of strand breaks. These alkylation sites however may be relatively innocuous.

Diverse alkylation phenomena, involving different types of repair processes, could lead to discrepancies between DNA strand break assays and genetic damage. Thus Brambilla *et al.* find that the potency of dimethylnitrosamine for the generation of SSBs in rat liver is 10 times that of methylnitrosourea,[9] yet the two compounds have comparable carcinogenic potencies, and methylnitrosourea is two orders of magnitude more potent than dimethylnitrosamine as a mutagen in the *Salmonella*/microsome test. This type of discrepancy might be resolved by taking into consideration the production of ALSs or unrepaired SSBs.

Interstrand Cross-Links

It is well known that bifunctional compounds that can bind to DNA at two sites so as to form cross-links are much more cytotoxic than monofunctional compounds, and in some cases are useful as antitumor drugs. The biological effects of DNA-cross-linking agents have recently been reviewed.[36,37,48] Most classes of bifunctional agents can produce both inter- and intrastrand cross-links, and both may be cytotoxic by blocking DNA replication or RNA transcription. Interstrand cross-linking can be sensitively measured in mammalian cells by alkaline elution, as well as by appropriate adaptations of the DNA denaturation/reassociation method.[49] There is no general technique however for measuring intrastrand cross-links in cells.

The major classes of compounds that produce interstrand cross-links in mammalian cells are (1) bifunctional alkylating agents, including nitrogen mustards and mitomycins; (2) *cis*-platinum(II) complexes; (3) chloroethylating agents, such as chloroethylnitrosoureas; and (4) photoreactive furocoumarins (psoralens) (see Kohn for review).[37] Current experience indicates that, in general, compounds that form interstrand cross-links would be expected to have predominantly cytotoxic effects. Most compounds that form interstrand cross-links however also generate monofunctional DNA adducts in large excess, and these may have strong mutagenic and carcinogenic effects.

DNA-Protein Cross-Links

A wide variety of agents, most of them known carcinogens, mutagens, or cytotoxins, have been found to produce DPCs. Many bifunctional agents, including alkylating agents, chloroethylnitrosoureas, and *cis*-Pt(II) complexes, produce a variety of cross-links in mammalian cells, including interstrand and intrastrand DNA cross-links, as well as DPCs (see Kohn for review).[37] The high cytotoxicity of the interstrand or intrastrand cross-links produced by these compounds in most, but perhaps not all, cases overshadows the effects of the DPCs. Ionizing and ultraviolet radiations also produce DPCs as minor lesions, the predominant lesions being base damage and, in the case of ionizing radiation, strand breaks; hence, the biological significance of the DPCs again is difficult to ascertain. In the case of the polycyclic carcinogens *N*-AcO-acetylaminofluorene and dimethylbenz[*a*]anthracene, Fornace and Little detected low levels of DPCs along with SSBs in human and mouse fibroblasts.[25] There are however compounds that produce DPCs as the most prominent macromolecular DNA lesions, and these will be considered next. DPCs produced by intercalating agents have already been considered.

DPCs were found to be the major DNA lesions, other than simple DNA adducts or base modifications, in mammalian cells treated with *trans*-Pt(II), formaldehyde, and Cr and Ni salts.

trans-*Platinum(II)*

trans-Pt(II) produces large numbers of DPCs, and, unlike *cis*-Pt(II), produces little or no interstrand cross-linking.[38] The DPCs are observed promptly after a 1-hour treatment of cells, and the great majority of these are usually removed over a 24-hour period, presumably by a repair process.[49,56,62,63,65] In mouse leukemia L1210 cells exposed to *trans*-Pt(II) for 1 hour at 37°C, the frequency of DPCs was measured to be 1.4×10^8 nucleotides/μM *trans*-Pt(II) and was proportional to the concentration of *trans*-Pt(II).[38] The cytotoxicity of *trans*-Pt(II) is much lower than that of *cis*-Pt(II); 100 μM *trans*-Pt(II), which generates 1.4 DPCs per 10^6 nucleotides (about 25,000 DPCs per cell), still allowed 50% survival of colony-forming ability.[65] *trans*-Pt(II) produced no detectable mutation frequency in V79 cells, even after treatment with 600 μM for 2 hours.[63] *trans*-Pt(II) does however produce sister chromatid exchanges[5] and malignant transformation of rodent cells.[26] These biological effects might be due to the DPCs or to Pt-DNA monoadducts. It would be of interest to study a monofunctional Pt(II) complex that could form monoadducts but not DPCs or other cross-links.

Although *trans*-Pt(II) does not ordinarily produce measurable levels of DNA strand breaks, easily measurable SSB frequencies do accumulate when cells are incubated in the presence of arabinosylcytosine plus hydroxyurea during the repair period.[21] These drugs inhibit DNA-repair synthesis, so that SSBs that are transiently formed during excision repair accumulate. However, it is not known whether this reflects excision repair of DPCs or of Pt-DNA monoadducts. Xeroderma pigmentosum cells, which differ from normal human cells in being deficient in some excision repair processes, did not accumulate SSBs even in the presence of the DNA synthesis inhibitors. DPCs were found to be more persistent in xeroderma cells than in normal human cells.[22] Hence these DPCs probably are removed by an excision repair process. Xeroderma cells have an increased sensitivity to *trans*-Pt(II), which could be due to deficient excision repair of DPCs and/or Pt-DNA monoadducts.

Formaldehyde

The findings reported with formaldehyde are in many respects similar to those found with *trans*-Pt(II). Formaldehyde produces large numbers of DPCs with little or no interstrand cross-linking, and ordinarily no detectable levels of SSBs. From the data of Ross and Shipley,[52] it can be estimated that L1210 cells treated with 100 μM formaldehyde for 2.5 hours acquired 0.5–1.0 DPCs per 10^6 nucleotides, a frequency quite similar to the case of *trans*-Pt(II). This treatment with formaldehyde caused little or no loss of colony-forming ability. The cells removed most of the formaldehyde-induced DPCs within 6 hours and essentially all of them within 24 hours. As in the case of *trans*-Pt(II), SSBs, presumably formed in the course of excision repair, accumulate when DNA-repair synthesis is inhibited by arabinosylcytosine plus hydroxyurea.[21] Thus the formation, repair, and perhaps biological significance of DNA damage produced by formaldehyde and by *trans*-Pt(II) appear to be remarkably similar.

Chromium Salts

Chromate salts are recognized carcinogens; they are strong mutagens and transform cells in culture (see references given by Fornace *et al.*).[27] Fornace *et al.* found that potassium chromate produced large numbers of DPCs in human and mouse cells.[27] There was little or no interstrand cross-linking. The frequency of DPCs produced by chromate appeared to be about one-half that produced by an equal concentration of *trans*-Pt(II). The active chemical species is probably Cr(III), since these ions produced DPCs both in cells and in isolated nuclei, whereas chromate generated these lesions only in intact cells. Cells are presumed capable of reducing chromate to Cr(III). In possible contradistinction to *trans*-Pt(II), the levels of chromate-induced DPCs were found to be unchanged for 12 hours of incubation of the cells in chromate-free medium. Some SSBs were observed, but these were completely removed within 4 hours. Large numbers of SSBs accumulated when arabinosylcytosine plus hydroxyurea was present during the repair period. These SSBs however accumulated in xeroderma (complementation group A) cells as well as in normal cells.[21]

Tsapakos *et al.* found DPCs as the predominant DNA lesions detected by alkaline elution in liver and kidney of chromate-treated rats.[59] The DPC production was dose dependent and was accompanied by very low levels of SSBs. In kidney, there may in addition have been a low frequency of interstrand cross-links (ISCs).

Nickel Compounds

Nickel(II) compounds are recognized carcinogens which produce chromosome aberrations and stimulate cell transformation (see references given by Ciccarelli *et al.*).[13,14] Ciccarelli and colleagues injected nickel carbonate intraperitoneally into rats and measured DNA damage in nuclei isolated from various tissues.[13,14] A dose-dependent production of DPCs and SSBs was noted in kidney, which was the major site of localization of nickel. SSBs were detected, although to a much smaller extent, in lung but not in liver or thymus. DPCs were detected only in kidney. Following injection of nickel carbonate (10 mg/kg), DPCs rose to a peak at 2–3 hours and then fell at 8 hours to about 40% of the peak; the DPCs then remaining persisted for 48 hours. SSBs appeared after a lag of 3–4 hours and then were completely removed over the next 15 hours. The timing of the SSB appearance and disappearance is appropriate for a manifestation of excision repair, since the peak occurred during the time when DPC levels were falling. However, the frequency of SSBs was higher than can be accounted for on the basis of repair of the DPCs alone. Some (or even all) of the SSBs might arise from repair of Ni-DNA adducts not involving protein. The class of DPCs that persist in kidney for over 48 hours may be particularly damaging. Some ISCs were also observed, especially at the higher doses; these completely disappeared within 20 hours.

In summary, it should be noted that DPCs are likely to be of widely different types, depending on the identity of the proteins linked, the sites of linkage to the DNA, and the chemical nature of the linkage. The different types of DPCs may have widely different biological significance. Some types of DPCs, such as those generated by *trans*-Pt(II) and formaldehyde, are efficiently repaired and are associated with only low potencies for genetic or cytotoxic damage. Other types of DPCs, such as those generated by Cr and Ni salts, are not repaired completely, or are repaired only after a long delay, and thus are more potent producers of significant biological damage.

EVALUATION OF DNA DAMAGE

Tests to Distinguish among Types of DNA Lesions

Both SSBs and ALSs can be detected by alkaline elution at relatively high pH, e.g., pH 12.8 (although the sensitivity for the detection of SSBs is greater at a lower pH, such as pH 12.2). Because of the probable importance of ALSs, a high pH assay would be desirable in order to optimize the likelihood of detecting these lesions. In order to detect SSBs arising due to excision repair of possibly hazardous DNA lesions, inhibitors of DNA-repair synthesis could be used to allow such SSBs to accumulate; this would greatly increase the sensitivity for the detection of excision repair. The assays for SSBs and ALSs should be conducted using proteinase K to eliminate the obscuring effect of any DPCs that may be present. If SSBs are detected, other assays should be performed to evaluate the significance of the lesions. Some recent reports pertinent to these considerations will be noted.

Cavanna *et al.* found that *N*-diacetylglycine produces SSBs in mouse tissues; the SSBs tended to persist for 24 hours and showed an alkali-labile component.[11] One may hypothesize that such findings would be indicative of serious potential for genetic damage.

Cavanna *et al.* found that cycasin, a carcinogen that was ranked as a false negative in the Ames mutagenesis test, produced dose-dependent SSBs in rat and mouse liver, and these SSBs were to a large extent unrepaired over an 18-hour period.[12]

Parodi *et al.* found that aniline, which is a weak carcinogen that has been found to be nonmutagenic in bacteria and yeast, produced SSBs in rat liver and kidney.[43] The SSBs in liver persisted for 24 hours, after which they declined.

Bolognesi *et al.* found that benzidine and 2-naphthylamine produced SSBs in mouse kidney and liver, and many of these lesions persisted for at least 12 hours.[1]

Experience such as this suggests that the ability of SSBs to be repaired in perhaps a 24-hour period is an important consideration. More data are needed however to test the validity of this hypothesis.

Another follow-up assay that should be considered is one omitting proteinase K, in order to determine whether the SSBs may be of the protein-associated type that are produced by intercalating agents.

The quantitative assessment of potential for hazard thus requires that the evaluation be made according to the type of DNA damage detected. The diversity of assays that would be needed for an adequate evaluation of a DNA-damaging compound probably will only be feasible in cell-culture systems.

Intact Animals versus Cell-Culture Systems

Attempts to correlate SSB assays in intact animals with the carcinogenic potencies of various compounds have had mixed success.[8,9,42,44–46] Some studies have found poor or nonexistent correlations. For example, Parodi *et al.* failed to find any correlation between SSBs in rat liver and carcinogenicity in a series of 16 aromatic amines and azo-derivatives.[45] Aside from the problems that could arise from diverse mechanisms of strand-break formation, the use of intact animals raises additional pharmacologic problems. In this study, most of the compounds were insoluble and were administered intraperitoneally as a suspension. The absorption of the compounds into the circulation therefore could have been quite variable. The rates of metabolism and excretion of the compounds also could have been variable. Hence the time course over which drug action was extended could have varied greatly. Assays of DNA lesions at a given time thus may not adequately reflect the total lesions conferred by a single injection. Furthermore, the doses used were chosen as one to two times the five-day LD_{50}, which may not be commensurate with doses used in chronic tumor-production assays. The authors are undoubtedly well aware of these problems, but the limitation is in what is feasible in a screening program.

Another difficulty is suggested in the finding that dimethylbenzanthracene and benzpyrene, which had given positive results for SSBs in V79 cells (after activation with a microsomal fraction),[57] showed only weak or nonexistent responses in rat or mouse liver.[44] The SSBs produced by these compounds presumably arise in the course of excision repair, which may be slow and prolonged, especially in rodent tissues.

Problems of this type may be avoidable by the use of suitable cell-culture systems. Dosage and timing could be more precisely controlled, and a greater number and variety of assays could be conducted. In particular, inhibitors of repair synthesis could be used to accumulate excision-induced SSBs and would greatly increase the sensitivity for the detection of carcinogens such as polycyclic hydrocarbons.

In order to take into account possible metabolic activation of test compounds, it is possible to use a mixture of liver cells and suitable target cells *in vitro*.[55] Rat hepatocytes can, for example, be used to activate potential DNA-damaging compounds, and the activity of the metabolites could be assayed in selected human cell types that have been prelabeled with radioactive thymidine. The use of human target cells has the advantage that the DNA-repair mechanisms, which can vary greatly from species to species, would be of the human type. Furthermore, repair-deficient human cell types might be used to enhance the sensitivity of particular DNA-damage assays and to distinguish between types of DNA lesions. The assay system can consist of a large excess of unlabeled hepatocytes mixed with a relatively small number of labeled human test cells. Measurement of radioactivity then would specifically determine DNA damage in the target cells, while fluorometric measurement of DNA mass[17,19] would reveal DNA damage in the hepatocytes themselves. DNA damage could occur in the hepatocytes without occurring in the target cells if a DNA-damaging metabolite is formed within the hepatocytes and is not released into the medium, either because of an inability to pass through the cell membrane or because of very rapid decomposition within the cell.

FINAL COMMENTS ON SCREENING FOR DNA DAMAGE

The first stage in screening should aim to detect any possible type of DNA damage; false positives can to some degree be tolerated, but the assays should be designed to minimize false negatives. False negatives could be minimized by the use of a battery of highly sensitive tests covering the widest possible range of DNA-lesion types. Two filter assays could be included in such a battery of tests: (1) alkaline elution assay for single-strand breaks plus alkali-labile sites, and (2) alkaline elution assay for DNA-protein cross-links plus interstrand cross-links.

A positive outcome in one or more of these initial tests however does not prove that there is significant DNA-damaging hazard. The degree of hazard could be further evaluated by a second stage of testing to determine the type of DNA damage produced and its capacity to be repaired by various human cell strains. Although much work must still be done to validate guidelines on the basis of which degree of hazard could be judged, there is evidence to support some useful hypotheses: (1) SSBs and DPCs that are promptly repaired do not necessarily indicate a high degree of genetic hazard, but delayed or incomplete repair of such lesions may indicate substantial hazard; (2) the presence of alkali-labile sites may indicate mutagenic hazard due to alkylation of DNA bases; (3) the detection of nucleotide excision repair would indicate that possibly carcinogenic DNA adducts have been formed; and (4) the formation of protein-associated strand breaks would suggest that DNA intercalation has taken place in the cell and that frameshift mutation or other genetic effects may occur.

REFERENCES

1. BOLOGNESI, C. F., C. F. CESARONE & L. SANTI. 1981. Evaluation of DNA damage by alkaline elution technique after *in vivo* treatment with aromatic amines. Carcinogenesis **4**: 265–268.
2. BRADLEY, M. O. & L. C. ERICKSON. 1981. Comparison of the effects of hydrogen peroxide and X-ray irradiation on toxicity, mutation and DNA damage/repair in mammalian cells (V-79). Biochim. Biophys. Acta **654**: 135–141.
3. BRADLEY, M. O., L. C. ERICKSON & K. W. KOHN. 1976. Normal and DNA strand rejoining and absence of DNA crosslinking in progeroid and aging human cells. Mutat. Res. **37**: 279–292.
4. BRADLEY, M. O., L. C. ERICKSON & K. W. KOHN. 1978. Non-enzymatic DNA strand breaks induced in mammalian cells by fluorescent light. Biochim. Biophys. Acta **520**: 11–20.
5. BRADLEY, M. O., I. C. HSU & C. C. HARRIS. 1979. Relationship between sister chromatid exchange and mutagenicity, toxicity and DNA damage. Nature **282**: 318–320.
6. BRADLEY, M. O. & K. W. KOHN. 1979. X-ray-induced DNA double strand break production and repair in mammalian cells as measured by neutral filter elution. Nucleic Acids Res. **7**: 793–804.
7. BRADLEY, M. O. & N. SHARKEY. 1977. Mutagenicity and toxicity of visible fluorescent light to cultured mammalian cells. Nature **266**: 724–725.
8. BRAMBILLA, G., M. CAVANNA, P. CARLO, R. FINOLLO, L. SCIABA, S. PARODI & C. BOLOGNESI. 1979. DNA damage and repair induced by diazoacetyl derivatives of amino acids with different mechanism of cytotoxicity. Correlations with mutagenicity and carcinogenicity. J. Cancer Res. Clin. Oncol. **94**: 7–20.
9. BRAMBILLA, G., M. CAVANNA, A. PINO & L. ROBBIANO. 1981. Quantitative correlation among DNA damaging potency of six N-nitroso compounds and their potency in inducing tumor growth and bacterial mutations. Carcinogenesis **2**: 425–429.
10. BURGER, R. M., J. PEISACH & S. B. HORWITZ. 1981. Mechanism of bleomycin action: in vitro studies. Life Sci. **28**: 715–727.
11. CAVANNA, M., S. PARODI, L. ROBBIANO, A. PINO, L. SCIABA & G. BRAMBILLA. 1980. Alkaline elution assay as a potentially useful method for assessing DNA damage induced *in vivo* by diazoalkanes. Gann **71**: 251–259.
12. CAVANNA, M., S. PARODI, M. TANINGHER, C. BOLOGNESI, L. SCIABA & G. BRAMBILLA. 1979. DNA fragmentation in some organs of rats and mice treated with cycasin. Br. J. Cancer **39**: 383–390.
13. CICCARELLI, R. B., T. H. HAMPTON & K. W. JENNETTE. 1981. Nickel carbonate induces DNA-protein crosslinks and DNA strand breaks in rat kidney. Cancer Lett. **12**: 349–354.
14. CICCARELLI, R. B. & K. E. WETTERHAHN. 1982. Nickel distribution and DNA lesions induced in rat tissues by the carcinogen nickel carbonate. Cancer Res. **42**: 3544–3549.
15. DUCORE, J. M., L. C. ERICKSON, L. A. ZWELLING, G. LAURENT & K. W. KOHN. 1982. Comparative studies of DNA cross-linking and cytotoxicity in Burkitt's lymphoma cell lines treated with cis-diaminedichloroplatinum(II) and L-phenylalanine mustard. Cancer Res. **42**: 897–902.
16. DUNN, W. C. & J. D. REGAN. 1979. Inhibition of DNA excision repair in human cells by arabinofuranosyl cytosine: effect on normal and xeroderma pigmentosum cells. Mol. Pharmacol. **15**: 367–374.
17. ERICKSON, L. C., M. O. BRADLEY & K. W. KOHN. 1980. Mechanisms for the production of

DNA damage in cultured human and hamster cells irradiated with light from fluorescent lamps. Biochim. Biophys. Acta **610:** 105–115.

18. ERICKSON, L. C., R. OSIEKA & K. W. KOHN. 1978. Differential repair of 1-(2-chloroethyl)-3-(4-methylcyclohexyl)-1-nitrosourea DNA damage in two human colon tumor cell lines. Cancer Res. **38:** 802–808.

19. ERICKSON, L. C., R. OSIEKA, N. A. SHARKEY & K. W. KOHN. 1980. Measurement of DNA damage in unlabeled mammalian cells analyzed by alkaline elution and a fluorometric DNA assay. Anal. Biochem. **106:** 169–174.

20. FORNACE, A. J., JR. 1982. Measurement of M. luteus endonuclease-sensitive lesions by alkaline elution. Mutat. Res. **94:** 263–276.

21. FORNACE, A. J., JR. 1982. Detection of DNA single-strand breaks produced during the repair of damage by DNA-protein cross-linking agents. Cancer Res. **42:** 145–149.

22. FORNACE, A. J., K. W. KOHN & H. E. KANN. 1976. DNA single-strand breaks during repair of UV damage in human fibroblasts and abnormalities of repair in xeroderma pigmentosum. Proc. Nat. Acad. Sci. USA **76:** 39–43.

23. FORNACE, A. J., K. W. KOHN & H. E. KANN. 1978. Inhibition of the ligase step of excision repair by 2-chloroethyl isocyanate, a decomposition product of 1,3-bis(2-chloroethyl)-nitrosourea. Cancer Res. **38:** 1064–1069.

24. FORNACE, A. J. & J. B. LITTLE. 1977. DNA crosslinking induced by X-rays and chemical agents. Biochim. Biophys. Acta **447:** 343–355.

25. FORNACE, A. J. & J. B. LITTLE. 1979. DNA-protein cross-linking by chemical carcinogens in mammalian cells. Cancer Res. **39:** 704–710.

26. FORNACE, A. J. & J. B. LITTLE. 1980. Malignant transformation by the DNA-protein crosslinking agent *trans*-Pt(II) diamminedichloride. Carcinogenesis **1:** 989–994.

27. FORNACE, A. J., JR., D. S. SERES, J. F. LECHNER & C. C. HARRIS. 1981. DNA-protein cross-linking by chromium salts. Chem. Biol. Interact. **36:** 345–354.

28. GILONI, I., M. TAKESHITA, F. JOHNSON, C. IDEN & A. P. GROLLMAN. 1981. Bleomycin-induced strand-scission of DNA. Mechanism of deoxyribose cleavage. J. Biol. Chem. **256:** 8608–8615.

29. GROLLMAN, A. P. & M. TAKESHITA. 1980. Interaction of bleomycin with DNA. Adv. Enzyme Regul. **18:** 67–83.

30. IQBAL, Z. M., K. W. KOHN, R. A. G. EWIG & A. J. FORNACE. 1976. Single-strand scission and repair of DNA in mammalian cells treated with bleomycin. Cancer Res. **36:** 3834–3838.

31. KANN, H. E., B. A. BLUMENSTEIN, A. PETKAS & M. A. SCHOTT. 1980. Radiation synergism by repair-inhibiting nitrosoureas in L1210 cells. Cancer Res. **40:** 771–775.

32. KANN, H. E., K. W. KOHN & J. M. LYLES. 1974. Inhibition of DNA repair by the 1,3-bis(2-chloroethyl)-1-nitrosourea breakdown product, 2-chloroethylisocyanate. Cancer Res **34:** 398–402.

33. KANN, H. E., M. A. SCHOTT & A. PETKAS. 1980. Effects of structure and chemical activity on the ability of nitrosoureas to inhibit DNA repair. Cancer Res **40:** 50–55.

34. KOHN, K. W. 1979. DNA as a target in cancer chemotherapy: measurements of macromolecular DNA damage produced in mammalian cells by anticancer agents and carcinogens. Methods Cancer Res. **16:** 291–345.

35. KOHN, K. W. 1981. DNA damage in mammalian cells. BioScience **31:** 593–597.

36. KOHN, K. W. 1981. Molecular mechanisms of crosslinking by alkylating agents and platinum complexes. *In* Molecular Actions and Targets for Cancer Chemotherapeutic Agents. A. C. Sartorelli, Ed.: 3–16. Academic Press, Inc. New York, N.Y.

37. KOHN, K. W. Biological aspects of DNA damage by crosslinking agents. *In* Molecular Aspects of Anticancer Drug Action. S. Neidle & M. Waring, Eds. Macmillian Press Ltd. London, England. (In press.)

38. KOHN, K. W. & R. A. G. EWIG. 1979. DNA-protein crosslinking by *trans*-platinum (II) diamminedichloride in mammalian cells, a new method of analysis. Biochem. Biophys. Acta **562:** 32–40.

39. KOHN, K. W., R. A. G. EWIG, L. C. ERICKSON & L. A. ZWELLING. 1981. Measurement of strand breaks and crosslinks by alkaline elution. *In* DNA Repair, a Laboratory Manual of Research Procedures. E. C. Friedberg & P. C. Hanawalt, Eds. **1:** 379–401. Marcel Dekker. New York, N.Y.

40. MEHTA, J. R., D. B. LUDLUM, A. RENARD & W. G. VERLY. 1981. Repair of O^6-ethylguanine in DNA by a chromatin fraction from rat liver: transfer of the ethyl group to an acceptor protein. Proc. Nat. Acad. Sci. USA **78**: 6766–6770.
41. OLSSON, M. & T. LINDAHL. 1980. Repair of alkylated DNA in *Escherichia coli*. Methyl group transfer from O^6-methylguanine to a protein cysteine residue. J. Biol. Chem. **255**: 10569–10571.
42. PARODI, S., S. DE FLORA, M. CAVANNA, A. PINO, L. ROBBIANO, C. BENNICELLI & G. BRAMBILLA. 1981. DNA-damaging activity *in vivo* and bacterial mutagenicity of sixteen hydrazine derivatives as related quantitatively to their carcinogenicity. Cancer Res. **41**: 1469–1482.
43. PARODI, S., M. PALA, P. RUSSO, A. ZUNINO, C. BALBI, A. ALBINI, F. VALERIO, M. R. CIMBERLE & L. SANTI. 1982. DNA damage in liver, kidney, bone marrow and spleen of rats and mice treated with commercial and purified alkaline elution assay and sister chromatid exchange induction. Cancer Res. **42**: 2277–2283.
44. PARODI, S., M. TANINGHER, M. PALA & L. SANTI. 1981. Alkaline DNA fragmentation in vivo: borderline or negative results obtained respectively with 7,12-dimethylbenz[*a*]anthracene and benzo[*a*]pyrene. Tumori **67**: 87–93.
45. PARODI, S., M. TANINGHER, P. RUSSO, M. PALA, M. TAMARO & C. MONTI-BRAGADIN. 1981. DNA-damaging activity *in vivo* and bacterial mutagenicity of sixteen aromatic amines and azo-derivatives, as related quantitatively to their carcinogenicity. Carcinogenesis **2**: 1317–1326.
46. PETZOLD, G. L. & J. A. SWENBERG. 1978. Detection of DNA damage induced *in vivo* following exposure of rats to carcinogens. Cancer Res. **38**: 1589–1594.
47. POMMIER, Y., D. KERRIGAN, R. SCHWARTZ & L. A. ZWELLING. 1982. The formation and resealing of intercalator-induced DNA strand breaks in isolated L1210 cell nuclei. Biochem. Biophys. Res. Commun. **107**: 576–583.
48. ROBERTS, J. J. 1978. The repair of DNA modified by cytotoxic, mutagenic and carcinogenic chemicals. Adv. Radiat. Biol. **7**: 212–436.
49. ROBERTS, J. J. & F. FRIEDLOS. 1981. Quantitative aspects of the formation and loss of DNA interstrand crosslinks in Chinese hamster cells following treatment with *cis*-diamminedichloroplatinum(II). I. Proportion of DNA platinum reactions involved in DNA cross-linking. Biochim. Biophys. Acta **655**: 146–151.
50. ROSS, W. E., D. L. GLAUBIGER & K. W. KOHN. 1978. Protein-associated DNA breaks in cells treated with adriamycin or ellipticine. Biochim. Biophys. Acta **519**: 23–30.
51. ROSS, W. E., D. GLAUBIGER & K. W. KOHN. 1979. Qualitative and quantitative aspects of intercalator-induced DNA strand breaks. Biochim. Biophys. Acta **562**: 41–50.
52. ROSS, W. E. & N. SHIPLEY. 1980. Relationship between DNA damage and survival in formaldehyde-treated mouse cells. Mutat. Res. **79**: 277–283.
53. SAUSVILLE, E. A., R. W. STEIN, J. PEISACH & S. B. HORWITZ. 1978. Properties and products of the degradation of DNA by bleomycin and iron (II). Biochemistry **17**: 2746–2754.
54. SOGNIER, M. A. & W. N. HITTELMAN. 1979. The repair of bleomycin-induced DNA damage and its relationship to chromosome aberration repair. Mutat. Res. **62**: 517–527.
55. STAIANO, N., L. C. ERICKSON & S. S. THORGEIRSSON. 1980. Bacterial mutagenesis and host cell DNA damage by chemical carcinogens in the salmonella/hepatocyte system. Biochem. Biophys. Res. Commun. **94**: 837–842.
56. STRANDBERG, M. C., E. BRESNICK & A. EASTMAN. 1982. The significance of DNA cross-linking to *cis*-diamminedichloroplatinum(II)-induced cytotoxicity in sensitive and resistant lines of murine leukemia L1210 cells. Chem. Biol. Interact. **15**: 169–180.
57. SWENBERG, J. A., G. L. PETZOLD & P. R. HARBACH. 1976. *In vitro* DNA damage/alkaline elution assay for predicting carcinogenic potential. Biochem. Biophys. Res. Commun. **72**: 732–738.
58. THOMAS, C. B., R. OSIEKA & K. W. KOHN. 1978. DNA cross-linking by in vivo treatment with 1-(2-chloroethyl)-3-(4-methylcyclohexyl)-1-nitrosourea of sensitive and resistant human colon carcinoma xenograms in nude mice. Cancer Res. **38**: 2448–2458.
59. TSAPAKOS, M. J., T. H. HAMPTON & K. W. JENNETTE. 1981. The carcinogen chromate induces DNA cross-links in rat liver and kidney. J. Biol. Chem. **256**: 3623–3626.
60. TSE, Y. C., K. KIRKEGAARD & J. C. WANG. 1980. Covalent bonds between protein and

DNA. Formation of phosphotyrosine linkage between certain DNA topoisomerases and DNA. J. Biol. Chem. **25:** 5560–5565.

61. WANG, J. C. 1982. DNA topoisomerases. Sci. Am. **247:** 94–97.
62. ZWELLING, L. A., T. ANDERSON & K. W. KOHN. 1979. DNA-protein and DNA interstrand crosslinking by *cis-* and *trans*-platinum(II)diamminedichloride in L1210 cells and relation to cytotoxicity. Cancer Res **39:** 365–369.
63. ZWELLING, L. A., M. O. BRADLEY, N. S. SHARKEY, T. ANDERSON & K. W. KOHN. 1979. Mutagenicity, cytotoxicity and DNA crosslinking in V79 Chinese hamster cells treated with *cis-* and *trans*-Pt(II)diamminedichloride. Mutat. Res. **67:** 271–280.
64. ZWELLING, L. A., D. KERRIGAN & Y. POMMIER. 1982. Inhibitors of poly-(adenosine diphosphoribose) synthesis slow the resealing rate of X-ray induced DNA strand breaks. Biochem. Biophys. Res. Commun. **104:** 897–902.
65. ZWELLING, L. A., K. W. KOHN, W. E. ROSS, R. A. G. EWIG & T. ANDERSON. 1978. Kinetics of formation and disappearance of DNA crosslinks in L1210 cells treated with *cis-* and *trans*-Pt(II). Cancer Res. **38:** 1762–1768.
66. ZWELLING, L. A., S. MICHAELS, L. C. ERICKSON, R. S. UNGERLEIDER, M. NICHOLS & K. W. KOHN. 1981. Protein-associated DNA strand breaks in L1210 cells treated with the DNA intercalating agents, 4′-(9-acridinylamino)-methanesulfon-*m*-anisidide (m-AMSA) and adriamycin. Biochemistry **20:** 6553–6563.
67. ZWELLING, L. A., S. MICHAELS, D. KERRIGAN, Y. POMMIER & K. W. KOHN. 1982. Protein-associated deoxyribonucleic acid strand breaks produced in mouse leukemia L1210 cells by ellipticine and 2-methyl-9-hydroxyellipticinium. Biochem. Pharmacol. **31:** 3261–3267.

THE USE OF CELLS FROM RAT, MOUSE, HAMSTER, AND RABBIT IN THE HEPATOCYTE PRIMARY CULTURE/DNA-REPAIR TEST*

Charlene A. McQueen and Gary M. Williams

The Naylor Dana Institute for Disease Prevention
American Health Foundation
Valhalla, New York 10595

INTRODUCTION

Liver-culture systems have been developed for toxicologic and pharmacologic studies in which the capacity of the liver for the biotransformation of diverse xenobiotics is utilized. Intact hepatocytes can be isolated and maintained as suspension or monolayer cultures.[1-3] These cells, at least initially, retain a broad metabolic capability. Comparison of intact cells and homogenates has revealed the metabolism of hepatocytes to more closely resemble metabolism *in vivo*.[4-8]

Cultured hepatocytes have been used as a screening test to assess the cytotoxic or genotoxic potential of chemicals.[9-14] Genotoxicity, defined as damage to DNA,[15] is often monitored as excision DNA repair, a specific response to DNA damage. DNA-repair synthesis can be demonstrated by unscheduled DNA synthesis (UDS) which is DNA synthesis occurring outside the replicative phase of the cell cycle.

A variety of culture conditions and methods of measuring DNA repair have been proposed.[9-11,16-21] The hepatocyte primary culture (HPC)/DNA-repair test[9,10] uses monolayer cultures of hepatocytes, and DNA-repair synthesis is measured by autoradiographic incorporation of ³H-thymidine. This approach has several advantages. Monolayer cultures are prepared so that at the time of exposure virtually all the cells are viable.[22] This is in contrast to suspension cultures, which consist of a mixture of viable and nonviable cells. In primary cultures of hepatocytes, less than 0.1% of the cells enter S-phase,[23] obviating the need for inhibitors of DNA synthesis. Additionally, with autoradiography, these few S-phase cells can be readily identified by their extensive incorporation of ³H-thymidine.

RAT HEPATOCYTE PRIMARY CULTURE/DNA-REPAIR TEST

Hepatocytes are isolated by a two-step *in situ* perfusion of the liver of adult animals.[23,24] In order to insure that only hepatocytes of high initial viability are used, the number of cells excluding trypan blue is determined. Monolayer cultures are initiated on coverslips by allowing the cells to attach for 1.5 to 2 hours. The cultures are washed and simultaneously exposed to ³H-thymidine (TdR) and the test chemical. After 18 hours the coverslips are processed for autoradiography.[9,24] Nuclear and background counts are made on an Artek counter. Net nuclear grain counts are calculated by subtracting the highest cytoplasmic count from the nuclear count.[9,24]

*This work was supported by National Institutes of Health Contract NOI-CP-55705 and Grants RR-05775-05 and CA33144 and U.S. Environmental Protection Agency Contract 68-01-6179. Although funded in part by EPA through a contract to Dr. Gary Williams, the research described in this article has not been subjected to agency review and therefore does not necessarily reflect the views of the agency, and no official endorsement should be inferred.

0077–8923/83/0407–0119 $01.75/0 © 1983, NYAS

The HPC/DNA-repair test depends on the ability of hepatocytes both to metabolize xenobiotics and to repair damage to DNA. These functions can be affected by cytotoxicity. Leakage of intracellular enzymes such as lactate dehydrogenase (LDH) has been used as a measure of toxicity in hepatocyte primary cultures.[12,13,25-27] The extracellular concentration of LDH is dependent on the concentration of the chemical and the length of exposure (FIGURE 1). Generally the concentration that induced maximum DNA repair resulted in little or no toxicity.[28] For example, 4,4'-methylene-bis-2-chloroaniline (MOCA) induced maximum repair at 10^{-5} M,[12,28] a concentration that resulted in an increase in extracellular LDH of 20% over the control level. This is not significantly different from the approximately 15% variation observed between control cultures.[12]

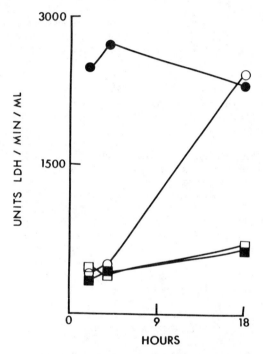

FIGURE 1. Cytotoxicity of 4,4'-methylenebis-2-chloroaniline (MOCA) in adult hepatocytes in primary culture. Lactate dehydrogenase (LDH) activity in the culture medium was measured following exposure to 10^{-3} M (filled circles), 10^{-4} M (open circles), 10^{-5} M (open squares), or control (filled squares) cultures. Values are the average of duplicate flasks.

In the HPC/DNA-repair test, excision repair of DNA is measured autoradiographically as UDS. In order to verify that this synthesis is occurring in nonreplicated DNA, density gradient centrifugation was used to separate replicated and nonreplicated DNA. Hepatocytes in primary culture were exposed to 2-acetylaminofluorene (2AAF) and ^{3}H-TdR in the presence of the density label bromodeoxyuridine (BUdR). The DNA was isolated and separated on a cesium chloride gradient.[29] The control samples displayed a single peak of absorbance at the density of nonreplicated DNA and a peak of ^{3}H-TdR at the density of replicated DNA (FIGURE 2A). The gradient

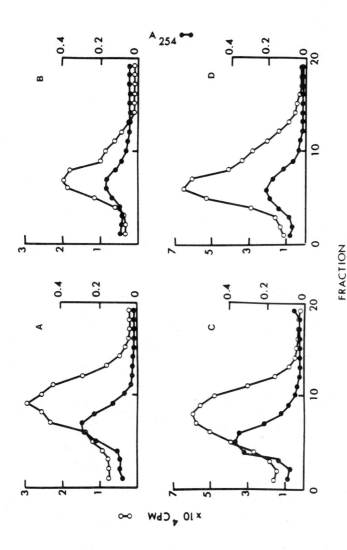

FIGURE 2. Neutral cesium chloride density gradient centrifugation of rat hepatocyte cultures exposed to 2-acetylaminofluorene (2AAF). The density of the gradient increases from left to right. (A) Control with ^{3}H-TdR. (B) 10^{-3} M 2AAF with ^{3}H-TdR. (C) Control with ^{3}H-CdR. (D) 10^{-3} M 2AAF with ^{3}H-CdR.

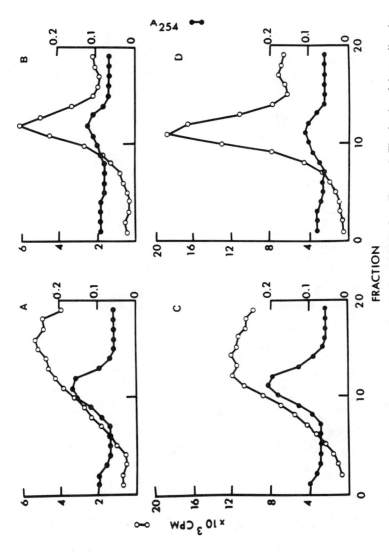

FIGURE 3. Alkaline cesium chloride density gradient reband of nonreplicated DNA from FIGURE 2. The density of the gradient increases from left to right. (A) Control with ³H-TdR. (B) 10⁻³ M 2AAF with ³H-TdR. (C) Control with ³H-CdR. (D) 10⁻³ M 2AAF with ³H-CdR.

from hepatocytes exposed to 2AAF also showed a single peak of absorbance at the density of nonreplicated DNA (FIGURE 2B). The peak of ^{3}H-TdR was displaced to the left at the density of nonreplicated DNA. Most of the incorporation was into nonreplicated DNA, indicating DNA repair.

The fractions containing nonreplicated DNA were recentrifuged on an alkaline gradient in order to confirm DNA repair. The control samples again showed a single peak of absorbance at the density of nonreplicated DNA with tailing of the ^{3}H-TdR at the bottom of the gradient (FIGURE 3A). The DNA from 2AAF-exposed hepatocytes had a single peak of absorbance at the density of nonreplicated DNA coincident with a single peak of ^{3}H-TdR (FIGURE 3B). This incorporation of ^{3}H-TdR into nonreplicated DNA is unequivocal evidence of repair synthesis. Similar results have also been reported with sodium iodide density gradients.[18]

TABLE 1

RAT HEPATOCYTE PRIMARY CULTURE/DNA-REPAIR TEST

Chemical Class	Carcinogen		Noncarcinogen	
	+	−	+	−
Alkylating agent	5	0	0	0
Aminoazo dyes	5	0	0	0
Azaaromatics	2	0	0	2
Monocyclic aromatic amines	3	1	0	0
Mycotoxins	7	1	0	1
Nitrosamines	7	1	0	3
Nitrosubstituted compounds	3	2	0	2
Polycyclic aromatic amines and amides	8	1	0	3
Polycyclic aromatic hydrocarbons	6	0	0	5
Pyrrolizidine alkaloids	4	0	0	0
Total	50	6	0	16

Bases other than TdR can be used for incorporation during DNA-repair synthesis.[30,31] Cesium chloride gradients of DNA from hepatocytes incubated with ^{3}H-deoxycytidine (CdR) showed a similar pattern to that seen with ^{3}H-TdR (FIGURES 2 and 3). However, although DNA repair could be demonstrated with either base, greater incorporation was seen with ^{3}H-CdR. An additional advantage of using ^{3}H-CdR is that higher concentrations of BUdR can be utilized, resulting in greater separation of replicated and nonreplicated DNA on the density gradient.[32]

The rat HPC/DNA-repair test has shown sensitivity to a wide variety of structural classes including alkylating agents, mycotoxins, aromatic amines, polycyclic aromatic hydrocarbons, nitrosamines, pyrrolizodine alkaloids, and aminoazo dyes.[9–11] Approximately 90% of the known carcinogens tested in our laboratory have been positive, and all of the known noncarcinogens have been negative (TABLE 1).

THE HPC/DNA-REPAIR TEST USING HEPATOCYTES OF VARIOUS SPECIES

Because of species differences in metabolic activation and *in vivo* susceptibility to chemical carcinogens,[33–35] it was desirable to extend the rat HPC/DNA-repair test to hepatocytes from other species. The procedures originally described for the rat

HPC/DNA-repair test have been adapted for use with mouse, hamster, and rabbit.[28,36-39] Validation of the mouse and hamster HPC/DNA-repair test was done by assaying carcinogens and noncarcinogens from six structural classes.[39] The carcinogenicity of these chemicals is well documented in the rat and mouse, but less information is available for the hamster (TABLE 2).

In the mouse HPC/DNA-repair test, all the carcinogens were positive and the noncarcinogens were negative (TABLE 2). Similarly, all the carcinogens elicited DNA repair in hamster hepatocytes. However, two of the presumed noncarcinogens, pyrene and aflatoxin G_2, were positive (TABLE 2). No carcinogenicity data in the hamster are available for these two chemicals, but the results suggest that pyrene and aflatoxin G_2 may be potential carcinogens in this species.

Although aflatoxin B_1 was positive in both mouse and hamster hepatocytes, hamster cells were more sensitive to the genotoxic effect of this chemical (FIGURE 4). The in vivo resistance of mouse to the tumorigenicity of this chemical[40] was reflected in the HPC/DNA-repair test. The results of this assay were consistent with the fact that mouse liver preparations form aflatoxin B_1-epoxide, the proposed ultimate metabolite.[41] Mouse hepatocytes and postmitochondrial liver supernatants, however, were more efficient in inactivating aflatoxin B_1 than were rat preparations.[41,42]

Multispecies testing of chemicals allows identification of genotoxic chemicals that may not be positive in the rat. Both safrole and procarbazine have been reported as negative in the rat HPC/DNA-repair test.[11] Positive results were obtained in mouse and hamster hepatocytes in response to safrole (FIGURE 5) and to procarbazine (FIGURE 6). Mouse hepatocytes were more sensitive to safrole than were hamster hepatocytes. Safrole induced maximum DNA repair at 10^{-4} M in mouse cells. Hamster hepatocytes had a maximum response at 10^{-3} M, a concentration that was toxic to mouse hepatocytes. However, hamster cells were more sensitive than mouse cells to procarbazine.

TABLE 2

VALIDATION OF THE HEPATOCYTE PRIMARY CULTURE/DNA-REPAIR TEST

	Carcinogenicity*			HPC/DNA-Repair Test		
Chemical	Rat	Hamster	Mouse	Rat	Hamster	Mouse
Methyl methanesulfonate	+		+	+	+	+
2-Acetylaminofluorene	+	+	+	+	+	+
Fluorene	−		−	−	−	−
Aflatoxin B_1	+		+	+	+	+
Aflatoxin G_2†				−	+	−
Benzo[a]pyrene	+	+	+	+	+	+
Pyrene	−		−	−	+	−
Nitrosopyrrolidine	+	+	+	+	+	+
Dimethylnitrosamine	+	+	+	+	+	+
Dimethylformamide	−			−	−	−
3'-Methyl-4-dimethyl-						
aminoazobenzene	+	+		+	+	+
p-Dimethylaminoazobenzene	+	+	+	+	+	+

*Results obtained primarily from International Agency for Research on Cancer Monographs on the Evaluation of the Carcinogenic Risk of Chemicals to Humans and U.S. Public Health Service Publication 149, Survey of Compounds Which Have Been Tested for Carcinogen Activity.

†Negative in trout.

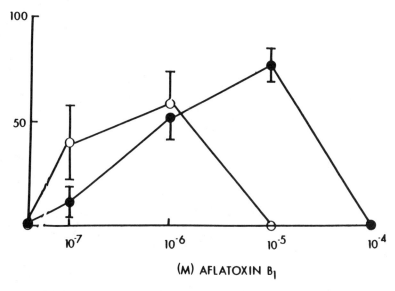

FIGURE 4. Genotoxicity of aflatoxin B_1 in mouse and hamster hepatocytes. DNA repair was measured by autoradiography. Each point represents the mean ± standard deviation of triplicate coverslips. Filled circles, mouse; open circles, hamster.

INTRASPECIES DIFFERENCES

The HPC/DNA-repair test can be used to investigate intraspecies differences in metabolism and their effect on genotoxicity. One step in the metabolism of aromatic amines and hydrazines is N-acetylation. Genetically determined polymorphic differences in this metabolic step are expressed in both humans and New Zealand white rabbits.[43–47] Individuals are classified as rapid or slow acetylators, with slow acetylation being the recessive trait. An association has been observed between the acetylator polymorphism and toxicity, including damage to DNA by aromatic amines and hydrazines.[48]

Rabbit hepatocytes in primary culture offer an *in vitro* system in which both N-acetylation and DNA damage can be investigated in the same cell. Hepatocytes from rapid and slow acetylator rabbits were shown to maintain in culture the differences in acetylation that are seen *in vivo*.[38] DNA damage in these cultures was indicated by the autoradiographic determination of DNA repair. Rabbit hepatocytes from both acetylator phenotypes were exposed to hydralazine (HDZ), an antihypertensive drug that is metabolized by *N*-acetyltransferase (NAT). A positive response was elicited only in hepatocytes from slow acetylators (TABLE 3). This response was concentration dependent.[38] HDZ has also been shown to induce DNA repair in Fisher F-344 rat hepatocytes, which have an acetylation rate similar to slow acetylator rabbits.[49]

Phenotype-dependent differences were also observed in response to 2-aminofluorene (2-AF). In contrast to the results with HDZ, hepatocytes from rapid acetylator rabbits were more sensitive to the effects of 2-AF than were hepatocytes from slow acetylators.[38] A concentration of 10^{-3} M 2-AF was toxic to the rapid phenotype but induced maximum repair in the slow acetylator hepatocytes (TABLE 3). Although the

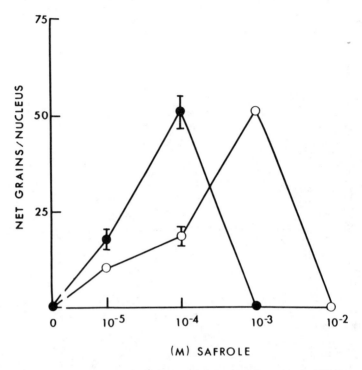

FIGURE 5. Genotoxicity of safrole in mouse and hamster hepatocytes. DNA repair was measured by autoradiography. Each point represents the mean ± standard deviation of triplicate coverslips. Filled circles, mouse; open circles, hamster.

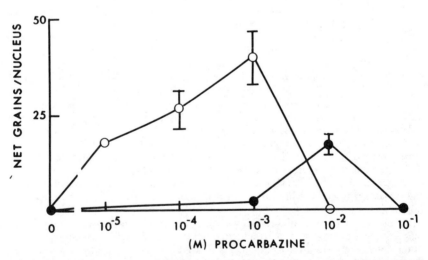

FIGURE 6. Genotoxicity of procarbazine in mouse and hamster hepatocytes. DNA repair was measured by autoradiography. Each point represents the mean ± standard deviation of triplicate coverslips. Filled circles, mouse; open circles, hamster.

TABLE 3

DNA REPAIR IN HEPATOCYTES FROM RAPID AND SLOW ACETYLATOR RABBITS

Chemical	Concentration (M)	Phenotype	
		Rapid	Slow
Hydralazine	10^{-3}	−	+
	5×10^{-3}	−/toxic	++
	10^{-2}	toxic	toxic
2-Aminofluorene	10^{-4}	++	+
	10^{-3}	toxic	++
	10^{-2}	toxic	toxic

acetylated derivative, 2-acetylaminofluorene, was positive, no correlation was observed between genotoxicity and the rate of acetylation.[50]

Benzidine was another aromatic amine that was tested in the rabbit HPC/DNA-repair test. Hepatocytes from a rapid acetylator showed maximum DNA repair at 10^{-5} M benzidine (FIGURE 7). At higher concentrations, the cytotoxic effect of benzidine decreased cellular DNA repair. With hepatocytes from a slow acetylator, however, the maximum response was seen at 10^{-3} M (FIGURE 7).

The demonstration of phenotype-dependent differences in genotoxicity of HDZ, 2-AF, and benzidine provides evidence for the role of genetic variability in N-

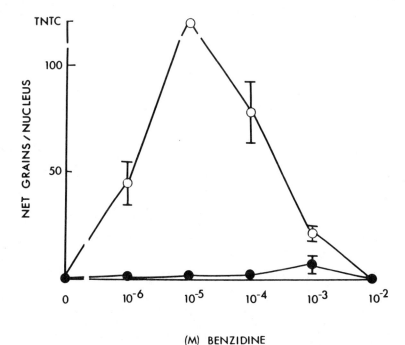

(M) BENZIDINE

FIGURE 7. Genotoxicity of benzidine in rabbit hepatocytes. DNA repair was measured by autoradiography in hepatocyte primary cultures from rapid (open circles) and slow (filled circles) acetylator rabbits. Each point represents the mean ± standard deviation of triplicate coverslips.

acetylating capacity in determining susceptibility to aromatic amines or hydrazine carcinogens.

CONCLUSIONS

The HPC/DNA-repair test has been developed to utilize both the metabolic capacity of intact liver cells and the specificity of DNA repair as an indicator of DNA damage. Extensive studies using this test have demonstrated that:

1. The HPC/DNA-repair test using rat hepatocytes is a reliable *in vitro* assay for identifying genotoxic chemicals.
2. The HPC/DNA-repair test can also be done with hepatocytes from rat, mouse, hamster, or rabbit.

The use of hepatocytes from a variety of species has demonstrated the advantages of multispecies testing. Chemicals such as procarbazine and safrole that have been reported as negative in the rat HPC/DNA-repair test were positive using mouse or hamster cells. Additionally, *in vivo* species differences in susceptibility to chemical carcinogens, such as the relative resistance of mouse to aflatoxin B_1, are reflected in the *in vitro* assay.

The HPC/DNA-repair test can also be used to assess the effect of intraspecies differences in metabolism on genotoxicity. Genetically determined differences in N-acetylation altered sensitivity to the genotoxic effect of HDZ, benzidine, and 2-AF. These results provide evidence for the role of the acetylator polymorphism as a factor in determining susceptibility to these genotoxic chemicals.

ACKNOWLEDGMENTS

We thank Dr. C. Tong and C. Maslansky for their collaboration and S. Dorn for preparing the manuscript.

REFERENCES

1. JEEJEEBHOY, K. N. & M. S. PHILIPS. 1976. Isolated mammalian hepatocytes in culture. Gastroenterology **71:** 1086–1096.
2. SIRICA, A. E. & H. C. PITOT. 1980. Drug metabolism and effects of carcinogens in cultured hepatic cells. Pharmacol. Rev. **31:** 205–228.
3. THURMAN, R. G. & F. C. KAUFMAN. 1980. Factors regulating drug metabolism in intact hepatocytes. Pharmacol. Rev. **31:** 229–251.
4. BILLINGS, R. E., R. E. MCMAHON, J. ASHMORE & S. R. WAGLE. 1977. The metabolism of drugs in isolated rat hepatocytes. Drug Metab. Dispos. **5:** 518–526.
5. DECAD, G. M., D. P. H. HSIEH & J. L. BYARD. 1977. Maintenance of cytochrome P-450 and metabolism of aflatoxin B_1 in primary hepatocyte cultures. Biochem. Biophys. Res. Commun. **78:** 279–287.
6. DYBING, E., E. SODERLUND, L. TIMM-HAUG & S. S. THORGEIRRSON. 1979. Metabolism and activation of 2-acetylaminofluorene in isolated rat hepatocytes. Cancer Res. **39:** 3268–3275.
7. VADI, H., P. MOLDEUS, J. CAPDEVILA & S. ORRENIUS. 1975. The metabolism of benzo[a]pyrene in isolated rat cells. Cancer Res. **35:** 2083–2091.
8. BATES, D. J., A. B. FOSTER & M. JARMAN. 1981. The metabolism of cyclophosphamide by isolated rat hepatocytes. Biochem. Pharmacol. **30:** 3055–3063.

9. WILLIAMS, G. M. 1977. The detection of chemical carcinogens by unscheduled DNA synthesis in rat liver primary cell cultures. Cancer Res. **37**: 1845–1851.
10. WILLIAMS, G. M. 1980. The detection of chemical mutagens/carcinogens by DNA repair and mutagenesis in liver cultures. *In* Chemical Mutagens. F. J. de Serres & A. Hollaender, Eds. **6**: 61–79. Plenum Press. New York, N.Y.
11. PROBST, G. S., R. E. MCMAHON, L. E. HILL, C. Z. THOMPSON, J. K. EPP & S. B. NEAL. 1981. Chemically-induced unscheduled DNA synthesis in primary rat hepatocyte cultures. Environ. Mutagen. **3**: 11–32.
12. MCQUEEN, C. A. & G. M. WILLIAMS. 1982. Cytotoxicity of xenobiotics in adult rat hepatocytes in primary culture. J. Fundam. Appl. Toxicol. **2**: 139–142.
13. ANUFORO, D. C., D. ACOSTA & R. V. SMITH. 1978. Hepatotoxicity studies with primary cultures of rat liver cells. In Vitro **14**: 981–987.
14. TYSON, C. A., C. MITOMA & J. KALIVODA. 1980. Evaluation of hepatocytes isolated by a nonperfusion technique in a prescreen for cytotoxicity. J. Toxicol. Environ. Health **6**: 197–205.
15. WILLIAMS, G. M. 1980. Classification of genotoxic and epigenetic hepatocarcinogens using liver culture assays. Ann. N.Y. Acad. Sci. **349**: 273–282.
16. HSIA, M. T. S. & B. L. KREAMER. 1979. Induction of unscheduled DNA synthesis in suspensions of rat hepatocytes by the environmental toxicant, 3,3',4,4'-tetrachloroazobenzene. Cancer Lett. **6**: 207–212.
17. MICHALOPOULOS, G., G. L. SATTLER, L. O'CONNOR & H. C. PITOT. 1978. Unscheduled DNA synthesis induced by procarcinogens in suspension and primary cultures of hepatocytes on collagen membranes. Cancer Res. **38**: 1866–1871.
18. YAGER, J. D. & J. A. MILLER. 1978. DNA repair in primary cultures of rat hepatocytes. Cancer Res. **38**: 4385–4394.
19. OLDHAM, J. W., D. A. CASCIANO & M. D. CAVE. 1980. Comparative induction of unscheduled DNA synthesis by physical and chemical agents in non-proliferating primary cultures of rat hepatocytes. Chem. Biol. Interact. **29**: 303–314.
20. BROUNS, R. E., M. POUT, R. DEVRIND, T. V. HOEK-KON & P. T. HENDERSON. 1979. Measurement of DNA excision repair in suspensions of freshly isolated rat hepatocytes after exposure to some carcinogenic compounds. Mutat. Res. **64**: 425–432.
21. ALTHAUS, F. R., S. D. LAWRENCE, G. L. SATTLER, D. G. LONGFELLOW & H. C. PITOT. 1982. Chemical quantification of unscheduled DNA synthesis in cultured hepatocytes as an assay for the rapid screening of potential chemical carcinogens. Cancer Res. **42**: 3010–3015.
22. WILLIAMS, G. M., E. BERMUDEZ & D. SCARAMUZZINO. 1977. Rat hepatocyte primary cell cultures. III. Improved dissociation and attachment techniques and the enhancement of survival by culture medium. In Vitro **13**: 809–817.
23. LAISHES, B. A. & G. M. WILLIAMS. 1976. Conditions affecting primary cell cultures of functional adult rat hepatocytes. II. Dexamethasone-enhanced longevity and maintenance of morphology. In Vitro **12**: 821–832.
24. WILLIAMS, G. M., M. F. LASPIA & V. C. DUNKEL. 1982. Reliability of the hepatocyte primary culture/DNA repair test in testing of coded carcinogens and noncarcinogens. Mutat. Res. **97**: 359–370.
25. ACOSTA, D., D. C. ANUFORO & R. V. SMITH. 1980. Cytotoxicity of acetamenophen and papaverine in primary cultures of rat hepatocytes. Toxicol. Appl. Pharmacol. **53**: 306–314.
26. SALOCKS, C. B., D. P. H. HSIEH & J. L. BYARD. 1981. Butylated hydroxytoluene pretreatment protects against cytotoxicity and reduces covalent binding of aflatoxin B₁ in primary hepatocyte cultures. Toxicol. Appl. Pharmacol. **59**: 331–345.
27. MITCHELL, D. B. & D. ACOSTA. 1981. Evaluation of the cytotoxicity of tricyclic antidepressants in primary cultures of rat hepatocytes. J. Toxicol. Environ. Health **7**: 83–92.
28. MCQUEEN, C. A., C. J. MASLANSKY, S. B. CRESCENZI & G. M. WILLIAMS. 1981. The genotoxicity of 4,4'-methylene-bis-2-chloroaniline in rat, mouse and hamster hepatocytes. Toxicol. Appl. Pharmacol. **58**: 231–235.
29. MCQUEEN, C. A. & G. M. WILLIAMS. 1981. Characterization of DNA repair elicited by carcinogens and drugs in the hepatocyte primary culture/DNA repair test. J. Toxicol. Environ. Health **8**: 463–477.

30. LIEBERMAN, M. W., R. N. BANEY, R. E. LEE, S. SELL & E. FARBER. 1971. Studies on DNA repair in human lymphocytes treated with proximate carcinogens and alkylating agents. Cancer Res. **31:** 1297–1306.
31. CLEAVER, J. E. 1973. DNA repair with purines and pyrimidines in radiation and carcinogen-damaged normal and xeroderma pigmentosum human cells. Cancer Res. **33:** 362–369.
32. ANDREA, U. & L. R. SCHWARZ. 1981. Induction of DNA repair synthesis in isolated rat hepatocytes by 5-diazouracil and other DNA damaging agents. Cancer Lett. **13:** 187–193.
33. MILLER, E. C., J. A. MILLER & M. ENOMOTO. 1964. The comparative carcinogenicity of 2-acetylaminofluorene and its N-hydroxy metabolite in mice, hamsters and guinea pigs. Cancer Res. **24:** 2018–2031.
34. IRVING, C. C. 1979. Species and tissue variation in the metabolic activation of aromatic amines. In Carcinogens: Identification and Mechanisms. A. C. Griffin & C. R. Shaw, Eds.: 221–227. Raven Press. New York, N.Y.
35. SELKIRK, J. K. 1980. Chemical carcinogenesis: a brief overview of the mechanism of action of polycyclic hydrocarbons, aromatic amines, nitrosamines and aflatoxins. In Carcinogenesis. T. J. Slaga, Ed. **5:** 1–31. Raven Press. New York, N.Y.
36. MASLANSKY, C. J. & G. M. WILLIAMS. 1981. Evidence for an epigenetic mode of action in organochlorine pesticide hepatocarcinogenicity: a lack of genotoxicity in rat, mouse and hamster hepatocytes. J. Toxicol. Environ. Health **8:** 121–130.
37. MASLANSKY, C. J. & G. M. WILLIAMS. 1982. Primary cultures and the levels of cytochrome P-450 in hepatocytes from mouse, rat, hamster and rabbit liver. In Vitro **18:** 663–693.
38. MCQUEEN, C. A., C. J. MASLANSKY, I. B. GLOWINSKI, S. B. CRESCENZI, W. W. WEBER & G. M. WILLIAMS. 1982. Relationship between the genetically determined acetylator phenotype and DNA damage induced by hydralazine and 2-aminofluorene in cultured rabbit hepatocytes. Proc. Nat. Acad. Sci. USA **79:** 1269–1272.
39. MCQUEEN, C. A., D. M. KREISER & G. M. WILLIAMS. 1983. The hepatocyte primary culture (HPC)/DNA repair assay using mouse or hamster hepatocytes. Environ. Mutagen. **5:** 1–8.
40. WOGAN, G. N. 1973. Aflatoxin carcinogenesis. Methods Cancer Res. **7:** 309–344.
41. DEGEN, G. H. & H. G. NEUMAN. 1981. Differences in aflatoxin B_1–susceptibility of rat and mouse are correlated with the capability in vitro to inactivate aflatoxin B_1-epoxide. Carcinogenesis **2:** 299–306.
42. DECAD, G. M., K. K. DOUGHERTY, D. P. H. HSIEH & J. L. BYARD. 1979. Metabolism of aflatoxin B_1 in cultured mouse hepatocytes: comparison with rat and effects of cyclohexene oxide and diethyl maleate. Toxicol. Appl. Pharmacol. **50:** 429–436.
43. KNIGHT, R. A., M. J. SELIN & H. W. HARRIS. 1959. Genetic factors influencing isoniazid blood levels in humans. Trans. Conf. Chemother. Tuberc. **18:** 52–60.
44. EVANS, D. A. P., K. A. MANLEY & V. A. MCKUSICK. 1960. Genetic control of isoniazid metabolism in man. Br. Med. J. **2:** 485–491.
45. FRYMOYER, J. W. & R. F. JACOX. 1963. Investigation of the genetic control of sulfadiazine and isoniazid metabolism in the rabbit. J. Lab. Clin. Med. **62:** 891–904.
46. FRYMOYER, J. W. & R. F. JACOX. 1963. Studies of genetically controlled sulfadiazine acetylation in rabbit livers: possible identification of the heterozygous trait. J. Lab. Clin. Med. **62:** 905–909.
47. GORDON, G. R., A. G. SHAFIZADEH & J. H. PETERS. 1973. Polymorphic acetylation of drugs in rabbits. Xenobiotica **3:** 133–150.
48. DRAYER, D. E. & M. M. REIDENBERG. 1977. Clinical consequences of polymorphic acetylation of basic drugs. Clin. Pharmacol. Ther. **22:** 251–258.
49. WILLIAMS, G. M., G. MAZUE, C. A. MCQUEEN & T. SHIMADA. 1980. Genotoxicity of the antihypertensive drugs hydralazine and dihydralazine. Science **210:** 329–330.
50. MCQUEEN, C. A. & G. M. WILLIAMS. 1982. Determination of host genetic susceptibility to genotoxic chemicals in hepatocyte cultures. IARC Sci. Publ. **39:** 413–420.

CYTOGENETIC METHODS FOR DETECTING EFFECTS OF CHEMICAL MUTAGENS

H. J. Evans

Medical Research Council
Clinical and Population Cytogenetics Unit
Western General Hospital
Edinburgh, EH4 2XU, United Kingdom

As an introduction to the session on chromosome effects, I was asked if I would provide an historical background and a rationale for the use of the cytogenetic approach for detecting the effects of chemical mutagens, so that subsequent speakers could address themselves to the details of specific test systems. I do not wish to dwell on different test systems, but neither do I wish to limit myself solely to remarks on history and on the basic rationale, and so I am going to spend some of my time at least in highlighting various problems associated with the cytogenetic analysis of chromosome damage induced by chemical mutagens. First, however, let me remind you that the discovery that chemical substances could act as mutagens in a manner analogous to x-rays dates back to the early 1940s and is rightly attributed to Charlotte Auerbach from her studies on the induction of mutations in *Drosophila* exposed to mustard gas, and to Oehklers for his independent demonstration that urethane produces chromosome aberrations in plant roots. As part of the British studies during the First World War, the late Peo Koller also demonstrated that mustard gas produced chromosome aberrations in *Tradescantia* pollen grains, and the parallel between these findings and the by this time well-known effects of ionizing radiations and of ultraviolet light in producing chromosome aberrations led to these mutagenic chemicals being referred to as "radiomimetic agents." In the succeeding 40 years a very large number of studies have shown that a wide variety of different types of chemical agent are capable of producing chromosome damage in whole organisms of plants, animals, and indeed man, as well as in cells taken from these organisms and exposed to these chemical substances in culture. From these studies have emerged a number of general principles as well as a number of problems, and I should like to consider some of these by addressing myself to a number of questions.

What Kinds of Chromosome Damage Can Be Analyzed in Appropriate Test Systems?

In essence we can distinguish between three different types of cytogenetic change in cells exposed to chemical mutagens: (1) alterations in chromosome structure, (2) sister chromatid exchange, and (3) alterations in chromosome number.

Alterations in Chromosome Structure

It is now well established that the continuity of the chromosome thread resides in its DNA and that a single duplex of DNA runs along the length of the chromosome from one end to the other. In terms of cytologically visible damage to the chromosome, a principal effect of chemical mutagens is to produce changes in chromosome structure

131

0077–8923/83/0407–0131 $01.75/0 © 1983, NYAS

that result in disruption of chromosome continuity to give free acentric fragments, or to rearrangements of segments of DNA within the chromosome, or to exchanges of DNA segments between chromosomes. Exchange events or fragments that involve sizable amounts of DNA may be readily visible under the microscope, and these changes in chromosome structure are loosely referred to as aberrations and can be classified into a number of distinct categories.[1,2]

Chemical substances or their metabolic derivatives that produce chromosome aberrations are reactive substances that interact with various components of the DNA. In most, but not all, cases, however, this interaction does not result in a direct breakage of the phosphodiester backbone of the DNA or chromosome, but such breakage or exchange follows as a consequence of misreplication at sites of damage in the DNA during the succeeding normal S phase of DNA replication. It is for this reason that most chemical mutagens primarily produce chromatid-type aberrations and not chromosome-type changes. This follows since, with the exception of very densely ionizing particulate radiations, the unit of breakage and rearrangement with virtually all mutagens is the single chromatid. Thus, events that occur during (or after) chromosome replication in S are seen to involve single chromatids when observed at the subsequent mitosis, whereas breakages and exchanges that involve unduplicated G_1 chromatids themselves become duplicated in S to give chromosome-type aberrations.[3]

The fact that the majority of chromatid-type aberrations induced by chemical mutagens are manufactured during S-phase DNA replication highlights two other interesting features of aberration induction. First is the fact that when proliferating cells are exposed to chemical mutagens, the highest aberration incidence is usually seen in those cells that were in the late G_1/early S phase of the cycle at the time of exposure.[4,5] A similar increased sensitivity to mutation induction occurs at this time, and one of the reasons for this is that the shorter the interval between treatment and replication the smaller the opportunity for lesion removal by DNA-repair processes. Another is that damage to nucleotides in the precursor pools occurs more readily than in the protected bases of the packaged DNA and such bases may then be incorporated during S phase and lead to mutations[6] and chromatid aberrations. Second, there have been many demonstrations that exposure of cells to chemical mutagens with exceedingly short half-lives results in chromatid-type aberrations not only at the first but also at the second and even subsequent mitoses following treatment. These findings show that if lesions persist in the DNA, and are not removed by the normal DNA-repair processes, then they may give rise to further aberrations during succeeding DNA replications.

Sister Chromatid Exchange

Sister chromatid exchanges (SCEs) comprise a special class of aberration that will be discussed in detail by others,[7] and so I shall limit my remarks to pointing out that they involve equal and symmetrical exchange between sister chromatids and hence do not result in an altered chromosome morphology and are not detectable except following special pretreatment and staining techniques. They may nevertheless be induced in very high frequency, and at very low concentrations, with certain classes of mutagen; are often, but perhaps not always, a consequence of lesions in DNA; are developed during S-phase replication; and involve a complete exchange between sister DNA duplexes. In these respects SCEs are similar to other chemically induced chromatid aberrations, but there are important differences and their relevance to mutation is yet to be unambiguously established.[8]

Alterations in Chromosome Number

In addition to what we might loosely term direct effects of mutagen exposure on chromosome structure, and which are usually a consequence of damage to the DNA within the chromosome or to nucleotides that could be later incorporated into the chromosome, there is a third class of cytogenetic effect that also results in mutational change in exposed cells or their descendants. This effect is one that influences the behavior of chromosomes at the time of cell division, resulting in a missegregation of chromosomes or chromatids to give progeny cells with abnormal chromosome numbers. These abnormal cells may have additional complete sets of chromosomes over and above the normal diploid number, i.e., they are polyploid, or they may have one, two, or more additional or missing chromosomes and are referred to as aneuploid cells. Both polyploidy and aneuploidy may result as a secondary consequence of a chromosomal structural change, but, in addition, there are a number of chemical agents that do not react with the DNA or protein structures within the chromosome, but rather interfere with the organization of the proteins involved in the formation of the spindle, which is the organelle responsible for the orderly segregation of chromosomes at cell division. These agents therefore will not be detectable in bacterial mutagenicity tests.

By definition, cytogenetic damage can only be observed cytologically in proliferating cell populations, but both the ability and the efficiency for detecting the three classes of damage referred to will vary according to the stage in development of the cells used to assay the damage. Structural changes are best observed at metaphase of mitosis, and sister chromatid exchanges can only be observed at this stage. Only a proportion of structural changes result in bridges and fragments at anaphase, so that anaphase analysis is less efficient and less informative, although having the advantage of being rather simpler and less time consuming than metaphase analysis. Acentric chromosome fragments, and indeed sometimes whole chromosomes, may be excluded from the daughter nuclei produced following mitosis and form additional small nuclei—micronuclei. The counting of micronuclei is a simple and very rapid procedure and is a useful end point for assaying chromosome damage in test systems.

Why Use Cytogenetic Effects to Detect Exposure to Chemical Mutagens?

Here there are really two questions: (1) How relevant is cytogenetic damage as an event to be measured in test systems? and (2) How specific and sensitive are our test systems?

Relevance

Cytogenetic variation resulting from changes in chromosome structure or chromosome number is an ever-present phenomenon in all eukaryotes, including man, in the absence of exposure to known external mutagens, but of course their frequency is much increased if cells are exposed to physical or chemical mutagens. The use of such changes as end points for detecting, and indeed measuring, exposure to mutagens is especially appropriate since a significant proportion of inherited genetic disease in man is directly attributable to chromosomal mutations involving changes in chromosome structure or number. Approximately 1 in every 170 live newborn babies has a chromosomal mutation that, in varying degree, may have an adverse effect on health and development in the affected child. A little over one-third of these abnormalities are due to chromosome structural rearrangements, and the remainder are a consequence of the presence of an additional chromosome or chromosomes (TABLE 1).[9] Many of the chromosomal abnormalities that occur in germ cells and in embryos are lethal, and as much as 60% of all human abortions are probably chromosomally abnormal.

In addition to the presence and consequence of chromosome changes in germ cells and embryos, we should not forget that there is now increasing evidence for the importance of chromosome rearrangement in somatic cells in the genesis of neoplastic change, and specific chromosome translocations are now clearly seen to be associated

TABLE 1

INCIDENCE OF CONSTITUTIONAL CHROMOSOME ANOMALIES IN CONSECUTIVE LIVE-BORN BABIES*

Numerical		45, X	0.07		4.14
		47, XXX	0.54	1.64	
		47, XXY	0.61		
		47, XYY	0.49		
	sex chromosome aneuploids	other sex chromosome anomalies (mostly mosaics)	0.96		
	autosomal trisomics	47, +D	0.12	1.45	
		47, +E	0.19		
		47, +G	1.14		
	triploidy		0.02		
Structural	balanced	D/D centric fusions	0.79	1.93	2.21
		D/G centric fusions	0.21		
		translocations; inversions	0.93		
	unbalanced	translocations; inversions; deletions	0.28		
Unclassified			0.51		
Total			6.86 (1 in 146)		

*Data summarized from a number of surveys (see Reference 9).

with lymphoproliferative malignancies such as chronic myeloid leukemia and various lymphomas.[10,11] Indeed, with improvement in our techniques for analyzing chromosomes, and the introduction of high-resolution prophase banding,[12] it is becoming even more evident that many human cancers involve or have associated with them some specific chromosomal abnormality.[13] The relevance of chromosome damage to the inheritance of genetic defects in both germ cells and somatic cells is therefore fairly clear, and since it is possible to visualize and measure chromosome damage in a wide variety of plant, mammalian, and human cells *in vivo* and *in vitro,* it is in this context a very appropriate end point to be utilized in test systems to detect chemical mutagens.

Specificity and Sensitivity

When we speak of specificity in mutagen testing systems, we are concerned with the ability of a system to detect different classes of mutagen. I think it is fair to say that all substances that produce chromosomal structural changes also produce what are loosely referred to as point mutations. This is not surprising, since the sizes of deletions and transpositions induced following exposure will range from those involving a few nucleotides in the DNA up to those involving larger tracts that are visible under the microscope. Frameshift and deletion mutants therefore form part of the family that the microscopist sees as aberrations. Indeed, it is not inappropriate here to point out that in recent years a number of genetic diseases in man that are considered to involve single gene effects have now been shown to be associated with deletions that are, in some cases, visible under the microscope. Now although all mutagens that induce structural rearrangements also induce point mutations, the reverse does not necessarily follow. Those mutagens that produce base changes that do not interfere with DNA replication will be inefficient in producing aberrations. However, few chemical mutagens produce no aberrations.

The sensitivity to the induction of chromosome damage in a given cell type and system will differ for different mutagens and classes of mutagens, and different cells will, for a given agent, show different sensitivities. For instance, among the alkylating agents polyfunctional compounds are more potent in inducing chromosome damage than are monofunctional agents, and those with high Swain-Scott substrate constant (s) values[14] are more potent than those with low s values. With regard to cell type, it has been shown that small lymphocytes may sometimes have a somewhat higher sensitivity to aberration induction than do cultured lymphoblastoid or fibroblast cells when exposed to direct-acting mutagens, but, because of their relatively diminished metabolic ability, they may be less sensitive to aberration induction by indirectly acting mutagens.

Sensitivity is obviously dictated by a variety of factors, but in the context of testing systems what we are really asking is, Does the end point of chromosome damage enable us to detect levels of exposure that are meaningful in the context of risk? The crude answer here is yes. To take an extreme example of the very potent damaging agent mitomycin C (MMC), we can detect a significant increase of background aberration frequencies in human lymphocytes or Chinese hamster fibroblasts exposed *in vitro* for one hour to MMC at a concentration of 1×10^{-7} M if we analyze a few hundred metaphase cells, and to 1×10^{-9} M if we score for SCE. SCEs are easier to observe and easier to score than are aberrations, and they provide an even more sensitive end point, so much so that we can detect the very small effects of substances that are claimed to be "weak mutagens" such as saccharine[15] or vitamin C[16] using a standard SCE test.

In relative terms, the SCE test is as sensitive as some of the more sensitive "Ames-type" bacterial tests. For example, we have recently, in collaboration with Dr. Bartsch, studied the response of Chinese hamster fibroblasts to SCE induction by a variety of pyrolysates of opium and its derivatives. In our hands we find that we can detect a significant increase in SCE frequency in cells exposed to an opium pyrolysate extract at a concentration of 10–50 μg/ml. In tests using *Salmonella* strains TA98 and 100, significant increases in reversion frequency are observed with concentrations of from 25 to 125 μg/ml.[17] As in the bacterial testing systems, it is of course possible to increase the sensitivity to chromosome damage in mammalian cells by utilizing cells deficient in various DNA-repair systems; examples of the latter include a number of well-known cell strains from man, the EM9 strain of Chinese hamster (CHO) cells,[18] and the sensitive strains of rat Yoshida sarcoma cells[19] to name but a few examples.

What Kinds of Factors May Influence the In Vitro *Response?*

The list of factors that may influence the frequency of chromosome damage following exposure to chemical mutagens is large, and many of these will be discussed by other speakers and include the requirement for, and control of, metabolic activation in the case of indirect mutagens; the influence of DNA repair and of metabolic inhibitors that affect DNA replication; and the effect of substances that scavenge free radicals as well as various other factors.[20] One factor that is not often controlled as well as it should be in tests for mutagenicity or chromosome-damaging effects in mammalian cells is the composition of the culture medium. Exposures in such tests are usually undertaken in the absence of serum, because of its protective effect and because it comprises a somewhat variable and ill-defined component of the medium. However, less attention is usually paid to other medium components, and to demonstrate how important they could be I should like to refer to two examples.

It has been known for over a decade that a gross imbalance in the pools of thymidine and cytidine triphosphates results in increasing mutation frequency in mammalian cells, and studies by Davidson and others have shown that culturing cells in the presence of excess thymidine—a nonessential component of many tissue-culture

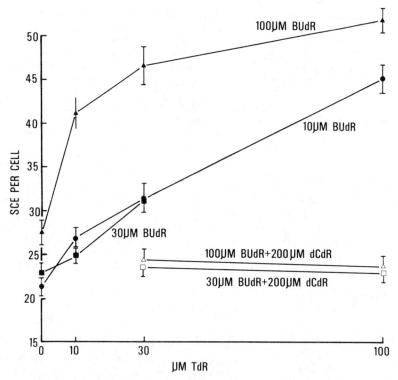

FIGURE 1. Incidence of SCE in CHO cells exposed to 1×10^{-7} M MNNG and various concentrations of thymidine (TdR) and BrdUrd. Note that the addition of deoxycytidine (dCdR) completely cancels out the potentiating effect of TdR.

TABLE 2

EFFECT OF DMFO ON SCE INDUCTION IN CULTURED RAT TUMOR CELLS
EXPOSED TO BCNU OR *cis*-DDP*

	SCEs/Metaphase	
Treatment	BCNU (1 μM)	*cis*-DDP (0.5 μM)
Control ± DFMO ± putrescine	13.4 ± 3.1	12.5 ± 3.0
Agent (BCNU or *cis*-DDP)	27.2 ± 6.1	56.5 ± 12.4
Agent + putrescine	26.0 ± 4.4	54.3 ± 12.6
Agent + DFMO	51.4 ± 11.0	29.7 ± 7.0
Agent + DFMO + putrescine	26.1 ± 4.4	50.0 ± 11.1

*After Tofilon *et al.*[23] BCNU is 1,3-bis-(2-chloroethyl)-1-nitrosourea. *cis*-DDP is *cis*-diamminedichloroplatinum II.

media—can also result in an increased frequency of mutation.[21] Incorporation of the thymidine analogue bromodeoxyuridine (BrdUdr), which is itself used to demonstrate SCEs, also results in mutation and SCE induction, and the presence of excess thymidine or its analogues enhances the response of cells to mutation induction by alkylating agents.[22] Indeed, it appears that nucleoside pool imbalance will generally enhance the mutation frequency in cells exposed to alkylating agents, perhaps by facilitating aberrant replication of alkylated DNA. Does such an imbalance also influence the SCE response of the cells following exposure to mutagens? The answer is yes. FIGURE 1 shows the results of one experiment on CHO cells exposed to 1×10^{-7} M *N*-methyl-*N'*-nitro-*N*-nitrosoguanidine (MNNG) where it can be seen that the SCE frequency is doubled at high nucleoside levels and that this effect is negated in the presence of excess deoxycytidine. The nucleoside levels to demonstrate this effect are much higher than those normally found in serum or in tissue-culture media, but the results underline the fact that factors that modify replication may also modify the cell's response to mutagens.

My second example is concerned with the influence of polyamines in the cell on its response to mutagens. Cationic polyamines play some role in maintaining DNA stability, and their levels are reduced under conditions that inhibit the enzyme ornithine decarboxylase (ODC). Cultured rat cells exposed to α-difluormethylornithine (DFMO), and in which levels of the polyamines spermidine and putrescine are reduced to 5% of normal, show a doubling of sensitivity to SCE induction by certain alkylating agents (TABLE 2) and a halving of the response to others, these effects being abolished if putrescine is added to the cultures.[23] These two examples serve to emphasize the importance of using well-defined media and conditions of culture for all kinds of tests involving mammalian cells.

Is There a Role for Automated Systems to Score Chromosome Damage?

I have referred to the exquisite sensitivity of the SCE induction test as a simple and rapid test where effects can be easily scored using a few tens of cells. However, the relevance of SCE with respect to mutation has yet to be clearly established. Structural changes, on the other hand, are clearly relevant but are less frequent, and the numbers of cells to be scored to obtain significant results may be of an order of magnitude more than is required for SCE analysis, and this is a major problem which adds considerable expense to mutagen tests. One answer to this problem is, of course, to score

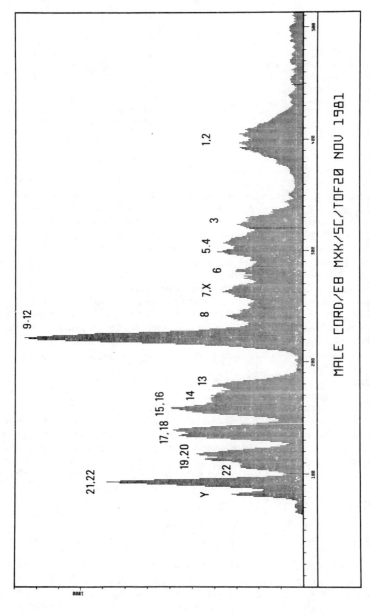

FIGURE 2. A normalized flow profile ("flow karyotype") of ethidium bromide–stained human chromosomes from cultured blood lymphocytes. Chromosomes and groups corresponding to the histogram peaks are numbered.

micronuclei, and this approach was, in fact, successfully used in early studies on relative biological efficiency (RBE) and dose response to ionizing radiations[24] and has recently been used extensively as an end point in *in vitro* and *in vivo* studies on chemical mutagens.[25] An alternative approach is to see whether we can count aberrations by machine.

If suspensions of ethidium bromide–stained chromosomes, rather than cells, are passed through an appropriate flow cytometer, it is possible to obtain a frequency distribution where each peak of fluorescence intensity corresponds to a particular chromosome or group of chromosomes in the complement, the position of each chromosome in the flow karyotype being a function of its size (FIGURE 2). As Carrano *et al.* originally demonstrated using Chinese hamster cells, the presence of chromosome aberrations following previous exposure to x-rays may significantly disturb the flow karyotype.[26] Such a disturbance is to be expected because the presence of chromosome fragments will reduce the heights of the peaks contributing to the larger chromosomes, increase the incidence of chromosome objects of smaller size, and increase the background noise level. We have pursued this approach using our own chromosome sorting machine to study x-ray-induced damage in chromosomes of cultured human lymphocytes and have been able to demonstrate changes in flow karyotype in cells exposed to relatively low doses of x-rays. Our results are extremely promising, for we obtain a dose-response curve from flow profiles that superimposes on that obtained from the conventional scoring of dicentric and fragment aberrations on metaphase preparations from the same cell samples (FIGURE 3). Although we can

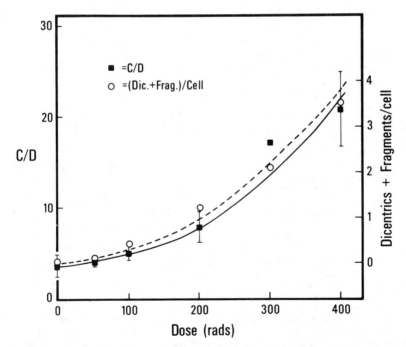

FIGURE 3. A comparison of x-ray dose response for chromosome damage in human lymphocytes measured by standard metaphase analysis of dicentrics and fragments in microscope preparations (circles), and as flow karyotype profile distortion (ratios of peaks to troughs) in chromosome samples from the same irradiated lymphocyte population (squares).

detect an exposure of as low as 25 rads by machine, there is much scope for improvement of the system to give increased discrimination in the detection of chromosome anomalies. However, it is worth pointing out that in *in vitro* studies on induced chromosome damage in human lymphocytes, a trained technician can score something like 200–300 metaphase cells per day, whereas with the flow machine the chromosomes of approximately 1,000 cells can be analyzed in the time of one minute. The potential for machine analysis is therefore really quite enormous and surely offers us a way to detect low levels of damage and to examine very large numbers of cells.

REFERENCES

1. EVANS, H. J. 1974. Effects of ionizing radiation on mammalian chromosomes. *In* Chromosomes and Cancer. J. German, Ed.: 191–237. John Wiley. New York, N.Y.
2. SAVAGE, J. R. K. 1976. Annotation: classification and relationships of induced chromosomal structural changes. J. Med. Genet. **13:** 103–122.
3. EVANS, H. J. 1977. Molecular mechanisms in the induction of chromosome aberrations. *In* Progress in Genetic Toxicology. D. Scott, B. A. Bridges & F. H. Sobels, Eds.: 57–74. Elsevier/North-Holland Biomedical Press. Amsterdam, the Netherlands.
4. EVANS, H. J. & D. SCOTT. 1964. Influence of DNA synthesis on the production of chromatid aberrations by X-rays and maleic hydrazide in Vicia faba. Genetics **49:** 17–38.
5. EVANS, H. J. & D. SCOTT. 1969. The induction of chromosome aberrations by nitrogen mustard and its dependence on DNA synthesis. Proc. R. Soc. London Ser. B **173:** 491–512.
6. TOPAL, M. D., C. A. HUTCHISON & M. S. BAKER. 1982. DNA precursors in chemical mutagenesis: a novel application of DNA sequencing. Nature **298:** 863–865.
7. WOLFF, S. 1983. Sister chromatid exchange as a test for mutagenic carcinogens. Ann. N.Y. Acad. Sci. (This volume.)
8. MORRIS, S. M., R. H. HEFLICH, D. T. BERANEK & R. L. KODELL. 1982. Alkylation-induced sister-chromatid exchanges correlate with reduced cell survival, not mutations. Mutat. Res. **105:** 163–168.
9. EVANS, H. J. 1980. How effects of chemicals might differ from those of radiations in giving rise to genetic ill-health in man. *In* Progress in Environmental Mutagenesis. M. Alacevic, Ed.: 3–21. Elsevier/North-Holland Biomedical Press. Amsterdam, the Netherlands.
10. EVANS, H. J. 1982. Cytogenetics of heritability in cancer. *In* Host Factors in Human Carcinogenesis. B. Armstrong & H. Bartsch, Eds.: 35–56. International Agency for Research on Cancer (IARC). Lyon, France.
11. ROWLEY, J. D. 1981. Nonrandom chromosome changes in human leukemia. *In* Genes, Chromosomes, and Neoplasia. F. E. Arrighi, P. N. Rao & E. Stubblefield, Eds.: 273–296. Raven Press. New York, N.Y.
12. ISCN. 1981. An International System for Human Cytogenetic Nomenclature—High Resolution Banding. Birth Defects Orig. Artic. Ser. **17:** 5.
13. YUNIS, J. J. 1981. Chromosomes and cancer: new nomenclature and future directions. Hum. Pathol. **12:** 494–503.
14. SWAIN, C. G. & C. B. SCOTT. 1953. Quantitative correlation of relative rates. Comparison of hydroxide ion with other nucleophilic reagents toward alkyl halides, esters, epoxides and acyl halides. J. Am. Chem. Soc. **75:** 141–147.
15. WOLFF, S. & B. RODIN. 1978. Saccharin-induced sister chromatid exchanges in Chinese hamster and human cells. Science **200:** 543–545.
16. GALLOWAY, S. M. & R. B. PAINTER. 1979. Vitamin C is positive in the DNA synthesis inhibition and sister-chromatid exchange tests. Mutat. Res. **60:** 321–327.
17. MALAVEILLE, C., M. FRIESEN, A.-M. CAMUS, L. GARREN, A. HAUTEFEUILLE, J.-C. BEREZIAT, P. GHADIRIAN, N. E. DAY & H. BARTSCH. 1982. Mutagens produced by the pyrolysis of opium and its alkaloids as possible risk factors in cancer of the oesophagus. Carcinogenesis **3:** 577–587.

18. THOMPSON, L. H., K. W. BROOKMAN, L. E. DILLEHAY, A. V. CARRANO, J. A. MAZRIMAS, C. L. MOONEY & J. L. MINKLER. 1982. A CHO-cell strain having hypersensitivity to mutagens, a defect in DNA strand–break repair, and an extraordinary baseline frequency of sister-chromatid exchange. Mutat. Res. **95:** 427–440.

19. SCOTT, D. 1977. Chromosome aberrations, DNA post-replication repair and lethality of tumour cells with a differential sensitivity to alkylating agents. *In* Chromosomes Today. A. de la Chapelle & M. Sorsa, Eds. **6:** 391–401. Elsevier/North-Holland Biomedical Press. Amsterdam, the Netherlands.

20. WOLFF, S. & W. F. MORGAN. 1982. Modulating factors in sister chromatid exchange induction by mutagenic carcinogens. *In* Sister Chromatid Exchange. S. Wolff, Ed.: 515–533. Alan Liss. New York, N.Y.

21. KAUFMANN, E. R. & R. L. DAVIDSON. 1979. Bromodeoxyuridine mutagenesis in mammalian cells is stimulated by purine deoxyribonucleosides. Somat. Cell Genet. **5:** 653–663.

22. PETERSON, A. R. & H. PETERSON. 1979. Facilitation by pyrimidine deoxyribonucleosides and hypoxanthine of mutagenic and cytotoxic effects of monofunctional alkylating agents in Chinese hamster cells. Mutat. Res. **61:** 319–331.

23. TOFILON, P. J., S. M. OREDSSON, D. F. DEEN & L. J. MARTON. 1982. Polyamine depletion influences drug-induced chromosomal damage. Science **217:** 1044–1046.

24. EVANS, H. J., G. J. NEARY & F. S. WILLIAMSON. 1959. The relative biological efficiency of single doses of fast neutrons and gamma rays in *Vicia faba* roots and the effect of oxygen. II. Chromosome damage: the production of micronuclei. Int. J. Radiat. Biol. **1:** 216–229.

25. HEDDLE, J. A., C. B. LUE, F. SAUNDERS & R. D. BENZ. 1978. Sensitivity to five mutagens in Fanconi's anemia as measured by the micronucleus method. Cancer Res. **38:** 2983–2988.

26. CARRANO, A. V., M. A. VAN DILLA & J. W. GRAY. 1979. Flow cytogenetics: a new approach to chromosome analysis. *In* Flow Cytometry and Sorting. M. R. Melemed, P. F. Mullaney & M. L. Mendelssohn, Eds.: 421–451. John Wiley. New York, N.Y.

SISTER CHROMATID EXCHANGE AS A TEST FOR MUTAGENIC CARCINOGENS*

Sheldon Wolff

*Laboratory of Radiobiology and Environmental Health
and Department of Anatomy
University of California
San Francisco, California 94143*

Beginning in the 1940s with the work of Karl Sax,[1] it became increasingly clear that if cells were irradiated, cytogenetic changes manifested as chromosome aberrations could be induced in a dose-related manner (for review, see Reference 2). The dose response became so well characterized that it was possible to use the induction of aberrations in the plant *Tradescantia* as a biological dosimeter to estimate the physical dose.[3] Subsequently, the number of chromosome aberrations induced in the long-lived peripheral lymphocytes of humans exposed to ionizing radiation was found to be an excellent indicator of the dose received by people in radiation accidents,[4,5] and currently is the method of choice to estimate exposure in cases where the actual physical dose is unknown. The method is also useful for the detection of exposure to clastogenic (chromosome-breaking) chemicals (for review, see Reference 6).

For most chemicals, which are S-dependent agents, however, another cytogenetic end point, the sister chromatid exchange (SCE), has proved to be a far more sensitive indicator of whether or not a cell has been exposed. SCEs are visualized in chromosomes from cells that have been treated so that the sister chromatids of a chromosome are physically[7] or chemically[8-10] different, and thus distinguishable from one another. If cells are exposed to chemicals at concentrations as low as 1/100th that necessary to produce chromosome aberrations, large numbers of SCEs can be seen,[11-13] i.e., the system is exquisitely sensitive. Furthermore, it is capable of detecting both direct-acting chemicals[13] and those that require metabolic activation, which can be accomplished *in vitro*[14,15] as well as *in vivo*[16,17] (for review, see Reference 18 and various chapters in Reference 19).

Because of its sensitivity and ease of scoring, the SCE test has been used increasingly as a short-term test for the detection of mutagenic carcinogens, and an excellent correspondence has been found between chemicals that induce SCEs and those that induce mutations, as determined by a battery of other short-term tests.[20] It should be noted, however, that the actual lesions in DNA that induce SCEs need not be the ones that induce mutations, for it has been shown that the ratio of SCEs to mutations varies among different chemicals that induce both end points.[21] Such results indicate either that some of the multitude of lesions induced in DNA can induce SCEs and some can induce mutations, or that only a small subset of the lesions can induce both SCEs and mutations. The induction of SCEs in human lymphocytes *in vivo* has also been used as a method to monitor the exposure of people to hazardous chemicals.

When the induction of SCEs is used as a test to determine whether or not a chemical agent is a potential mutagenic carcinogen, two philosophically different approaches are taken depending on the purpose of the test. If the test is to be used simply to see if a chemical could be a mutagenic carcinogen that might be dangerous to an exposed population, then a sensitive cell system is needed. This is analogous to the

*This work was supported by the U.S. Department of Energy.

142

TABLE 1

EFFECT OF SACCHARIN ON THE INDUCTION OF SCEs IN CHO CELLS*

Concentration (%)	Number of SCEs	Number of Chromosomes	SCEs/Cell
0	875	2027	8.75
0.1	953	2003	9.53
0.5	995	2027	9.95
1.0	1246	2028	12.46
5.0†			
0	845	1967	8.45
1.0	1294	1982	12.94
≧1.5†			
0	855	1947	8.55
0.5 (purified)	1021	1969	10.21
1.0 (purified)	1121	2006	11.21
≧1.5†			
0	872	1979	8.72
0.8	1105	1987	11.05
1.0	1196	1996	11.96
≧1.2†			

*One-hundred cells per point. (Table modified from Reference 26.)
†No growth.

development of special strains of *Salmonella* that are repair defective to enhance the ability of cells to mutate. In fact, some repair-defective cells have been shown to be more sensitive to the induction of SCEs than are normal cells.[22-24] Positive results in such tests are warnings that the chemical is potentially dangerous.

If, however, the purpose of the test is to obtain an approximation, or estimate, of the degree of hazard to man of a given chemical, then a cell system with a sensitivity of normal human cells should be used; most often, human peripheral lymphocytes are chosen.

As mentioned, the SCE test is exceedingly sensitive, and thus has proved its utility in the detection of cytogenetic effects of suspected carcinogens—such as saccharin, phorbol esters, and benzene—that are often missed in conventional tests for mutagenicity.

TABLE 2

EFFECT OF SACCHARIN ON THE INDUCTION OF SCEs IN CULTURED HUMAN LYMPHOCYTES*

Concentration (%)	Number of SCEs	Number of Chromosomes	SCEs/Cell
0	981	4588	9.81
0.1	1169	4592	11.69
0.5	1711	4594	17.11
≧1.0†			
0	950	4592	9.50
0.3	1311	4595	13.11
0.5	1607	4598	16.07
≧0.6†			
0.3 (purified)	1322	4590	13.22
0.5 (purified)	1745	4596	17.45
≧0.6†			

*One hundred cells per point. (Table modified from Reference 26.)
†No growth.

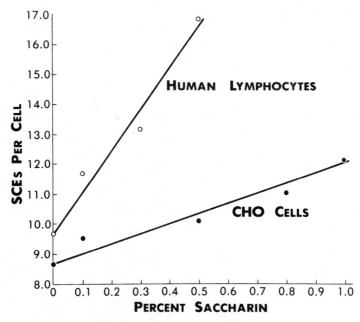

FIGURE 1. Induction of SCEs by saccharin. (Reprinted from Reference 26 with permission. Copyright 1978 by the American Association for the Advancement of Science.)

Saccharin, which has been reported to be a weak carcinogen that induces stomach cancer in rats exposed to high doses,[25] can reproducibly induce SCEs in Chinese hamster ovary (CHO) cells[26,27] and human lymphocytes.[26] When saccharin is administered to cells at high concentrations approaching cytotoxicity, a dose-related increase in SCEs can be found (TABLES 1 and 2; FIGURE 1). The doses necessary to see this effect, however, are so high that the compound at worst can only be considered a very weak genotoxin. Similar results have also been obtained with the artificial sweetener sodium cyclamate (and cyclohexylamine HCl) (TABLES 3 and 4). The irony of the situation is that compounds of one type, the cyclamates, have been banned, whereas the other has not.

With other weak compounds the results are not always so straightforward. Thus the tumor promoter 12-O-tetradecanoylphorbol-13-acetate (TPA) has been reported

TABLE 3

EFFECT OF SODIUM CYCLAMATE ON THE INDUCTION OF SCES IN CHO CELLS*

Concentration (%)	Number of SCEs	Number of Chromosomes	SCEs/Cell
0	2722	5936	9.07
0.25	2895	5940	9.65
0.50	3273	5963	10.41
0.75	3436	5897	11.45
≥1.0†			

*Three experiments; 100 cells per point per experiment.
†No growth.

TABLE 4

EFFECT OF SODIUM CYCLAMATE AND CYCLOHEXYLAMINE HYDROCHLORIDE ON THE
INDUCTION OF SCEs IN CULTURED HUMAN LYMPHOCYTES

Concentration	Number of Cells	Number of SCEs	Number of Chromosomes	SCEs/Cell
Sodium cyclamate (%)				
0	100	1040	4596	10.4
0.10	100	1018	4598	10.2
0.25	100	1068	4598	10.7
0.50	100	1229	4598	12.3
0.75	100	1640	4598	16.4
1.00	100	1722	4594	17.2
Cyclohexylamine HCl (M)				
0	200	1855	9182	9.3
10^{-4}	200	2293	9186	11.5
10^{-3}	100	1202	4593	12.0

either to induce SCEs[28–31] or to be inactive.[32–35] In a series of experiments carried out in our laboratory with both CHO cells and human fibroblasts, it was found that a 20–50% increase in SCEs[31] (TABLES 5 and 6) was always obtained. This increase, which was independent of the time at which the cells were fixed after treatment, was not due to an impurity present in the TPA, for when the stock solution (97% TPA) was purified by high-pressure liquid chromatography, the purified fractions still retained the activity.

TABLE 5

EFFECT OF TPA ON THE INDUCTION OF SCEs IN CHO CELLS*

Treatment	Number of Cells	Number of SCEs	SCEs per Cell (mean ± standard error)	Percent Increase
None	100	1094	10.9 ± 0.3	
1% Ethanol	100	1099	11.0 ± 0.3	
1 μg/ml TPA	100	1361	13.6 ± 0.4	24.8
None	50	474	9.5 ± 0.4	
1% Ethanol	50	440	8.8 ± 0.4	
1 μg/ml TPA	50	668	13.4 ± 0.5	52.3
None	50	518	10.4 ± 0.5	
1 mg/ml TPA	50	646	12.9 ± 0.5	24.5
10 mg/ml TPA	50	707	14.1 ± 0.5	35.6
100 mg/ml TPA	50	635	12.7 ± 0.5	22.1
1 μg/ml TPA	50	682	13.6 ± 0.5	30.8
None	50	476	9.5 ± 0.4	
1% Ethanol	50	477	9.5 ± 0.4	
1 mg/ml TPA	50	575	11.5 ± 0.5	21.1
10 mg/ml TPA	50	502	10.0 ± 0.4	5.3†
100 mg/ml TPA	50	574	11.5 ± 0.5	21.1
1 μg/ml TPA	50	593	11.4 ± 0.5	20.0
1% Ethanol	50	524	10.5 ± 0.5	
1 μg/ml TPA	50	679	13.6 ± 0.5	29.5

*Table modified from Reference 31.
†Not significantly different.

TABLE 6

EFFECT OF TPA ON THE INDUCTION OF SCES IN SV40-TRANSFORMED HUMAN FIBROBLASTS*

Treatment	Number of SCEs	Number of Chromo-somes	Number of Cells	SCEs/Cell (mean ± SE)	Percent Increase†
Normal cells (GM637)					
None	316	1256	20	15.8 ± 0.9	
1 μg/ml TPA	420	1346	20	21.0 ± 1.0	32.9
None	440	1614	25	17.6 ± 0.8	
1 μg/ml TPA	596	1640	25	23.8 ± 1.0	35.2
Xeroderma pigmentosum cells (XP12R0)					
None	356	1426	20	19.6 ± 1.0	
1 μg/ml TPA	466	1305	20	23.3 ± 1.1	18.9
None	393	1426	20	19.6 ± 1.0	
1 μg/ml TPA	499	1501	20	25.0 ± 1.1	27.6

*Table modified from Reference 31.
†All values were significantly different from respective controls by two-tailed Student's t-test (p < 0.005).

Since bromodeoxyuridine (BrdUrd), which is used to make the sister chromatids different from one another, can lead to the formation of SCEs,[36] other experiments were carried out to rule out the possibility that the increase was related to a difference in the amount of BrdUrd incorporated into DNA. When CHO cells were grown for 28 hours in medium supplemented with 10 μM BrdUrd, in the presence or absence of TPA, cesium chloride density gradient profiles of the DNA showed no differences in incorporation.

It should be emphasized that although TPA does bring about an increase in SCEs, the mechanism by which this occurs is still unknown. In addition to being independent of the factors just mentioned, the effect is related neither to TPA toxicity nor to the excision repair capability of the cells. Cells that were proficient (GM637), mildly deficient (CHO), or severely deficient (XP12RO) in excision repair[22,37] all had similar increases in SCE frequency after treatment with TPA. TPA does have a wide range of effects on cells, but the relation of these effects to SCE frequency is as yet unknown, as is the relation between SCE induction and the other effects of TPA, including tumor promotion.

On the other hand, the results obtained with some of the other agents that are detected by the SCE test transcend the results obtained with saccharin and TPA in that they often give new information and lead to biological insights. An example of this can be seen in experiments in which human lymphocytes were exposed to diethylstilbestrol (DES). DES is a synthetic estrogen that has been implicated in the induction of adenocarcinoma of the vagina in young women who had been exposed *in utero*.[38] Although the administration of DES is highly correlated with carcinogenicity, DES has proved to be negative in the *Salmonella* test.[39,40] Its metabolites, however, have been found to induce SCEs in cultured human fibroblasts.[41] When experiments were carried out in which human lymphocytes were exposed to various concentrations of DES (FIGURE 2), it was found that cells from pregnant women were more susceptible than were cells from men or postmenopausal women. Lymphocytes from nonpregnant, premenopausal women had an intermediate sensitivity. Furthermore, the results obtained from three different men were homogeneous, as were the results obtained from three postmenopausal women. The results obtained from pregnant women at all concentrations of DES, however, were heterogeneous, as were the results obtained at the highest concentrations of DES administered to cells of nonpregnant, premenopausal women. These results indicated that the response of cells to DES was dependent on the hormonal status of the individual. Furthermore, the effect seems to be the result of an intrinsic difference in the cells rather than a diffusible factor in women's blood that interacts with DES. This was shown in an experiment in which blood from a pregnant woman and blood from a man were cocultured. The mixed blood was exposed to DES for the duration of the culture period, and the Y chromosome–bearing cells from the man were scored separately from the XX cells of the woman. The results obtained from lymphocytes from pregnant women and from men were the same when grown in mixed culture as when grown separately (FIGURE 3).[42]

It has been claimed that DES is preferentially a clastogen,[43] i.e., that it only breaks

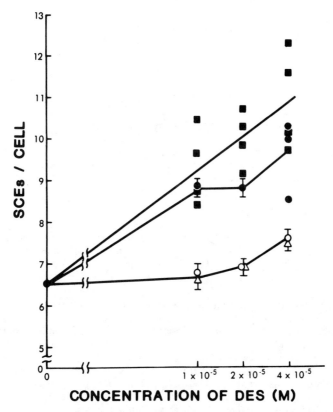

FIGURE 2. Induction of SCEs by DES in lymphocytes from four different pregnant women (squares), premenopausal women (filled circles), postmenopausal women (open circles), and men (triangles). When data from different individuals were homogeneous, they were pooled. Points with error bars represent the average from three different individuals ± standard error (SE). (Reprinted from Reference 42 by permission of the American Association for Cancer Research.)

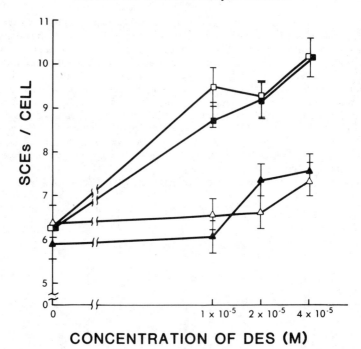

FIGURE 3. Induction of SCEs by DES in cocultured lymphocytes from a man and a pregnant woman. Lymphocytes from the pregnant woman (control) (filled squares), pregnant woman (cocultured) (open squares), man (control) (filled triangles), and man (cocultured) (open triangles). Bars, SE. (Reprinted from Reference 42 by permission of the American Association for Cancer Research.)

chromosomes. The results of Hill and Wolff, however, show that this is not so because DES induces SCEs,[42] which are not clastogenic events. Recent results have also shown that although DES does not induce mutations in the *Salmonella* system, it can induce mutations in yeast provided it is administered in the presence of strong oxidizing agents.[44]

Because men and postmenopausal women have significantly lower amounts of estrogens and progesterones than do pregnant and premenopausal women, and because DES induced only a small increase in the numbers of SCEs in lymphocytes from men and postmenopausal women, the data suggest that cells containing high levels of female hormones interact with DES to produce SCEs. The variability seen in individual responses of lymphocytes from premenopausal and pregnant women might then be related to the variability of estrogen and progesterone concentrations that is known to exist among women. Furthermore, because low SCE frequencies were seen in the presence of high testosterone levels as well as in the absence of testosterone, the data indicate that testosterone does not play a protective role.

Experiments on the induction of SCEs by the weak carcinogen benzene and its metabolites—phenol, hydroquinone, and catechol—have shown that although benzene itself and phenol do not induce SCEs in human lymphocytes, hydroquinone and catechol are very potent compounds in this respect.[45] If the cells are exposed to benzene in the presence of S-9 mix containing microsomes from polychlorinated biphenyl (Aroclor)–induced rat liver, to bring about metabolic activation, then benzene itself can induce SCEs. When similar experiments were carried out with both hydroquinone and catechol in the presence of S-9 mix, a somewhat greater increase was found than when these compounds were administered by themselves (FIGURE 4).[46] These results indicate that the phenolic metabolites of benzene can be enzymatically converted further to produce metabolites even more potent in inducing those DNA lesions that lead to SCE formation. The results obtained with SCE studies indicate that the formation of catechol and hydroquinone, and their further biotransformation, could be a likely cause of benzene's toxicity and presumably even of its leukemogenicity.

The same characteristics that have made the SCE an extremely valuable system for determining whether a compound is potentially genotoxic have made it attractive to think of using the observation of SCEs in cultured lymphocytes to monitor populations exposed to chronic low doses of hazardous compounds (for review, see Reference 47). Several sources of variability must still be understood before we will be certain about the utility of such a monitoring system.[48] One of the major problems has been that in general the SCE frequencies decrease with time after treatment,[47] as was originally found in animal experiments.[16] Studies on animals[49] and people[50] exposed to one potent chemical, ethylene oxide, however, indicate that this might not be a problem; the

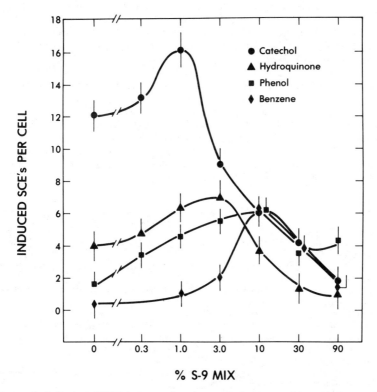

FIGURE 4. Induction of SCEs in human lymphocytes by benzene and its oxidative metabolites in the presence of a metabolic activation system. (Modified from Reference 46.)

induced lesions could be long-lived and lead to an increased yield of SCEs in peripheral blood lymphocytes long after exposure has ceased. When rabbits were exposed to ethylene oxide in exposure chambers, significant increases in SCEs were obtained at 50 and 250 parts per million. The yields stayed high for a period of about four months after exposure ceased.[49] Increases in SCEs have also been found in lymphocytes obtained from hospital workers exposed to ethylene oxide, which is used as a sterilant.[51,52] Pertinently, preliminary results have been reported in which a dose-related response was noted among industrial workers exposed to ethylene oxide.[50] In one heavily exposed group of workers, the yields remained high six months after exposure ceased. Such persistence is reminiscent of the persistence found for the induction of chromosome dicentrics in human lymphocytes by radiation and augurs well for the development of the SCE test as a method to monitor exposure of workers to potentially hazardous chemicals.

CONCLUSION

When cells are exposed to genotoxins either *in vitro* or *in vivo*, SCEs are readily induced. This induction can be used to detect direct-acting compounds as well as those requiring metabolic activation. The SCE test, which is exquisitely sensitive, can detect chromosomal effects of weak carcinogens such as saccharin, or other compounds, such as phorbol esters or DES, that are often missed in genetic tests. The system also can give biological insights. For instance, the response of human lymphocytes to DES is influenced by the cells' hormonal status, i.e., cells from pregnant women are more sensitive than are cells from men and postmenopausal women. Furthermore, although benzene can induce SCEs only when metabolically activated, its active metabolites hydroquinone and catechol are somewhat more effective after further metabolic activation.

In several places the *in vivo* induction of SCEs in peripheral blood lymphocytes is being tested as a possible method to monitor human exposure to toxic substances. Many sources of variability still need to be worked out before we will be certain about the utility of such a system. Its potential, however, is indicated by reports that a dose response has been noted in people exposed to ethylene oxide and, in some cases, that the effect has lasted for up to six months after exposure ceased.

REFERENCES

1. SAX, K. 1940. An analysis of X-ray-induced chromosomal aberrations in Tradescantia. Genetics **25:** 41–68.
2. LEA, D. E. 1946. Actions of Radiations on Living Cells. Cambridge University Press. Cambridge, England.
3. CONGER, A. D. 1954. Radiobiological studies with *Tradescantia* at nuclear test detonations. Am. Nat. **88:** 215–224.
4. GOOCH, P. C., M. BENDER & M. L. RANDOLPH. 1964. Chromosome aberrations induced in human somatic cells by neutrons. *In* Biological Effects of Neutron and Proton Irradiations: 325–342. International Atomic Energy Agency. Vienna, Austria.
5. BENDER, M. & P. C. GOOCH. 1967. Chromosome aberrations in irradiated humans. Excerpta Med. Int. Congr. Ser. No. 105: 1421–1425.
6. EVANS, H. J. & D. C. LLOYD, Eds. 1978. Mutagen-Induced Chromosome Damage in Man. Edinburgh University Press. Edinburgh, Scotland.
7. TAYLOR, J. H. 1958. Sister chromatid exchanges in tritium-labeled chromosomes. Genetics **43:** 515–529.

8. ZAKHAROV, A. F. & N. A. EGOLINA. 1972. Differential spiralization among mammalian mitotic chromosomes. I. BUdR-revealed differentiation in Chinese hamster chromosomes. Chromosoma **38**: 341–365.
9. LATT, S. A. 1973. Microfluorometric detection of deoxyribonucleic acid replication in human metaphase chromosomes. Proc. Nat. Acad. Sci. USA **70**: 3395–3399.
10. IKUSHIMA, T. & S. WOLFF. 1974. Sister chromatid exchanges induced by light flashes to 5-bromodeoxyuridine and 5-iododeoxyuridine substituted Chinese hamster chromosomes. Exp. Cell Res. **87**: 15–19.
11. LATT, S. A. 1974. Sister chromatid exchanges, indices of human chromosome damage and repair: detection by fluorescence and induction by mitomycin C. Proc. Nat. Acad. Sci. USA **71**: 3162–3166.
12. SOLOMON, E. & M. BOBROW. 1975. Sister chromatid exchanges—a sensitive assay of agents damaging human chromosomes. Mutat. Res. **30**: 273–278.
13. PERRY, P. & H. J. EVANS. 1975. Cytological detection of mutagen-carcinogen exposure by sister chromatid exchange. Nature London **258**: 121–125.
14. STETKA, D. G. & S. WOLFF. 1976. Sister chromatid exchange as an assay for genetic damage induced by mutagen-carcinogens. II. In vitro test for compounds requiring metabolic activation. Mutat. Res. **41**: 343–350.
15. NATARAJAN, A. T., A. D. TATES, P. P. W. VAN BUUL, M. MEIJERS & N. DE VOGEL. 1976. Cytogenetic effects of mutagens/carcinogens after activation in a microsomal system in vitro. I. Induction of chromosome aberrations and sister chromatid exchanges by diethylnitrosamine (DEN) and dimethylnitrosamine (DMN) in CHO cells in the presence of rat-liver microsomes. Mutat. Res. **37**: 83–90.
16. STETKA, D. G. & S. WOLFF. 1976. Sister chromatid exchange as an assay for genetic damage induced by mutagen-carcinogens. I. In vivo test for compounds requiring metabolic activation. Mutat. Res. **41**: 333–342.
17. ALLEN, J. W. & S. A. LATT. 1976. In vivo BrdU-33258 Hoechst analysis of DNA replication kinetics and sister chromatid exchange formation in mouse somatic and meiotic cells. Chromosoma **58**: 325–340.
18. WOLFF, S. 1977. Sister chromatid exchange. Annu. Rev. Genet. **11**: 183–201.
19. WOLFF, S., Ed. 1982. Sister Chromatid Exchange. John Wiley & Sons. New York, N.Y.
20. Report of the Committee on Chemical Environmental Mutagens. 1982. The Identification and Impact of Chemical Mutagens. National Academy of Sciences, NAS Press. Washington, D.C.
21. CARRANO, A. V., L. H. THOMPSON, P. A. LINDL & J. L. MINKLER. 1978. Sister chromatid exchange as an indicator of mutagenesis. Nature **271**: 551–553.
22. WOLFF, S., B. RODIN & J. E. CLEAVER. 1977. Sister chromatid exchanges induced by mutagenic carcinogens in normal and xeroderma pigmentosum cells. Nature **265**: 347–349.
23. PERRY, P. E., M. JAGER & H. J. EVANS. 1978. Mutagen-induced sister chromatid exchanges in xeroderma pigmentosum and normal lymphocytes. In Mutagen-Induced Chromosome Damage in Man. H. J. Evans & D. C. Lloyd, Eds.: 201–207. Edinburgh University Press. Edinburgh, Scotland.
24. WENT, M., A. SZORENYI, A. POLAY & N. SIMON. 1981. Investigation of the sister chromatid exchange (SCE) frequency in one patient with xeroderma pigmentosum. Arch. Dermatol. Res. **271**: 407–409.
25. ARNOLD, D. L., C. A. MOODIE, C. H. GRICE, S. M. CHARBONNEAU, B. STAVRIC, B. F. COLLINS, P. F. MCGUIRE, Z. Z. CZAWIDZKA & I. C. MUNRO. 1980. Long-term toxicity of orthotoluene sulfonamide and sodium saccharin in the rat. Toxicol. Appl. Pharmacol. **52**: 113–152.
26. WOLFF, S. & B. RODIN. 1978. Saccharin-induced sister chromatid exchanges in Chinese hamster and human cells. Science **200**: 543–545.
27. ABE, S. & M. SASAKI. 1977. Chromosome aberrations and sister chromatid exchanges in Chinese hamster cells exposed to various chemicals. J. Nat. Cancer Inst. **58**: 1635–1641.
28. KINSELLA, A. R. & M. RADMAN. 1978. Tumor promoter induces sister chromatid exchanges: relevance to mechanisms of carcinogenesis. Proc. Nat. Acad. Sci. USA **75**: 6149–6153.

29. NAGASAWA, H. & J. B. LITTLE. 1979. Effect of tumor promoters, protease inhibitors, and repair processes on x-ray-induced sister chromatid exchanges in mouse cells. Proc. Nat. Acad. Sci. USA **76:** 1943–1947.

30. GENTIL, A., G. RENAULT & A. MARGOT. 1980. The effect of the tumour promoter 12-O-tetradecanoyl-phorbol-13-acetate (TPA) on UV- and MNNG-induced sister chromatid exchanges in mammalian cells. Int. J. Cancer **26:** 517–521.

31. SCHWARTZ, J. L., M. J. BANDA & S. WOLFF. 1982. 12-O-Tetradecanoylphorbol-13-acetate (TPA) induces sister-chromatid exchanges and delays in cell progression in Chinese hamster ovary and human cell lines. Mutat. Res. **92:** 393–409.

32. LOVEDAY, K. S. & S. A. LATT. 1979. The effect of tumor promoter, 12-O-tetradecanoyl-phorbol-13-acetate (TPA), on sister-chromatid exchange formation in cultured Chinese hamster cells. Mutat. Res. **67:** 343–348.

33. CARRANO, A. V., L. H. THOMPSON, D. G. STETKA, J. L. MINKLER, J. A. MAZRIMAS & S. FONG. 1979. DNA crosslinking, sister-chromatid exchange and specific-locus mutations. Mutat. Res. **63:** 175–188.

34. FUJIWARA, Y., Y. KANO, M. TATSUMI & P. PAUL. 1980. Effects of a tumor promoter and an anti-promoter on spontaneous and UV-induced 6-thioguanine-resistant mutations and sister-chromatid exchanges in V79 Chinese hamster cells. Mutat. Res. **71:** 243–251.

35. POPESCU, N. C., S. C. AMSBAUGH & J. A. DIPAOLO. 1980. Enhancement of N-methyl-N'-nitro-N-nitrosoguanidine transformation of Syrian hamster cells by a phorbol diester is independent of sister chromatid exchanges and chromosome aberrations. Proc. Nat. Acad. Sci. USA **77:** 7282–7286.

36. WOLFF, S. & P. PERRY. 1974. Differential Giemsa staining of sister chromatids and the study of sister chromatid exchanges without autoradiography. Chromosoma **48:** 341–353.

37. GOTH-GOLDSTEIN, R. 1977. Repair of DNA damaged by alkylating carcinogens is defective in xeroderma pigmentosum–derived fibroblasts. Nature **267:** 81–82.

38. GREENWALD, P., P. C. NASCA, W. S. BURNETT & A. POLAN. 1973. Prenatal stilbestrol experience of mothers of young cancer patients. Cancer **31:** 568–572.

39. MCCANN, J., E. CHOI, E. YAMASAKI & B. N. AMES. 1975. Detection of carcinogens as mutagens in the *Salmonella*/microsome test: assay of 300 chemicals. Proc. Nat. Acad. Sci. USA **72:** 5135–5139.

40. GLATT, H. R., M. METZLER & F. OESCH. 1979. Diethylstilbestrol and 11 derivatives: a mutagenicity study with *Salmonella typhimurium*. Mutat. Res. **67:** 113–121.

41. RUDIGER, H. W., F. HAENISCH, M. METZLER, F. OESCH & H. R. GLATT. 1979. Metabolites of diethylstilbestrol induce sister chromatid exchange in human cultured fibroblasts. Nature **281:** 392–394.

42. HILL, A. & S. WOLFF. 1982. Increased induction of sister chromatid exchange by diethylstilbestrol in lymphocytes from pregnant and premenopausal women. Cancer Res. **42:** 893–896.

43. ASHBY, J. 1982. Screening chemicals for mutagenicity: practices and pitfalls. *In* Mutagenicity: New Horizons in Genetic Toxicology. J. A. Heddle, Ed.: 1–33. Academic Press, Inc. New York, N.Y.

44. MEHTA, R. D. & R. C. VON BORSTEL. 1982. Genetic activity of diethylstilbestrol in *Saccharomyces cerevisiae*. Enhancement of mutagenicity by oxidizing agents. Mutat. Res. **92:** 49–61.

45. MORIMOTO, K. & S. WOLFF. 1980. Increase of sister chromatid exchanges and perturbations of cell division kinetics in human lymphocytes by benzene metabolites. Cancer Res. **40:** 1189–1193.

46. MORIMOTO, K., S. WOLFF & A. KOIZUMI. 1983. Induction of sister chromatid exchanges in human lymphocytes by microsomal activation of benzene metabolites. Mutat. Res. Lett. (In press.)

47. LAMBERT, B., A. LINDBLAD, K. HOLMBERG & D. FRANCESCONI. 1982. The use of sister chromatid exchange to monitor human populations for exposure to toxicologically harmful agents. *In* Sister Chromatid Exchange. S. Wolff, Ed.: 149–182. John Wiley & Sons. New York, N.Y.

48. WOLFF, S. 1981. Cytogenetic analysis at chemical disposal sites: problems and prospects. *In* Assessment of Health Effects at Chemical Disposal Sites. W. W. Lowrance, Ed.: 61–80. Rockefeller University. New York, N.Y.
49. YAGER, J. W. & R. D. BENZ. 1982. Sister chromatid exchanges induced in rabbit lymphocytes by ethylene oxide after inhalation exposure. Environ. Mutagen. **4:** 121–134.
50. Johnson & Johnson Co. 1982. Preliminary results from a pilot research chromosome study. *In* Current Report, Occupational Safety, Health Reporter. Bureau of National Affairs. Washington, D.C.
51. GARRY, V. F., J. HOZIER, D. JACOBS, R. L. WADE & D. G. GRAY. 1979. Ethylene oxide: evidence of human chromosomal effects. Environ. Mutagen. **1:** 375–382.
52. YAGER, J. W., C. J. HINES & R. C. SPEAR. 1982. Exposure to ethylene oxide at work increases sister chromatid exchanges in human peripheral lymphocytes. Science. (In press.)

THE Salmonella MUTAGENICITY ASSAY: PROMISES AND PROBLEMS

Genetic Toxicology Branch
Food and Drug Administration
Washington, D.C. 20204

DEVELOPMENT OF THE AMES Salmonella TEST

The ability of certain chemicals to cause mutations in bacteria has been known for at least 30 years. It was only in the 1970s, however, that chemical mutagenesis in bacteria changed from an interesting and useful scientific tool into a subject of concern to those interested in the safety of chemicals.

This transition was the direct result of the development of techniques for metabolizing chemical mutagens to their active form. As it was becoming apparent that many chemical carcinogens require metabolic "activation" to exert their carcinogenic effect, Heinrich Malling, then working at the Oak Ridge National Laboratory, demonstrated that such activation was required for mutagens as well. By using the nonenzymatic Udenfriend hydroxylation system, Malling was able to convert the carcinogens dimethylnitrosamine (DMN) and diethylnitrosamine into mutagens for the fungus *Neurospora crassa*.[1] These nitrosamines were not, in themselves, mutagenic to *Neurospora* or to other microorganisms. At about the same time, David Smith, then at the National Institutes of Health, enzymatically hydrolyzed the carcinogenic plant glycoside cycasin, purified the resulting aglycone, methylazoxymethanol, and showed that this compound is mutagenic to the bacteria *Salmonella typhimurium*.[2]

In searching for a more general mammalian enzyme-mediated method for activating chemical mutagens, Michael Gabridge and Marvin Legator, then at the Food and Drug Administration, developed the "host-mediated assay."[3] In this assay, bacteria were injected into the intraperitoneal cavity of a mouse and DMN was given intramuscularly to the same mouse. The bacteria were later removed from the peritoneal cavity, and mutant cells were counted. It was found that DMN passing through the tissues of the mouse had, in fact, been metabolized to a chemical that could cause mutations in the bacteria. Although this host-mediated assay did not prove to be useful in detecting a wide range of chemical mutagens or carcinogens,[4] it did serve to demonstrate the importance of linking mutagen-sensitive bacterial cells with a mammalian metabolic activation system.

The key to making bacterial mutagenesis tests useful was to mix the bacterial cells *in vitro* with an enzymatic activating system based on mammalian liver. Malling was the first investigator to demonstrate that a mouse liver extract, rather than a whole living mouse, was sufficient to activate DMN to a bacterial mutagen.[5] Bruce Ames at Berkeley had, meanwhile, been working on developing a set of bacterial strains that are particularly sensitive to chemical mutagens. Ames and his coworkers adapted the *in vitro* activation method developed by Malling into a rapid protocol now referred to as the "Ames test."[6,7]

The Ames test utilizes bacteria (*Salmonella typhimurium*) that are unable to synthesize the amino acid histidine and therefore require exogenous histidine for growth. In this test, about 10^8 of these bacteria and the test chemical are mixed

154

0077-8923/83/0407-0154 $01.75/0 © 1983, NYAS

together in the presence of rat liver S-9 (the 9,000 × g supernatant from homogenized liver) and appropriate cofactors in a test tube containing molten soft agar. The mixture is then plated on petri plates containing agar medium deficient in histidine. After two days of incubation at 37°C, the colonies on the plates are counted. Only those bacteria that have had mutations that result in their regaining the ability to synthesize histidine will grow into visible colonies. Thus, if a plate treated with the test chemical contains more colonies than does a control plate without test chemical, the chemical is a mutagen.

USES OF THE AMES TEST

In general, the Ames *Salmonella* assay has been applied to chemicals for two distinct safety evaluation purposes. The first is to screen chemicals to determine whether or not they may be capable of inducing heritable mutations in humans. The second use of the Ames test is to screen chemicals for potential carcinogenicity.

A number of legitimate criticisms can be made against using a test such as this for these two purposes. One of the criticisms is that the *in vitro* metabolic activation system used does not accurately reproduce *in vivo* mammalian metabolism. Thus, certain chemicals may not be activated to mutagens in the Ames test even though they may be carcinogenic or mutagenic to humans or other mammals. The breaking of liver cells by homogenization, use of liver as the only organ for preparing the metabolic activation system, selection of a particular level of S-9, standardization of cofactors and the fixing of cofactor concentrations, use of aerobic rather than anaerobic incubation conditions, and other particular conditions in the assay may prevent detection of some carcinogens or mammalian mutagens.

Another criticism of bacterial tests is that bacteria are so different from mammals that the information obtained is not relevant to safety evaluation. In general, the fact that the genetic material, DNA, is chemically similar in all organisms, including bacteria and humans, means that a chemical that attacks bacterial DNA is also likely to attack human DNA. However, there would appear to be validity to the argument that some chemicals may affect humans by attacking the chromosomal structures to which the DNA is bound rather than by attacking the DNA itself. Since bacterial DNA is not arranged on chromosomes as is that of higher organisms, it is possible that chemicals that cause exclusively chromosomal changes could be missed in bacterial mutagenesis assay systems.

Similarly, it has been argued that the Ames *Salmonella* assay should fail to detect some mutagens because it is an assay for reverse, rather than for forward, mutations. Since the mutation being assayed is one that changes a nonfunctional gene product into a functional one, there should be a certain degree of specificity in the types of genetic events that will be detected in this assay. Forward mutational assays that detect changes from functional gene products to nonfunctional ones would be expected to be more general in nature and thus to detect a wider spectrum of mutagens. Although this argument is plausible, there is no evidence to indicate that forward mutational assays in bacteria are more sensitive or more general in detecting mutagens than is the Ames assay.[8]

In spite of all the theoretical arguments that can be raised against the Ames test, it remains the most widely used method for screening chemicals for potential carcinogenicity. There are several reasons for this. First, there have been several independent large-scale studies on the correlation between mutagenicity in *Salmonella* and carcinogenicity in mammals.[9-14] In addition, the Ames test is relatively rapid, simple, and inexpensive to perform. Its rapidity, sensitivity, and low cost have made this test a useful tool in more basic scientific research as well as in toxicological evaluations of chemicals. One such use, for example, has been in the identification of biologically active metabolites of mutagens and carcinogens.

In the large studies that have been published on this subject, high levels of correlation have been reported between mutagenicity and carcinogenicity. It appears, however, that the numerical correlations arrived at may be biased on two counts: (1) the proportion of all carcinogens that are actually mutagens is likely to be less than the percentages indicated in the studies, and (2) the proportion of mutagens that are carcinogens is likely to exceed those reported. That is, the Ames test probably has, in actuality, more false negatives and fewer false positives than one would conclude it has from the available published studies.

To understand why the published correlations are probably numerically inaccurate, it is necessary to consider how the chemicals used in correlation studies are usually selected. In general, a large number of chemicals generally accepted to be carcinogenic are identified. These chemicals tend, of course, to fall into one of the major classes of known chemical carcinogens: the aromatic amines and nitro compounds, the N-nitrosamines, the polycyclic hydrocarbons, and the direct-acting alkylating agents. Then a series of noncarcinogens is selected, usually containing as many chemicals as possible that are closely related in structure to the carcinogens being assayed. Thus, aromatic amines and nitro compounds, N-nitroso compounds, and polycyclic hydrocarbons tend to be overrepresented in the "noncarcinogen" class in these studies.

This method of selection introduces two types of bias into the study. First, the Ames test is good for detecting aromatic amines and nitro compounds, polycyclic hydrocarbons, and direct alkylating agents as mutagens. If more carcinogens outside of these classes were tested, then the predictive value of positive mutagenicity test results would probably appear to fall. For example, the Ames test is not capable, in general, of detecting highly chlorinated organic carcinogens or hormonally active carcinogens[15] as mutagens. Thus, it appears likely that as more carcinogens outside of the well-known chemical classes of carcinogens are tested in the Ames test, the percentage of "false negatives" will increase.

In looking at the problem of "false positives," we must consider the method of selection of "noncarcinogens" in the correlation studies. Since this class of compounds tends to include chemicals that are closely related in chemical structure to known carcinogens, positive results on some of these chemicals may, in fact, not be "false." This is because the so-called noncarcinogens may be contaminated with enough of a structurally related carcinogen to be active in the *Salmonella* assay.[16] Also, the chemical under test is more likely than a randomly selected chemical to be misclassified as a noncarcinogen; it may simply be too weak a carcinogen to be detected in the whole-animal bioassays, which are relatively insensitive and often difficult to evaluate. Therefore, the testing of a truly random selection of noncarcinogens is likely to result in a lower rate of "false positives" in the Ames test than has been indicated by published correlation studies.

There are a number of "false negatives" and even some "false positives" in the *Salmonella* assay that have been shown to arise because of deficiencies in the test protocol used. Modification of the protocol, sometimes in ways that could be selected based upon the structure of the test chemical, can rectify "false" results in some cases. Appreciation and understanding of the cases in which a standard test protocol fails to give correct results may assist us in interpreting results for chemicals for which the "correct" result is unknown.

PROTOCOL MODIFICATIONS

In discussions of the use of mutagenicity testing for predicting the carcinogenicity of chemicals, the case of the food additive furylfuramide, or AF-2, is often cited. This nitrofuran preservative was widely used in many popular foods in Japan. By 1974 it had been found to be genetically active in bacteria,[17–19] *Neurospora* and yeast,[20] silkworms,[21] and human lymphocytes in culture.[22] At that time, however, AF-2 was nonmutagenic in the Ames *Salmonella* assay although it was positive in the closely related bacterial species *Escherichia coli*. This false-negative result in the *Salmonella* assay was consistent with the results of chronic feeding studies in rats and mice in which AF-2 had been found to be noncarcinogenic at levels up to 0.2% in the diet.[23] The mutagenicity of AF-2 in a variety of test systems prompted a reevaluation of the negative findings in both laboratory rodents and in the *Salmonella* assay. Subsequent studies in laboratory animals confirmed that AF-2 is carcinogenic.[24–28] The *Salmonella* assay had to be modified by the introduction of the plasmid pKM101 into the tester strains[29] before AF-2 and other carcinogenic nitrofurans could be shown to be mutagenic in this test.[9,30] Thus, in the case of AF-2, the use of mutagenicity tests other than the *Salmonella* assay was required in order to bring enough attention to this compound for its safety to be questioned. It should be pointed out, however, that the chemical structure of AF-2 made it a highly suspect chemical, even in the absence of mutagenicity data, because many nitrofuran compounds are carcinogenic.[31]

The flame retardant tris(2,3-dibromopropyl) phosphate, which was widely used in children's sleepwear in the United States, is another example of a chemical for which a demonstration of mutagenicity[32,33] correctly predicted its carcinogenicity.[34,35] This compound was mutagenic in the standard Ames *Salmonella* mutagenicity assay. However, a closely related chemical that has been used as a substitute for the carcinogenic flame retardant was negative in the standard assay with Aroclor 1254–induced rat liver S-9 in the metabolic activation system.[32] This compound, tris(1,3-dichloro-2-propyl) phosphate, or "tris-CP" (commercial name: Fyrol FR-2), was positive in the Ames *Salmonella* assay when rats or mice induced with phenobarbital were used for the S-9 rather than the standard Aroclor 1254–induced rats.[36,37]

Because tris-CP has not been tested for carcinogenicity, it is unclear whether this compound is a "false negative" with Aroclor 1254–induced rat S-9 or a "false positive" with phenobarbital-induced S-9. The results, however, do demonstrate that apparently minor changes in the test protocol can make the difference between a positive and a negative result.

The problem of establishing a quantitative correlation between mutagenicity and carcinogenicity can be complicated by difficulties in evaluating whole-animal chronic carcinogenicity tests. This is illustrated by results of testing the phenylenediamines. Sontag recently reevaluated the carcinogenicity data for the phenylenediamines that have been tested by the National Cancer Institute's bioassay program.[38] All of those that have been tested for mutagenicity in *Salmonella* have been reported to be positive. Of the nine reported mutagens for which detailed tumor-incidence data were available, Sontag concluded that five were carcinogenic. However, for four of the remaining mutagens, which the National Cancer Institute had not reported as carcinogenic, Sontag concluded that there were possible treatment-related responses in at least one sex of one species. Thus, by Sontag's interpretation of the carcinogenicity data, these four compounds (2,5-diaminotoluene, 2,6-diaminotoluene, *o*-nitrophenylenediamine, and *p*-phenylenediamine) are not necessarily "false positives" in the Ames assay. The difficulties inherent in interpreting carcinogenicity bioassays in rodents, as illustrated by results obtained with these phenylenediamines, make evaluation of the correlation between carcinogenicity and mutagenicity particularly difficult. For example, Sontag considered as suspicious an increase in tumors of types that were induced by one or more of the carcinogenic phenylenediamines, even if the increase was not statistically significant. Considerations such as this can raise questions about negative conclusions drawn on some whole-animal studies.

One of the phenylenediamines reported by the National Cancer Institute to be noncarcinogenic was p-phenylenediamine,[39] which has been reported to be mutagenic in the *Salmonella* assay.[40,41] However, this apparently "false-positive" mutagen is, in fact, nonmutagenic in the *Salmonella* assay if it is dissolved in water rather than in the commonly used solvent dimethylsulfoxide (DMSO). Burnett and his colleagues at the Clairol Research Laboratories found that p-phenylenediamine becomes mutagenic when dissolved in DMSO and tested several hours later.[42] Thus, of the four "false-positive" results for the phenylenediamines evaluated by Sontag, three may be due to inadequacies in the whole-animal carcinogenicity test while one is due to an inappropriate choice of solvent in the mutagenicity test.

This conclusion is in contrast with that of Golberg, who reported that five mutagenic phenylenediamines used as hair-dye ingredients were false positives in *Salmonella*.[43] His findings of noncarcinogenicity for these chemicals were based upon the conclusions drawn in the original research reports. Of the five false-positive compounds reported by Golberg, Sontag found that 2,5-diaminotoluene and 4-nitro-o-phenylenediamine resulted in tumor increases that may have been treatment related but were not statistically significant. The carcinogenicity of N-phenyl-p-phenylenediamine is not discussed by Sontag, but the *Salmonella* mutagenicity data cited by Golberg for this chemical are only marginally positive at best.[44] Sontag found that detailed tumor incidence data were not available for reanalysis of the National Cancer Institute–sponsored carcinogenicity study on Golberg's fourth *Salmonella* false positive, m-phenylenediamine, while the fifth false positive, p-phenylenediamine, is actually nonmutagenic to *Salmonella,* as discussed above. The discrepant conclusions that can be drawn concerning the mutagenicity and carcinogenicity of phenylenediamines demonstrate how different methods of evaluating test data can alter the apparent correlation between carcinogenicity and mutagenicity.

The "false-negative" result for the carcinogen DMN in the standard Ames *Salmonella* assay shows the importance of minor protocol modifications as well as solvent effects. Yahagi and her colleagues developed a modification of the standard Ames protocol called the "preincubation assay."[45] In the preincubation assay, the test chemical, S-9 mix, and bacteria are incubated for 20–30 minutes in a test tube before the molten soft agar is added and the mixture poured on the selective medium in the petri plate. With preincubation, rather than the usual procedure of mixing all the ingredients in the molten agar and pouring the mixture immediately, DMN is mutagenic in *Salmonella*. Changing the order of addition of ingredients and delaying the addition of agar appear to be effective because the test chemical, the S-9 mix, and the bacteria are all about four times more concentrated during the preincubation step than they are when added to the molten agar in the standard Ames protocol.[46]

DMN can also be converted from a "false negative" to a "true positive" by a different minor modification of the standard Ames protocol. If S-9 derived from Syrian golden hamster liver is used rather than that from rat liver, the mutagenicity of DMN can be demonstrated without any other modification of the Ames procedure.[46] The negative finding with DMN in the presence of rat liver S-9 appears to be due, at least in part, to the presence of some substance in the rat S-9 that interferes with activation of DMN to a mutagen.[47] This was shown by the fact that a mixture of rat S-9 or microsomes with hamster S-9 failed to activate DMN to a mutagen, although hamster S-9 alone activated it well. This same inhibitor was found in mouse liver S-9 and microsomes. Rat S-9 also inhibited the mutagenicity of the carcinogenic aromatic amine benzidine in the presence of hamster S-9.[47] However, no such inhibitor in rat S-9 was found for phenacetin, which was also mutagenic when tested with hamster S-9 but not with rat S-9.[48]

DMN can also be used to demonstrate one of the hazards of using DMSO as a solvent. Both the preincubation assay with rat liver S-9 and the standard Ames assay with hamster liver S-9 fail to show that DMN is mutagenic when DMSO is used as the solvent.[49,50] This is probably due to the close structural similarity between DMSO and DMN, as shown by the fact that DMSO is a *competitive* inhibitor of DMN demethylase,[50] which is thought by many investigators to be the sole enzyme involved in the activation of DMN. Thus, DMSO probably interferes with DMN mutagenesis by acting as a structural analogue of DMN at the active site of the enzyme required to metabolize it to a mutagen.

In some cases, consideration of the structure of test chemicals immediately suggests the type of modification of test protocol that may be appropriate. For example, many flavonols that are found in plants exist as glycosides. Upon ingestion by a human or other animal, these glycosides will be hydrolyzed by intestinal microflora. However, this hydrolytic reaction does not occur to any significant extent in the standard Ames *Salmonella* protocol. Therefore, in testing such compounds, glycosidases can be added to the test system. These glycosidases may be derived from any of a variety of sources, such as the mold *Aspergillus niger*,[51] the snail *Helix pomatia*,[52] the cecal content of rats,[52] or even human feces.[53] Under these conditions, rutin, which is a glycoside of the mutagenic flavonol quercetin, can be shown to be mutagenic. The published data concerning the carcinogenicity of quercetin in rodents are conflicting.[54-57] It is unclear whether quercetin is a "false positive" in the standard Ames *Salmonella* assay or if rutin is a "false negative" that becomes a "true positive" upon addition of glycosidase.

Thus, the standard metabolic activation system used in the *Salmonella* assay and many other mutagenicity tests fails to mimic certain metabolic pathways that exist in the gastrointestinal tract. Another example of this would be the reduction of azo compounds. Because many azo compounds can be reduced in the intestine to give rise to carcinogenic and mutagenic aromatic amines, Sugimura and his colleagues proposed that the *Salmonella* assay be modified to permit such azo reduction to occur.[58] They made several modifications to accomplish this, including the use of a preincubation step described previously and the addition of riboflavin to the S-9 mix as a reducing agent. Under these conditions, certain azo compounds that were negative in the standard *Salmonella* protocol were reported to be mutagenic. For at least some azo dyes, detection as mutagens also requires the use of hamster liver S-9 rather than rat liver S-9.[59]

CONCLUSIONS

The Ames *Salmonella* assay has been a useful tool in the safety evaluation of chemicals to which humans are exposed. The speed and low cost of this test, combined with its apparent correlation with whole-animal carcinogenicity tests, have resulted in the application of the assay to a very large number of chemicals, certainly numbering in the thousands. Our confidence in relying on the results of such large-scale screening is, however, diminished by a number of factors. First, the ability of the *Salmonella* assay to predict carcinogenicity with accuracy has been most clearly demonstrated for a relatively small number of chemical classes, i.e., those classes that contain many known carcinogens. Outside of these classes, negative results may not be an adequate indication of noncarcinogenicity. Difficulties in interpreting the results of chronic whole-animal carcinogenicity tests make the correlation of mutagenicity and carcinogenicity even more uncertain. Some types of chemicals, such as glycosides and certain types of azo compounds, require specific modifications of the standard protocol. Other chemicals, such as DMN and phenacetin, can only be shown to be mutagenic if the metabolic activation system is modified by using liver from species other than the rat. The use of a particular solvent, such as DMSO, can change a nonmutagen such as p-phenylenediamine into a "false positive" or, conversely, a carcinogenic mutagen such as DMN into a "false negative."

These problems serve to underscore the necessity of evaluating the mutagenicity of any chemical in a variety of test systems. This is particularly important when there is any suspicion concerning the carcinogenicity or mutagenicity of a chemical due to its structure. Unfortunately, in reality, most of the chemicals that have been tested in the *Salmonella* assay and found to be negative will never be tested in other mutagenicity assays. This is a serious problem that can easily lead to a false sense of security, particularly among decision makers who are not aware of the shortcomings of using only one assay to evaluate mutagenicity. Since a number of the limitations of the *Salmonella* assay are due to the nature of the *in vitro* metabolic activation system, there is a danger inherent in applying more than one test to a chemical when all the tests applied use similar activation systems. We should strive to use tests that not only have different types of end points, but also have different types of systems for activating the test agents.

REFERENCES

1. MALLING, H. V. 1966. Mutagenicity of two potent carcinogens, dimethylnitrosamine and diethylnitrosamine, in *Neurospora crassa*. Mutat. Res. **3:** 537–540.
2. SMITH, D. W. E. 1966. Mutagenicity of cycasin aglycone (methylazoxymethanol), a naturally occurring carcinogen. Science **192:** 1273–1274.
3. GABRIDGE, M. G. & M. S. LEGATOR. 1969. A host-mediated microbial assay for the detection of mutagenic compounds. Proc. Soc. Exp. Biol. Med. **130:** 831–834.
4. SIMMON, V. F., H. S. ROSENKRANZ, E. ZEIGER & L. POIRIER. 1979. Mutagenic activity of chemical carcinogens and related compounds in the intraperitoneal host-mediated assay. J. Nat. Cancer Inst. **62:** 911–918.
5. MALLING, H. V. 1971. Dimethylnitrosamine: formation of mutagenic compounds by interaction with mouse liver microsomes. Mutat. Res. **13:** 425–429.
6. AMES, B. N., W. E. DURSTON, E. YAMASAKI & F. D. LEE. 1973. Carcinogens are mutagens: a simple test system combining liver homogenates for activation and bacteria for detection. Proc. Nat. Acad. Sci. USA **70:** 2281–2285.
7. AMES, B. N., J. MCCANN & E. YAMASAKI. 1975. Methods for detecting carcinogens and mutagens with the Salmonella/mammalian-microsome mutagenicity test. Mutat. Res. **31:** 347–364.
8. SKOPEK, T. R., H. L. LIBER, D. A. KADEN & W. G. THILLY. 1978. Relative sensitivities of forward and reverse mutation assays in *Salmonella typhimurium*. Proc. Nat. Acad. Sci. USA **75:** 4465–4469.
9. MCCANN, J., E. CHOI, E. YAMASAKI & B. N. AMES. 1975. Detection of carcinogens as mutagens in the *Salmonella*/microsome test: assay of 300 chemicals. Proc. Nat. Acad. Sci. USA **72:** 5135–5139.
10. MCCANN, J. & B. N. AMES. 1976. Detection of carcinogens as mutagens in the *Salmonella*/microsome test: assay of 300 chemicals: discussion. Proc. Nat. Acad. Sci. USA **73:** 950–954.
11. COMMONER, B., A. J. VITHAYATHIL, J. I. HENRY, J. C. GOLD & M. J. REDING. 1976. Reliability of Bacterial Mutagenesis Techniques to Distinguish Carcinogenic and Noncar-

cinogenic Chemicals. U.S. Environmental Protection Agency Report EPA-600/1-76-022. National Technical Information Service. Springfield, Va.

12. SUGIMURA, T., S. SATO, M. NAGAO, T. YAHAGI, T. MATSUSHIMA, Y. SEINO, M. TAKEUCHI & T. KAWACHI. 1976. Overlapping of carcinogens and mutagens. *In* Fundamentals in Cancer Prevention. P. N. Magee, S. Takayama, T. Sugimura & T. Matsushima, Eds.: 191–215. University Park Press. Baltimore, Md.
13. PURCHASE, I. F. H., E. LONGSTAFF, J. ASHBY, J. A. STYLES, D. ANDERSON, P. A. LEFEVRE & F. R. WESTWOOD. 1978. An evaluation of 6 short-term tests for detecting organic chemical carcinogens. Br. J. Cancer **37**: 873–959.
14. SIMMON, V. F. 1979. In vitro mutagenicity assays of chemical carcinogens and related compounds with *Salmonella typhimurium*. J. Nat. Cancer Inst. **62**: 893–899.
15. LANG, R. & U. REDMANN. 1979. Non-mutagenicity of some sex hormones in the Ames Salmonella/microsome mutagenicity test. Mutat. Res. **67**: 361–365.
16. DONAHUE, E. V., J. MCCANN & B. N. AMES. 1976. Detection of mutagenic impurities in carcinogens and noncarcinogens by high-pressure liquid chromatography and the *Salmonella*/microsome test. Cancer Res. **38**: 431–438.
17. KADA, T. 1973. *Escherichia coli* mutagenicity of furylfuramide. Jpn. J. Genet. **48**: 301–305.
18. KONDO, S. & H. ICHIKAWA-RYO. 1973. Testing and classification of mutagenicity of furylfuramide in *Escherichia coli*. Jpn. J. Genet. **48**: 295–300.
19. YAHAGI, T., M. NAGAO, K. HARA, T. MATSUSHIMA, T. SUGIMURA & G. T. BRYAN. 1974. Relationships between the carcinogenic and mutagenic or DNA-modifying effects of nitrofuran derivatives, including 2-(2-furyl)-3-(5-nitro-2-furyl)acrylamide, a food additive. Cancer Res. **34**: 2266–2273.
20. ONG, T. & M. M. SHAHIN. 1974. Mutagenic and recombinogenic activities of the food additive furylfuramide in eukaryotes. Science **184**: 1086–1087.
21. TAZIMA, Y. 1973. Some problems in environmental mutagens: from the experiences on AF-2. Kagaku **43**: 745–749. (In Japanese.) (As cited in Reference 19.)
22. TONOMURA, A. & M. S. SASAKI. 1973. Chromosome aberrations and DNA repair synthesis in cultured human cells exposed to nitrofurans. Jpn. J. Genet. **48**: 291–294.
23. MIYAJI, T. 1971. Acute and chronic toxicity of furylfuramide in rats and mice. Tokohu J. Exp. Med. **103**: 331–369.
24. SANO, T., T. KAWACHI, N. MATSUKURA, K. SASAJIMA & T. SUGIMURA. 1977. Carcinogenicity of a food additive, AF-2, in hamsters and mice. Z. Krebsforsch. **89**: 61–68.
25. IKEDA, Y., S. HORUICHI, T. FURUYA, O. UCHIDA, K. SUZUKI & J. AZEGAMI. 1974. Induction of gastric tumors in mice by feeding of furylfuramide. Food Sanitation Study Council, Ministry of Health and Welfare, Japan, August, 1974. (In Japanese.) (As cited in Reference 24.)
26. UCHIDA, Y., T. FURIYA, K. KAWAMATA, S. HORIUCHI & Y. IKEDA. 1975. Study about toxicity of furylfuramide (AF-2). II. Study of carcinogenicity. Folia Pharmacol. Jpn. **71**: 80–81. (In Japanese.) (As cited in Reference 24.)
27. NOMURA, T. 1975. Carcinogenicity of the food additive furylfuramide in foetal and young mice. Nature **258**: 610–611.
28. TAKAYAMA, S. & N. KUWABARA. 1977. The production of skeletal muscle atrophy and mammary tumors in rats by feeding 2-(2-furyl)-3-(5-nitro-2-furyl)acrylamide. Toxicol. Lett. **1**: 11–16.
29. MCCANN, J., N. E. SPINGARN, J. KOBORI & B. N. AMES. 1975. Detection of carcinogens as mutagens: bacterial tester strains with R factor plasmids. Proc. Nat. Acad. Sci. USA **72**: 979–983.
30. YAHAGI, T., T. MATSUSHIMA, M. NAGAO, Y. SEINO, T. SUGIMURA & G. T. BRYAN. 1976. Mutagenicity of nitrofuran derivatives on a bacterial tester strain with an R factor plasmid. Mutat. Res. **40**: 9–14.
31. IARC. 1974. IARC Monographs on the Evaluation of Carcinogenic Risk of Chemicals to Man **7**: 143–196. International Agency for Research on Cancer. Lyon, France.
32. PRIVAL, M. J., E. C. MCCOY, B. GUTTER & H. S. ROSENKRANZ. 1977. Tris(2,3-dibromopropyl) phosphate: mutagenicity of a widely used flame retardant. Science **195**: 76–78.

33. BLUM, A. & B. N. AMES. 1977. Flame-retardant additives as possible cancer hazards. Science **195:** 17–23.
34. National Cancer Institute. 1978. Bioassay of Tris(2,3-dibromopropyl) phosphate for Possible Carcinogenicity. Technical Report Series No. 76. DHEW Publication No. (NIH) 78-1326. National Cancer Institute. Bethesda, Md.
35. VAN DUUREN, B. L., G. LOEWENGART, I. SEIDMAN, A. C. SMITH & S. MELCHIONNE. 1978. Mouse skin carcinogenicity tests of the flame retardants tris(2,3-dibromopropyl) phosphate, tetrakis (hydroxymethyl) phosphonium chloride, and polyvinyl bromide. Cancer Res. **38:** 3236–3240.
36. GOLD, M. D., A BLUM & B. N. AMES. 1978. Another flame retardant, tris(1,3-dichloro-2-propyl)phosphate and its expected metabolites are mutagens. Science **200:** 785–786.
37. BRUSICK, D., D. MATHESON, D. R. JAGANNATH, S. GOODE, H. LEBOWITZ, M. REED, G. ROY & S. BENSON. 1980. A comparison of the genotoxic properties of tris(2,3-dibromopropyl) phosphate and tris(1,3-dichloro-2-propyl) phosphate in a battery of short-term bioassays. J. Environ. Pathol. Toxicol. **3:** 207–226.
38. SONTAG, J. M. 1981. Carcinogenicity of substituted-benzenediamines (phenylenediamines) in rats and mice. J. Nat. Cancer Inst. **66:** 591–602.
39. National Cancer Institute. 1979. Bioassay of p-Phenylenediamine dihydrochloride for Possible Carcinogenicity. Technical Report Series No. 174. DHEW Publication No. (NIH) 79-1730. National Cancer Institute. Bethesda, Md.
40. GARNER, R. C. & C. A. NUTMAN. 1977. Testing of some azo dyes and their reduction products for mutagenicity using Salmonella typhimurium TA1538. Mutat. Res. **44:** 9–19.
41. DEGAWA, M., Y. SHOJI, K. MASUKO & Y. HASHIMOTO. 1979. Mutagenicity of metabolites of carcinogenic aminoazo dyes. Cancer Lett. **8:** 71–76.
42. BURNETT, C., C. FUCHS, J. CORBETT & J. MENKART. 1982. The effect of dimethylsulfoxide on the mutagenicity of the hair dye p-phenylenediamine. Mutat. Res. **103:** 1–4.
43. GOLBERG, L. 1980. Rapid tests in animals and lower organisms as predictors of long-term toxic effects. In Current Concepts in Cutaneous Toxicity. V. A. Drill & P. Lazar, Eds.: 171–212. Academic Press, Inc. New York, N.Y.
44. YOSHIKAWA, K., H. UCHINO & H. KURATA. 1976. Studies on the mutagenicity of hair dye. Eisei Shikensho Hokoku **94:** 28–32. (In Japanese.)
45. YAHAGI, T., M. DEGAWA, Y. SEINO, T. MATSUSHIMA, M. NAGAO, T. SUGIMURA & Y. HASHIMOTO. 1975. Mutagenicity of carcinogenic azo dyes and their derivatives. Cancer Lett. **1:** 91–96.
46. PRIVAL, M. J., V. D. KING & A. T. SHELDON, JR. 1979. The mutagenicity of dialkylnitrosamines in the Salmonella plate assay. Environ. Mutagen. **1:** 95–104.
47. PRIVAL, M. J. & V. D. MITCHELL. 1981. Influence of microsomal and cytosolic fractions from rat, mouse, and hamster liver on the mutagenicity of dimethylnitrosamine in the Salmonella plate incorporation assay. Cancer Res. **41:** 4361–4367.
48. WEINSTEIN, D., M. KATZ & S. KAZMER. 1981. Use of a rat/hamster S-9 mixture in the Ames mutagenicity assay. Environ. Mutagen. **3:** 1–9.
49. YAHAGI, T., M. NAGAO, Y. SEINO, T. MATSUSHIMA, T. SUGIMURA & M. OKADA. 1977. Mutagenicities of N-nitrosamines on Salmonella. Mutat. Res. **48:** 121–130.
50. PRIVAL, M. J. & V. D. MITCHELL. 1983. Dimethylnitrosamine demethylase and the mutagenicity of dimethylnitrosamine: effects of rodent liver fractions and dimethylsulfoxide. In Topics in Chemical Mutagenesis. N-Nitrosamines. T. K. Rao, W. Lijinski & J. Epler, Eds. **1.** Plenum Press. New York, N.Y. (In press.)
51. NAGAO, M., N. MORITA, T. YAHAGI, M. SHIMIZU, M. KUROYANAGI, M. FUKUOKA, K. YOSHIHIRA, S. NATORI, T. FUJINO & T. SUGIMURA. 1981. Mutagenicities of 61 flavonoids and 11 related compounds. Environ. Mutagen. **3:** 401–419.
52. BROWN, J. P. & P. S. DIETRICH. 1979. Mutagenicity of plant flavonols in the Salmonella/mammalian microsome test. Activation of flavonol glycosides by mixed glycosidases from rat cecal bacteria and other sources. Mutat. Res. **66:** 223–240.
53. TAMURA, G., C. GOLD, A. FERRO-LUZZI & B. N. AMES. 1980. Fecalase: a model for activation of dietary glycosides to mutagens by intestinal flora. Proc. Nat. Acad. Sci. USA **77:** 4961–4965.

54. PAMUKCU, A. M., S. YALCINER, J. F. HATCHER & G. T. BRYAN. 1980. Quercetin, a rat intestinal and bladder carcinogen present in bracken fern (*Pteridium aquilinum*). Cancer Res. **40:** 3468–3472.
55. SAITO, D., A. SHIRAI, T. MATSUSHIMA, T. SUGIMURA & I. HIRONO. 1980. Test of carcinogenicity of quercetin, a widely distributed mutagen in food. Teratogen. Carcinogen. Mutagen. **1:** 213–221.
56. HIRONO, I., I. UENO, S. HOSAKA, H. TAKANASHI, T. MATSUSHIMA, T. SUGIMURA & S. NATORI. 1981. Carcinogenicity examination of quercetin and rutin in ACI rats. Cancer Lett. **13:** 15–21.
57. MORINO, K., N. MATSUKURA, T. KAWACHI, T. OHGAKI, T. SUGIMURA & I. HIRONO. 1982. Carcinogenicity test of quercetin and rutin in golden hamsters by oral administration. Carcinogenesis **3:** 93–97.
58. SUGIMURA, T., M. NAGAO, T. KAWACHI, M. HONDA, T. YAHAGI, Y. SEINO, S. SATO, N. MATSUKURA, T. MATSUSHIMA, A. SHIRAI, M. SAWAMURA & H. MATSUMOTO. 1977. Mutagen-carcinogens in food, with special reference to highly mutagenic pyrolytic products in broiled foods. *In* Origins of Human Cancer. H. H. Hiatt, J. D. Watson & J. A. Winsten, Eds. (Book C): 1561–1577. Cold Spring Harbor Laboratory. Cold Spring Harbor, N.Y.
59. PRIVAL, M. J. & V. D. MITCHELL. 1982. Analysis of a method for testing azo dyes for mutagenic activity in *Salmonella typhimurium* in the presence of flavin mononucleotide and hamster liver S9. Mutat. Res. **97:** 103–116.

MUTAGENICITY AND CARCINOGENICITY CORRELATIONS BETWEEN BACTERIA AND RODENTS

David Brusick

*Litton Bionetics, Inc.
Kensington, Maryland 20895*

INTRODUCTION

Most chemicals are operationally identified as mutagens using test results from bacterial assays. For example, more than 800 chemically defined substances have been reported in the open literature as mutagens in the Ames *Salmonella*/microsome assay (GeneTox *Salmonella* working group). Many of the same materials have also been detected as mutagens in *Escherichia coli*[9] or other bacteria assays.

Because the number of chemicals identified as mutagens is rather large and because the bacterial test results represent the sole basis for designating the chemical a mutagen, it has become critical to understand the significance of these findings.

Are chemicals identified as mutagens on the basis of bacteria tests implicated in the induction of chronic toxicity in mammalian species? During the past 10 years this question has focused on the predictive ability of bacteria tests to distinguish animal carcinogens. Answers to the question have been derived by correlation studies assessing the concordance between mutagens and carcinogens and between nonmutagens and noncarcinogens. This latter comparison is stated with full appreciation for the problems involved in defining nonmutagens and noncarcinogens.

This report is an attempt to review the results of 10 years of correlation studies and to assess their findings.

SELECTION OF TEST SYSTEMS

The number of methods capable of measuring mutation in bacteria is too large to cover adequately in this report; however, the job becomes much more manageable when one realizes that over 90% of all mutagenicity results have been derived from two species and from only a relatively few techniques using these two species. TABLE 1 identifies the systems from which most data are available. The tests are divided by both species and selective system (either forward or reverse mutation). There is a considerable range in the size of data bases representing each assay and, especially, correlation studies. Large-scale systematic attempts at defining correlations have only been conducted for *S. typhimurium* (*his*⁻) and *E. coli* (*trp*⁻) bacteria.[16]

The Environmental Protection Agency GeneTox working group concluded in its report that *E. coli* WP$_2$ contributed few if any unique responses not obtainable from the Ames *Salmonella* assay.[9] Data from additional strains such as WP67 and TM1 might improve the utility of the *E. coli* test, but the data base for these strains is not yet sufficient.

Purchase classified the developmental process of a test system into three stages: (1) a developing test, (2) a developed test, and (3) an established predictive test.[16] Only the *Salmonella*/microsomal assay meets the criteria of an established predictive test (TABLE 2).

164

0077–8923/83/0407–0164 $01.75/0 © 1983, NYAS

TABLE 1

DESCRIPTION OF MICROBIAL MUTAGENESIS ASSAYS COMMONLY USED IN GENETIC SCREENING

Organism	Detected	Locus	System	Reference
Salmonella typhimurium	Reverse mutation in several auxotrophic tester strains, TA1535, TA1537, TA1538, TA98, and TA100.	his G, his C, and his D	Minimal basal agar medium without histidine	Ames et al.[3]
	Forward mutation in a single strain TM677	XGPRT (xanthine-guanine phosphoribosyltransferase)	Minimal basal agar medium supplemented with 8-azaguanine	Skopek et al.[20]
Escherichia coli	Reverse mutation in several auxotrophic tester strains, WP$_2$, WP$_2$ uvrA, CM611, WP67, and TM1	Trytophan $trp^- \rightarrow trp^+$	Minimal basal agar medium without tryptophan	Brusick et al.[9]

TABLE 2

FEATURES OF THE *SALMONELLA*/MICROSOMAL ASSAY

Criteria*	Documented for the Ames Assay
Used in a number of laboratories	Tests conducted in over 2000 laboratories worldwide†
Data from at least 1000 unique substances	GeneTox as of 1979 identified 803 chemicals. Since that time there have been new publications. These will easily yield more than 200 new chemicals.
An understanding of the extrapolative value of the test beginning to develop	The predictive value of the test to identify both rodent and human carcinogens has been calculated and published. Potency correlations have also been made.
An understanding of the limitations of the test	Chemical classes that do not do well in the test are known.

*From Purchase.[16]
†Ashby, J. Personal communication.

Consequently, there appears to be little to gain from a critical analysis of tests that have less than 100 chemicals in the published data base (a "developing test" according to Purchase) or a test such as *E coli* WP_2, which has been reviewed and appears to parallel the Ames test for the most part.

Other attributes of the Ames test that make it unique among short-term test methods concern available knowledge about the operational aspects of the test, which can affect the outcome of tests. Very few short-term methods (with the possible exception of *Drosophila*) have been sufficiently studied to also have this information:

● Chemical class specificity has been documented in numerous reports.[18] Examples are given in TABLE 3.
● Interlaboratory variability has been reviewed.[10]
● Sensitive areas of the protocol have been identified, and the consequences of protocol changes involving these sensitive areas can be reasonably well predicted.[15]

TABLE 3

CHEMICAL GROUPS CONTAINING HIGH PROPORTIONS OF FALSE NEGATIVES

Antimetabolites	Alter synthetic pathway necessary for normal nucleic acid synthesis and stabilization.
Azo compounds	Require nonhepatic metabolism, often gut flora.
Carbamyls and thiocarbamyls	Complex metabolic activation pathways and phylogenetic specificities (several *Salmonella* negatives are positive in cultured mammalian cells or *Drosophila*).
Halogenated compounds	Complex metabolic activation pathways and volatility (several positive in closed system).
Steroids	Not initiators.
Cross-linking agents	Effects are primarily lethal and not mutagenic; can be detected in repair-sufficient substrains.
Inorganic compounds	Solubility and bioavailability, many considered to lack initiating properties.
Promoters	Lack initiating activity.

The remaining sections of the report therefore will examine several published and unpublished reports with the intent of determining the usefulness of the Ames test as a predictor of rodent carcinogens.

PREDICTIVE PERFORMANCE OF THE AMES TEST FOR CARCINOGENS

Realization that the plate-test reverse-mutation method employing *Salmonella typhimurium* could serve as an indicator of certain classes of carcinogens was acknowledged by Ames in 1971.[1] The applicability and relevance of the method were enhanced by the addition of an exogenous mammalian hepatic metabolic activation system[4] and the introduction of the plasmid pKM101 into the tester strains.[13]

Shortly after these final improvements in the assay, McCann *et al.* published the results of a carcinogen correlation ("validation") study with 300 chemicals.[12] This analysis in 1975 stimulated other similar "validation" studies extending over the next eight years. The findings, specifically the extent of correlation between carcinogens and mutagens, fell into a broad range of almost 50% to 95%, and the results from selected studies were used both to support and to criticize the use of *in vitro* testing (TABLE 4).

This report has been prepared to summarize and, to a limited degree, analyze the findings of a series of these comparative studies, and attempts are made to explain some of the variation observed among the correlation coefficients reported.

Studies were selected for the summary only if they included the minimal set of tester strains (TA1535, TA1537, TA98, and TA100) and employed metabolic activation (i.e., the "standard" Ames test). Most of the results reported in this summary were obtained from laboratories with demonstrated capabilities in conducting the Ames test, and deficiencies resulting from method nonfamiliarity probably represent a minor contribution to the range of variability.

The summary is given in TABLE 4. Values are given in TABLE 4 for the sensitivity (mutagenic carcinogens/total carcinogens), specificity (nonmutagenic noncarcinogens/total noncarcinogens), and predictive value (mutagenic carcinogens/total mutagens).

Values for all three categories could not be calculated for every study. In addition, two studies compared the Ames test responses against results from very special *in vivo* data bases. The ICPEMC report compared bacterial mutagenesis with *in vivo* germ cell mutagenesis,[11] and the report by Waters *et al.* used the human carcinogen data base for comparison with Ames results.[22]

As seen in TABLE 4, the range of sensitivities (number of carcinogens detected as a mutagen in the Ames test) and predictive values (number of Ames mutagens that are also carcinogens) varied considerably, which might be interpreted in a way to discredit the use of the Ames test as a predictive test for carcinogens.

A separate but related issue involved is that of "false-positive" and "false-negative" responses. These responses are difficult to analyze reliably because the accuracy of the rodent standard is questionable. In comparing cancer responses for a given set of chemicals in rodent species, one encounters unique responses; for example, the two rodent species do not agree for all chemicals and an Ames result may be "false" for one species but "true" for another. Rodents also fail to detect several carcinogens known to induce tumors in humans; these might then be considered false negatives for rodents but false positives for the Ames test if they fail to agree with the rodent responses. False responses, therefore, occur due to inadequacies in both the microbial predictive model and the rodent target organism used as the standard. TABLE 5 provides some explanations for apparent false responses.

TABLE 4

ASSAY PERFORMANCE OF THE AMES *SALMONELLA* REVERSE-MUTATION ASSAY*

Ames Assay Assessment	Year Reported	Chemicals in the Sample	Sensitivity	Specificity	Predictive Value	Basis of Chemical Selection
McCann et al.[13]	1975	300	90%§§	87%§§	92%	investigator selected
Sugimura et al.[21]	1976	241	93%§§	77%§§	87%	investigator selected
Purchase et al.[17]	1978	120	91%§§	93%§§	93%	investigator selected
Simmon, V.[19]	1979	92	64%	78%	90%	committee selected
Andersen et al.[5]	1979	221	90%	100%	100%	investigator selected
Rinkus and Legator[18]	1979	271	77%¶	NA‖	—	literature data base
Bartsch et al.[7]	1980	89	76%§§	57%§§	95%	investigator selected
Bridges et al.[8]	1981	42	45–70%	60–80%	—	investigator selected
EPA GeneTox†	1982	122	63%	—	—	literature data base
NIEHS/NTP[23]	1982	218	54%	—	76%	selection committee
Waters, et al.[22]	1982	17	65%	NA	100%	human carcinogens
ICPEMC[11]	1982	51	64%	35%	58%	animal germ cell mutagens
NCI Comparative‡	1982	58	78%	67%	81%	selection committee

*All values reported are approximate and based upon evaluations of results by the authors of the reports or individuals giving their personal communications.

†From A. Auletta, EPA GeneTox Coordinating Committee.

‡V. C. Dunkel, personal communication.

§From I. F. H. Purchase.[16]

¶Reanalysis of the report by B. Ames and J. McCann showed an 82% correlation.[2]

‖NA means not applicable.

Other factors can also influence the outcomes of the individual comparative studies. TABLE 6 addresses several factors that appear to have biased the predictive values shown in TABLE 4. These factors were interpreted as follows:

1. *Method by which the chemical sample used in the comparison was selected.* This factor appears to be extremely important in determining the outcome of the comparative exercise. Studies in which the investigator selected the chemicals to be included in the sample show significantly higher correlations than do studies using chemical samples that were not influenced by the investigator.

2. *Tests on coded and uncoded samples.* The correlations obtained from tests with coded samples were lower than those found in studies in which the investigators knew the identity of the materials at the time of testing or at the time of data interpretation.

3. *Tests conducted in a single laboratory or in multiple laboratories.* Test results from multiple laboratory data tend to produce lower correlation values because of protocol variations and because of the need for consensus responses. A special variation

TABLE 5

EXPLANATIONS FOR THE OCCURRENCE OF APPARENT FALSE RESPONSES IN THE AMES TEST

False negatives
  1. Inadequate bioassay evaluation or interpretation that suggests positive effects without independent confirmation in another species.
  2. *In vitro* metabolic activation system that results in a failure to generate the appropriate carcinogenic intermediate under the test conditions employed.
False positives
  1. Inadequate bioassay due to improper species selection or suboptimal exposure conditions.
  2. Bacterial activation of chemicals to mutagens not formed at biologically active levels in mammalian metabolic systems.
  3. Repair system differences between indicator cells in the model system and the target organism *in vivo*.
  4. Mutagenic impurities in the test sample that are detected in the model system but not by the *in vivo* target organism.
  5. Inadequate *in vitro* metabolic activation, which may generate active intermediates that are not adequately detoxified in a fashion similar to that occurring *in vivo*.

of this problem was inherent in those studies that relied on published data because few reports itemize the responses from all trials conducted during the evaluation of each chemical in the sample. Consequently, the true reproducibility of a reported response may not be reflected in the final manuscript.

4. *Correlation results of the Ames test with restricted data sets.* The Ames test did not appear to correlate well enough with the mammalian germ cell data base to consider the assay predictive for this end point. The correlation with human carcinogens is not particularly good in this report; however, a review by Bartsch *et al.* shows a 71% sensitivity.[6]

In general, then, many of the correlation studies showing the highest predictive values might be questioned concerning bias introduced via mechanisms identified in items 1, 2, or 3. Conversely, this conclusion should not imply that the true correlation values are those represented by the lowest values. In fact, closer inspection of the spectrum of studies has revealed information valuable in determining overall accuracy.

TABLE 6

ASSAY PERFORMANCE OF THE AMES *SALMONELLA* REVERSE-MUTATION ASSAY

Study	Composition of Chemical Sample			Restricted to		Evaluations Based on		
	Chemicals Selected by Investigators	Literature Review	Selection by Committee	Human Carcinogens	Animal Germ Cell Mutagens	Multiple Laboratory Results	Coded Samples	Noncoded Samples
McCann et al.[13]	X							X*
Sugimura et al.[21]	X							X*
Purchase et al.[17]	X							X*
Simmon, V.[19]			X				?	
Andersen et al.[5]	X							X*
Rinkus and Legator[18]		X						
Bartsch et al.[7]	X							X*
Bridges et al.[8]		X	X			X (12)†	X	
EPA GeneTox		X				X some		X
NIEHS/NTP[23]			X			X (3)†	X	
Waters, et al.[22]		X		X		X some		X
ICPEMC[11]		X			X			X
NCI Comparative			X			X (4)†	X	
Mean								
% Sensitivity (range)	88 (76–93)	67 (63–77)	67 (45–78)	65	64	63 (45–78)	67 (54–78)	79 (63–93)
% Specificity (range)	83 (57–100)	35	75 (67–70)	—	35	75 (67–70)	74 (67–80)	64 (35–100)
% Predictive (range)	93 (87–100)	79 (58–100)	82 (76–90)	100	58	79 (76–81)	79 (76–81)	89 (58–100)

*Presumed to be noncoded.
†Number of laboratories involved.

ANALYSIS OF SPECIFIC STUDIES

Based on statements in the previous section, results reported by McCann et al.,[12,13] Sugimura et al.,[21] Purchase et al.,[16,17] Andersen et al.,[5] and Bartsch et al.[6] might be biased in favor of overestimating the positive correlation. Closer attention is warranted for the remaining studies to ensure that, because of confounding factors, the evaluation is not underestimating positive correlations.

1. Simmon.[19] This study consisted of chemicals selected from National Cancer Institute–tested compounds. The chemical classes selected for this correlation study consisted of several known to produce problems in the Ames assay. For example, there were seven compounds from the amide/urea/acylating class, five compounds from the antimetabolite class, three metallic compounds, and three promoters. Consequently, the correlation values from this study would be expected to be lower than those of McCann et al.,[12,13] Purchase et al.,[16,17] and Sugimura et al.,[21] in which the number of known problem chemical classes was smaller.

2. Rinkus and Legator.[18] This report created considerable controversy when initially published because it was the first review to question seriously the high positive correlations reported by McCann et al.,[12,13] Sugimura et al.,[21] and Purchase et al.[16,17] The most critical point of this manuscript is the reliability of the carcinogenicity data. Many of the chemicals identified as carcinogens by the authors were disputed by other experts. A reanalysis of the manuscript by Ames and McCann produced an 82% correlation (sensitivity),[2] bringing the predictive efficiency of the assay to a more respectable level.

3. Bridges et al.[8] This evaluation consisted of studies by approximately 12 laboratories on 42 chemicals evenly divided between carcinogens and noncarcinogens. The correlation values were low because of several reasons, such as (a) the intentional selection of chemicals known to be difficult to assess in the Ames test, (b) the selection of several chemicals lacking good data defining their lack of carcinogenicity, (c) the lack of any attempt to standardize the test conditions among the 12 laboratories, and (d) the use of coded samples.

4. National Institute of Environmental Health Sciences/National Toxicology Program.[23] The data shown for this program represent the present status of the NTP-sponsored testing program. The chemicals selected for the sample are of industrial origin and include a high percentage of chlorinated organics already known to be biased toward false-negative results. In fact, the data base in general is biased away from chemical classes known to do well in the Ames assay.[23] This program is an ongoing one, and the data base is likely to change within the next year or two. A second analysis will be warranted at that time.

5. EPA GeneTox. This evaluation represents the largest data base in the summary. Virtually no selection was involved, and results from more than 800 chemicals were available for comparison with carcinogenicity data. The low positive correlation may rest with a number of variables, which cannot be controlled when searching published literature; however, the most probable cause rests with the fact that the vast majority of the early reports (prior to 1974) contain data from studies that did not use an exogenous metabolic activation system (S-9 mix or equivalent). This fact is certain to bias the correlations toward lower levels of concordance. The data base must be reanalyzed excluding results where S-9 mix was not used.

6. National Cancer Institute Comparative. This data set may well represent one of the least biased correlation studies and as such may be the best representative of the true predictive power of the Ames test for rodent carcinogens. The details of the studies are briefly summarized in TABLE 7. Over 100 chemicals were tested during the course

of this study, but results for a large group have already been reported by Dunkel.[10] The remaining data are summarized in TABLE 8. The strength of this correlation study rests with the fact that the chemicals were tested in multiple laboratories, under code, using the samples employed in the rodent bioassays.

The mutation response and tumor responses from the NCI collaborative study were subjected to considerable analysis. Specific points indicate the power of the Ames test in this evaluation.

1. The incidence of complete positive or negative concordance across the four participating laboratories was about 95%. Criteria used in the evaluation of mutagenic activity have been defined by Dunkel.[10]

2. Among chemicals in which both rodent species gave qualitatively similar responses (both showed increased tumor levels or both were not affected), the Ames test provided good sensitivity (88% for positives) and specificity (67% for negatives). The predictive values would, by definition, be 100% for this group.

3. A comparison of concordance between mice and rats for the 58 chemicals showed that 72% (42/58) of the substances produced the same qualitative responses in both species. The Ames tests when compared to the rodent results showed an equally good (or bad, depending upon one's preconceived expectations) concordance in that 74% (43/58) of the samples produced qualitatively similar responses. The fact that the mouse and rat differed as to whether or not the compound was a carcinogen approximately 30% of the time is of interest.

4. The study produced eight "false negatives," among which nitrilotriacetic acid (NTA), reserpine, cinnamyl anthranilate, m-cresidine, and Dapsone are generally regarded as nongenotoxic carcinogens. Styrene has not been shown to be mutagenic in bacteria, but its oxide is.[14]

5. Seven compounds were identified as "false positives." Among this group at least three agents, 4-nitro-o-phenylenediamine, p-phenylenediamine, and 3-nitropropionic

TABLE 7

NCI-SPONSORED COLLABORATIVE STUDY

● Four laboratories involved
● Three activation systems (noninduced and induced)
Mouse
Rat
Hamster
● Five *Salmonella* strains
TA 1535
TA 1537
TA 1538
TA 98
TA 100
● One *E. coli* strain
WP$_2$ *uvr A*
● Total of 139 chemicals (several were used as internal positive controls)
● Dose range 0.3 μg to 3333 μg/plate
● Triplicate plates per dose
● Seven concentrations per activation condition and for nonactivation run
● Concurrent positive and negative controls with each test
● All samples coded during testing and data evaluation
● All chemical samples were from the same batches that were employed in the animal tests

TABLE 8

SUMMARY OF THE RESULTS FROM 58 CHEMICALS EVALUATED UNDER CODE BY FOUR
LABORATORIES USING THE AMES ASSAY

Carcinogen Classification	Ames Positives in Sample Tested*	Chemicals Not Correctly Predicted by Ames Test
Positives		
Chemicals with sufficient evidence in two species	88% (14/16)	NTA reserpine
Chemicals with sufficient evidence in one species	67% (6/9)	cinnamyl anthranilate, m-cresidine, Dapsone
Chemicals with limited evidence in animals	83% (5/6)	daminozide
Chemicals with marginal evidence	67% (4/6)	acetohexamide, styrene
Negatives		
Chemicals without evidence of carcinogenicity in animals	33% (7/21)	4'-(chloroacetyl)-acetaniline 2,4-dimethoxyaniline HCl 4-nitro-o-phenylenediamine p-phenylenediamine 2,5-toluenediamine sulfate 1-nitronaphthalene 3-nitropropionic acid

*Sensitivity = (no. mutagenic carcinogens)/(no. carcinogens) = 29/37 = 78%. Specificity = (no. nonmutagenic noncarcinogens)/(no. noncarcinogens) = 14/21 = 67%. Predictive value = (no. mutagenic carcinogens)/(no. mutagens) = 29/36 = 81%.

acid, produce positive effects in other *in vitro* assays for mutation, unscheduled DNA synthesis (UDS), and cell transformation. It may be useful to reexamine the rodent bioassay protocols used in the evaluation of these agents for carcinogenicity.

The spectrum of chemicals covered in this evaluation along with the power of the Ames protocol that was employed lend a high degree of credibility to the conclusion that a correlation value of about 80% may well be typical for the Ames test. It is clear that the early reports of 90–95% correlation were not realistic and that correlations of 50–70% are usually associated with atypical conditions likely to bias against good positive concordance.

TESTING RECOMMENDATIONS

Recommendations applicable to conducting the Ames test fall into three categories.

1. *Recommendations associated with a minimal protocol acceptable for screening.* The minimal protocol is sufficient to satisfy the need for rapid, inexpensive screening of large numbers of chemicals. The margin of error with respect to accurate carcinogen detection would be moderate. The protocol should include both nonactivation and activation test conditions with at least seven concentrations of the material in each. A minimum of four tester strains (TA1535, TA1537, TA98, and TA100) should be used with at least duplicate plates per dose group. Positive, negative, and solvent controls must be conducted in duplicate concurrently with the test material. Dose

levels should be selected on the basis of the toxicity of the test material and should be spaced at intervals no greater than half-log. The S-9 mix should be prepared from Arochlor 1254–induced Sprague-Dawley rats or Syrian hamsters. The study design should clearly describe the criteria for determination of a positive response. Evidence of increases in reversion frequencies that do not meet the criteria for a positive response should be repeated over a narrower dose range encompassing the dose levels of suspected activity in the initial test. Criteria for evaluating the repeat test results should also be defined in the study design.

2. An expanded and more powerful testing approach would include the following recommended modifications to the minimal protocol: (a) Toxicity testing should be performed both with and without S-9 mix to set appropriate dose ranges for the two test conditions. (b) Both rat- and hamster-induced S-9 mix should be used. (c) All dose groups and controls should be conducted in at least triplicate. (d) The entire assay should be repeated as an independent confirmation. (e) Five strains including TA1538 should be used.

3. The last set of recommendations consists of things to consider whether the minimal or expanded test design is used. (a) In setting the maximum dose level, the test system must determine the maximum tolerated dose (MTD), preferably by clear evidence of toxicity. Solubility or maximum limit concentrations may not be adequate. (b) Attempts should be made to determine the purity of the sample and the identity (if possible) of any impurities. (c) Set an acceptable range (upper and lower limits) for the spontaneous background for each strain, and do not use cultures falling outside the preset range. This procedure facilitates evaluation of data from year to year. (d) Perform sterility controls for the S-9 mix, the test material, the overlay agar, and the solvent. (e) Determine the maximum nontoxic concentration of the solvent or vehicle employed before using it in studies. (f) Perform a simple volatility test by comparing the evaporation time of 50 μl of test material to 50 μl of water and 50 μl of 95% ethanol. Agents with volatilities close to ethanol should be tested by preincubation or in closed chamber conditions rather than by the agar overlay method. If there is any doubt, use the preincubation method.

CONCLUSIONS

A reasonably critical analysis of the performance of the "standard" Ames *Salmonella*/microsome assay as a predictor of rodent carcinogenesis suggests that when the technique is applied properly to a nonbiased spectrum of chemicals, the predictive accuracy for mouse and rat carcinogens will be approximately 80%. Studies showing correlations substantially higher (87–93%) and lower (50–70%) appear to be biased in one direction or the other by deficiencies in study design or by unusual frequencies of atypical chemicals in the sample used in the comparison.

After 10 years of application and numerous validation programs, the actual performance of the "standard" Ames test appears to be less predictive than initially hoped for, but not poor enough that the test should be abandoned. Judicious application of the method in specific testing situations can be a highly efficient way to screen large numbers of chemicals for their carcinogenic potential.

SUMMARY

Detection of mutation in bacteria has acquired the status of an accepted procedure in genetic toxicology programs. The methods presently employed in such programs include both forward and reverse mutation-induction techniques in strains of *Salmonella typhimurium* and *Escherichia coli*.

The specific strains used in these techniques have been selected over the years on the basis of their sensitivity to a broad range of chemical mutagens. In addition, it has been reported that chemical carcinogens can be presumptively identified on the basis of these assays, and bacterial testing has been generally considered the front-line test procedure for the identification of presumptive mutagenic carcinogens. An analysis of correlative studies both retrospective and cross-sectional shows a range of predictive capabilities depending on features such as chemical class, carcinogenic mechanism, and requirements for specific metabolic toxification processes.

The greatest limitations associated with the use of bacteria mutation testing is the real and/or perceived issue of the test or a misinterpretation of the correlation coefficients under conditions of routine application. Concerns related to the performance (reliability, reproducibility, and predictability) and relevance of bacteria assays perpetuate controversy surrounding their application to hazard assessment.

A review of several studies comparing mutation induction and tumor induction indicates that the Ames test can be useful in screening large numbers of chemicals, but the true correlation coefficient is only about 80% when compared to tumor responses in mice and rats.

REFERENCES

1. AMES, B. N. 1971. The detection of chemical mutagens with enteric bacteria. *In* Chemical Mutagens: Principles and Methods for Their Detection. A. Hollaender, Ed. 1: 267–282. Plenum Press. New York, N.Y.
2. AMES, B. N. & J. MCCANN. 1981. Validation of the *Salmonella* test: a reply to Rinkus and Legator. Cancer Res. 41: 4192–4203.
3. AMES, B. N., J. MCCANN & E. YAMASAKI. 1975. Methods for detecting carcinogens and mutagens with the *Salmonella*/microsome mutagenicity tests. Mutat. Res. 31: 347–364.
4. AMES, B. N., W. DURSTON, E. YAMASAKI & F. LEE. 1973. Carcinogens are mutagens: a simple test system combining liver homogenates for activation and bacteria for detection. Proc. Nat. Acad. Sci. USA 70: 2281–2285.
5. ANDERSON, M., M. L. BINDERUP, P. KIEL & H. LARSEN. 1979. Evaluation of the Ames test. Ugeskr. Laeg. 14: 305–307.
6. BARTSCH, H., L. TOMATIS & C. MALAVEILLE. 1982. Mutagenicity and carcinogenicity of environmental chemicals. Regulatory Toxicol. Pharmacol. 2: 94–105.
7. BARTSCH, H., C. MALAVEILLE, A. M. CAMUS, G. MARTEL-PLANCHE, G. BRUN, A. HUNTEFEUILLE, N. SABADIE, A. BARBIN, T. KUROKI, C. DREVON, C. PICCOLI & R. MONTESANO. 1980. Validation and comparative studies on 180 chemicals with *S. typhimurium* strains and V79 Chinese hamster cells in the presence of various metabolizing systems. Mutat. Res. 76: 1–50.
8. BRIDGES, B. A., D. MACGREGOR & E. ZEIGER. 1981. Summary report on the performance of bacterial mutation assays. *In* Evaluation of Short-Term Tests for Carcinogens. F. J. de Serres & J. Ashby, Eds.: 49–67. Elsevier/North-Holland. New York, N.Y.
9. BRUSICK, D. J., V. F. SIMMON, H. S. ROSENKRANZ, V. A. RAY & R. S. STAFFORD. 1980. An evaluation of the *Escherichia coli* WP₂ and WP₂ *uvr*A reverse mutation assay. Mutat. Res. 76: 169–190.
10. DUNKEL, V. C. 1979. Collaborative studies on the *Salmonella*/microsome mutagenicity assay. J. Assoc. Off. Anal. Chem. 62: 874–882.
11. ICPEMC. 1982. Report from Committee 1 on Short-Term Testing Strategies. Mutat. Res. (In press.)
12. MCCANN, J., E. CHOI, E. YAMASAKI & B. N. AMES. 1975. Detection of carcinogens as mutagens in the *Salmonella*/microsome test: assay of 300 chemicals. Proc. Nat. Acad. Sci. USA 72: 5135–5139.

13. MCCANN, J., N. E. SPINGARN, J. KOBORI & B. N. AMES. 1975. Detection of carcinogens as mutagens: bacterial tester strains with R factor plasmids. Proc. Nat. Acad. Sci. USA 72: 979–983.
14. MILVEY, P. & A. J. GARRO. 1976. Mutagenic activity of styrene oxide (1,2-epoxyethylbenzene), a presumed styrene metabolite. Mutat. Res. 40: 15–18.
15. PRIVAL, M. J. 1983. The Salmonella mutagenicity assay: promises and problems. Ann. N.Y. Acad. Sci. (This volume.)
16. PURCHASE, I. F. H. 1982. An appraisal of predictive tests for carcinogenicity. Mutat. Res. 99: 53–71.
17. PURCHASE, I. F., E. LONGSTAFF, J. ASHBY, J. A. STYLES, D. ANDERSON, P. A. LEFEVRE & F. R. WESTWOOD. 1978. An evaluation of 6 short-term tests for detecting organic chemical carcinogens. Br. J. Cancer 37: 873–903.
18. RINKUS, S. J. & M. S. LEGATOR. 1979. Chemical characterization of 465 known or suspected carcinogens and their correlation with mutagenic activity in the Salmonella typhimurium system. Cancer Res. 39: 3289–3318.
19. SIMMON, V. F. 1979. In vitro mutagenicity assays of chemical carcinogens and related compounds with Salmonella typhimurium. Nat. Cancer Inst. 62: 893–899.
20. SKOPEK, T. R., H. L. LIBER, J. J. KROWLEWKSI & W. G. THILLY. 1978. Quantitative forward mutation assay in Salmonella typhimurium using 8-azaguanine resistance as a genetic marker. Proc. Nat. Acad. Sci. USA 75: 410–414.
21. SUGIMURA, T., S. SATO, M. NAGAO, T. YAHAGI, T. MATSUSHIMA, Y. SEINO, M. TAKEUCHI & T. KAWACHI. 1976. Overlapping of carcinogens and mutagens. In Fundamentals in Cancer Prevention. P. N. Magee, S. Takayoma, T. Sugimura & T. Matsushima, Eds.: 191–215. University Park Press. Baltimore, Md.
22. WATERS, M. D., N. E. GARRETT, C. M. COVONE-DE SERRES, B. E. HOWARD & H. F. STACK. 1982. Genetic bioassay data on some known or suspected human carcinogens. In Chemical Mutagens: Principles and Methods for Their Detection. F. J. de Serres & A. Hollaender, Eds. 8. Plenum Press. New York, N.Y.
23. ZEIGER, E. 1983. An overview of genetic toxicity testing in the National Toxicology Program. Ann. N.Y. Acad. Sci. (This volume.)

THE ROLE OF *NEUROSPORA* IN EVALUATING ENVIRONMENTAL CHEMICALS FOR MUTAGENIC ACTIVITY

Frederick J. de Serres* and Henrich V. Malling†

*Office of the Director
†Laboratory of Biochemical Genetics
National Institute of Environmental Health Sciences
Research Triangle Park, North Carolina 27709

INTRODUCTION

Testing of environmental chemicals to evaluate their mutagenic activity is performed in three distinct phases: (1) hazard identification, (2) hazard evaluation, and (3) risk estimation. In hazard identification, we are interested in determining whether a chemical possesses mutagenic activity. In hazard evaluation, we try to determine the spectrum of genetic damage that will be produced in eukaryotic organisms as well as to obtain information on relative potency. In risk estimation, we try to determine the impact of transmitted genetic damage on subsequent generations. Because of the fact that genetically active chemicals can produce different types of genetic damage, the types of assay systems used for each phase of evaluation can be quite different. For example, in hazard identification, the *Salmonella* assay for reverse mutation developed by Ames and his colleagues is widely used,[1,2] but this assay only provides information on the induction of point mutations. In hazard evaluation, we try to determine whether the chemical that is active in a prokaryotic organism is also active in eukaryotic organisms. We try also to determine the spectrum of genetic damage produced (i.e., chromosome rearrangements, abnormal chromosome numbers, gene mutation by means of point mutation as well as small chromosomal or multilocus deletions, etc.) In hazard evaluation, it is also useful to use assays in such eukaryotic organisms as fungi, *Drosophila,* and mammalian cells in culture. For risk estimation, assays are made on whole animals, not only to determine relative potency but also to determine whether the agent produces genetic damage in germ cells that will be passed on to subsequent generations. For gene mutation it is known, for example, that more serious deleterious effects can be found in the F_1 progeny when the gene mutations result from multilocus deletions.[26,27] Thus chemicals that produce gene mutations predominantly by multilocus deletion would be expected to have a much greater impact on the F_1 generation than would those chemicals that produce gene mutations predominantly by point mutation.

THE ROLE OF *NEUROSPORA* IN TESTING ENVIRONMENTAL CHEMICALS FOR MUTAGENIC ACTIVITY

Neurospora crassa has been used extensively to evaluate the mutagenicity of environmental chemicals and to provide useful information on their mutagenic activity. The earliest role was the use of a *Neurospora* reverse-mutation assay in hazard identification. A *Neurospora* reverse-mutation assay developed by M. Westergaard and H. K. Mitchell (unpublished) was one of the first assays used to test environmental chemicals for mutagenic activity. This assay utilized a colonial,

177

0077–8923/83/0407–0177 $01.75/0 © 1983, NYAS

adenine-requiring strain (the double mutant 70007, 38701) to measure reverse mutation from adenine requiring to adenine independence.[19] The 38701 allele has a low spontaneous reversion frequency (0.8×10^{-7} conidia) so that effects of very weak mutagens could be readily detected. This strain was used in the late 1940s and early 1950s by several research groups to investigate the mutagenicity of a wide range of chemicals both alone and in combination.[11,14,17,18,20] In an early review of some of this work, data on 24 chemicals were tabulated to illustrate these early attempts to relate chemical structure with mutagenic activity.[29] In addition, the dose-response curves for reverse mutation of 38701 made it possible to evaluate relative potency. It was clear from this work that chemical mutagens could mimic the mutagenic effects of radiation on this strain. In addition, the dose-response curves for reverse mutation of 38701 made it possible to evaluate relative potency. For example, formaldehyde had an induced reverse-mutation frequency of 0.3×10^{-6} treated conidia, whereas that for diepoxybutane was 85.0×10^{-6}.

The development of a double-mutant strain that was both adenine requiring and inositol requiring (38701, 37401)[17] made it possible to assay for reverse mutation at two loci. Those experiments showed striking evidence for allele specificity.[17,29] For example, ultraviolet (UV) treatment gave a higher frequency of inositol revertants than adenine revertants, whereas the reverse was true after treatment with diethyl sulfate. In experiments with chloroethyl sulfate, there was a high frequency of adenine revertants induced but no inositol revertants.

From these studies, it became clear that induced revertant frequencies were allele specific and that the reverse-mutation pattern of a particular allele was probably dependent upon how the gene was originally damaged. This allele specificity was a problem of great concern since any attempt to determine dose-response curves and relative potency would be allele specific. It became obvious that it was possible to determine whether a chemical was mutagenic with this double-mutant strain but not to study induction kinetics and relative potency. Clearly, this assay was useful for hazard identification but, unfortunately, it required more resources than assaying for reversion with bacteria.[12] Because of this, the *Neurospora* assays developed with 38701 and with the double mutant 38701, 37401 were never widely used.

Neurospora plays a more important role, however, in the assessment process by providing data useful for hazard evaluation. Data from forward-mutational assays are particularly useful to provide information on relative potency in eukaryotic organisms as well as information on the characteristics of the specific locus mutations (i.e., whether they result from point mutation and/or multilocus deletion).

The first direct method for detecting forward mutations that occur at specific loci was based on pigment accumulation by the *ad-3A* and *ad-3B* mutants.[5] Mutants at these two loci not only have a requirement for adenine but they accumulate a reddish purple pigment in the vacuoles of the mycelium. By using a colonial, albino strain and incubating conidia in soft agar (0.15%) in large florence flasks with aeration, all conidia develop into colonies about 2 mm in diameter after about seven days of incubation. Wild-type conidia form white colonies, whereas *ad-3* mutants form pigmented colonies. The development of this technique made it possible to collect large numbers of *ad-3* mutants and to determine their frequencies precisely.

In a typical experiment, four to five different treatments are given and the resulting data can be expressed in terms of a dose-response curve for the induction of *ad-3* mutations. The genetic characterization of these purple *ad-3* mutants makes it possible to develop separate dose-response curves for the *ad-3A* locus and for the *ad-3B* locus.[5] The original technique was then modified so that a standard wild-type strain could be used (rather than the colonial, albino strain used in the original experiment), and the treatment procedure was standardized.[3,8]

The next major advance was the development of a two-component heterokaryon that is heterozygous in the *ad-3* region ($+/ad-3A$, $+/ad-3B$). With this strain, we can not only detect *ad-3* mutants that result from point mutation at the *ad-3A* and *ad-3B* loci but also multilocus deletions. Because the *ad-3A* and *ad-3B* loci are closely linked, these multilocus deletions can cover the *ad-3A* locus or the *ad-3B* locus or both loci simultaneously.[28] This approach provides not only precise dose-response curves for mutation induction in the *ad-3* region but also information on relative potency (TABLE 1). When the *ad-3* mutations recovered are characterized genetically,[4] then the relative distribution of mutants into various subclasses (i.e., point mutations versus multilocus deletions) provides information that is useful for helping to evaluate the impact of this damage on the F_1 and subsequent generations.

To illustrate the utility of the *ad-3* assay in *Neurospora*, we can examine data on the mutagenicity of 1,2 dibromoethane [or ethylene dibromide (EDB)].

TABLE 1

CHARACTERIZATION OF CHEMICAL MUTAGENS ON THE BASIS OF
INDUCED *ad-3* MUTATION FREQUENCY

Mutation Frequency $\times$ 10^{-6} Survivors	Relative Activity
1–10	weak
10–100	moderate
100–1,000	strong
1,000–10,000	potent

MUTAGENICITY OF EDB IN *NEUROSPORA*

The mutagenicity of EDB in *Neurospora* was first reported by us in 1969,[21] and a preliminary note on the genetic analysis of EDB-induced *ad-3* mutation was reported in 1970.[6] This chemical is in widespread distribution in the environment because of its use as a gasoline antiknock additive and as a soil and grain fumigant. As a result of these uses, human exposure can be expected; occupational exposures may include gasoline station workers, agricultural and grain storage workers, as well as workers in oil refineries and plants producing EDB. EDB has been found to be carcinogenic in both rats and mice.[13]

The results of an experiment in which EDB-treated conidia of heterokaryon 12 (H-12) were assayed for inactivation and the induction of *ad-3* mutations are given in FIGURE 1. In this experiment, conidia were treated for three hours with various concentrations of EDB (1.25 to 1.63 pl per ml) of a 0.067 M phosphate buffer adjusted to pH 7.0 containing 10% dimethyl sulfoxide.

One particularly interesting aspect of these data is the very narrow range of concentrations required to decrease survival from 100% down to 30%. We experienced considerable variation in the dose-response curves for inactivation as well as mutation induction from experiment to experiment, and a number of experiments were performed to obtain the dose-response curves plotted in FIGURE 1.

Another interesting aspect of the experiment is the nonlinear dose-response curve for mutation induction. The spontaneous *ad-3* mutation frequency is 0.38×10^{-6} survivors, and the lowest EDB exposure gave an induced frequency of 19.3×10^{-6} survivors or roughly a 50-fold increase over the spontaneous frequency. The main point here is that if we were to subtract the spontaneous frequency from the induced frequency, the dose-response curve would not become linear.

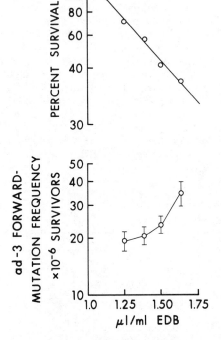

FIGURE 1. Dose-response curves for inactivation and induction of *ad-3* mutations in heterokaryon 12 of *Neurospora crassa* (experiment 12-125).

From these data we can see that EDB would be classified as a moderate mutagen (see TABLE 1) since the maximum forward-mutation frequency found falls into the range 10 to 100 × 10^{-6} survivors.

The next phase of the assessment is the genetic characterization of the *ad-3* mutations to determine the spectrum of EDB-induced genetic alterations. As discussed in previous publications, *ad-3* mutations can result from either point mutations in the *ad-3* region at the *ad-3A* or *ad-3B* locus (*ad-3A^R* or *ad-3B^R*) or multilocus deletions covering one or both loci as well as the *nic-2* (nicotinic acid–requiring) locus located some 4–8 map units to the right.[2,4,8,9] Multilocus deletions can result from any one of the following genotypes: *ad-3A^{IR}*, *ad-3B^{IR}*, (*ad-3A, ad-3B*)^{IR}, (*ad-3A ad-3B nic-2*)^{IR}, or (*ad-3B nic-2*)^{IR}. Several other double *ad-3* mutant genotypes are possible, but these usually occur very infrequently: *ad-3A^R ad-3B^R*, *ad-3A^{IR} ad-3B^R*, *ad-3A^R ad-3B^{IR}*. The probability of these latter mutants, if the events that give rise to them occur independently of one another, should be the product of their individual frequencies. Thus if *ad-3A^R* = 1.0 × 10^{-5} and *ad-3B^{IR}* = 2.0 × 10^{-6}, then *ad-3A^R ad-3B^{IR}* should be 2.0 × 10^{-11}. In the sample sizes generally used to characterize *ad-3* mutants (250–300 mutants), we never expect to encounter mutants of this genotype with forward-mutation frequencies in the range 50 to 100 × 10^{-6} survivors.

Ad-3 mutations that are picked up on the basis of pigment accumulation can be characterized by simple heterokaryon tests as *ad-3A* or *ad-3B*. In addition to this type of analysis, *ad-3^R* mutations show allelic complementation and specify a linear complementation map of 17 complementation groups, or *complons*.[10] On this map, mutants have three types of complementation patterns: complementing with either nonpolarized (NP) or polarized (P) patterns or noncomplementing (NC). NP mutants are believed to result from missense mutations that specify complete polypeptide chains with only a single erroneous amino acid. P and NC mutants result from nonsense or frameshift mutations that specify polypeptide fragments or, in the case of NC mutants, no polypeptide chain at all.[22,23] If the percentages of *ad-3B* mutants that show allelic complementation (NP + P) are high, then we know that the majority of mutants result from base-pair substitution. For example the percentage of complementing mutants induced by 2-aminopurine is 77.1%[3] and *N*-methyl-*N'*-nitro-*N*-nitrosoguanidine (MNNG) is 82.8%.[24] However, the percentage of complementing mutants is 25.5% when the *ad-3B* mutants are derived from a frameshift mutagen like ICR-170 {2-methoxy-6-chloro-9-[(ethyl-2-chloroethyl) amino propylamino] acridine dihydrochloride}.[7]

Equally interesting are the data on recessive lethal (RL) damage at loci other than the *ad-3A* locus or the *ad-3B* locus. The dikaryon and trikaryon tests[4] enable us to determine whether additional damage has been produced outside of the *ad-3* region elsewhere in the genome (*ad-3^R* + *RL*) or whether this RL damage is closely linked to the *ad-3* region (*ad-3^R* + *RL^{CL}*). This latter type of damage would be expected to occur very infrequently, if the two sites of damage occur independently of one another. If they occur frequently, this should tell us something about the mutagenic activity of the agent under analysis.

CHARACTERIZATION OF EDB-INDUCED *ad-3* MUTATIONS

The results of the genetic characterization of EDB-induced *ad-3* mutations are given in TABLE 2. In a total of 389 *ad-3* mutations characterized, 84% were point

TABLE 2

DISTRIBUTION AND FREQUENCY OF EDB-INDUCED *ad-3* MUTATIONS

Genotype	Number	Subset	
		Fraction	Percent
ad-3	388	388/388	100.0
ad-3^R	327	327/388	84.3
ad-3^{IR}	61	61/388	15.7
ad-3A^R	103	103/388	26.6
ad-3B^R	224	224/388	57.7
ad-3B^{NP}	64	64/388	16.5
ad-3B^P	30	30/388	7.7
ad-3B^{NC}	130	130/388	33.5
ad-3A^{IR}	16	16/388	4.1
ad-3B^{IR}	8	8/388	2.1
(*ad-3A ad-3B*)^{IR}	32	32/388	8.3
(*ad-3A ad-3B nic-2*)^{IR}	4	4/388	1.0
(*ad-3B nic-2*)^{IR}	0	0/388	0.0
ad-3A^{IR} ad-3B^R	1	1/388	0.3
ad-3A^R ad-3B^{IR}	0	0/388	0.0
ad-3^R + *RL*	53	53/327	16.2
ad-3^R + *RL^{CL}*	16	16/327	4.9

mutations $(ad\text{-}3^R)$ and 16% were multilocus deletions $(ad\text{-}3^{IR})$. The 327 point mutations can be further subdivided into 31.5% $ad\text{-}3A^R$ and 68.5% $ad\text{-}3B^R$ mutants. The ratio $ad\text{-}3B/ad\text{-}3A$ indicates that there are twice as many $ad\text{-}3B$ mutants as $ad\text{-}3A$ mutants, which is in agreement with our data on x-rays[28] as well as our data with UV.[15] When heterokaryon tests for allelic complementation were performed on the $ad\text{-}3B$ mutants, a total of 42% were found to be complementing. This spectrum is similar to that produced by UV,[15] which has been shown to produce $ad\text{-}3B$ mutants as a result of both base-pair substitutions and frameshift mutations.[16]

When the 62 mutants resulting from multilocus deletion $(ad\text{-}3^{IR})$ were characterized, the majority (52%) were found to be $(ad\text{-}3A\ ad\text{-}3B)^{IR}$ with 13% $ad\text{-}3B^{IR}$ and 26% $ad\text{-}3A^{IR}$. Four, or 6.5%, were large deletions, which also covered the $nic\text{-}2$ locus $(ad\text{-}3A\ ad\text{-}3B\ nic\text{-}2)^{IR}$. This distribution is different from the spectrum obtained with either x-rays $(0.025 < p < 0.05)$ or UV $(p < 0.005)$ as shown in TABLE 3. There are usually two to three times as many $ad\text{-}3B^{IR}$ mutants as $ad\text{-}3A^{IR}$ mutants. In the EDB sample, this situation is reversed for reasons that are not immediately obvious. In general, the EDB sample (with the exception of the $ad\text{-}3B/ad\text{-}3A$ ratio) is more like that produced by x-rays than by UV in that the majority of the mutants are deletions that cover two or more loci in the $ad\text{-}3$ and adjacent regions.

TABLE 3

GENOTYPES OF MULTILOCUS DELETION MUTATIONS IN THE $ad\text{-}3$ REGION
AS A FUNCTION OF MUTAGENIC ORIGIN

Mutagen	Total $ad\text{-}3^{IR}$ Sample	$ad\text{-}3A^{IR}$		$ad\text{-}3B^{IR}$		$ad\text{-}3A$ $ad\text{-}3B$ $ad\text{-}3B^{IR}$		$ad\text{-}3A$ $ad\text{-}3B$ $nic\text{-}2^{IR}$		$ad\text{-}3B$ $nic\text{-}2^{IR}$	
		No.	%	No.	%	No.	%	No.	%	No.	%
EDB	62	16	25.8	8	12.9	32	51.6	4	6.5	0	0.0
UV[15]	48	7	14.6	22	45.8	19	39.6	0	0.0	0	0.0
X-Rays[28]	201	34	16.9	68	33.8	87	43.3	10	5.0	2	1.0

It is also of interest that one of the EDB-induced multilocus deletions results from genetic damage that is expected to occur very infrequently: $ad\text{-}3A^{IR}\ ad\text{-}3B^R$. This double mutant is basically $ad\text{-}3^{IR} + RL^{CL}$ where the closely linked site of recessive lethal damage is located at the $ad\text{-}3B$ locus. Further evidence for double mutants comes from the dikaryon and trikaryon tests, which show that 69 of the 327 $ad\text{-}3^R$ mutations have a separate site of recessive lethal damage: $ad\text{-}3^R + RL$. What is totally unexpected is that in 16/69 such mutants, the separate site of RL damage is in the $ad\text{-}3$ region $(ad\text{-}3^R + RL^{CL})$.

IMPLICATION OF DATA FROM THE $ad\text{-}3$ BIOASSAY FOR HAZARD EVALUATION

The data from forward-mutational assays with the $ad\text{-}3$ test system in two-component heterokaryons is especially relevant to hazard evaluation where we are trying to determine whether an agent is mutagenic in a eukaryotic organism, where we try to get some idea of relative potency (in this case for inducing gene mutations by forward mutation), and where we want to be able to evaluate the production of the same spectrum of genetic damage that would occur in the whole animal. The $ad\text{-}3$

forward-mutation assay in two-component heterokaryons provides the same capability, for example, as the *Drosophila* sex-linked recessive lethal assay and the specific locus assay in mammalian cells in culture. All three assays provide useful information with regard to relative potency as well as a measure of mutagenic activity in a eukaryotic organism.

Where the *ad-3* assay system in *Neurospora* is especially useful is to provide a genetic characterization of the specific locus mutations recovered in terms of the extent of the genetic damage in each mutant. The relative frequencies of point mutations and multilocus deletions can be readily determined. This assay provides the same type of information as do mutations in the dilute (d) short ear (se) region of mice, where mutants can also be recovered at two closely linked loci.[26] The albino (c) locus in mice permits a similar type of analysis.[27] With the discovery by Russell that many of the *d se* mutations as well as the *c* locus mutations that result from multilocus deletion have deleterious effects in heterozygous condition on the F_1 progeny, the significance of this type of analysis has become readily apparent. It is clear that in evaluating the relative risk of the production of gene mutations, we must have information on the relative frequencies of point mutations and multilocus deletions. The relative risk of mutations of the latter type is much greater than the former. For example, if we have two agents of comparable potency, but in the first case the agent produces specific locus mutations exclusively by point mutation whereas in the second case the agent produces specific locus mutations by multilocus deletions, the risk to future generations would be completely different. An example of a mutagen of the first type is MNNG. Essentially 100% of the MNNG-induced *ad-3* mutations result from point mutation at the *ad-3A* or *ad-3B* locus.[24] On the other hand, *ad-3* mutations induced by hycanthone, lucanthone, and four structurally related analogues result predominantly from multilocus deletion. In these cases, some 82–88% of the *ad-3* mutations are of genotype $(ad\text{-}3A \ ad\text{-}3B \ nic\text{-}2)^{IR}$ or $(ad\text{-}3B \ nic\text{-}2)^{IR}$, showing that not only are multilocus deletions produced but they tend to be, in the main part, the longest that can be recovered in this region.[25] Assuming comparable transmission of both kinds of mutations through the germ line (and there is no evidence to the contrary), the specific locus mutations resulting from multilocus deletions would provide a serious hazard to the F_1 progeny whereas the risk from those arising from point mutation would be essentially negligible by comparison. The main point of interest with regard to risk estimation is that gene mutations can be qualitatively different from one another and as a result can produce very different effects on the F_1 and subsequent generations.

In summary, fungal systems offer no clear-cut advantage over bacterial systems for hazard identification. There are chemicals that are positive in fungi but negative in bacteria and vice versa, but the data base is not large enough to make an objective evaluation. However, the *Neurospora* assay of forward mutation in the *ad-3* region in two-component heterokaryons does provide a unique data base for hazard evaluation. The same spectrum of genetic alterations can be recovered as gives rise to specific locus mutations in higher organisms. Also the direct method for recovery of *ad-3* mutations makes it possible to obtain quantitative estimates of the induction of forward mutations at two specific loci that provide data on relative potency. The ability to characterize the *ad-3* mutants by a series of genetic tests makes it possible to generate a data base that can supplement the data on whole animals used in risk estimation.

ACKNOWLEDGMENT

The authors are indebted to Dr. Christopher Portier of the Biometry and Risk Assessment Program for providing the statistical analysis of the EDB data.

REFERENCES

1. AMES, B. N. 1971. The detection of chemical mutagens with enteric bacteria. *In* Chemical Mutagens; Principles and Methods for Their Detection. A. Hollaender, Ed. **1:** 267–282. Plenum Press. New York, N.Y.
2. AMES, B. N., J. McCANN & E. YAMASAKI. 1975. Methods for detecting carcinogens and mutagens with the salmonella/mammalian-microsome mutagenicity test. Mutat. Res. **31:** 347–364.
3. BROCKMAN, H. E. & F. J. DE SERRES. 1963. Induction of *ad-3* mutants of *Neurospora crassa* by 2-aminopurine. Genetics **48:** 597–604.
4. DE SERRES, F. J. 1981. Induction and genetic characterization of specific locus mutations in the *ad-3* region in two-component heterokaryons of *Neurospora crassa*. *In* Short-Term Tests for Chemical Carcinogens. H. F. Stich & R. H. C. San, Eds.: 175–186. Springer-Verlag. New York, N.Y.
5. DE SERRES, F. J. & H. G. KØLMARK. 1958. A direct method for determination of forward-mutation rates in *Neurospora crassa*. Nature **182:** 1249–1250.
6. DE SERRES, F. J. & H. V. MALLING. 1970. Genetic analysis of *ad-3* mutants of *Neurospora crassa* induced by ethylene dibromide. EMS Newsletter **3:** 36–37.
7. DE SERRES, F. J. & H. V. MALLING. Unpublished.
8. DE SERRES, F. J. & H. V. MALLING. 1971. Measurement of recessive lethal damage over the entire genome and at two specific loci in the *ad-3* region of a two-component heterokaryon of *Neurospora crassa*. *In* Chemical Mutagens; Principles and Methods for Their Detection. A. Hollaender, Ed. **2:** 311–342. Plenum Press. New York, N.Y.
9. DE SERRES, F. J. & R. S. OSTERBIND. 1962. Estimation of the relative frequencies of X-ray-induced variable and recessive lethal mutations in the *ad-3* region of *Neurospora crassa*. Genetics **47:** 793–796.
10. DE SERRES, F. J., H. E. BROCKMAN, W. E. BARNETT & H. G. KØLMARK. 1971. Mutagen specificity in neurospora. Mutat. Res. **12:** 129–142.
11. DICKEY, F. H., G. H. CLELAND & C. LOTZ. 1949. The role of organic peroxides in the induction of mutations. Proc. Nat. Acad. Sci. USA **35:** 581–586.
12. HEMMERLY, J. & M. DEMEREC. 1955. Tests of chemicals for mutagenicity. Cancer Res. **15**(Suppl.): 69–75.
13. IARC. 1977. Some Fumigants, the Herbicides 2,4-D and 2,4,5-T, Chlorinate Dibenzodioxins and Miscellaneous Industrial Chemicals. IARC Monographs on the Evaluation of the Carcinogenic Risk of Chemicals to Man **15:** 195–209. International Agency for Research on Cancer. Lyon, France.
14. JENSEN, K. A., H. G. KØLMARK & M. WESTERGAARD. 1949. Back mutation in *Neurospora crassa* induced by diazomethane. Hereditas **35:** 521–525.
15. KILBEY, B. J. & F. J. DE SERRES. 1967. Quantitative and qualitative aspects of photoreactivation of premutational ultraviolet damage at the *ad-3* loci of *Neurospora crassa*. Mutat. Res. **4:** 21–29.
16. KILBEY, B. J., F. J. DE SERRES & H. V. MALLING. 1971. Identification of the genetic alteration at the molecular level of ultraviolet-light-induced *ad-3B* mutants in *Neurospora crassa*. Mutat. Res. **12:** 47–56.
17. KØLMARK, H. G. 1953. Differential response to mutagens as studied by the neurospora reverse-mutation test. Genetics **39:** 270–276.
18. KØLMARK, G. & N. H. GILES. 1955. Comparative studies of monoepoxides as inducers of reverse-mutation in neurospora. Genetics **40:** 890–902.
19. KØLMARK, H. G. & M. WESTERGAARD. 1949. Induced back-mutations in a specific gene of *Neurospora crassa*. Hereditas **35:** 490–506.
20. KØLMARK, G. & M. WESTERGAARD. 1953. Further studies on chemically induced reversions at the adenine locus of neurospora. Hereditas **39:** 209–224.
21. MALLING, H. V. 1969. Ethylene dibromide: a potent pesticide with high mutagenic activity. Genetics **61:** s39.
22. MALLING, H. V. & F. J. DE SERRES. 1967. Relation between complementation patterns and genetic alterations in nitrous acid–induced *ad-3B* mutants of *Neurospora crassa*. Mutat. Res. **4:** 425–440.

23. MALLING, H. V. & F. J. DE SERRES. 1968. Identification of genetic alterations induced by ethyl methane sulfonate in *Neurospora crassa*. Mutat. Res. **6:** 181–193.
24. MALLING, H. V. & F. J. DE SERRES. 1970. Genetic effects of *N*-methyl-*N'*-nitro-*N*-nitrosoguanidine in *Neurospora crassa*. Molec. Gen. Genet. **106:** 195–207.
25. ONG, T. & F. J. DE SERRES. 1980. Genetic analysis of *ad-3* mutants induced by hycanthone, lucanthone and their indazole analogs in *Neurospora crassa*. J. Environ. Pathol. Toxicol. **4:** 1–8.
26. RUSSELL, L. B. 1971. Definition of functional units in a small chromosomal segment of the mouse and its use in interpreting the nature of radiation induced mutations. Mutat. Res. **11:** 107–123.
27. RUSSELL, L. B., W. L. RUSSELL & E. M. KELLY. 1979. Analysis of the albino-locus region of the mouse. I. Origin and variability. Genetics **91:** 127–139.
28. WEBBER, B. B. & F. J. DE SERRES. 1965. Induction kinetics and genetic analysis of X-ray-induced mutations in the *ad-3* region of *Neurospora crassa*. Proc. Nat. Acad. Sci. USA **53:** 430–437.
29. WESTERGAARD, M. 1957. Chemical mutagenesis in relation to the concept of the gene. Experientia **13:** 224–234.

MUTAGENICITY SCREENING WITH FUNGAL SYSTEMS

Friedrich K. Zimmermann*

Genetics
Biology Department
Technical University
D-6100 Darmstadt, Federal Republic of Germany

INTRODUCTION

Adverse effects induced in the genome of man can be created by the induction of point mutations, chromosomal structural aberrations, and aneuploidy. Experimental systems to evaluate the mutagenic potential[1] of a given agent to man require genetic tests that cover all these effects. However, tests commonly used in environmental mutagenicity screening are based on reactions that result from chemical alterations induced directly in DNA or interfere with DNA metabolism.[2] However, proper segregation of chromosomes requires an intact spindle fiber function, and there the molecular target and essential component is tubulin.[3] Consequently, tests based on the induction of point mutations or on differences in DNA-repair systems between normal cells and repair-deficient derivatives need not reveal the mutagenic activity of chemicals that induce only chromosomal malsegregation because they interfere with tubulin but not with DNA. Specific tests for chromosomal malsegregations are available in fungi (FIGURE 1). The use of fungi in mutagenicity testing requires exogenous sources for metabolic activation. This was recognized very early, and the first use of mouse liver microsomes was reported by Malling in experiments with *Neurospora crassa*.[4] Since then a wide spectrum of sources of exogenous activation have been combined with fungal mutagenicity test systems.[5] Consequently, fungal systems appear to be particularly suited for mutagenicity testing.

The importance of fungi in a comprehensive testing battery is documented by two recent reports. Parry and associates tested a number of tumor promoters and related substances for induction of mitotic aneuploidy in yeast.[6] This strain allows simultaneously the detection of point mutation, mitotic recombination, and mitotic chromosome loss. Several tumor promoters, among them 12-*O*-tetradecanoylphorbol-13-acetate, were positive only in the induction of mitotic aneuploidy and not in any other type of genetic effect. Most interesting was the observation that two fatty acids, lauric and oleic acids, induced mitotic chromosome loss. It is most unlikely that fatty acids can directly alter DNA; however, they may very well interfere with the integrity and function of the cellular membrane systems.[7] The other contribution, by Gualandi and Bellincampi, is based on the use of a strain of *Aspergillus nidulans*,[8] which monitors the full spectrum of genetic eukaryotic changes. The authors tested known membrane-active chemicals like ethanol, amphotericin B, and miconazole and two agents known to interfere with the spindle fiber apparatus, benomyl and isopropyl-3-chlorophenyl carbamate.[9] Again, there was a specific induction of mitotic aneuploidy but no other type of genetic effect. Consequently, both reports document clearly that there are agents that would be missed in standard genetic toxicology testing based on effects induced directly in DNA or indirectly by interference with enzymes of DNA

*The author's research is supported by Contract No. NO1-ES-1-5005, Development of a Yeast Aneuploidy Test System, from the National Institute of Environmental Health Sciences, Research Triangle Park, N.C.

a. Meiotic Systems

Neurospora crassa[27], a set of two haploid strains of opposite

mating types: a <u>arg-1   centromere   +   ad-3B   + </u>
               A   +   centromere ad-3A   +   nic-2

Saccharomyces cerevisiae

Strain D9J2[40]

$\dfrac{\text{ade3} \quad +}{+ \quad \text{ade6}}$ centromere $\dfrac{\text{leu1} \quad + \quad \text{cyh2} \quad + \quad \text{aro2} \quad + \quad \text{ade5}}{+ \quad \text{trp5} \quad + \quad \text{met13} \quad + \quad \text{lys5} \quad +}$

Strain DIS13[38]

$\dfrac{\text{can} \quad +}{+ \quad \text{ura3}}$ centromere $\dfrac{+ \quad \text{his1} \quad + \quad \text{ilv1} \quad + \quad \text{met5}}{\text{hom3} \quad + \quad \text{arg6} \quad + \quad \text{trp2} \quad +}$

b. Mitotic Systems

Aspergillus nidulans

Strain P[14]

$\dfrac{\text{su adE} \quad \text{riboA1} \quad + \quad + \quad \text{proA1}}{+ \quad + \quad \text{fpaA1 anA1} \quad +}$ centr $\dfrac{+ \quad + \quad \text{adE20} \quad \text{biA1}}{\text{pabaA1} \quad \text{yA2} \quad + \quad +}$

Saccharomyces cerevisiae

Strain D6[41]

$\dfrac{\text{ade3}}{+}$ centromere $\dfrac{\text{leu1} \quad \text{trp5} \quad \text{cyh2} \quad \text{met13}}{+ \quad + \quad + \quad +}$ $\dfrac{\text{ade2-40}}{\text{ade2-40}}$

Strain D61.M

$\dfrac{\text{ade6}}{+}$ centromere $\dfrac{\text{leu1} \quad + \quad \text{cyh2}}{+ \quad \text{trp5} \quad +}$ $\dfrac{\text{ade2-40}}{\text{ade2-40}}$

FIGURE 1. The relevant genotypes of fungal strains used to detect induction of aneuploidy. (Superscripts following strain designations indicate references.)

metabolism. It is precisely this situation that makes fungi very important components of a testing battery.

TEST ORGANISMS AND METHODS

Fungi are eukaryotes with defined chromosome numbers contained in a typical nucleus. Mitotic nuclear division proceeds without the breakdown of the nuclear membrane. Meiosis occurs under specific conditions and leads to the formation of haploid spores contained in an ascus in the case of ascomycetous fungi. Analysis of segregation of genetic markers in such asci clearly shows a proper Mendelian segregation. Fungal chromosomes are unambiguously defined as formal genetic elements but are too small to be identified under the light microscope. This is the reason why no cytogenetic studies have been reported in connection with mutagenicity testing. However, it is possible to construct special strains that monitor induction of chromosomal structural aberrations. Fungal species have been used to test for the induction of point mutations (forward and reverse mutation), mitotic recombination, and only recently aneuploidy in mitotic and meiotic cells.

Aspergillus nidulans. A general reference for the genetics of this mold has been provided by Clutterbuck.[10] Present methods and techniques have been described in several articles.[11-14]

Saccharomyces cerevisiae. This is probably the best investigated eukaryotic organism in respect to genetics and molecular biology. DNA in yeast chromosomes is organized in nucleosomes and there associated with typical histones H2A, H2B, H3, and H4. Histone H1 has not been found yet.[15] Chromosomal cytology and behavior during mitosis and meiosis have been reviewed,[16] and the spindle fiber apparatus is under intensive investigation.[17] Formal genetic analysis and electron-microscopic studies have established a haploid chromosome number of 17. Aspects of mutation induction[18] and mitotic recombination[19] have been reviewed recently.

Schizosaccharomyces pombe. This yeast has been used by a number of investigators for the study of induced forward and reverse mutation, but it has not found wide acceptance. Its use in forward-mutation assays has been described by Loprieno.[20]

Neurospora crassa. This is actually the classical microorganism for mutation induction. A very elaborate "heterokaryon test" developed by de Serres allows the study and identification of the full range of eukaryotic mutational events[21] (see also de Serres and Malling, this volume).

METABOLIC ACTIVATION

Fungal systems can monitor all kinds of genetic changes typical of eukaryotic organisms. However, fungal metabolism can differ from that of the various types of mammalian cells and, therefore, there is a problem with false negatives—mutagens that are not activated in fungi. On the other hand, fungi seem to lack certain detoxification systems typical of mammals and thus generate false positives. Dichlorvos is a prime example of such a false positive.[22] Moreover, there is a distinct possibility that fungi activate certain chemicals to ultimate mutagens whereas the same chemicals may not be metabolized in mammals.

Fungal cells are quite robust, and they can be combined with quite a variety of activation systems (reviewed in Reference 5) and can even be used in an intrasanguineous host-mediated assay as developed by Fahrig.[23] Consequently, a broad range of metabolic activation systems can be combined with fungal tester strains.

FUNGAL MUTAGENICITY TESTING

The most obvious test is a point mutation assay, and this is becoming more and more accepted because of the success of the *Salmonella* reverse-mutation assay. There have been recent attempts to establish equivalent tests for fungi in *Aspergillus*[12] and also in *Saccharomyces cerevisiae*.[24]

Most of the testing has been done using the induction of mitotic recombination in *Aspergillus* and in *Saccharomyces cerevisiae*. The Gene-Tox report for yeast lists 521 chemicals as tested.[25] Of these, 257 gave a positive response, and most of the results were obtained in mitotic recombination assays. Mitotic recombination has to be considered as an expression of repair activity in response to primary genetic damage inflicted upon DNA. A general discussion of mitotic recombination and its biological relevance has been presented by Resnick.[26]

FUNGAL SYSTEMS TO DETECT CHROMOSOMAL MALSEGREGATION

Bellincampi *et al.* set out to "establish whether damage to cellular membranes in fungi can result in a disturbance of mitotic division ultimately leading to a non-disjunction through disorientation of the spindle."[13] They tested a number of membrane-active antimycetic agents and found that amphotericin B, pimaricin, fenarimol, miconazole, and econazole were efficient inducers of mitotic nondisjunction in *Aspergillus nidulans*. As shown in TABLE 1, there are indeed a number of chemicals that induce only this type of damage with no other genetic changes being induced (such chemicals are shown in italics).

There are several systems in fungi that can be used to study meiotic malsegregation. They are shown in FIGURE 1. The *Neurospora* system is based on two haploid strains, which can be crossed.[27] Each strain carries a number of nutritional markers in coupling. Only multiple crossing-over events will result in the generation of prototrophs, and this is much rarer than the formation of disomic spores. This test system used on 48 agents has identified 10 as definite inducers of meiotic nondisjunction.

There are two systems in yeast where meiotic chromosomal malsegregation can be studied. The basic principle is a chromosome pair that shows an alternating array of recessive nutritional markers. There is Parry *et al.*'s strain D9J2 and Sora *et al.*'s DIS13 (see FIGURE 1 for genotypes).[40,38] Strain DIS13 is also designed to distinguish effects during meiosis I from those in meiosis II. The difficulty in yeast is that vegetative cells must be killed at very high efficiency, otherwise they will obscure the formation of the selected prototrophic disomics. Only a few agents have been tested in the two systems (TABLE 1). However, it is possible to compare the spontaneous frequencies of meiotic nondisjunctional events in the two species on a per-chromosome basis. In *Neurospora* the frequencies are about 5×10^{-5}, in *Saccharomyces* strain D9J2 around $2\text{--}8 \times 10^{-5}$, and in DIS13 almost 1×10^{-4}. Taking these frequencies at face value suggests that meiotic chromosomal segregation is quite precise in fungi.

Mitotic chromosomal malsegregation has been a test in *Aspergillus* for many years, and many data derive from work with this mold.[14] The basic principle there and in *Saccharomyces cerevisiae* is the same. There are recessive markers flanking the centromere of one chromosome. The markers are in coupling (see FIGURE 1). There is strain P of *Aspergillus nidulans* and strains D6 and D61.M of *Saccharomyces cerevisiae*. All three strains have in common that they not only signal induction of mitotic chromosome loss but also that of point mutation, mitotic recombination, and deletion of chromosomal fragments. It is the expression of recessive markers on both sides of the centromere, especially the ones very close to it, that indicates mitotic chromosome loss as opposed to other events. This situation is fortunate since it provides each strain with an internal control for classical types of genetic damage originating from direct or indirect DNA damage under the same conditions that induce chromosomal malsegregation. As shown in TABLE 1 for published data and TABLE 2 for original data, there are indeed numerous agents that induce mitotic chromosome loss in the two fungal species. It is difficult to compare at a quantitative level the

TABLE 1

SOME CHEMICALS SHOWN TO INDUCE CHROMOSOMAL MALSEGREGATION IN MITOTIC
AND/OR MEIOTIC FUNGAL TEST SYSTEMS*

Chemical	Test
Acetone	yeast mit.†
2-Acetylaminofluorene	yeast mit.[32]
4-Acetylaminofluorene	yeast mit.[32]
Actinomycin D	Asp. mit.[14]
Amethopterine	Neur. mei.[27]
Aminotriazole	Asp. mit.;[14] yeast mit.[32]
Amphotericin B	Asp. mit.[27]
Anthraline	yeast mit.[6]
Atrazine	Neur. mei.[27]
Benomyl (benlate)	Asp. mit.;[14] yeast mei.[38] yeast mit.;[39]†
Benzidine	yeast mit.[32]
Benzo[*a*]pyrene	yeast mit.[32]
Caffeine	Neur. mei.;[27] yeast mit.[40]
Carbendazim (bavistan)	Asp. mit.;[14] yeast mit.†
Chloroform	yeast mit.[32]
Chloroneb	Asp. mit.[14]
Cyclophosphamide	yeast mit.[32]
Decanoic acid	yeast mit.†
2,4-Diaminoanisole sulfate	Neur. mei.[27]
2,4-Diaminotoluene	Neur. mei.[27]
Dicloran	Asp. mit.[14]
Dichlorvos	Asp. mit.[14]
Diethylstilbestrol	yeast mit.[6]
Digitonin	yeast mit.†
9,10-Dimethylanthracene	yeast mit.[32]
Dimethylformamide	yeast mit.[32]
Diphenylnitrosamine	yeast mit.[32]
Econazole	Asp. mit.[14]
Epichlorhydrin	yeast mit.[32]
Ethanol	Asp. mit.[14]
Ethionine	yeast mit.[32]
Ethylenethiourea	yeast mit.[32]
Ethylmethanesulfonate	yeast mit.[40]
Fenarimol	Asp. mit.[13]
p-Fluorophenylalaine	Neur. mei.;[27] yeast mei.;[40] Asp. mit.;[14] yeast mit.[40]
Formaldehyde	Asp. mit.[14]
Griseofulvin	Asp. mit.[14]
Hexamethylphosphoramide	yeast mit.[32]
Hydrazine sulfate	yeast mit.[32]
8-Hydroquinoline	Neur. mei.[27]
Iodoacetic acid	yeast mit.[6]
O-Isopropyl-*N*-3-chlorophenyl carbamate	Asp. mit.[8]
Lanthanum nitrate	yeast mit.†
Lauric acid (dodecaonic acid)	yeast mit.[6]
Mercurochrome	oeast mit.[40]
2(3-Methoxycarbonyl thioureidoaniline)	Asp. mit.[14]
4,4'-Methylenebis(bischloroaniline)	yeast mit.[32]
Methylthiophanate	osp. mit.[14]
Methylmethanesulfonate	yeast mei.;[38] Asp. mit.[14]
Mitomycin C	yeast mit.[40]
3-Methyl-4-nitroquinoline-*N*-oxide	yeast mit.[32]

TABLE 1 (*continued*)

Chemical	Test
Miconazole	Asp. mit.[13]
Nitrogen mustard	Asp. mit.[14]
4-Nitroquinoline-*N*-oxide	yeast mit.[32]
Octanoic acid	yeast mit.†
Oleic acid	yeast mit.[6]
Pentachloronitrobenzine	Asp. mit.[14]
Phenmedipham	Asp. mit.[14]
Phorbol-12-13-didecanoate	yeast mit.[6]
Pimaricin	Asp. mit.[13]
β-Propiolactone	yeast mit.[32]
Saccharin	yeast mit.[6]
Safrole	yeast mit.[32]
Silver nitrate	yeast mit.†
Sulfacetamide	Neur. mei.[27]
Tetrachloronitrobenzene	Asp. mit.[14]
Tetradecanoic acid	yeast mit.†
12-O-Tetradecanoylphorbol-13-acetate	yeast mit.[6]
Thiabenzazole	Asp. mit.[14]
Thiocid	yeast mit.[40]
Thiophanate	Asp. mit.[14]
o-Toluidine	yeast mit.[32]
Trifluraline	Neur. mei.[27]
Trimethoprime	Neur. mei.[14]
Urethane	yeast mit.[40]

*Asp mit. = *Aspergillus nidulans* mitotic aneuploidy; Neur. mei. = *Neurospora crassa* meiotic nondisjunction; yeast mei. = yeast meiotic nondisjunction; yeast mit. = yeast mitotic aneuploidy. Chemicals that have definitely been reported as nonmutagenic in the same organism are italicized.
†See TABLE 2.

TABLE 2

CHEMICALS INDUCING MITOTIC ANEUPLOIDY IN STRAIN D61.M
OF *SACCHAROMYCES CEREVISIAE**

Chemical	Most Effective Concentration	White Colonies	Presumptive Monosomics	Per 10^6 Cells
Digitonin	control	8	8	0.69
	0.25 ppm	23	23	2.15
Fatty acids	control	0	0	0.13
Octanoic acid	5.0 ppm	14	13	1.22
Decanoic acid	14.5 ppm	102	45	6.12
Tetradecanoic acid	2.5 ppm	15	15	2.02
Acetone	control	1	1	0.25
	4.76%	20	20	1.39
Silver nitrate	control	0	0	<0.15
	0.14 ppm	12	11	1.78
Lanthanum nitrate	control	0	0	0.24
	33.3 ppm	23	22	3.96
Fungicides	control	0	0	0.10
Carbendazim				
CAS Nr. 10605-21-7	25.0 ppm	100	99	20.50
Benomyl (benlate)				
CAS Nr. 17804-35-2	123.0 ppm	102	102	19.73

*Treatment conditions: growing cultures in a synthetic medium for 17 hours at 28°C, 4–5 generations in the presence of chemical. White colonies express two recessive markers, cyh2 on the left and ade6 on the right-hand side of centromere VII. Presumptive monosomics express in addition to those markers centromere-linked recessive leu1 (causing a leucine requirement).

frequencies of spontaneous mitotic chromosome loss. In yeast, events are related to plating units usually in cultures that have grown for several generations in the presence of the agent. In *Aspergillus*, aneuploid sectors are counted in mycelial colonies and it is not immediately possible to determine frequencies on a per-nucleus basis. Spontaneous frequencies of such sectoring colonies are around 5 per 10,000. In yeast, spontaneous frequencies are much lower. Lower values in Parry's work are around 1 per 10^6, and this probably reflects the spontaneous rate. In strain D61.M, the frequencies are rather around 1 per 10^7 cells. All in all, the two species differ considerably with respect to the stability of mitotic chromosome numbers. It is possible that this may also reflect the difference between a natural diploid, yeast, and an exceptional diploid condition, as in *Aspergillus*.

THE RELEVANCE OF INDUCED FUNGAL ANEUPLOIDY

The agents listed in TABLE 1 comprise quite a variety of chemicals. Some are "classical" mutagens, and their registering positive in fungal aneuploidy tests would not provoke a need for specific aneuploidy tests.[28] However, some of the chemicals tested are tumor promoters and they registered positive only in the aneuploidy test. They are negative using the classical genetic end points.[6] Also positive in the aneuploidy test were membrane-active compounds[13] and the benzimidazole carbamate derivatives.[8] The specificity of inducing only aneuploidy but no other type of genetic effect raises the question of the molecular target.

Proper chromosomal segregation in eukaryotes is mediated by the spindle fiber apparatus, and it is there that the molecular targets have to be identified. The detailed mechanism of chromosomal segregation is still controversial even for the mitotic type of division.[2] However, there is general agreement that microtubules are a major component of the spindle apparatus, and at the molecular level, tubulin is a prominent constituent. Benomyl, one of the major agents inducing fungal aneuploidy, has been used to isolate resistant mutants in *Aspergillus*. Sheir-Neiss et al. could demonstrate that the resistance was based on altered β-tubulin.[29] This clearly proves that benomyl indeed acts on tubulin and on microtubules and that the genetic effects induced with this popular pesticide are indeed mediated through its interaction with the machinery needed for proper chromosomal segregation. Of course, this does not mean that this is also what happens in other eukaryotes. However, if tubulin is the target, then it has to be considered that tubulin is a highly conserved protein among eukaryotes[30,31] and, therefore, many eukaryotes may well be susceptible to the pesticide and respond with the same genetic effects. Another carbamate fungicide is isopropyl-3-chlorophenyl carbamate, CIPC, which induces specifically aneuploidy in *Aspergillus*.[8] This agent also has effects on mouse 3T3 fibroblasts *in vitro* when applied at concentrations common for pest control. It caused the disappearance of cytoplasmic microtubules, multipolar spindles, and unequal partitioning of nuclei between daughter cells. And the authors, Oliver et al., stated: "Fibroblast cultures exposed to CIPC and then returned to drug-free medium may provide chromosome deficient cells of value for genetic mapping and recombinant studies. On the other hand if similar responses to CIPC exposure and withdrawal occur in vivo then the herbicide may pose a genetic risk to man that is not revealed by usual testing procedures."[9] This makes it clear that even though the data presented in TABLE 1 refer to effects induced in fungi, they indicate at least a potential and serious threat of the same agents to man. Moreover, another carbamate derivative, parbendazole, has been shown to interfere with microtubules in tissue-culture cells and to bind to tubulin.[33]

It may not be justified to state that tubulin is tubulin as DNA is DNA. The high conservation of tubulin on the one hand would support this view. On the other hand, the experiments of Sheir-Neiss *et al.* suggest that one or two amino acid substitutions are sufficient to cause resistance to drugs interfering with microtubule assembly and function.[29] Consequently, we have to consider a certain degree of organism specificity. Nevertheless, fungal systems do indeed provide reliable genetic tests for the induction of aneuploidy and the identification of agents that induce aneuploidy. The predictive value of fungal tests for mammals can only be assessed in comparative studies with large numbers of agents.

There are other agents in TABLE 1 where the primary target is probably not tubulin and microtubules—the solvents acetone and dimethylformamide, the fatty acids, and some antimycetic pharmaceuticals like amphotericin B, pimaricin, econazole, and miconazole. It is known for the latter group of agents that they interfere with membranes.[34] However, membranes are an extremely important element of intracellular structure in eukaryotes. It is conceivable that agents interacting with membranes may induce secondary effects. Hoover *et al.* showed that *cis*-unsaturated fatty acids had an effect on the organization of cytoskeletal elements in lymphocytes and concluded: "It is suggested that the perturbation of the plasma membrane with unsaturated fatty acids alters the interaction of surface receptors with the cytoskeleton, which in turn affects cytoplasmic distribution of proteins."[7] After all, telomeres of chromosomes are attached to the nuclear membrane and, consequently, chromosomal malsegregation could very well be increased by primary membrane changes.

It is not so very obvious why lanthanum nitrate would lead to chromosomal malsegregation (TABLE 2). However, this agent was chosen because it has been used to contrast DNA for electron microscopy.[35] This agent turned out to induce only aneuploidy but no other type of genetic change. It remains to be seen what other kinds of heavy metal salts will have the same effects.

CONCLUSION

Fungi have played a certain role in environmental mutagenicity testing using as genetic end points the induction of point mutation and mitotic recombination. They also have identified a number of mutagens that turned out to be negative in blind testing in bacterial tests.[36] However, this was not always felt compelling enough to include fungal tests in a comprehensive testing battery and the conclusion was, "There seems to be no evidence that assays of recombinagenicity or aneuploidy were capable of detecting classes of carcinogens not detected by the prokaryote mutation assay."[28] This conclusion by Brookes and de Serres was based on the results of blind testing of 42 chemicals in the International Collaborative Program.[37] However, this program was based on the classical type of mutagens and carcinogens versus their inactive counterparts. The only notable exception was the inclusion of the solvent dimethylformamide, which registered positive, actually as a "false positive," in the aneuploidy test of Parry and Sharp.[32] In the future, fungi certainly will play a very important role in short-term tests for the detection of aneugenic agents that induce genetic effects via targets that are neither DNA nor enzymes involved in DNA metabolism. Consequently, they are not an optional alternative to all the other mutagenicity/short-term carcinogenicity tests, but a mandatory component in all comprehensive testing batteries.

SUMMARY

Several fungal species have been used for mutagenicity screening: *Aspergillus nidulans, Saccharomyces cerevisiae,* and *Neurospora crassa.* The eukaryotic nature of these organisms with typical chromosomes in a nucleus and their mitotic and meiotic mode of nuclear division have been the basis for the development of test systems that cover the full spectrum of genetic changes typical for eukaryotes. It is possible to detect simple point mutations and also grosser structural chromosomal alterations. Mitotic recombination as a repair test has found wide application. In recent years, the induction of mitotic and also meiotic chromosomal malsegregation has been investigated. It turned out that there are numerous chemicals that specifically induce only aneuploidy but no other type of genetic change. Among such chemicals are well-known tumor promoters, membrane-active agents, and others that have been shown to interact with tubulin and interfere with microtubule formation and function in mammalian cells. Such agents will be classified as nonmutagenic in any of the presently used mutagenicity test batteries, which monitor only changes that result from primary effects exerted either directly on DNA or indirectly via interference with DNA metabolizing enzymes. Consequently, fungal tests for induction of aneuploidy do not represent an optional alternative to other tests, but they are a mandatory part of any test battery aimed at uncovering all kinds of mutagenic agents.

REFERENCES

1. ZIMMERMANN, F. K. 1982. Can we determine mutagenicity or only a mutagenic potential? Mutat. Res. **92:** 3–7.
2. HAYNES, R. H., J. G. LITTLE, B. A. KUNZ & B. J. BARCLAY. 1981. Non-DNA primary targets for the induction of genetic change. *In* Environmental Mutagens and Carcinogens. T. Sugimura, S. Kondo & H. Takebe, Eds.: 121–128. University of Tokyo Press. Tokyo, Japan.
3. PICKET-HEAPS, J. D., D. H. TIPPIT & K. R. PORTER. 1982. Rethinking mitotis. Cell **29:** 729–744.
4. MALLING, H. V. 1971. Dimethylnitrosamine formation of mutagenic compounds by interaction with mouse liver microsomes. Mutat. Res. **13:** 425–429.
5. ZIMMERMANN, F. K., V. W. MAYER & J. M. PARRY. 1982. Genetic toxicology studies using *Saccharomyces cerevisiae*. J. Appl. Toxicol. **2:** 1–10.
6. PARRY, J. M., E. M. PARRY & J. C. BARRETT. 1981. Tumor promoters induce mitotic aneuploidy in yeast. Nature London **294:** 263–265.
7. HOOVER, R. L., K. FUJIMARA, R. D. KLAUSNER, D. K. BHALLA, R. TUCKER & M. J. KRANOSKY. 1981. Effects of free fatty acids on the organization of cytoskeletal elements in lymphocytes. Mol. Cell. Biol. **1:** 939–948.
8. GUALANDI, G. & D. BELLINCAMPI. 1981. Induced gene mutation and mitotic non-disjunction in *A. nidulans*. Toxicol. Lett. **9:** 389–394.
9. OLIVER, J. M., J. A. KRAWIEC & R. D. BERLIN. 1978. A carbamate herbicide causes microtubule and microfilament disruption and nuclear fragmentation in fibroblasts. Exp. Cell Res. **116:** 229–237.
10. CLUTTERBUCK, A. J. 1974. *Aspergillus nidulans*. *In* Handbook of Genetics. R. C. King, Ed. **1:** 447–510. Plenum Press. New York, N.Y.
11. DEMOPOULOS, N. A., A. KAPPAS & M. PELECANOS. 1982. Recombinagenic and mutagenic effects of the antitumor antibiotic bleomycin in *Aspergillus nidulans*. Mutat. Res. **102:** 51–57.
12. BIGNAMI, M., A. CARERE, G. CONTI, L. CONTI, R. CREBELLI & M. FABRICI. 1982. Evaluation of different genetic markers for the detection of frameshift and missense mutagens in *A. nidulans*. Mutat. Res. **97:** 293–302.
13. BELLINCAMPI, D., G. GUALANDI, E. LAMONICA, C. POLEY & G. MORPURGO. 1980. Membrane-damaging agents cause mitotic nondisjunction in *A. nidulans*. Mutat. Res. **79:** 169–172.
14. MORPURGO, G., D. BELLINCAMPI, G. GUALANDI, L. BADINELLI & O. S. CRESCENZI. 1979. Analysis of mitotic nondisjunction with *Aspergillus nidulans*. Environ. Health Perspect. **31:** 81–95.

15. FANGMAN, W. L. & V. A. ZAKIAN. 1981. Genome structure and replication. *In* The Molecular Biology of the Yeast *Saccharomyces*. Life Cycle and Inheritance. J. N. Strathern, E. W. Jones & J. R. Broach, Eds. **1**: 27–58. Cold Spring Harbor Laboratories. Cold Spring Harbor, N.Y.

16. BYERS, B. 1981. Cytology of the yeast life cycle. *In* The Molecular Biology of the Yeast *Saccharomyces*. Life Cycle and Inheritance. J. N. Strathern, E. W. Jones & J. R. Broach, Eds. **1**: 59–96. Cold Spring Harbor Laboratories. Cold Spring Harbor, N.Y.

17. KING, S. M., J. S. HYAMS & A. LUBA. 1982. Absence of microtubule sliding and an analysis of spindle formation and elongation in isolated spindles from the yeast *Saccharomyces cerevisiae*. J. Cell Biol. **94**: 341–349.

18. HAYNES, R. H. & B. A. KUNZ. 1981. DNA repair and mutagenesis in yeast. *In* The Molecular Biology of the Yeast *Saccharomyces*. Life Cycle and Inheritance. J. N. Strathern, E. W. Jones & J. R. Broach, Eds. **1**: 371–414. Cold Spring Harbor Laboratories. Cold Spring Harbor, N.Y.

19. KUNZ, B. A. & R. H. HAYNES. 1981. Phenomenology and genetic control of mitotic recombination in yeast. Annu. Rev. Genet. **15**: 57–89.

20. LOPRIENO, N. 1981. Screening of coded carcinogenic/noncarcinogenic chemicals by a forward-mutation system with the yeast *Schizosaccharomyces pombe*. *In* Evaluation of Short-Term Tests for Carcinogens. Report of the International Collaborative Program. F. J. de Serres & J. Ashby, Eds.: 424–433. Eleseiver/North-Holland, Inc. New York, N.Y.

21. DE SERRES, F. J. & H. V. MALLING. 1971. Measurement of recessive lethal damage over the entire genome and at two specific loci in the ad-3 region of a two-component heterokaryon of *Neurospora crassa*. *In* Chemical Mutagens, Principles and Methods for Their Detection. A. Hollaender, Ed. **2**: 311–342. Plenum Press. New York, N.Y.

22. IARC. 1979. IARC Monographs on the Evaluation of the Carcinogenic Risk of Chemicals to Humans **20**: 97–127. International Agency for Research on Cancer. Lyon, France.

23. FAHRIG, R. 1975. Development of host-mediated mutagenicity tests—yeast systems. II. Recovery of yeast cells out of testes, liver, lung and peritoneum of rats. Mutat. Res. **31**: 381–389.

24. MEHTA, R. D., U. G. G. HENNIG, R. C. VON BORSTEL & L. G. CHATTEN. 1982. Genetic activity in *Saccharomyces cerevisiae* and thin-layer chromatographic comparisons of medical grades of pyrvinium pamoate and monopyrvinium salts. Mutat. Res. **102**: 59–69.

25. ZIMMERMANN, F. K., R. C. VON BORSTEL, J. M. PARRY, D. SIEBERT, G. ZETTERBERG, E. S. VON HALLE, R. BARALE & N. LOPRIENO. Gene-Tox Workshop Program: report on *Saccharomyces cerevisiae*. Mutat. Res. (Submitted).

26. RESNICK, M. A. 1979. The induction of molecular and genetic recombination in eukaryotic cells. Adv. Radiat. Biol. **8**: 175–217.

27. GRIFFITHS, A. J. F. 1979. Neurospora prototroph selection system for studying aneuploidy production. Environ. Health Perspect. **31**: 75–80.

28. BROOKES, P. & F. J. DE SERRES. 1981. Overview of assay system performance. *In* Evaluation of Short-term Tests for Carcinogens. Report of the International Collaborative Program. F. J. de Serres & J. Ashby, Eds.: 96–111. Elsevier/North-Holland, Inc. New York, N.Y.

29. SHEIR-NEISS, G., M. H. LAI & N. R. MORRIS. 1978. Identification of a gene for β-tubulin in *Aspergillus nidulans*. Cell **16**: 437–442.

30. LUDUENA, R. F. & M. LITTLE. 1981. Comparative structure and chemistry of tubulins from different eukaryotes. Biosystems **14**: 239–246.

31. LITTLE, M., E. KRAUS & H. POSTINGL. 1981. Tubuline sequence conservation. Biosystems **14**: 239–246.

32. PARRY, J. M. & D. SHARP. 1981. Induction of mitotic aneuploidy in the yeast strain D6 by 42 coded compounds. *In* Evaluation of Short-term Tests for Carcinogens. Report of the International Collaborative Program. F. J. de Serres & J. Ashby, Eds.: 468–480. Elsevier/North-Holland, Inc. New York, N.Y.

33. HAVERCROFT, J. C., R. A. QUINLAN & K. GULL. 1981. Binding of parbendazole to tubuline and its influence on microtubules in tissue culture cells as revealed by immunofluorescence microscopy. J. Cell Sci. **49**: 195–204.

34. KERN, R. & F. K. ZIMMERMANN. 1976. Über den Wirkungsmechanismus des Antimyzetikums Econazol. Mykosen **20**: 133–146.

35. STOCKENIUS, W. 1961. Electron microscopy of DNA molecules stained with heavy metal salts. J. Biophys. Biochem. Cytol. **11:** 297–310.
36. ASHBY, J. 1981. Overview of study and test chemical activities. *In* Evaluation of Short-term Tests for Carcinogens. Report of the International Collaborative Program. F. J. de Serres & J. Ashby, Eds.: 112–171. Elsevier/North-Holland, Inc. New York, N.Y.
37. DE SERRES, F. J. & J. ASHBY, Eds. 1981. Evaluation of Short-term Tests for Carcinogens. Report of the International Collaborative Program. Elsevier/North-Holland, Inc. New York, N.Y.
38. SORA, S., G. LUCCHINI & G. E. MAGNI. 1982. Meiotic diploid progeny and meiotic nondisjunction in *Saccharomyces cerevisiae.* Genetics **101:** 17–33.
39. MORTIMER, R. K., R. CONTOPOULO & D. SCHILD. 1981. Mitotic chromosome loss in a radiation-sensitive strain of the yeast *Saccharomyces cerevisiae.* Proc. Nat. Acad. Sci. USA **78:** 5578–5582.
40. PARRY, J. M., D. SHARP & E. M. PARRY. 1979. Detection of mitotic and meiotic aneuploidy in the yeast *Saccharomyces cerevisiae.* Environ. Health Perspect. **31:** 97–111.
41. PARRY, J. M. & F. K. ZIMMERMANN. 1976. The detection of monosomic colonies produced by mitotic chromosome non-disjunction in the yeast *Saccharomyces cerevisiae.* Mutat. Res. **36:** 49–66.

THE VERSATILITY OF *Drosophila melanogaster* FOR MUTAGENICITY TESTING*

Ruby Valencia

Zoology Department
University of Wisconsin
Madison, Wisconsin 53706

The fruit fly *Drosophila melanogaster* is a familiar laboratory organism. Much of our basic knowledge of genetics came from studies with the fly, and it has been used in genetics classrooms routinely over the years. *Drosophila* mutagenesis tests were used extensively to study the genetic effects of ionizing radiations. Since World War II *Drosophila* testing for chemical mutagenesis has become routine. The problems are somewhat different for chemicals than for radiation. Exposure is not as easy or as certain, and we are dealing not with just one agent but with a very large number of different agents. Thus a major problem is screening the myriad of possible mutagens.

One test, the sex-linked recessive lethal assay, has proved to be the best all-around *Drosophila* test for screening chemicals. There are, however, other *Drosophila* mutagenesis tests, and in addition, the sex-linked recessive lethal assay can be used in ways other than primary screening or it can be adapted to the purpose of the screening.

The aim of this presentation, then, is to look at an array of suitable *Drosophila* mutagenesis tests and show how it is possible to coordinate exposure methods, germ cell sampling procedures, genetic schemes, and genetic end points to design optimum test protocols.

CHARACTERISTICS OF THE *DROSOPHILA* SYSTEM

The major advantages of *Drosophila* for studies in genetics have been enumerated many times: it is a sexually reproducing eukaryote, with germ cell stages paralleling mammals; it has only three major chromosomes; it is genetically well known; there are many mutations with visible effects that can be used as markers; there are many special chromosomes, with combinations of markers and rearrangements, especially inversions, to work with; the generation time is short; large numbers of progeny can be obtained, etc.

The mutagenesis assays screen for mutations in germ cells, which are transmitted to the progeny, low-level effects can be detected (such as 0.2% sex-linked recessive lethals), the tests are generally adaptable for large-scale screening programs, several different mutagenic end points can be used, and several different cytogenetic mutational events can be detected.

An additional characteristic of *Drosophila* is that it has been found to be able to activate indirect-acting mutagens (promutagens), and the activation occurs not only in the gut and other tissues, but usually within the gonadal tissue as well.[1,2]

There are also available a number of different types of repair-deficient strains, which are being studied for their possible enhancement of the sensitivity of chemical

*Supported by the U.S. Environmental Protection Agency, the U.S. National Institute of Environmental Health Sciences, and Raltech Scientific Services.

197

mutagenesis tests. They seem to be especially effective in chromosome-loss tests,[3] although sex-linked recessive lethals have also been observed to be increased.[4]

TESTS, END POINTS, AND CYTOGENETIC EVENTS

Many genetic mutagen testing schemes have been devised and used in genetic studies with *Drosophila*. At least five of them can be discussed as possible tests for use in chemical evaluation. They are:

The sex-linked recessive lethal test (SLRL).
The heritable translocation test (HT).
The chromosome loss test (CL).
The nondisjunction test (ND).
The specific locus test (SL).

TABLE 1

DROSOPHILA MUTAGENESIS TESTS

Sex-linked recessive lethal test (SLRL)	lethal effect	point mutations; deletions (small); gross rearrangements
Heritable translocation test (HT)	linkage of phenotypic markers	chromosome breakage and eucentric reunion, resulting in exchange of parts between two chromosomes
Chromosome loss test (CL)	loss of phenotypic markers	chromosome breakage and no reunion or incorrect reunion; chromosome lost
Nondisjunction (ND)	loss or gain of phenotypic markers	meiotic malfunction, resulting in nullo and diplo gametes
Specific locus (SL)	visible phenotype	point mutations; small deletions; position effects of gross rearrangements

TABLE 1 lists these five tests and indicates the end point that is observed in each case as well as the basic cytogenetic events that give rise to the end points. Some of the tests (HT, CL, and ND) are very specific for limited types of genetic damage while the others (SLRL and SL) detect a variety of types of damage. It should be noted that there are also some good somatic mutation assays available in *Drosophila*.

It is not within the scope of this paper to describe in any detail how these different assays are conducted, or to recommend specific procedures. In most cases, there are choices of genetic markers and chromosomes that can be used. For all except the specific locus test, the Gene-Tox reviews on the sex-linked recessive lethal test[5] and on chromosome mutations[6] are recommended. It is necessary, however, to point out a few characteristics of each one that are pertinent to the present discussion.

The SLRL assay screens some 800 loci[7] and detects lethal effects associated with either intragenic mutations or chromosome rearrangements.[8] The spontaneous mutation rates observed per locus are comparable to rates found for human loci.[5] The test is easy to adapt to routine large-scale testing programs, although large numbers of chromosome tests may be needed (depending on the purpose of the test) and concurrent controls should usually be conducted.

The HT test is much more laborious than the SLRL test, but the spontaneous background mutation rate is essentially zero. Thus historic controls can be used and the number of chromosome tests can be relatively small. It is a very "clean" and clear-cut proof of chromosome breakage and transmitted chromosome rearrangement.

The CL test is a much easier and more rapid test for clastogenesis than is the HT test, and it can be sensitized greatly by the use of a ring X chromosome in the male (treated) and a repair-deficient mutation in the female (untreated). Concurrent controls are needed, since the background rate is often variable and may be quite high, if repair-deficient strains are used.

Nondisjunction, a genetic effect that is very important to human health, is not detected in any of the other four mutagenesis tests. It must, therefore, be tested for in a separate experiment. Germ cell stage sampling must be very strictly controlled, and it must be kept in mind that chromosome gain is a clear indicator of nondisjunction but that chromosome loss, which can also occur as a result of chromosome breakage, can be used as an indicator only with caution.

The specific locus test is in a different category from the four discussed above. It is a one-generation test but it is very time consuming and yields data on a very limited number of loci. It is not considered appropriate for screening large numbers of chemicals. It does provide data that are useful for refined genetic studies of mutagenic agents of special interest. It is probably best used for comparative mutagenicity studies, where only specific locus data are available from other species for comparison.

GERM CELL STAGE SAMPLING

It is possible to treat and sample any germ cell developmental stage in *Drosophila,* but it does require knowledge of when these stages occur and the appropriate management of experiments. First, the desired stage must be present in the fly at the time of treatment. Second, the mating and brooding procedure must be such that the desired stage is identified and tested in the progeny. Third, the germ cell stage developmental pattern is very different in the male and in the female. TABLE 2 shows the fly developmental stages, the germ cell stages present in each, and the nuclear division stage that would be in progress in each stage.

Treating male or female first and second instar larvae would mean treating gonia in either sex, but beginning with third instar larvae, the germ cell stages present in the two sexes differ greatly.

Meiosis begins in the third instar larva of the male, and all stages from gonia to mature sperm are already present in the male pupa.[9] Thus, when any of the male developmental stages is exposed to a mutagen, all germ cell stages may be affected. By planning the matings and rematings (brooding) of treated males appropriately, specific germ cell stages can be sampled.[10] For example, when adult males are treated, mated in a proportion of one male to three virgin females, and remated every two or three days, it can be expected that in brood 3 or 4 sperm will be ejaculated that were treated as spermatocytes. On the other hand, if third instar larvae are treated, the very first sperm ejaculated, in the first brood, would have been treated in the spermatocyte stage. Thus the fly stage treated and the sperm sampled must be coordinated in order to sample germ cell stages correctly.

In the female, meiosis may begin in the pupa,[11] but it proceeds very slowly and is then arrested in prophase I[12] while the oocyte matures. Finally, in the most mature oocyte (stage 14), the nuclear division proceeds to metaphase I. Meiosis continues only after fertilization of the oocyte, or egg, just before it is laid. Note that the stages most often studied in the male, the postmeiotic germ cells, do not exist in the adult female. Thus in comparing mutagenesis results of treating adult male and female flies, it must be kept in mind that a very different germ cell population might be compared.

One other important point to consider in germ cell staging after chemical exposure is that the exposure cannot be delimited accurately, as it was possible to do with

TABLE 2

FLY DEVELOPMENT STAGES, GERM CELL STAGES, AND NUCLEAR STAGES

Sex	Germ Cell Stage	Nuclear Stage	Fly Stage
Male	gonia	mitosis	first and second instar
Female	gonia	mitosis	larvae
Male	gonia	mitosis	third instar larvae
	first spermatocytes	meiosis	
Female	gonia	mitosis	
Male	gonia	mitosis	pupae
	spermatocytes	meiosis	
	spermatids	postmeiosis	
	small amount of sperm	postmeiosis	
Female	gonia (immature oocytes?)	mitosis (early prophase of meiosis I)	
Male	gonia	mitosis	adult
	spermatocytes	meiosis	
	spermatids	postmeiosis	
	sperm	postmeiosis	
Female	gonia	mitosis	recently emerged adult
	immature oocytes (up to stage 7)	prophase of meiosis I	
Female	gonia	mitosis	aged adult
	immature oocytes	prophase I	
	stage 14 oocytes	metaphase I	
	fertilized egg	anaphase I	
		meiosis II	

radiation exposure. A chemical compound might enter the gonad and remain there to affect later stages. This property would be expected to differ for different types of compounds.

ROUTES OF ADMINISTRATION

There are four routes of administration of chemical compounds that are easily available to any testing program. They are (1) adult feeding, (2) adult injection, (3) inhalation, and (4) larval exposure. The feeding procedure is the easiest to perform and has been used routinely in many testing programs. The other routes of administration, however, are not too difficult to use routinely, once the protocol is in place in a laboratory.

The bases for choice of route of administration are several. Most important is the physicochemical nature of the compound to be tested. For example, if it is unstable in an aqueous solution or mixture, it may not be suitable for the feeding protocol, which requires an extended period of time in aqueous sugar solution. Injection could get the compound into the fly sooner. If it is highly volatile, inhalation might be a more suitable exposure method. A material that cannot be put into solution or into a very fine suspension might be presented to larvae, which can engulf particles.

A second consideration in choosing the exposure method is the route by which humans might be exposed. If that route of administration is available (given the above-mentioned limitations), it might be desirable to choose the human exposure route.

In a comparative mutagenesis study, on the other hand, an effort should probably be made to obtain data from a comparable route of administration. Perhaps feeding results in *Drosophila* cannot validly be compared with injection results in mice.

TESTING PROTOCOLS

There are three very different purposes for which *Drosophila* testing protocols may be designed. These are (1) screening, (2) risk evaluation, and (3) comparative mutagenesis.

For screening, the SLRL assay is by far the best choice, for the reasons enumerated above and also because test results have shown it to be more sensitive than chromosome loss or translocation tests.[1] It is also convenient that, for screening, it is usually necessary to test only postmeiotic germ cells, since they are often the most sensitive. The experiment size may be adjusted to the test sensitivity desired (see below).

A screening program should also include a test for nondisjunction, since the SLRL test does not detect meiotic malfunction, and the knowledge that a chemical could induce aneuploidy is a very important human health consideration. Much work is in progress at this time to develop and validate acceptable protocols for aneuploid testing. It is not within the scope of this paper to discuss the pros and cons of the different genetic schemes and procedures being studied.

For risk evaluation, it would be useful to have, in addition to the postmeiotic SLRL data and the ND data, information on mutagenesis in gonia, the germ cells most at risk in human exposures. A gonial assay can best be done in a separate experiment, i.e., not incorporated in the original screen. A clastogenesis test, either HT or CL, would complete the mutagenic profile.

If the testing is being done as part of a comparative mutagenesis program, the protocol should be designed to provide the best comparative data. It will not always be possible to have all parameters comparable, but this should at least be considered and addressed where possible.

SLRL TEST SENSITIVITY

In conducting a sex-linked recessive lethal assay, the size of the experiments (i.e., the number of treated and control X chromosomes tested for lethal mutations) can be varied over a wide range. The number that is appropriate to test depends upon the test sensitivity desired and the control mutation frequency.

If the purpose is to detect not only strong mutagens but also weaker ones, then the test can be run to detect as low as 0.2% induced mutations. Most *Drosophila* screening tests are conducted routinely at that level of sensitivity. If, however, the goal is simply to pick up potent mutagens, the experiment size can be drastically reduced, which of course would make it possible to screen much larger numbers of chemicals in a much shorter time.

The control frequency greatly affects the number of chromosomes to be tested for a given mutation increment to be detected with statistical significance. TABLE 3 shows the test numbers needed, for treated and control, assuming equal numbers of each, given different control rates and different induced mutation increments. These numbers are based on the Kastenbaum-Bowman statistical test[13] and were prepared by Dr. B. Margolin.[14] It can be seen that the higher the control rate the more limited is the test sensitivity, given that test numbers over 10,000 are not practical for screening experiments. Most *Drosophila* laboratory stock mutation frequencies are in the range

TABLE 3
SENSITIVITY RANGE OF SLRL ASSAY*

Induced Increment (%)	Control Rate (%)		
	0.05	0.10	0.20
0.05	35,156	57,274	100,392
0.10	11,899	17,575	28,628
0.15	6,642	9,170	14,265
0.20	4,547	5,948	8,784
0.30	2,739	3,320	4,583
0.40	1,933	2,273	2,973
0.50	1,485	1,715	2,141
0.60	1,203	1,369	1,659
0.70	1,010	1,134	1,350
0.80	870	966	1,136

*Selected from tables of B. Margolin.[14]

of 0.1% to 0.2%, so that sample sizes of 10,000 or less allow for considerable variability of test sensitivities, down to mutation increments of 0.15%.

EFFECTS OF PROTOCOL VARIATIONS ON MUTAGENESIS RESULTS

Following are some samples of data that demonstrate how the SLRL mutation frequency is affected by the treating and sampling protocol employed.

First, observe (TABLE 4) the frequency pattern in different germ cell stages for different compounds. Four compounds were chosen, which were tested at different times but with the same brooding protocol, so that comparison is easier. Ethyleneimine gave a peak effect in brood 1, with a slight drop in brood 2.[15] Thus mature sperm and spermatids were most sensitive and earlier stages were much less affected.

Ethylmethanesulfonate (EMS) was about equally mutagenic throughout the postmeiotic stages, but the gonial rate decreased markedly.[15] (The drop is partly explained, however, by germinal selection, discussed below.)

TABLE 4

GERM CELL STAGE SPECIFICITY PATTERNS*

	Days Postmating			
	1,2,3,4	5,6,7	8,9,10	11,12,13,14
Ethyleneimine[15] 30 ppm	280–2582	79–1019	15–1219	20–2382
	10.8%	7.8%	1.2%	0.8%
Ethylmethanesulfonate[15] 400 ppm	353–1939	267–1340	107–618	18–801
	18.2%	19.9%	17.3%	2.3%
Tris(2,3-dibromopropyl)-phosphate[15] 1000 ppm	53–3837	87–1212	17–1287	23–2241
	1.4%	7.2%	1.3%	1.0%
Platinum chloride[16] 100 ppm	6–2121	30–1618	6–871	4–1205
	0.3%	1.9%	0.7%	0.3%
Platinum chloride[16] 500 ppm	7–1902	23–1838	14–1301	2–1253
	0.4%	1.3%	1.1%	0.2%

*Adult males, fed.

The flame retardant tris(2,3-dibromopropyl)phosphate showed a definite peak effect in brood 2, but was still a fairly strong mutagen in the other stages.[15] The effectiveness of platinum chloride also peaked in brood 2, but it was a weaker mutagen.[16] It might have been difficult to detect it as a mutagen had spermatids and spermatocytes not been tested.

TABLE 5 demonstrates three points. First, when adult males were fed tris, a peak of mutagenicity was found in spermatids (here better sampled) and other stages were considerably less affected. (These data may look as though they do not agree with those shown in TABLE 4, but in fact they agree very well. Note that the brooding days were different.) When larvae were exposed, by mixing tris into the culture medium and rearing the larvae in the mixture, the highest mutagenicity appeared in brood 1. This is to be expected, since the first sperm to be ejaculated by these males would have been exposed as spermatocytes or early spermatids. What was surprising was that the mutagenic effect persisted into the later broods, which in the case of larval exposure should represent exposed gonia. The conclusion must be that either the gonia were better exposed in this procedure, or that larvae activate this mutagen more effectively than do the adults, or that tris stayed in the gonads and was therefore present to affect later stages, not present at the time of the exposure.

TABLE 6 shows how the route of administration can affect results. We tested

TABLE 5

ADULT VERSUS LARVAL EXPOSURE*

	Days Postmating			
	1,2,3	4,5	6,7	8,9
Adults	4–529	18–296	5–256	8–1361
	0.76%	6.08%	2.00%	0.59%
Larvae	53–465	16–179	5–223	6–310
	11.40%	8.94%	2.24%	1.94%

*Larval means entire larval life (in medium). Tris(2,3-dibromopropyl)phosphate (500 ppm) fed to males. (Data from Reference 15.)

TABLE 6

SLRL INDUCED BY β-PROPIOLACTONE IN ADULT MALES*

	Lethals/ Chromosomes Tested	Percent Lethals	Percent Induced†
Fed			
0 ppm	16/3018	0.53	—
250 ppm	18/3122	0.58	0.05
1000 ppm	27/3377	0.80	0.27
3000 ppm	33/3375	0.98	0.45
Injected			
0 ppm	40/8960	0.45	—
1000 ppm	61/3024	2.02	1.56
3000 ppm	118/1524	7.74	7.36

*See Reference 15.
†Experiments were done with an "FM6" stock giving a high spontaneous mutation rate. Thus "percent induced" (treated-control) is shown.

beta-propiolactone in a blind test,[15] but we were told that the compound was unstable in aqueous solution. We therefore tested it first by injection and got, as can be seen, a highly positive, dose-related effect. Then, just to test protocols, we tested it by adult feeding. It would have been detected only as a weak mutagen by this route of administration, and then perhaps only at the higher dose.

It has been noted for some time that there is often a lack of effect, or at least relatively less effect, when females are exposed to chemicals, as compared to exposed males. It was thought that this was probably related to the nature of the mature oocyte, with its large amount of cytoplasm surrounding the nucleus. Past work with radiation effects on the female germ cells, including the mature (stage 14) oocytes and the recently fertilized eggs, tended to indicate, however, that the observations with chemical exposures had more to do with the nuclear stage at the time of exposure.[17,18] As shown in TABLE 2 and discussed above, the progression of the meiotic stages is very different in the two sexes. Some experiments were done, therefore, in which special effort was made to sample the most mature oocytes. Since this is quite difficult to do, the number of chromosomes tested was sometimes quite small. Comparison was made in parallel experiments when possible, with results from spermatocytes in exposed males.

TABLE 7 shows the results with four compounds. Spermatocytes yielded, with 100 ppm EMS, about 6% lethals. With 10 times that dose, 1,000 ppm, to females, only a little over 1% lethals were obtained.[15] A similar experiment with tris,[15] using 500 ppm in both sexes, yielded about 5% from spermatocytes and only 0.2% from oocytes.

Dimethylnitrosamine (DMN) was both fed and injected in both males and females in concurrent experiments.[15] The dose, however, was 2.5 ppm for feeding and 500 ppm for injection, due to differences in toxicity and sterility by the two routes of administration. Feeding for 24 hours yielded nearly 10 times as many lethals in spermatocytes as in oocytes. The differential was even greater when the flies were injected. Spermatocytes gave 22% lethals, while oocytes gave only 1.4%.

An experiment was then carried out by S. Abrahamson to obtain oogonial mutation data from methyl nitrosourea (MNU).[19] Rather than discard the earlier broods, data were collected from them also. Clearly, MNU was as mutagenic in oocytes as in postmeiotic male germ cells. Actually, MNU killed spermatocytes so effectively that the sample was mostly later stages. Similarly, the female germ cell sample was not the most mature oocytes, because they were also killed. Nevertheless, the results show that MNU does indeed reach and effectively mutate maturing oocytes, whereas the other chemicals did not. Bleomycin was earlier found by Traut to be mutagenic in females.[20]

The gonial data for MNU show another very important effect, which must be borne in mind when collecting gonial mutation data. As can be seen, the oogonia yielded far more mutations than did the spermatogonia. Now, both of these frequencies are inflated by clusters of mutations, i.e., groups of sperm or oocytes bearing identical lethals induced in gonial cells that then reduplicated. There is, therefore, a large error in both cases. Nevertheless, the results show that oogonia yielded more lethals than did spermatogonia. The reason is germinal selection, a phenomenon that has been long recognized.[21]

In oogonia, there are two X chromosomes. Thus any recessive lethals occurring in one X are covered by a normal allele in the other. The male, however, has a single X chromosome, and a lethal mutation in a vital gene may kill the metabolically active gonial cell. It is for this reason that we strongly recommend that gonial sampling be done using autosomes or, preferably (since autosomal genetic schemes are difficult), using females. Some experimentation is needed, and is being done in our laboratory, to determine the optimum exposure method (for example, adults vs. larvae) and to validate an overall gonial test protocol.

Time prohibits discussing still other parameters here. For example, duration of feeding exposure may affect results, as shown by data from dimethylnitrosamine fed for 24 hours or 72 hours.[15] In inhalation studies, duration of exposure and concentration of the gas in the air work together to affect the mutations obtained, as well demonstrated by studies of dibromoethane (DBE).[22] In translocation tests, whether or not the sperm is aged in the female may greatly affect results.

TABLE 7

LETHALS INDUCED IN SPERMATOCYTES AND OOCYTES

	Lethals/ Chromosomes Tested	Percent Lethals
EMS (fed)*		
Spermatocytes (100 ppm)	24/384	6.25
Oocytes (1000 ppm)	16/1432	1.12
Tris(2,3-dibromopropyl)phosphate (fed; 500 ppm)*		
Spermatocytes	63/1242	5.07
Oocytes	5/2399	0.21
Dimethylnitrosamine (fed; 25 ppm)*		
Spermatocytes	54/1108	4.87
Oocytes	43/7869	0.55
Dimethylnitrosamine (injected; 500 ppm)*		
Spermatocytes	81/364	22.25
Oocytes	17/1259	1.35
Methylnitrosourea (fed; 500 ppm)†		
Postmeiotic male germ cells	81/533	15.20
Oocytes	23/149	15.44
Spermatogonia	69/1065	6.48
Oogonia	133/403	33.00

*Data from Reference 15.
†Data from References 15 and 19.

SUMMARY

I trust that I have accomplished my aim in this presentation, to demonstrate that *Drosophila* mutagenesis testing protocols are varied and are adaptable. The choices are many, at different levels.

First, the route of administration can be chosen that is most appropriate for the chemical compound or to duplicate the human exposure route or to obtain data needed for some particular purpose, such as comparative mutagenesis.

Second, the genetic test scheme or combination of schemes can be chosen to suit the purpose of the testing (screening, risk evaluation, or comparative mutagenesis).

Third, by treating the appropriate fly development stage and managing the mating and brooding procedures, the germ cell stage can be sampled that is the critical one for the mutagenic effect of that compound and the correct one for the end point to be observed.

Fourth, in screening, using the SLRL assay, the experiment size can be adjusted in accordance with the test sensitivity desired, the projected use of the data, and time and cost restrictions.

In summary, *Drosophila melanogaster* has long been known as a most versatile organism for genetic studies of all kinds—and we now know that this versatility extends to the evaluation of chemical compounds for mutagenic effects. If it were only a mammal!

ACKNOWLEDGMENTS

The valuable assistance of Kathleen Houtchens and Douglas White in producing data and developing test protocols is much appreciated, as are the valuable discussions with Seymour Abrahamson, in whose laboratory most of the work has been done.

REFERENCES

1. VOGEL, E. & F. H. SOBELS. 1976. *In* Chemical Mutagens, Principles and Methods for Their Detection. A. Hollaender, Ed. **4**: 93–142. Plenum Press. New York, N.Y.
2. VOGEL, E. 1981. *In* Short-Term Tests for Chemical Carcinogens. H. F. Stich & R. H. C. San, Eds.: 379–398. Springer-Verlag. New York, N.Y.
3. ZIMMERING, S. 1982. Induced chromosome loss with nitrosopiperidine in the male *Drosophila melanogaster*. Environ. Mutagen. **4**: 521–524.
4. GRAF, U., M. M. GREEN & F. E. WURGLER. 1979. Mutat. Res. **63**: 101–112.
5. LEE, W. R., S. ABRAHAMSON, R. VALENCIA, E. VON HALLE, F. E. WURGLER & S. ZIMMERING. 1983. The sex-linked recessive lethal test for mutagenesis in *Drosophila melanogaster*. A report of the U.S. Environmental Protection Agency Gene-Tox Program. Mutat. Res. (In press.)
6. VALENCIA, R., S. ABRAHAMSON, E. VON HALLE, W. R. LEE, R. C. WOODRUFF, F. E. WURGLER & S. ZIMMERING. Chromosome mutation tests for mutagenesis in *Drosophila melanogaster*. A report of the U.S. Environmental Protection Agency Gene-Tox Program. Mutat. Res. (In preparation.)
7. ABRAHAMSON, S., F. E. WURGLER, C. DEJONGH & H. UNGER MEYER. 1980. How many loci on the X-chromosome of *Drosophila melanogaster* can mutate to recessive lethals? Environ. Mutagen. **2**: 447–453.
8. VALENCIA, R. 1970. A cytogenetic study of radiation damage in entire genomes of Drosophila. Mutat. Res. **10**: 207–219.
9. COOPER, K. W. 1950. Normal spermatogenesis in Drosophila. *In* Biology of Drosophila. M. Demerec, Ed.: 1–61. Hafner Publishing Co. New York, N.Y. (Reprinted in 1965.)

10. CHANDLEY, A. C. & A. J. BATEMAN. 1962. Timing of spermatogenesis in *Drosophila melanogaster* using tritiated thymidine. Nature **193**(4912): 299–300.
11. BODENSTEIN, D. 1950. The postembryonic development of Drosophila. *In* Biology of Drosophila. M. Demerec, Ed.: 275–367. Hafner Publishing Co. New York, N.Y. (Reprinted in 1965.)
12. PURO, J. & S. NOKKALA. 1977. Meiotic segregation of chromosomes in *Drosophila melanogaster* oocytes. Chromosoma **63**: 273–286.
13. KASTENBAUM, M. A. & K. O. BOWMAN. 1970. Tables for determining the statistical significance of mutation frequencies. Mutat. Res. **9**: 527–549.
14. MARGOLIN, B. National Toxicology Program, Research Triangle Park, N.C. Personal communication, from tables prepared for use in mutagenesis testing program.
15. VALENCIA, R. Unpublished data.
16. WOODRUFF, R. C., R. VALENCIA, R. F. LYMAN, B. A. EARLE & J. T. BOYCE. 1980. The mutagenic effect of platinum compounds in *Drosophila melanogaster*. Environ. Mutagen. **2**: 133–138.
17. VALENCIA, R. M. & J. I. VALENCIA. 1964. The radiosensitivity of mature germ cells and fertilized eggs in Drosophila melanogaster. *In* Mammalian Cytogenetics and Related Problems in Radiobiology. C. Pavan, C. Chagas, O. Frota-Pessoa & L. R. Caldas, Eds.: 345–360. Pergamon Press. New York, N.Y.
18. VALENCIA, R. M. 1965. The radiosensitivity of stage 14 oocytes. Genetics **52**: 481–482.
19. ABRAHAMSON, S. Unpublished data.
20. TRAUT, H. 1980. Mutagenic effects of bleomycin in Drosophila melanogaster. Environ. Mutagen. **2**(1): 89–96.
21. ABRAHAMSON, S., H. UNGER MEYER, E. HIMOE & G. DANIEL. 1966. Further evidence demonstrating germinal selection in early premeiotic germ cells of Drosophila males. Genetics **54**(2): 687–693.
22. KALE, P. & J. W. BAUM. 1979. Sensitivity of Drosophila melanogaster to low concentrations of the gaseous 1,2-dibromoethane. I. Acute exposures. Environ. Mutagen. **1**(1): 15–18.

APPROACHES TO COMPARATIVE MUTAGENESIS IN HIGHER EUKARYOTES: SIGNIFICANCE OF DNA MODIFICATIONS WITH ALKYLATING AGENTS IN *DROSOPHILA MELANOGASTER**

E. W. Vogel

Department of Radiation Genetics and Chemical Mutagenesis
Sylvius Laboratories
State University of Leiden
2333 AL Leiden, the Netherlands

INTRODUCTION

Speaking of most recent facets in the development of genetic bioassays for genotoxic agents, it would seem appropriate to point out the advantages and limitations of *Drosophila* within the frame of genetic bioassays, and to confine attention to the improvements that have been achieved during the past few years in understanding metabolism of aromatic amines and polycyclic hydrocarbons in this system.[1-4] Another relevant topic worth discussing could be to trace and review the progress that has been made in developing several somatic mutation assays suitable for testing large numbers of chemicals in *Drosophila*.[5,6] These rapid methods, once evaluated against a broad array of reference carcinogens, may provide an efficient means of overcoming the limitations imposed on mutagenicity testing in *Drosophila* by limited capacities for large-scale experiments. However, I have deliberately refrained from discussing in more detail these new methods currently under way for monitoring genotoxic agents in *Drosophila*. Instead, I shall limit my discussion here to the presentation of a concept more bearing on quantitative aspects of chemical mutagenesis in *Drosophila*.

THE CONCEPT

The great success of short-term mutagen screening has led to the identification as genotoxic agents of a wide variety of synthetic and naturally occurring compounds belonging to nearly all chemical classes. However, an area where we are rather ignorant and which has aroused much interest in recent years concerns the question of whether general characteristics and relationships can be identified for the genetic effects of chemical reagents that are valid beyond the particular species, strain, tissue, cell stage, and set of environmental conditions under which the correlations were obtained, and thus would permit extrapolation from one species to another.

In order to approach these problems, one prerequisite is the identification of the actual DNA sites and the types of DNA lesions responsible for the different kinds of genetic alterations. Thus, DNA-binding studies should be coupled with a powerful system for the genetic analysis of all or at least most of the genetic alterations: viable

*This work was supported by the National Institute of Environmental Health Sciences (U.S.A.), Contract No. ESO 1027-05/07, and the "Stichting Koningin Wilhelmina Fonds" (the Netherlands), Contract No. 81.90. Part of this study also received support from the Association Contract No. 139-77-1 ENV N between the European Communities (Environmental Research Program) and the State University of Leiden.

0077–8923/83/0407–0208 $01.75/0 © 1983, NYAS

and lethal gene mutations, deficiencies, and various categories of chromosomal aberrations. Obviously, the eukaryotic chromosome is the prime candidate to be chosen for such structure-effect studies. The purpose of this paper is to examine the possible role that a complex eukaryote like *Drosophila* might play in this research area, but it is outside the scope of this overview to review exhaustively all pertinent and important documentation available. Certainly, the reader is referred to the original literature that contains the data.

At the present time, *Drosophila* constitutes one of the few eukaryotic *in vivo* systems that would enable us to detect, in the same population of treated cells, practically all the known changes capable of occurring in the genetic material. This provides us with a tool of specifying the mutagenic properties of a reagent in terms of its ability to cause various types of genetic damage. If the determination of the mutational profile is paralleled by DNA-binding studies or carried out with model compounds for which the DNA alteration pattern is already known, it is possible to search for links between the mode of initial interaction with DNA and the kinds of genetic damage produced.

Choice of Model Mutagens

As a group, alkylating agents (AAs) are particularly suited for approaches aimed at identifying links between primary DNA damage and the genetic consequences, because detailed information is available on the actual distribution of alkylation at various DNA sites.[7-11] But the most stringent argument for choosing this class of mutagens is that the relative distribution in DNA of the same type of alkyl group can be manipulated, depending on the alkylating substrate selected. There has yet been no other class of known mutagens that would enable the production of a similar variation in the distribution of DNA adducts. As examples of these modifications, we may quote differential ethylation of guanine in DNA at the 7 position as opposed to the O^6 site and phosphotriester formation (TABLE 1). All this makes AAs particularly attractive as model compounds for the analysis of basic mechanisms in chemical mutagenesis and, at a more advanced stage, for interspecies comparisons.

Choice of Biological End Points

Thus far, seven biological parameters have been elaborated to search in germ cells or somatic tissue of *Drosophila* for links between initial alkylation pattern and the kinds of genetic effects observed. These are, in germ cells:

1. Mutation induction relative to cytotoxicity.
2. The proportion of chromosomal aberrations versus that of recessive lethal mutations.
3. The ability of AAs to cause delayed mutations, expressed by the ratio of F_2 lethals to F_3 lethals.
4. The role of repair processes.
5. The influence of expression time on the formation of chromosomal aberrations.
6. The induction of mutations relative to quantitative DNA binding.

And in somatic cells:

7. The proportion of somatic recombination versus somatic mutations.

TABLE 1

RELATIONSHIP BETWEEN CYTOTOXICITY AND RELATIVE MUTAGENICITY OF ALKYLATING AGENTS IN *DROSOPHILA**

Substrate	s	Ratio O^6/N-7 Alkylation of Guanine	k_{AA}/k_{EMS} at n =		$\dfrac{(LC_{50})_{EMS}}{(LC_{50})_{AA}}$	$\dfrac{(CM_4)_{EMS}}{(CM_4)_{AA}}$ †	$\dfrac{CM_4}{LC_{50}} = Q_{AA}$ ‡
			5.1	2.5			
MMS	0.86	0.004	16	4	25	5	0.128
DMS	0.86	0.003	16?	4?	26	5	0.130
EMS	0.67	0.03	(1.0)	(1.0)	(1.0)	(1.0)	0.025
DES	0.64	0.02/0.03					
MNU	0.42	0.11			7.6	2.9	0.069
iPMS	0.31	0.30			2.7	1.0	0.069
ENU	0.26	0.6–0.7			7.0	8.3	0.021
DEN	—	0.5–0.6					
ENNG	0.26	0.4–1.1					

*For original literature see the overview by Vogel and Natarajan;[14] k_{AA}/k_{EMS}, reaction rates relative to EMS at 20°C at n = 5.1 and n = 2.5 (from Reference 12); LC_{50}, 50% survival; CM_4, concentration that produced 4% recessive lethals.

†Mutagenic effectiveness.

‡Mutagenic efficiency.

COMPARISON OF MULTIPLE GENETIC END POINTS

Mutagenic Efficiency in Relation to Cytotoxicity

The advantage of comparing in the same cell population various biological parameters is that this enables us to relate relative mutagenic efficiency of AAs, even without knowing the actual molecular dose in the target tissue. Confining our first example to the relationship between relative mutagenicity and cytotoxicity, i.e., the proportion of X-linked recessive lethals/LC_{50} (50% survival), one would expect the relative mutagenic efficiency at equal survival to parallel the DNA-alkylation/protein-alkylation ratios (and also the O^6/N-7 alkylation of guanine) to some extent, because not only alkylation of DNA but also extensive protein alkylation can lead to cell death.[1,2]

TABLE 1 is a compilation of data on mutation induction and cytotoxicity (50% survival of adult males) following exposure to some alkylating agents.[13] Comparisons were made of the CM_4 values (the exposure concentration producing 4% recessive lethals) versus LC_{50}. Analysis on the basis of the Q values ($CM_4/LC_{50} = Q$) reveals a decrease in relative *mutagenic efficiency* in the sequence N-ethyl-N-nitrosourea (ENU) > ethylmethanesulfonate (EMS) > isopropylmethanesulfonate (iPMS) = N-methyl-N-nitrosourea (MNU) > methylmethanesulfonate (MMS) = dimethylsulfate.[13,14] Evidently, supermutagens like ENU and EMS are characterized by having low Q values (TABLE 1).

Mutagenic efficiency should be clearly distinguished from *mutagenic effectiveness* in such semiquantitative considerations.[14–16] When calculations are made on a molar base, MMS is fivefold more *effective* as a mutagen than is EMS (TABLE 1). The simple explanation is the fourfold higher reaction rate of MMS relative to EMS at $n = 2.5$, the average nucleophilicity (n) of the DNA molecule.[12,17] Thus, comparisons on a molar base are potentially misleading, because they do not account for existing differences between AAs in reaction rates, reaction mechanism, and the Swain-Scott substrate constant s.

Chromosomal Aberrations Relative to Mutation Induction

A good example of the need for such considerations is the second parameter, the proportion T:M, i.e., of reciprocal translocations to mutations at different mutation frequencies ("doses"). As outlined, the percentages of recessive lethals can serve as a relative biological dosimeter for the efficiency of the reagent to induce, for instance, reciprocal translocations between the two large autosomes (2-3 translocations).

The methylating MMS provides a typical example of a reagent with high chromosome-breaking ability. Thus, following administration of MMS, reciprocal translocations occurred at relatively low MMS doses producing a fivefold increase in the percentage of lethals.[13] However, the results also showed that with an ENU-type mutagen, the recovery of chromosomal aberrations can be expected only at an extremely high degree of DNA ethylation, i.e., at a dose leading to about a 100- to 300-fold increase in the yield of recessive lethals. N-Ethyl-N'-nitro-N-nitrosoguanidine (ENNG) and diethylnitrosamine (DEN) showed a behavior similar to ENU, while diethylsulfate (DES) and EMS took an intermediate position between these reagents and MMS. AAs can therefore be graded in a series from MMS through to ENU, ENNG, and DEN according to their ability to induce chromosome aberrations relative to mutations: MMS > DMS = MNU > dimethylnitrosamine (DMN) > EMS = DES > iPMS > ENNG = ENU = DEN.[14] Obviously, this is nearly the reverse sequence in relative mutagenic efficiency, as compared to that established for mutation induction (see above).

In comparisons of the mutagenic efficiencies of these agents, another crucial question is whether indications of the biological importance of specific adducts in DNA may be obtained by comparing the relative amounts of various alkylation products formed in relation to their potency as mutagens. Only a few of the nearly 20 sites at which DNA can be alkylated may be significant for mutagenesis.[10] Actually, few data exist at present that could enable such detailed assessments to be made. It therefore seems to us that the search for "negative correlations" might be the best approach in preventing misinterpretation. Such a negative correlation arises for the ethylating reagents ENU, ENNG, and DEN when we contrast their preference for O ethylation with their remarkably low capacity for chromosome aberrations in *Drosophila*. To begin with, the striking similarities in the mutational spectra observed for ENU, ENNG, and DEN are consistent with their very similar ethylation patterns in DNA: over 80% of DNA modifications by these agents are on oxygen sites.[10,11,18] The corresponding values for the DES-EMS pair, having a distinctly higher chromosome-breaking efficiency in postmeiotic germ cells of *Drosophila*, are clearly lower, namely, 8–20%. Thus it is tempting to postulate that, in view of the marked specificity of DEN, ENU, and ENNG for point mutations, the molecular events that characterize the action on DNA of these reagents, O ethylation, in particular the formation of ethylphosphotriesters, which are the predominant products, are not the critical primary lesions leading to chromosomal aberrations in *Drosophila*.

Delayed Mutations

Another striking feature of these AAs, which is instructive to examine, is the lower yield of delayed mutations in passing from MMS, EMS, DES through ENU, and DEN.[13] Recessive lethals that arise with a delay give rise to mosaic gonads in F_1 and cannot be detected with the conventional F_2 recessive lethal test. An F_3 generation must be set up to determine these effects, and the ratio of F_2 to F_3 lethals can serve to express the relative efficiency of a reagent in causing delayed mutations.

It is striking that the agents relatively inefficient in inducing chromosome breakage, ENU and DEN, gave high F_2-lethal to F_3-lethal ratios. MMS, EMS, DES, and DMS at various levels of exposure tended to produce more delayed mutations, this being reflected by lower F_2-lethal to F_3-lethal ratios.[13] Thus, for the AAs considered here, ability to induce delayed mutations and efficiency to cause chromosome breakage seem to be positively related. In other words, ability to produce delayed mutations seems to correlate with preference for alkylation of base nitrogens, especially the guanine N-7, the major product of DNA reacted with MMS, EMS, or DES.[11,19,20] Alkylation at the 7 position and the 3 position of purine bases on nucleic acids is known to cause depurination from DNA.[19,21] Depurination has been equated with strand breakage, a potentially lethal lesion.[22,23] Shearman and Loeb showed a possible relationship between depurination and mutagenesis through incorporation of noncomplementing nucleotides during DNA synthesis *in vitro*.[24] It has also been suggested that, besides its involvement in lethal effects, depurination may contribute to chromosome breakage and occasional base-pair deletions, which may occur as delayed effects.[25,26,27]

Modification of Mutational Spectra by Repair Processes

In the preceding discussion, we were considering the possible genetic significance of some initial DNA-alkylation products. It was early recognized that alkylation damage in DNA reacted with AAs might be subjected to repair processes.[19] Differential repair of DNA lesions may lead to changes in the frequency of genetic end points and consequently may alter the proportion with which they occur. Thus, DNA repair would be an important means of investigating mechanisms in alkylation mutagenesis in *Drosophila*.

In recent years, a number of recombination-defective and DNA-repair-defective mutants have been isolated in *Drosophila* to study the genetic control of mutagenesis

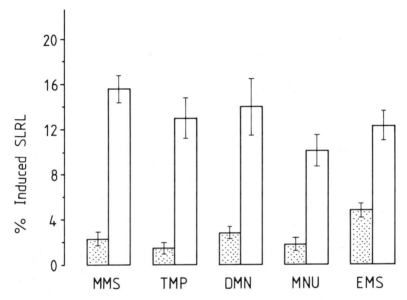

FIGURE 1. The influence of maternal repair on the induction of recessive lethal mutations following mutagen treatment of postmeiotic germ cells in *Basc* males. Crosses performed with either Berlin K females (dotted columns) or *mei-9^LI* females (white columns).

in this eukaryote.[28] Some of these mutants such as *mei-9* have been reported to be defective in excision repair.[29] We have employed one of the *mei-9* alleles,[30] the repair-defective stock *mei-9^LI*, to investigate its response to alkylation-induced damage. In these experiments, postmeiotic germ cells from repair-deficient *Basc* males were treated and mated to either *mei-9^LI* or repair-proficient (Berlin K) females to explore the effect on mutation induction of a deficient maternal repair.

FIGURE 1 demonstrates the lowered ability of *mei-9^LI* to carry out repair of DNA damage inflicted by methylating or ethylating reagents with preference for N alkylation. Following administration of moderate or high doses, alkylation of DNA by MMS, trimethylphosphate (TMP), DMN, MNU, and EMS led to markedly increased percentages of X-linked recessive lethals in oocytes with an excision-repair deficiency (*mei-9*), as opposed to wild-type oocytes. This pattern is in line with the hypermutability observed for the *mei-9* locus in other laboratories with MMS, EMS, and MNU.[31,32] In the interpretation of these repair experiments, a distinction must be made between the effects of high doses, causing a significantly greater effect in *mei-9^{LI}* cells than in normal ones, and the low dose range, where such differences were not observed (Vogel, unpublished observations). These studies raise the possibility that low quantities of alkylation damage can be handled by these mutants, possibly indicating the operation at low doses of a "leaky" repair system.

What is, however, of greater importance to the present discussion is that with ENU, DEN, and ENNG, when administered as a series of high and moderate doses, reduction in DNA-excision-repair capacity had no apparent consequences: the effects of ENU, DEN, and ENNG treatments are virtually the same in the two types of females (FIGURE 2). We may tentatively interpret these results as suggesting that in *Drosophila*, ENU, DEN, and ENNG exert their major mutagenic effects through induction of direct miscoding, a mechanism early proposed by Loveless.[33]

A second proposed mechanism, then, denoted as indirect miscoding induced by misrepair of DNA,[34] may account for the marked increases in mutation yield by MMS, TMP, DMN, MNU, and EMS. Generally speaking two major mechanisms of alkylation mutagenesis can so far be distinguished in *Drosophila*.

On the other hand, the AAs did not split into two groups when assayed for chromosomal aberrations. With regard to induced chromosome losses and partial losses, absence of *mei-9^+* resulted in an enhanced response to *all* the AAs. This is consistent with the results reported for various AAs by Zimmering and coworkers.[35-37] It suggests that in DNA reacted with ENU, DEN, and ENNG, the primary lesions responsible for the production of chromosomal aberrations and those giving rise to mutations may not be the same.

It must be clearly stated, however, that the repair-deficient, mutant strains with *mei-9* (*mei-9^a*, *mei-9^{ATI}*, *mei-9^{LI}*) are the only mutants that have yet been investigated

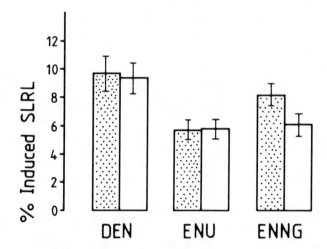

FIGURE 2. Absence of an effect of maternal excision-repair deficiency on mutation induction in males with DEN, ENU, or ENNG. Crosses performed with either Berlin K females (dotted columns) or *mei-9^{LI}* females (white columns).

TABLE 2

ABSENCE OF A STORAGE EFFECT FOR 2-3 TRANSLOCATIONS INDUCED BY HMPA

Treatment	Genetic End Point	Storage*	Gametes Tested	Percent Lethals or Translocations
0.01% × 48 hours	recessive lethals	U	2221	7.4
	translocations	U	2206	0.27
	translocations	S	7553	0.38
0.02% × 48 hours	recessive lethals	U	3217	11.0
	translocations	U	3188	0.72
	translocations	S	9568	0.79

*U, unstored; S, storage period up to 17 days.

extensively. Much more research on the role of processes in the induction of mutations by AAs, utilizing other stocks with a reduced excision capacity, is required before firm conclusions can be drawn.

Significance of Expression Time

A characteristic feature of chemically induced DNA breaks in *Drosophila* is their delayed formation.[38-41] The mechanism of this "storage phenomenon" remains unknown, but Auerbach has suggested that "depurination leading to strand breakage is a conceivable candidate for the potential lesions that, in the chromosomes of *Drosophila* and plants, mature gradually into chromosome breaks."[25] Thus, the hypothesis on the link between alkylation damage at base nitrogen and the occurrence of delayed mutations is also applicable to the delayed formation of chromosomal aberrations. Moreover, this whole concept most recently has received further experimental support from work with hexamethylphosphoramide (HMPA).

The mutational pattern of HMPA is, in many respects, entirely different from that seen for alkylating agents. When HMPA-treated males are mated to untreated females, the percentage of 2-3 translocations is the same whether the females are allowed to lay eggs at once (unstored) or whether they are prevented from doing so for several days or weeks (TABLE 2). This indicates no delayed formation of exchanges from HMPA-induced breaks. When the same type of experiment is carried out with MMS, for example, a gradual increase in the percentage of translocations (or chromosome losses) is observed in the later broods.[13,42] HMPA, and again this stands in contrast to the picture seen with MMS-type mutagens, is also not very efficient in the production of delayed mutations, measured as F_3 recessive lethals. The frequencies of recessive lethals in the F_2 were four- to sevenfold higher than the percentages of mutations that could be detected by an analysis of the F_3 test (Vogel, unpublished observation).

Preliminary, rather than definitive, experiments were carried out to examine whether specific interactions with DNA could explain the peculiar mutagenic profile of HMPA. Two experiments were designed in which *Drosophila* DNA was reacted *in vivo* with ^{14}C-HMPA (0.23%) for two hours (van Zeeland, unpublished observation). By far the major part of radioactive label was at a position of unidentified products, possibly the pyrimidines, whereas HMPA did not bind to a measurable extent with the O^6 or the N-7 sites of guanine. The negative correlation we may draw here is absence of the storage phenomenon for a mutagen that does not seem to interact measurably with guanine in DNA. Further knowledge of the actual DNA sites attacked by HMPA is clearly desirable for the understanding of the molecular processes relevant to the delayed formation of breaks.

DNA Dosimetry

EMS has yet been the only chemical mutagen for which an exact dosimetry has been developed.[43] Aaron and Lee determined the dose-response curve for EMS with dose measured as ethylation of DNA per sperm cell and response measured as the relative frequency of X-linked recessive lethals induced in sperm cells of *Drosophila*.[44] The relative frequency of mutations was found to increase linearly with the dose over a range in dose of 2.1×10^{-4} to 1.4×10^{-2} ethylations per nucleotide. This linear relation suggests that no change in mechanism of mutagenesis occurred from low to high doses in *Drosophila*.[44] Therefore, Lee proposed to use relative frequency of X-linked recessive lethals as a biological dosimeter for comparing different studies in EMS mutagenesis in *Drosophila*.[43]

Future work will have to incorporate other AAs, to be able to extrapolate with confidence among different reagents. Clearly, the use of DNA dosimetry appears the best approach currently available to bridge the methodological gap from one species to another in the quantitative assessment of adverse effects of chemical mutagens.

Mutagenic Properties of AAs in Somatic Tissue

In view of the attractiveness of a genetic mechanism for the initiation of carcinogenesis, we felt that a brief consideration of the effects caused by AAs in somatic cells of *Drosophila* could be of interest in the present context, the possible relationship between alkylation-induced DNA damage and the resulting genetic consequences. Becker has described a system that measures mosaic single spots (light and dark segments) and mosaic twin spots in the eyes of $w^{co} sn^2/w +; se h/se h$ females following treatment of larvae.[45] We chose this method to compare the ability of MMS, EMS, DES, ENU, DEN, and ENNG to produce somatic mutations and recombinational events. All six AAs were efficient in producing twin spots (TS) and single spots (MS) in females treated as larvae. However, the proportion of TS versus MS varied considerably, depending on the type of alkylation damage introduced.[46] MMS produced the highest TS to MS ratio (1.16) followed by EMS and DES, which gave similar indices. DEN, ENNG, and ENU were the least effective reagents in the

TABLE 3

INDUCTION OF SOMATIC MUTATION AND RECOMBINATION BY MMS AND ENU
IN FEMALE w^{co}/w LARVAE*

	MMS	ENU	ENU	ENU	ENU	Controls
Concentration (mM)	5.0	0.1	0.5	2.5	5.0	—
Number of experiments	2	2	2	2	2	6
Eyes tested	754	1074	1160	1120	1120	3550
% TS	6.50	0.28	0.28	1.88	3.30	0.08
% ML	7.29	1.02	2.16	6.07	8.66	0.23
TS:ML†	0.91	0.25	0.10	0.31	0.38	—

*Summary table from data to be published elsewhere. TS, twin spots; ML, light spots.
†Induced TS:induced ML.

TABLE 4

MUTAGENIC PROFILES OF ALKYLATING AGENTS IN *DROSOPHILA*

	MMS	EMS DES	ENU DEN ENNG
Preferential reactivity in DNA	N	N > 0	0
Postmeiotic male germ line			
Mutation induction relative to cytotoxicity	low	high	high
Ability to cause chromosome breakage	high	moderate	low
Efficiency for delayed mutation	high	moderate	low
Response to *mei-9⁻*			
Mutations	increase	increase	not detectable
Chromosomal aberrations	increase	increase	increase
Somatic cells in females*			
Ratio of somatic			
recombination to mutation	high	moderate	low

*w/w^{co} system following treatment of first instar larvae.

induction of twin spots. This was obvious from the low TS to MS indices determined for these three AAs.[46] Also, it was found that for the reagents that alkylate DNA in a similar fashion, the TS:MS indices were also very similar: for the EMS-DES pair (0.76 versus 0.78) and for DEN (0.23), ENNG (0.29), and ENU (0.36).

Mosaic twin spots are supposed to reflect somatic recombination events and chromosomal abnormalities; mosaic single spots may be due to gene mutations.[45] Both mosaic twin spots and single spots occur in clones, i.e., they form characteristic segments in the eyes, corresponding with the direction in which the developing eye tissue differentiates. However, this typical clone formation was not apparent for a substantial part of the mosaic dark spots. This suggests that dark spots are mostly nongenetic in origin. Therefore, it was decided to no longer include them in the calculations.

Another possible pitfall that must be accounted for when comparing these ratios concerns the meaning of dose. In recent experiments with four concentrations of ENU, the TS:ML (mosaic light spots) indices showed a tendency to decrease with decreasing dose (TABLE 3). Yet there was, even at the highest ENU dose, no overlap with the high TS:ML ratios established for MMS, indicating that the differences between these AAs are real. In other words, a reagent such as MMS, which in germ cells is an efficient inducer of chromosomal aberrations, gives relatively high twin spot:single spot indices in somatic tissue.

CONCLUSIONS

The rather complex picture of alkylation-induced mutagenesis that has been drawn suggests a model function for *Drosophila* that may prove of increasing value in the future. By determining the mutagenic profile of a given genotoxic reagent as completely as possible, employing the seven parameters currently available for such an integrated analysis in *Drosophila*, it should be feasible to link certain types of DNA-alkylation adducts with various genetic end points.

The broad spectrum of genetic alterations seen with AAs in postmeiotic germ cells of males clearly documents that "mutagenic potency" is a very relative term, the ranking order largely depending on the type of genetic end point selected for the measurements (TABLE 4). Thus, there is a general, direct correlation between chromosome-breaking efficiency, the occurrence of delayed genetic damage, and N alkylation and a general inverse association between the latter parameter and the ability of AAs to induce point mutations. Conversely, those AAs acting more extensively at the oxygen atoms, while being less active with regard to the production of breaks, are more potent as inducers of point mutations (TABLE 4).

The picture is further complicated by the demonstration that DNA-repair processes can selectively modify the mutagenic efficiency. Nevertheless, I believe analysis of multiple genetic parameters provides a means of gaining further insight into possible links between chemical structure and mutagenic mechanisms. Although some of these studies are still in an initial stage, they have already demonstrated that events at the molecular level can be related with genetic effects in whole organisms.

REFERENCES

1. BAARS, A. J., M. JANSEN & D. D. BREIMER. 1979. Xenobiotica-metabolizing enzymes in *Drosophila melanogaster*. Activities of epoxide hydratase and glutathione S-transferase compared with similar activities in rat liver. Mutat. Res. **62:** 279–291.

2. FORBES, C. 1980. Sex-linked lethal mutations induced in Drosophila melanogaster by 7,12-dimethylbenz[a]anthracene. Mutat. Res. **79:** 231–237.

3. HÄLLSTRÖM, I. & R. GRAFTSTRÖM. 1981. The metabolism of drugs and carcinogens in isolated subcellular fractions of *Drosophila melanogaster*. II. Enzyme induction and metabolism of benzo[a]pyrene. Chem. Biol. Interact. **34:** 145–159.

4. VOGEL, E. 1981. Recent achievements with Drosophila as an assay system for carcinogens. *In* Short-Term Tests for Chemical Carcinogens. H. F. Stich & R. H. C. San, Eds.: 379–398. Springer Verlag. New York, N.Y.

5. RASMUSON, B., H. SVAHLIN, A. RASMUSON, I. MONTELL & H. OLOFSSON. 1978. The use of a mutationally unstable X-chromosome in Drosophila melanogaster for mutagenicity testing. Mutat. Res. 54: 33–38.

6. VOGEL, E., W. G. H. BLIJLEVEN, P. M. KLAPWIJK & J. A. ZIJLSTRA. 1980. Some current perspectives of the application of Drosophila in the evaluation of carcinogens. *In* The Predictive Value of Short-Term Screening Tests in Carcinogenicity. G. M. Williams *et al.*, Eds.: 125–147. Elsevier/North-Holland Biomedical Press. Amsterdam, the Netherlands.

7. LAWLEY, P. D., D. J. ORR & H. JARMAN. 1975. Isolation and identification of products from alkylation of nucleic acids: ethyl- and isopropyl-purines. Biochem. J. **145:** 73–84.

8. LAWLEY, P. D. & S. A. SHAH. 1972. Reaction of alkylating mutagens and carcinogens with nucleic acids: detection and estimation of a small extent of methylation of O-6 of guanine in DNA by MMS *in vitro*. Chem. Biol. Interact. **5:** 286–288.

9. PEGG, A. E. & J. W. NICOLL. 1976. Nitrosamine carcinogenesis: the importance of the persistence in DNA of alkylated bases in the organotropism of tumour induction. IARC Sci. Publ. **12:** 571–592.

10. SINGER, B. 1976. All oxygens in nucleic acids react with carcinogenic ethylating agents. Nature **264:** 333–339.

11. SUN, L. & B. SINGER. 1975. The specificity of different classes of ethylating agents toward various sites of Hela cell DNA in vitro and in vivo. Biochemistry **14:** 1795–1802.

12. OSTERMAN-GOLKAR, A., L. EHRENBERG & C. WACHTMEISTER. 1970. Reaction kinetics and biological action in barley of monofunctional methanesulfonic esters. Radiat. Bot. **10:** 303–327.

13. VOGEL, E. & A. T. NATARAJAN. 1979. The relation between reaction kinetics and mutagenic action of mono-functional alkylating agents in higher eukaryotic systems. I. Recessive lethal mutations and translocations in Drosophila. Mutat. Res. **62:** 51–100.

14. VOGEL, E. & A. T. NATARAJAN. 1982. The relation between reaction kinetics and

mutagenic action of monofunctional alkylating agents in higher eukaryotic systems: interspecies comparisons. *In* Chemical Mutagens. F. J. de Serres & A. Hollaender, Eds. **7:** 295–336. Plenum Publishing Corp. New York, N.Y.

15. EHRENBERG, L., K. D. HIESCHE, S. OSTERMAN-GOLKAR & I. WENNBERG. 1974. Evaluation of genetic risks of alkylating agents: tissue doses in the mouse from air contaminated with ethylene oxide. Mutat. Res. **24:** 83–103.

16. HUSSEIN, S. & L. EHRENBERG. 1975. Prophage inductive efficiency of alkylating agents and radiations. Int. J. Radiat. Biol. **27:** 355–362.

17. WALLES, S. & L. EHRENBERG. 1969. Determination of the rate constants for alkylation of DNA *in vitro* with methanesulfonic esters. Acta Chem. Scand. **23:** 1080–1084.

18. SINGER, B. 1979. *N*-Nitroso alkylating agents: formation and persistence of alkyl derivatives in mammalian nucleic acids as contributing factors in carcinogenesis. J. Nat. Cancer Inst. **62:** 1329–1339.

19. LAWLEY, P. D. & P. BROOKES. 1963. Further studies on the alkylation of nucleic acids and their constituent nucleotides. Biochem. J. **89:** 127–138.

20. SWENSON, D. H. & P. D. LAWLEY. 1978. Alkylation of deoxyribonucleic acid by carcinogens dimethyl sulphate, ethyl methanesulfonate, *N*-ethyl-*N*-nitrosourea and *N*-methyl-*N*-nitrosourea. Biochem. J. **171:** 575–587.

21. LAWLEY, P. D. 1974. Some chemical aspects of dose-response relationships in alkylation mutagenesis. Mutat. Res. **23:** 283–295.

22. BROOKES, P. & P. D. LAWLEY. 1961. The reaction of mono- and difunctional alkylating agents with nucleic acids. Biochem. J. **80:** 496–503.

23. BROOKES, P. & P. D. LAWLEY. 1963. Effects of alkylating agents on T2 and T4 bacteriophages. Biochem. J. **89:**138–144.

24. SHEARMAN, C. W. & L. A. LOEB. 1977. Depurination decreases fidelity of DNA synthesis in vitro. Nature **270:** 537–538.

25. AUERBACH, C. 1976. Mutation Research: Problems, Results, and Perspectives. Chapman and Hall. London, England.

26. FREESE, E. 1971. Molecular mechanism of mutations. *In* Chemical Mutagens, Principles and Methods for Their Detection. A. Hollaender, Ed. **1:** 1–56. Plenum Press. New York, N.Y.

27. OESCHGER, N. S. & P. E. HARTMAN. 1970. ICR-induced frameshift mutations in the histidine operon of Salmonella. J. Bacteriol. **101:** 490–504.

28. BAKER, B. S., J. B. BOYD, A. T. C. CARPENTER, M. M. GREEN, T. D. NGUYEN, P. RIPOLL & P. D. SMITH. 1976. Genetic controls of meiotic recombination and somatic DNA metabolism in Drosophila melanogaster. Proc. Nat. Acad. Sci. USA **73:** 4140–4144.

29. BOYD, J. B., M. D. GOLINO & R. B. SETLOW. 1976. The mei-9^a mutant of Drosophila melanogaster increases mutagen sensitivity and decreases excision repair. Genetics **84:** 527–544.

30. GRAF, U., E. VOGEL, U. P. BIBER & F. E. WÜRGLER. 1979. A new allele at the mei-9 locus on the X chromosome. Mutat. Res. **59:** 129–133.

31. GRAF, U. & F. E. WÜRGLER. 1976. MMS-sensitive strains in Drosophila melanogaster. Mutat. Res. **34:** 251–258.

32. SMITH, P. D., R. L. DUSENBERY, S. F. COOPER & C. F. BAUMEN. 1982. Examining the mechanism of mutagenesis in DNA repair–deficient strains of Drosophila melanogaster. *In* Environmental Mutagens and Carcinogens. T. Sugimura, S. Kondo & H. Takebe, Eds.: 147–155. University of Tokyo Press. Tokyo, Japan.

33. LOVELESS, A. 1969. Possible relevance of O-6 alkylation of deoxyguanosine to the mutagenicity and carcinogenicity of nitrosamines and nitrosamides. Nature London **233:** 206–207.

34. LAWLEY, P. D. 1979. Approaches to chemical dosimetry in mutagenesis and carcinogenesis: the relevance of reactions of chemical mutagens and carcinogens with DNA. *In* Chemical Carcinogens and DNA. P. L. Grover, Ed. **1:** 1–36. CRC Press. Boca Raton, Fla.

35. ZIMMERING, S. 1981. Reviews of the current status of the mei-9^a test for chromosome loss in *Drosophila melanogaster:* an assay with radically improved detection capacity for chromosome lesions induced by methyl methanesulfonate (MMS), dimethylnitrosamine (DMN), and especially diethylnitrosamine (DEN) and procarbazine. Mutat. Res. **83:** 63–80.

220 Annals New York Academy of Sciences

36. ZIMMERING, S. 1982. Induced chromosome loss following treatment of postmeiotic cells of the *Drosophila melanogaster* male with MMS and DMN and matings with repair-proficient females and the repair-deficient females *mei-9ᵃ* and *st* mus 302. Mutat. Res. **94:** 79–86.
37. ZIMMERING, S., A. W. HARTMANN & A. S. W. SONG. 1981. The repair-deficient mei-9ᵃ Drosophila female potentiates chromosome loss induced in the paternal genome by diethylnitrosamine. Mutat. Res. **91:** 123–128.
38. AUERBACH, C. 1967. The chemical production of mutations. Science **158:** 1141–1147.
39. AUERBACH, C., J. M. ROBSON & J. G. CARR. 1947. The chemical production of mutations. Science **105:** 243–247.
40. HERSKOWITZ, I. H. 1956. Mutagenesis in mature *Drosophila* spermatozoa by "triazine" applied in vaginal douches. Genetics **41:** 605–609.
41. SCHALET, A. 1955. The relationship between the frequency of nitrogen mustard induced translocations in mature sperm of *Drosophila* and utilization of sperm by females. Genetics **40:** 594.
42. VOGEL, E. & A. T. NATARAJAN. 1979. The relation between reaction kinetics and mutagenic action of monofunctional alkylating agents in higher eukaryotic systems. II. Total and partial sex-chromosome loss in Drosophila. Mutat. Res. **62:** 101–123.
43. LEE, W. R. 1976. Molecular dosimetry of chemical mutagens. Determination of molecular dose to the germ line. Mutat. Res. **38:** 311–316.
44. AARON, C. S. & W. R. LEE. 1978. Molecular dosimetry of the mutagen ethyl methanesulfonate in *Drosophila melanogaster* spermatozoa: linear relation of DNA alkylation per sperm cell (dose) to sex-linked recessive lethals. Mutat. Res. **49:** 27–44.
45. BECKER, H. J. 1966. Genetic and variegation mosaics in the eye of Drosophila. Curr. Top. Dev. Biol. **1:** 155–171.
46. VOGEL, E. W., W. G. H. BLIJLEVEN, M. J. H. KORTSELIUS & J. A. ZIJLSTRA. 1982. A search for some common characteristics of the effects of chemical mutagens in Drosophila. Mutat. Res. **92:** 69–87.

MUTATION SYSTEMS IN CULTURED
MAMMALIAN CELLS*

Ernest H. Y. Chu

Department of Human Genetics
University of Michigan Medical School
Ann Arbor, Michigan 48109

In 1968 during the Thirteenth International Congress of Genetics in Tokyo, Heinrich Malling and I presented evidence for chemical induction of specific locus mutation in cultured mammalian cells.[1,2] In the same meeting, Kao and Puck also reported the success of chemical mutagenesis in Chinese hamster cells.[3,4] Following those initial experiments, there was a period of consolidation, mostly devoted to the demonstration that the heritable variations under study did include true mutations. Evidence for mutation has now been obtained in a great number of cell systems. This evidence has taken several forms, including quantitative predictions of the effects of ploidy, the efficacy of known mutagens in the induction of cell variants, and, most convincingly, the direct demonstration in mutant cells of altered gene products.[5,6] Although we can take satisfaction from the conclusion that mutations can be experimentally induced and analyzed in cultured mammalian cells, certainly many important questions remain. In this presentation I shall attempt to provide the background on and introduce a series of questions, uncertainties, and investigative possibilities in this field.

The significant problems that persist in mammalian cell mutagenesis may be grouped into three overlapping areas of research:

1. The investigation of basic mutational events that take place in mammalian cells.

2. The development of methods for the quantitation of mutation induced in cultured mammalian cells.

3. The application of these methods for the screening of environmental mutagens.

For convenience of presentation, these areas will be covered in sequence.

INVESTIGATION OF BASIC MUTATIONAL EVENTS

Mutation is much more than a chemical or configurational change in DNA; it is a biological process deeply enmeshed in the structural and biological complexities of the cell and modulated by both internal and external factors. We still know little about the causes of spontaneous mutations, but the experimental evidence leaves no doubt that mutations occur in replicating as well as nonreplicating genomes of both pro- and eukaryotes. Our knowledge of the origin and nature of induced mutations far outstrips that relating to spontaneous ones, because it is always easier to analyze processes that one can manipulate experimentally than naturally occurring ones. Nevertheless, the manifestation of primary DNA lesion to a substantial alteration in gene expression follows a complex process, which acts as a sieve that allows only a proportion of the DNA lesion to proceed toward the final product of a mutant clone. According to

*Research supported by Public Health Service Grant CA 26803.

221

Auerbach, the mutagenic process for induced mutations may include (1) penetration of the mutagen to DNA; (2) production of a premutational lesion; (3) death from unrepaired damage; (4) repair of lesion, restoring the normal gene; (5) fixation of the premutational lesion as a mutated gene; (6) formation of mutant cells; (7) death of mutant cells; and (8) formation of a clone of mutant cells.[7] Two steps in the mutagenic process, namely, the premutational lesion caused by mispairing and the misrepair of the lesion, are the most important and basic mutational mechanisms. Many of the enzymes of DNA replication and repair contribute to the fate of primary DNA lesions, and the specificity of these enzymes contributes to the diversity of the mutational response.

Mutational responses in cultured mammalian cells can be measured at the DNA, protein, and phenotype levels. The entire way in which we think about genetic events, especially mutation, in mammalian cells may soon take on new meaning as the significance of introns and exons and genomic fluidity begins to be understood. These problems of understanding some of the most fundamental aspects of the structural organization of eukaryotic genes are exciting ones, especially in the promise of new recombinant DNA and nucleic acid sequencing techniques to aid in solving them. Studies of mutations of mammalian cells at the DNA level, as measured by either restriction analysis or nucleotide sequencing, are just the beginning.

Mutation can also be identified and measured at the protein level. Methods for one- or two-dimensional gel electrophoresis have been developed that are potentially powerful tools for monitoring mutations both in human somatic cells, *in vitro* and *in vivo,* and in human populations. If a change in electrophoretic mobility or molecular weight of a protein is indicated, the protein or peptide may be isolated for determination of amino acid composition and sequence. A specific mutation at the gene level may thus be inferred.

The effect of spontaneous and induced mutations on the expression of hypoxanthine guanine phosphoribosyltransferase (HPRT) (resistance to the analogues of the purine bases) in cultured mammalian somatic cells has been studied extensively. Several investigators have reported abnormalities in the immunochemical, catalytic, and electrophoretic properties of the enzyme in mutagenized HPRT-deficient cell lines and their revertants (reviewed in Reference 8). Alterations in peptide maps of the enzyme have been detected in several variant cell lines.[9–11] These studies provide evidence for the existence of a variety of undefined mutations in the structural gene for HPRT in cultured mammalian cells.

In man, an inherited deficiency of HPRT is associated with two distinct clinical syndromes: a virtually complete deficiency of HPRT activity has been described in patients with Lesch-Nyhan syndrome, while a partial enzyme deficiency is found in some male patients presenting with hyperuricemia and a severe form of gout. Human HPRT exists, in its native state, as a tetramer of identical subunits, coded by a single X-linked gene locus.[12] The complete amino acid sequence of HPRT from human erythrocytes has been defined.[13] Each subunit of the enzyme is 217 amino acids long with a molecular weight equal to 24,470. Jolly *et al.* recently reported the isolation and preliminary characterization of cloned cDNA sequences of the human HPRT gene.[14] The amino acid sequence predicted from the nucleotide sequence of the cDNA is in complete agreement with that defined by protein sequencing except that the NH_2 terminal methionine coded for by the initiator codon is absent in the mature enzyme.

More recently, the molecular basis for HPRT deficiency in two unrelated gout patients has been defined as single amino acid substitutions.[15,16] One enzyme variant, called $HPRT_{London}$, has a serine-to-leucine substitution at position 109. This substitution can be explained by a single nucleotide change in the codon for serine (UCA →

UUA). Sequence analysis of a single peptide in the variant enzyme ($HPRT_{Toronto}$) of another gout patient revealed an arginine-to-glycine amino acid substitution at position 50 (codon for Arg_{50} changed from CGA to GGA). These significant developments not only demonstrate for the first time the molecular basis of human mutation at the HPRT locus but also suggest the feasibility of identifying the molecular defects of HPRT mutants, spontaneous and induced, that can be obtained in cultured human cells.

The most prevailing end point for assessing mutations in cultured mammalian cells is a change in the cellular phenotype, such as nutritional requirement, resistance to cytotoxic agents, or cell membrane alterations. At this level, it is particularly important to distinguish the nature of somatic variations arising *in vitro,* whether it is genetic or epigenetic. Recently, Morgan Harris showed that brief exposure of 5-bromodeoxyuridine-tolerant, thymidine kinase–deficient (TK^-) Chinese hamster cells to 5-azacytidine resulted in massive conversion to the HAT^+ state (ability to grow in medium containing hypoxanthine, aminopterin, and thymidine),[17] suggesting that the induction of revertants might have resulted from changes in DNA methylation patterns. Similarly, 5-azacytidine treatment has been used to reactivate the expression of the HPRT gene on the inactive X chromosome in Chinese hamster cells.[18]

Even though the change is genetic, an alteration in the phenotype could be the result of either a genic or chromosomal event. For instance, Cox and Masson have shown that x-ray-induced mutations to 6-thioguanine resistance in cultured human diploid fibroblasts consist mostly of chromosomal deletions or rearrangements.[19] Hozier and his associates have made a cytogenetic analysis of the $L5178Y/TK^{+/-} \rightarrow TK^{-/-}$ mouse lymphoma mutagenesis assay system.[20] Mutant $TK^{-/-}$ colonies form a bimodal frequency distribution of colony sizes for most mutagenic or carcinogenic test substances. Large-colony $TK^{-/-}$ mutants with normal growth kinetics appear karyologically identical within and among clones and with the $TK^{+/-}$ parental cells. In contrast, most slow-growing, small-colony $TK^{-/-}$ mutants have a readily recognizable chromosome rearrangement involving the mouse chromosome 11 on which the TK locus is located.

Cell lines have been developed in which point mutation, chromosome loss, or chromosome deletion leading to the disappearance of a phenotype could be directly assayed. Puck has developed a test system in which the marker is contained on a chromosome not itself necessary for cell reproduction.[21] One such example is a reduced human–Chinese hamster cell hybrid in which a single human chromosome 11 is retained and mutations at several loci on this chromosome, namely, the antigenic markers a_1, a_2, a_3, LDH-A, and AcP-2, can be induced with mutagens. Any possible modification of a human protein or enzyme and the presence or absence of chromosome 11 or its segment can be studied. Using two-dimensional gel electrophoresis, at least eight distinct spots unique to the cell containing chromosome 11 have been located on the electrophoretogram.[22]

DEVELOPMENT OF METHODS FOR QUANTITATION OF MUTATION
INDUCTION IN CULTURED MAMMALIAN CELLS

Choice of Cell Material

A general question regarding any study of cultured cells is whether it accurately reflects the cellular responses in the intact animal. This point may be addressed using the direct-acting mutagen testing systems in which peripheral lymphocytes in the intact animal are examined. Another question is, Can the results derived with rodent cells be extended to the cells of humans? There is evidence that the repair of DNA damage differs both in process and extent between human and rodent cells.[23] Clearly, our knowledge of the similarities and differences in DNA replication and repair of human and other mammalian cells is lacking because of the dearth of human cell mutants for these fundamental processes. Therefore, studies of mutagenesis with human diploid fibroblasts, lymphoblastoid cell lines, or other established near-diploid cell lines should reduce the necessity for extrapolation to man.

Development of Selectable Genetic Markers

On the basis of classical studies with corn and *Drosophila*, the rates of spontaneous mutation may differ among different loci. Evidence has been presented that in certain mutator strains of Chinese hamster ovary (CHO) cells, the spontaneous mutation rates were 5- to 50-fold higher than those in parental cells for two genetic markers (resistance to 6-thioguanine or to ouabain), but the rates for two other markers (reversion of proline auxotrophy to prototrophy and forward mutation to emetine resistance) were unaffected.[24] As shown by several investigators, certain agents such as x-rays and gamma-rays fail to induce ouabain-resistant mutants in mammalian cells,[25-27] although they are capable of inducing 6-thioguanine-resistant mutants in the same cell type. Therefore, it appears that mutation rates may be site specific for mammalian cells as well. From both the theoretical and practical points of view, one of the important tasks is to develop additional selectable genetic markers for comparative mutagenesis among loci and for more accurate assessment of mutagenic potential of environmental chemicals.

Metabolic Activation

Environmental chemical mutagens often are nonreactive and have to be enzymatically activated before they can manifest their biological effects. Many cell types, including CHO and mouse lymphoma L5178Y cells, are not capable of activating procarcinogens or promutagens. Mutagenesis in these cells has to be studied by the addition of enzymatically active liver subcellular fraction, or by cocultivation with intact hepatocytes (cell-mediated mutagenesis). Tong and Williams have developed a system in which adult rat liver epithelial cells are used to detect genotoxic compounds by mutations at the HPRT locus.[28]

The subcellular enzyme fractions, such as rat liver microsomal S-9 fractions, are simpler to use and have been shown to activate a wide variety of procarcinogens. However, it has been shown that the profile of the metabolites and DNA adducts formed after metabolism of various potent carcinogens is different from that of intact cells. This indicates that the use of subcellular fractions does not accurately simulate the *in vivo* situation. It is also known that for certain classes of chemicals, metabolic activation or inactivation is species specific. If it can be metabolically activated *in vitro*, will it also be activated in humans? The major dilemma here is that given a chemical about which little is known, it is difficult to ensure that the activation system used will detect its possible mutagenic potential. Furthermore, the concentration of the S-9 fraction used in the cell-mutagenesis experiments may affect cell viability, thus introducing another source of biological variability to quantitative mutagenesis.

Phenotypic Expression

The course of events from the initial mutagenic insult to a cell to the appearance of the cell's progeny as a mutant colony is complicated and largely undefined. The so-called phenotypic expression refers to this series of unknown processes leading to the eventual recognition of the mutant. Forward mutations, such as HPRT⁻ or TK⁻, which represents a loss of function, could be the result of a spectrum of mutational events including chromosome loss, chromosomal deletion, nucleotide base substitution, and base addition or deletion (frameshift changes). The lesion, such as a loss or change of a base, could be repaired without error, thus producing no mutation, or could be misrepaired, leading to a change (mutation) in the normal nucleotide sequence. In theory, two rounds of cell replication are needed for "error replication" resulting in an altered base pair. Furthermore, cell divisions are necessary to allow dilution or degradation of the parental mRNA and/or enzyme in the cell carrying the mutated gene. These are the reasons why a period of phenotypic expression between the removal of the mutagen and the addition of the selective agent is necessary. The length of this period has been experimentally determined for a maximum recovery of the HPRT⁻ and TK⁻ mutations, and different lengths of time have been shown for mutants at other loci.

For both test systems using CHO or L5178Y cells, a predetermined period for phenotypic expression is allowed. Li devised a simple technique for growth of CHO cells as unattached cultures.[29] The optimum expression time was shorter for unattached cells (six days) than for attached cells (nine days). However, it is conceivable that different newly arisen mutants may have different growth rates and different selective advantages, relative to each other and to the parental cells. Some mutants may have divided more frequently than the parental cells and are overrepresented; others may be very slow growers, which may be missed in an enumeration of mutant colonies at a fixed time. Still other mutants may not thrive at all. Using the respreading technique, the mutation frequency is therefore only an average estimate.

Cell Density (Metabolic Cooperation)

Another factor that may affect mutation frequency is cell density. It was shown, especially in the case of HPRT⁻ mutant selection, that intercellular communication upon contact will diminish the number of mutants recovered. This phenomenon, known as "metabolic cooperation," occurs during 6-thioguanine selection and entails a transfer of metabolites, such as 6-thioguanosine monophosphate, through the gap junctions on the cell membrane from the dying parental cells to mutants, killing the latter in the process. For CHO cells, reconstruction experiments have determined the maximum cell density ($2 \times 10^5/10$ cm dish) that would not affect HPRT⁻ mutant frequency.

The problem of metabolic cooperation may be less serious in L5178Y cells grown in suspension due to a small number of gap junctions on the cell surface. Metabolic cooperation may also be reduced or abolished in the presence of tumor promoters (P. Liu, personal communication). For certain membrane markers, such as resistance to ouabain or abrin,[30] there was no demonstrable metabolic cooperation. For still other genetic markers under different situations, cross feeding in crowded cultures may actually increase, rather than decrease, the mutation frequency. Harris showed that the incidence of spontaneous variation to puromycin resistance in pig kidney cells increased in proportion to population density.[31] Large colonies or aggregates of sensitive cells grew progressively in puromycin at concentrations that destroyed the same populations completely when present in dispersed form. Harris concluded that response to puromycin is conditioned by cellular interactions as well as by genetic susceptibility.

Back Mutation and Preexisting Mutants

It is generally assumed that there is no back mutation during the course of mutagenicity testing, but this possibility needs to be verified. If it does indeed occur, the frequency should be accounted for in an estimate of the spontaneous and induced forward mutations.

The existence of preexisting mutants of spontaneous origin may be troublesome in the evaluation of a weak mutagen. There are experimental procedures, however, such as recloning or pregrowth in counterselective media, to eliminate or reduce the number of preexisting mutants in cell populations prior to mutagenicity testing.

Effects at Low Doses

At very low doses of chemicals, the linearity of dose-effect curves has not been settled. In both CHO and the mouse lymphoma cell systems, dose-effect relationships within a certain range of concentrations have been obtained for many chemicals, such as ethyl methanesulfonate and ethyl nitrosourea. However, it is debatable whether a linear extrapolation to lower doses is either reasonable or desirable.

A number of biological factors may play important roles during the entry of a chemical (or its derivatives) into the cell and the nucleus before its interaction with the genetic material. The question in a testing situation is to find the lowest effective dose that can induce mutation, not the theoretical threshold. The practical problem is the chronic exposure to low doses, rather than the demonstration of mutation induction after a single acute exposure at a very low dose. It seems that a dose-response curve using a series of doses is more informative than the determination of a "doubling dose" of the spontaneous frequency.

APPLICATION OF MAMMALIAN CELL MUTATION ASSAY SYSTEMS FOR THE SCREENING OF ENVIRONMENTAL MUTAGENS

Choice of Test Systems

Numerous cell lines, markers, and selective systems are now available. The use of human versus nonhuman somatic cells in cultures for mutation assay may be considered on the basis of technical advantages and relevance to man. The extent to which the assays could or should be used is a matter of controversy. Some people may feel that the assays are functional and the available systems should now be employed as far as practicable in the process of hazard evaluation. To wait for the ideal system is, in fact, a decision to continue to expose people and the environment to chemicals of unknown activity. At the Banbury conference in 1979, the weight of opinion was that the present test systems used in conjunction with various other end points are better than nothing. It was considered better to have a few false positives and negatives rather than exposure to a whole lot of potentially hazardous chemicals.

Definition of Basic Parameters of a Mutagen Screening System

Many factors must be examined to distinguish between genetic and epigenetic events, and certain basic criteria must be met if one claims to have a true mutational system. The types of mutation that are measured, whether genic, chromosomal, or both, should be delineated. In fact, for screening of mutagenicity of environmental chemicals, forward mutations—which encompass a broad spectrum of mutational events—may be preferable to reverse mutations. On the other hand, reverse mutations can help to identify the mutagenic specificity of chemicals.

The fact that many have developed reliable mutation systems does not imply that the work is easy or without problems. Many problems must be addressed for each system. Laboratories adopting a system should standardize the assay conditions and demonstrate that consistent, reproducible results can be obtained. The use of positive and solvent controls, the use of appropriate activation systems, background mutation frequencies consistent with historical controls, dose-response curves, and other baseline information should be defined. It may also be helpful to establish the classes of carcinogens and mutagens that the system accurately detects.

Increase in the Mutability of Cells

Recent years have witnessed a sharply increased number of studies of mutator and antimutator mutations in a variety of organisms, the most penetrating analyses being conducted in prokaryotes and fungi. These studies have proceeded in parallel with analyses of mechanisms of DNA replication and repair, at the same time as genes responsible for these functions have been identified by conditional lethal and/or mutagen-sensitive mutations.

In mammalian cells, increased rates of spontaneous mutations have been demonstrated in mutator strains that have modified intracellular nucleoside triphosphate pools.[24,32,33] In addition to cell strains obtained from the repair-defective human syndromes, several laboratories have reported the isolation of mutants in cultured rodent cells that have increased sensitivity to killing by mutagens, such as ultraviolet (UV) light (e.g., References 34–36). In some of these UV-sensitive mutants, enhancement of UV mutagenic response has been demonstrated.[37] Conceivably, these and other hypermutable cell strains will be very useful in increasing the sensitivity for detecting certain classes of mutagens. It is of particular importance to obtain human cell mutants defective in DNA repair for such purposes.

Statistical Analysis

In quantitative measurements of cell survival and mutation frequency, there are a number of sources of errors that affect the final mutant yield, including cell counts, cell viability during cytotoxicity tests, growth rate, and competition among the newly arisen mutants. Appropriate statistical models to account for these sources of variability and the choice of statistical methods for the analysis of data are important considerations in screening tests of environmental mutagens and carcinogens. There seem to be many valid problems in this area that require further research and discussion. A workshop on statistical analysis of *in vitro* tests for mutagenicity was held in April 1981 in Chapel Hill, North Carolina, under the joint sponsorship of the National Institute of Environmental Health Sciences and the National Toxicology Program. A report of the workshop will be published.

SUMMARY

Experimental studies on the complex process of mutagenesis are conducted in the belief that they will increase our understanding of the effects of environmental agents on human populations. Cultured mammalian cells offer distinct advantages over other biological systems for studying this problem. Mutagenesis in cultured mammalian cells is unquestionably more relevant to man than are bacterial, fungal, or insect systems. At the same time, compared to whole-animal studies, cultured cells offer the advantage of ease of handling, low cost, and rapidity of assay.[38]

Mutation in cultured mammalian cells can be identified and measured at the phenotypic, protein, and DNA levels. Test systems have been developed to detect both gene and chromosome mutations. Direct evidence is available for true mutations occurring in mammalian cells, although some of the heritable variations could be epigenetic. It is now possible to identify the specific molecular alternations in mammalian cell mutants arising *in vitro*.

The development of methods for the quantitation of mutation in cultured mammalian cells has been briefly considered in this presentation, particularly with reference to the choice of cell material, development of selective markers, and the various factors that affect quantitative measurement of mutations. Although notable progress has been made, much still remains to be learned to make these measurements more reliable.

In applying the mammalian cell mutation assays for the screening of environmental mutagens, the choice of a system or systems should be based on their technical advantages and relevance to man. The basic parameters of the mutagen screening system must be well defined and controlled. The sensitivity for detection may be increased by the use of hypermutable cell strains. Appropriate statistical models to account for various sources of variability and the choice of statistical methods for the analysis of data are important considerations in screening environmental mutagens and carcinogens.

REFERENCES

1. CHU, E. H. Y. & H. V. MALLING. 1968. Chemical mutagenesis in Chinese hamster cells *in vitro*. *In* Proceedings of the 12th International Congress on Genetics. 1: 102. Science Council of Japan. Tokyo, Japan.
2. CHU, E. H. Y. & H. V. MALLING. 1968. Mammalian cell genetics. II. Chemical induction of specific locus mutations in Chinese hamster cells *in vitro*. Proc. Nat. Acad. Sci. USA 61: 1306–1312.
3. KAO, F.-T. & T. T. PUCK. 1968. Isolation of nutritionally deficient mutants of Chinese hamster cells. *In* Proceedings of the 12th International Congress on Genetics. 1: 157. Science Council of Japan. Tokyo, Japan.
4. KAO, F.-T. & T. T. PUCK. 1968. Genetics of somatic mammalian cells. VII. Induction and isolation of nutritional mutants in Chinese hamster cells. Proc. Nat. Acad. Sci. USA 60: 1275–1281.
5. CHU, E. H. Y. 1974. Induction and analysis of gene mutations in cultured mammalian somatic cells. Genetics 78: 115–132.
6. SIMINOVITCH, L. 1976. On the nature of heritable variation in cultured somatic cells. Cell 7: 1–11.
7. AUERBACH, C. 1976. Mutation Research. Problems, Results and Perspectives: 381–388. Chapman and Hall. London, England.
8. CASKEY, C. T. & G. D. KRUH. 1979. The HPRT locus. Cell 16: 1–9.
9. CAPECCHI, M. R., R. A. VON DER HAAR, N. E. CAPECCHI & M. M. SVEDA. 1977. The isolation of a suppressible nonsense mutant in mammalian cells. Cell 12: 371–381.

10. MILMAN, G., S. W. KRAUSS & A. S. OLSEN. 1977. Tryptic peptide analysis of normal and mutant forms of hypoxanthine phosphoribosyltransferase from HeLa cells. Proc. Nat. Acad. Sci. USA **74:** 926–930.
11. KRUH, C. D., R. G. FENWICK, JR. & C. T. CASKEY. 1981. Structural analysis of mutant and revertant forms of Chinese hamster hypoxanthine-guanine phosphoribosyltransferase. J. Biol. Chem. **256:** 2878–2886.
12. HOLDEN, J. A. & W. N. KELLEY. 1978. Human hypoxanthine-guanine phosphoribosyl-transferase: evidence for a tetrameric structure. J. Biol. Chem. **253:** 4459–4463.
13. WILSON, J. M., G. E. TARR, W. C. MAHONEY & W. N. KELLEY. 1982. Human hypoxanthine-guanine phosphoribosyltransferase: complete amino acid sequence of the erythrocyte enzyme. J. Biol. Chem. **257:** 10978–10985.
14. JOLLY, D. J., A. C. ESTY, H. U. BERNARD & T. FRIEDMANN. 1982. Isolation of a genomic clone partially encoding human hypoxanthine phosphoribosyltransferase. Proc. Nat. Acad. Sci. USA **79:** 5038–5041.
15. WILSON, J. M., G. E. TARR & W. N. KELLEY. 1983. Human hypoxanthine-guanine phosphoribosyltransferase: a single amino acid substitution in a mutant form of the enzyme isolated from a patient with gout. Proc. Nat. Acad. Sci. USA. (In press.)
16. WILSON, J. M., I. H. FOX, G. E. TARR & W. N. KELLEY. The molecular abnormality in a structural variant of hypoxanthine-guanine phosphoribosyltransferase (HPRT$_{Toronto}$) isolated from a patient with gout. J. Biol. Chem. (Submitted.)
17. HARRIS, M. 1982. Induction of thymidine kinase in enzyme-deficient Chinese hamster cells. Cell **29:** 483–492.
18. GRANT, S. G. & R. G. WORTON. 1982. 5-Azacytidine-induced reactivation of HPRT on the inactive X chromosome in diploid Chinese hamster cells. Am. J. Hum. Genet. **34:** 171A.
19. COX, R. & W. K. MASSON. 1978. Do radiation-induced thioguanine-resistant mutants of cultured mammalian cells arise by HGPRT gene mutation or X-chromosome rearrangement? Nature **276:** 629–630.
20. HOZIER, J., J. SAWYER, M. MOORE, B. HOWARD & D. CLIVE. 1981. Cytogenetic analysis of the L5178/TK$^{+/-}$ → TK$^{-/-}$ mouse lymphoma mutagenesis assay system. Mutat. Res. **84:** 169–181.
21. PUCK, T. T. 1979. Historical perspective on mutation studies with somatic mammalian cells. *In* Mammalian Cell Mutagenesis: The Maturation of Test System(s). Banbury Report 2. A. W. Hsie, J. P. O'Neill & V. K. McElheny, Eds.: 3–14. Cold Spring Harbor Laboratory. Cold Spring Harbor, N.Y.
22. SCOGGIN, C. H., E. GABRIELSON, J. N. DAVIDSON, C. JONES, D. PATTERSON & T. T. PUCK. 1981. Two dimensional electrophoresis of human-CHO cell hybrids containing human chromosome 11. Somat. Cell Genet. **7:** 389–398.
23. REGAN, J. D. & R. B. SETLOW. 1973. Repair of chemical damage to human DNA. *In* Chemical Mutagens: Principles and Methods for Their Detection. A. Hollaender, Ed. **3:** 151–170. Plenum Press. New York, N.Y.
24. MEUTH, M., N. L'HEUREUX-HUARD & M. TRUDEL. 1979. Characterization of a mutator gene in Chinese hamster ovary cells. Proc. Nat. Acad. Sci. USA **76:** 6505–6509.
25. ARLETT, C. F., D. TURNBULL, S. A. HARCOURT, A. R. LEHMANN & C. M. COLELLA. 1975. A comparison of the 8-azaguanine and ouabain-resistance systems for the selection of induced mutant Chinese hamster cells. Mutat. Res. **33:** 261–278.
26. CHANG, C. C., J. E. TROSKO & T. AKERA. 1978. Characterization of ultraviolet light–induced ouabain-resistant mutations in Chinese hamster cells. Mutat. Res. **51:** 85–98.
27. THACKER, J., M. A. STEPHENS & A. STRETCH. 1978. Mutation to ouabain-resistance in Chinese hamster cells: induction by ethyl methanesulfonate and lack of induction by ionizing radiation. Mutat. Res. **51:** 255–270.
28. TONG, C. & G. M. WILLIAMS. 1980. Definition of conditions for the detection of genotoxic chemicals in the adult rat-liver epithelial cell/hypoxanthine-guanine phosphoribosyl transferase (ARL/HGPRT) mutagenesis assay. Mutat. Res. **74:** 19.
29. LI, A. P. 1981. Simplification on the CHO/HGPRT mutation assay through the growth of Chinese hamster ovary cells as unattached cultures. Mutat. Res. **85:** 165–175.
30. LI, I.-C., D. A. BLAKE, I. J. GOLDSTEIN & E. H. Y. CHU. 1980. Modification of cell membrane in variants of Chinese hamster cells resistant to abrin. Exp. Cell Res. **129:** 351–360.

31. HARRIS, M. 1967. Phenotypic expression of drug resistance in cell culture. J. Nat. Cancer Inst. **38:** 185–192.
32. CHANG, C.-C., J. A. BOEZI, S. T. WARREN, C. L. K. SABOURIN, P. K. LIU, L. GLATZER & J. E. TROSKO. 1981. Isolation and characterization of a UV-sensitive hypermutable aphidicolin-resistant Chinese hamster cell line. Somat. Cell Genet. **7:** 235–253.
33. WEINBURG, G., B. ULLMAN & D. W. MARTIN, JR. 1981. Mutator phenotypes in mammalian cell mutants with distinct biochemical defects and abnormal deoxyribonu-cleoside triphosphate pools. Proc. Nat. Acad. Sci. USA **78:** 2447–2451.
34. KUROKI, T. & S. Y. MIYASHITA. 1976. Isolation of UV-sensitive clones from mouse cell lines by Lederberg style replica plating. J. Cell. Physiol. **90:** 79–90.
35. SHIOMI, T. & K. SATO. 1979. Isolation of UV-sensitive variants of human FL cells by a viral suicide method. Somat. Cell Genet. **5:** 193–201.
36. THOMPSON, L. H., J. S. RUBIN, J. E. CLEAVER, G. F. WHITMORE & K. BROOKMAN. 1980. A screening method for isolating DNA repair–deficient mutants of CHO cells. Somat. Cell Genet. **6:** 391–405.
37. BUSCH, D. B., J. E. CLEAVER & D. A. GLASER. 1980. Large scale isolation of UV-sensitive clones of CHO cells. Somat. Cell Genet. **6:** 407–418.
38. COX, R. 1982. Mechanisms of mutagenesis in cultured mammalian cells. *In* Environmental Mutagens and Carcinogens. T. Sugimura, S. Kondo & H. Takebe, Eds.: 157–166. A. R. Liss, Inc. New York, N.Y.

CHINESE HAMSTER CELL MUTAGENESIS:
A COMPARISON OF THE CHO AND V79 SYSTEMS

David F. Krahn

*Haskell Laboratory for Toxicology
and Industrial Medicine
E. I. du Pont de Nemours & Co., Inc.
Newark, Delaware 19711*

INTRODUCTION

Mutation assays involving cultured mammalian cells can provide valuable information regarding the potential mutagenic and carcinogenic activities of chemicals. Data from these and similar assays provide guidance for decision making early in the process of assessing the toxicity of chemicals. Even though results from mammalian cell mutation tests are not sufficient for classifying chemicals as mutagens or carcinogens in mammals, including humans, the results can be used to make important decisions. However, it is critical for users of these tests to understand their sensitivity and reliability as well as to have a feeling for their relevance for predicting effects in humans.

Two commonly used mammalian cell mutation assays involve Chinese hamster V79 and CHO (Chinese hamster ovary) cells. The uses of these cell types for identifying mutagenic chemicals were thoroughly reviewed and analyzed by the Environmental Protection Agency Gene-Tox Program.[1,2] These reviews include recommendations for assay protocols and should be consulted for a complete treatment of the relevant literature published through 1980. The intent of this review is to describe briefly and compare these two cell types with regard to their usefulness as tests for identifying and studying mutagenic chemicals.

CELL CHARACTERISTICS

The V79 cell line was derived from the lung of a male Chinese hamster in 1958,[3] and the Chinese hamster ovary cell line was first isolated in 1957.[4] The V79 and CHO cells used today are clones or subclones of those isolates. V79 and CHO cells have stable karyotypes with modal chromosome numbers of 21 ± 1 and 20, respectively. The widely used CHO-K_1 line has a modal chromosome number of 21. Both of these cells double every 12–16 hours and exhibit plating efficiencies of 75–95%.[1,2] Different clones have related but not necessarily identical karyotypes, and therefore, care must be taken when comparisons are made among them.

Many carcinogens and mutagens require metabolism to exert their carcinogenic and mutagenic activities.[5,6] However, V79 and CHO cells do not contain appreciable levels of the mixed function oxidase (MFO) activities required for these conversions. Marquardt *et al.* reported that *trans*-1,2-dihydro-1,2-dihydroxy-7-methylbenz[*a*]anthracene induced gene mutations at the hypoxanthine-guanine phosphoribosyl transferase (HGPRT) locus in V79 cells.[7] The *trans*-8,9-dihydro-8,9-dihydroxy derivative was weakly mutagenic, but 7-methylbenz[*a*]anthracene was not. Subsequently, Pal *et al.* reported that the 7,8-dihydriol derivative of benzo[*a*]pyrene (BaP) induced sister chromatid exchanges (SCEs) in CHO cells whereas other diols and benzo[*a*]pyrene itself showed only weak activities.[8] The data from these studies imply that V79 and

231

0077–8923/83/0407–0231 $01.75/0 © 1983, NYAS

CHO cells contain low levels of MFO activity if it is assumed that the induction of gene mutations in V79 cells and SCEs in CHO cells requires metabolic activation of the polycyclic aromatic hydrocarbons and their diol derivatives. More recently, Pal et al. demonstrated that low levels of BaP monooxygenase and epoxide hydrase activities are indeed present in V79 cells.[9] However, the marginal levels of MFO activities in V79 and CHO cells are insufficient, and an exogenous source of metabolizing enzymes is necessary to detect mutagens requiring metabolism.

Summer and Wiebel reported that both V79 and CHO cells contain glutathione (GSH) and GSH S-transferases.[10] These findings are relevant because of the involvement of GSH and its transferases in the detoxification of electrophilic chemicals. Differences in their levels in the two cells lines may affect the relative sensitivities of these cell types to mutagenic chemicals. The authors reported that GSH S-transferases (with 1-chloro-2,4-dinitrobenzene and 1,2-dichloro-4-nitrobenzene as substrates) were severalfold higher in V79 cells than in CHO cells, while the opposite relation was observed for GSH. These results should be interpreted carefully since the levels of these substances may be dependent on the growth phase, which was not indicated. In addition, the enzyme activities and GSH levels were reported as concentrations per milligram protein, which may not be directly convertible to concentration per cell.

GENETIC MARKERS

Numerous genetic markers have been used to identify and study chemically induced mutations in V79[1,11-13] and CHO[2,14] cells. However, the majority of the published work involves two gene loci, HGPRT and Na/K-ATPase. The HGPRT locus is used with both the V79 and CHO cells,[1,2] while the Na/K-ATPase locus is predominantly used with V79 cells.[2]

The purine analogues 6-thioguanine (6TG) and 8-azaguanine (8AG) are commonly used to select for cells with mutations at the HGPRT locus. The details of these selection systems have been adequately described.[1,2] There is considerable evidence indicating that cells resistant to 6TG or 8AG possess mutations at the HGPRT locus. The evidence is similar for both the V79 and CHO cell systems.[1,2] The evidence that guanine-analogue-resistant cells possess mutations at the HGPRT locus includes the following: (1) The frequency of resistant colonies is increased by treatment with mutagens. (2) Certain resistant cells are revertible by specific mutagens. (3) The resistant phenotype is stably expressed. (4) Resistant cells lack HGPRT activity or possess markedly reduced levels, fail to incorporate hypoxanthine into cellular macromolecules, and are sensitive to aminopterin. (5) HGPRT in resistant cells, if present, often shows altered heat sensitivity, electrophoretic mobility, or enzyme kinetics. (6) In the case of V79 cells, HGPRT activity was restored to HGPRT-deficient cells by gene transfer.

With V79 cells, ouabain is often used to select for cells with mutations in the membrane-bound Na/K-ATPase.[1] Ouabain inhibits Na/K-ATPase by binding to this enzyme. Since Na/K-ATPase is essential for cell survival, the mutations that ouabain selects for are probably base substitutions that affect its binding but allow the retention of normal Na/K-ATPase activity.

EXPRESSION TIME

The expression time is the period between treatment with the test compound and the addition of the selective agent. This period is necessary to allow mutations to be fixed and to provide time for constitutive HGPRT and its mRNA to degrade. A short expression time has the practical advantages of allowing a more rapid and inexpensive test.

With respect to the HGPRT locus, V79 cells appear to require a slightly shorter expression time for maximal recovery of HGPRT mutants than do CHO cells. The Gene-Tox review group concluded that the appropriate expression time for V79 cells is six days,[1] while CHO cells generally require seven to nine days.[2]

Several factors may influence expression time. Perhaps the most likely factor is an inherent difference in the half-life of HGPRT in the two cell types. The HGPRT half-life in V79 cells is about 24 hours,[15] but no data on the half-life of CHO cell HGPRT are available. A second factor is the rate at which cells divide, since division dilutes the level of HGPRT activity per cell. In this regard, when Bradley *et al.* observed a five-day expression time,[16] their V79 cells had a doubling time of 9–11 hours, as compared to times of 12–14 hours standardly observed with CHO cells.[2] They propose that the faster rate of division may be at least partially responsible for the short expression time. A third factor possibly affecting expression time may be the selective agent used. Bonatti and colleagues recently concluded that enzyme dilution is the main factor in determining the differential responses to mutant selection with 8AG and 6TG.[17] 8AG is clearly a poorer substrate for HGPRT than is 6TG.[18,19] Thus, a residual level of HGPRT, capable of leading to cell death when 6TG is present, may not result in cell death with 8AG. However, Bradley *et al.* and Shaw and Hsie carefully examined the expression time in V79 cells with 6TG as the selective agent and found that five days was a sufficient period to allow the complete recovery of mutants induced by ethyl methanesulfonate (EMS) and some nitrosoureas.[16,20]

Some modified procedures for handling cells during expression time have been assessed with CHO cells. A modification gaining wider use is the growth of cells unattached to cell-culture plates during the expression period.[21,22] This treatment does not affect the recovery of HGPRT mutants, but has a major advantage over conventional methods with regard to the amount of time and effort required. This technique may be applicable to V79 cells, but no data are available. Using CHO cells, O'Neill and Hsie demonstrated that phenotypic expression could occur when cells were held in a nondividing state.[23]

USE TO TEST DIVERSE CHEMICALS

A review of the literature indicates that many more chemicals have been tested for mutagenic activity in V79 cells than in CHO cells. A general breakdown by chemical classes is presented in TABLE 1. The numbers include the chemicals listed in the V79 and CHO Gene-Tox reviews[1,2] and articles published for V79[24–34] and CHO[35–44] cells since then. I did not evaluate the recently published studies on the basis of the criteria outlined in the Gene-Tox reviews.

It is difficult to compare the sensitivities of the V79 and CHO cell mutation assays. Purity of the test material, culture conditions, treatment and selection conditions, selective agent, selective agent concentration, and activation system composition can all dramatically influence the degree of observed mutagenic activity. The only study in which both cell types were studied in the same laboratory under near identical conditions indicated that EMS was about 2.5 times more active in CHO cells than in V79 cells.[20] However, this observation may not apply to other mutagens and, thus, it is not clear from the published data that one cell type is more sensitive than the other.

For cell-culture mutation assays, the ideal situation is to have cells that allow the quantitation of genetic effects and possess the capability to metabolize test chemicals in a manner similar to cells in mammals. Simulation of human metabolism is the ultimate goal, but a system that reliably and accurately mimics rodent metabolism would be a major step forward. The major deficiency of the V79, CHO, and other similar mammalian cell assays is the activation systems. The commonly used liver-homogenate-mediated activation systems have proven to be extremely useful for identifying mutagenic chemicals requiring metabolic activation.[1,2] However, the relevance of mutation data generated with the use of tissue homogenates must be carefully assessed because a balance of activating and detoxifying pathways is not necessarily maintained. In addition, tissue homogenates do not always generate the same metabolites or metabolic profiles as intact cells.[45–47] Cell-mediated activation systems address some of the deficiencies of tissue homogenates. Dr. R. Langenbach has addressed this subject in this conference.[48]

The selection of a particular system should involve a determination of whether the system is sensitive to chemicals similar to those planned for evaluations or study. The system with the most test results may or may not be the most appropriate system. For example, little is known about the activities of arylamines in the V79 and CHO systems. We reported that the mutagenic activity of 2-acetylaminofluorene (2AAF) can be observed in CHO cells treated in the presence of a liver homogenate activation

TABLE 1

PUBLISHED TEST RESULTS

	V79	CHO
Polycyclic aromatic hydrocarbons	119	1
Nitro and nitroso compounds	26	6
Alkyl sulfate and alkane sulfonates	2	5
Aryl and alkyl halides	13	5
Purine and pyrimidine derivatives	7	3
Heterocyclic compounds	5	19
Metal-containing compounds	—	6
Other	48	4
	220	49

system.[49] However, with the same test conditions that convert 2AAF to a CHO cell mutagen, 2-naphthylamine (2NA) and 4-aminobiphenyl are either inactive or barely detectable (unpublished observations). Interestingly, *Salmonella* strain TA98 is mutagenized by 2NA with the same treatment medium. Thus, CHO cells and *Salmonella* appear to participate to differing degrees in the activation or detoxification of these chemicals. The modification of treatment conditions or the use of liver from different species or from animals treated with different inducers may allow these carcinogens and bacterial mutagens to express their mutagenic activities in CHO cells. However, this example points out that the routine use of these or similar mammalian systems must be approached with caution.

ROLE IN GENETIC TOXICITY TESTING

The V79 and CHO cell mutation assays are easy to use and provide valuable information to aid in assessing the potential carcinogenic and mutagenic activities of chemicals. Specific uses of these and other mammalian cell mutation assays include (1) identifying potential carcinogenic and mutagenic chemicals early in the process of the development of new product candidates; (2) screening potential substitutes for chemicals demonstrated to be carcinogenic or mutagenic; (3) guiding the synthesis of nonmutagenic derivatives of mutagenic chemicals; (4) monitoring samples for mutagenic impurities; (5) setting priorities for testing chemicals in animal carcinogenesis studies; and (6) gaining insights into the mechanisms by which carcinogens act. In most cases, a bacterial mutagenesis assay is the first test to conduct to begin addressing these concerns. However, mammalian cells may provide confirmatory information and provide insight into mammalian cell effects.

The activation systems are the weakest aspect of these and similar assays. Thus, the most important criterion for choosing a test system is whether it accurately predicts the activities of chemicals with structures similar to the test compound. Unfortunately, for a substantial portion of test chemicals encountered in industry, this kind of information is not available.

As with most cell-culture systems, these cells can be used to measure multiple end points such as sister chromatid exchange[50] and chromosome aberrations.[51] This flexibility is a significant advantage if the objective of a study is to investigate the relative abilities of a chemical to induce different effects. However, in a screening program, it may be disadvantageous to use only one cell type to measure multiple end points because metabolic or other cell-specific biochemical properties may result in a lack of sensitivity that is confirmed by all of the end points measured.

There are important questions that these assays do not adequately address. First, they cannot be relied upon to identify nongenotoxic carcinogens that participate in carcinogenesis by mechanisms that do not involve the induction of mutations in exposed cells. This, of course, is widely acknowledged, and considerable effort is being directed to address this concern. Second, the cultured mammalian cell mutation assays do not provide sufficient information for predicting *in vivo* potency and, hence, they do not provide a basis for setting exposure levels or assessing risk. Work should continue to develop, understand, and refine cell-culture systems for the uses described earlier. However, to address the *in vivo* potency/risk assessment issue, especially for carcinogenesis, *in vivo/in vitro* assays in which genetic or biochemical effects are measured in cells isolated from treated animals are necessary.

ACKNOWLEDGMENT

I want to thank Dr. A. M. Sarrif for reviewing the manuscript and Dr. J. D. Irr for his helpful comments. The secretarial assistance of J. M. Skrotsky is also acknowledged.

REFERENCES

1. BRADLEY, M. O., B. BHUYAN, M. C. FRANCIS, R. LANGENBACH, A. PETERSON & E. HUBERMAN. 1981. Mutagenesis by chemical agents in V79 Chinese hamster cells: a review and analysis of the literature. Mutat. Res. **87:** 81–142.
2. HSIE, A. W., D. A. CASCIANO, D. B. COUCH, D. F. KRAHN, J. P. O'NEILL & B. L. WHITFIELD. 1981. The use of Chinese hamster ovary cells to quantify specific locus mutation and to determine mutagenicity of chemicals: a report of the Gene-Tox Program. Mutat. Res. **86:** 193–214.
3. FORD, D. K. & G. YERGANIAN. 1958. Observations on the chromosomes of Chinese hamster cells in tissue culture. J. Nat. Cancer Inst. **21:** 393–425.
4. PUCK, T. T., S. J. CIRCIURA & A. ROBINSON. 1958. Genetics of somatic mammalian cells.

III. Long-term cultivation of euploid cells from human and animal subjects. J. Exp. Med. **108:** 945–956.

5. HEIDELBERGER, C. 1973. Chemical carcinogenesis. Annu. Rev. Biochem. **44:** 79–121.
6. MILLER, J. A. 1970. Carcinogenesis by chemicals: an overview—G. H. A. Clowes Memorial Lecture. Cancer Res. **30:** 559–576.
7. MARQUARDT, H., S. BAKER, B. TIERNEY, P. L. GROVER & P. SIMS. 1977. The metabolic activation of 7-methylbenz[a]anthracene: the induction of malignant transformation and mutation in mammalian cells by non-K-region dihydrodiols. Int. J. Cancer **19:** 828–833.
8. PAL, K., B. TIERNEY, P. L. GROVER & P. SIMS. 1978. Induction of sister-chromatid exchanges in Chinese hamster ovary cells treated *in vitro* with non-K-region dihydrodiols of 7-methylbenz[a]anthracene and benzo[a]pyrene. Mutat. Res. **50:** 367–375.
9. PAL, K., P. L. GROVER & P. SIMS. 1980. The metabolism of benzo[a]pyrene by Chinese hamster cells in culture. Xenobiotica **10:** 25–31.
10. SUMMER, K. & F. J. WIEBEL. 1981. Glutathione and glutathione S-transferase activities of mammalian cells in culture. Toxicol. Lett. **9:** 409–413.
11. MORROW, J., D. SAMMONS & E. BARRON. 1980. Puromycin resistance in Chinese hamster cells: genetic and biochemical studies of partially resistant, unstable clones. Mutat. Res. **69:** 333–346.
12. WASMUTH, J. J., J. M. HILL & L. S. VOCK. 1980. Biochemical and genetic evidence for a new class of emetine-resistant Chinese hamster cells with alterations in the protein biosynthetic machinery. Somatic Cell Genet. **6**(4): 495–516.
13. HUBERMAN, E., C. K. MCKEOWN & J. FRIEDMAN. 1981. Mutagen-induced resistance to mycophenolic acid in hamster cells can be associated with increased inosine 5'-phosphate dehydrogenase activity. Proc. Nat. Acad. Sci. USA **78**(5): 3151–3154.
14. FUNANAGE, V. L. 1982. Isolation and characterization of 5,6-dichloro-1-β-D-ribofuranosyl-benzimidazole-resistant mutants of the Chinese hamster ovary cell line. Mol. Cell. Biol. **2:** 467–477.
15. CASKEY, C. T. & G. D. KRUH. 1979. The HGPRT locus. Cell **16:** 1–9.
16. BRADLEY, M. O., N. A. SHARKEY, K. W. KOHN & M. W. LAYARD. 1980. Mutagenicity and cytotoxicity of various nitrosoureas in V79 Chinese hamster cells. Cancer Res. **40:** 2719–2725.
17. BONATTI, S., A. ABBONDANDOLO, A. MAZZACCARO, R. FIORIO & L. MARIANI. 1980. Experiments on the effect of the medium in mutation tests with the HGPRT system in cultured mammalian cells. Mutat. Res. **72:** 475–482.
18. GILLIN, F. D., D. J. ROUFA, A. L. BEAUDET & C. T. CASKEY. 1972. 8-Azaguanine resistance in mammalian cells I. Hypoxanthine-guanine phosphoribosyltransferase. Genetics **72:** 239–252.
19. VAN DIGGELEN, O. P., T. F. DONAHUA & S. ILSHIN. 1979. Basis for differential cellular sensitivity to 8-azaguanine and 6-thioguanine. J. Cell. Physiol. **98:** 59–72.
20. SHAW, E. I. & A. W. HSIE. 1978. Conditions necessary for quantifying ethyl methanesulfonate–induced mutations to purine-analog resistance in Chinese hamster V79 cells. Mutat. Res. **51:** 237–254.
21. LI, A. P. 1981. Simplification of the CHO/HGPRT mutation assay through the growth of Chinese hamster ovary cells as unattached cultures. Mutat. Res. **85:** 165–175.
22. KRUSZEWSKI, F. H., H. E. SCRIBNER & K. L. MCCARTHY. 1982. Comparison of CHO/HGPRT mutation assay results after attached and unattached growth during expression. Environ. Mutagen. **4:** 393–394.
23. O'NEILL, J. P. & A. W. HSIE. 1979. Phenotypic expression time of mutagen-induced 6-thioguanine resistance in Chinese hamster cells (CHO/HGPRT system). Mutat. Res. **59:** 109–118.
24. GRUENER, N. & M. L. LOCKWOOD. 1980. Mutagenic activity in drinking water. Am. J. Public Health **70:** 276–278.
25. JONES, J. A., J. R. STARKEY & A. KLEINHOFS. 1980. Toxicity and mutagenicity of sodium azide in mammalian cell cultures. Mutat. Res. **77:** 293–299.
26. TAKANISHI, H., M. UMEDA & I. HIRONO. 1980. Chromosomal aberrations and mutations in cultured mammalian cells induced by pyrrolizidine alkaloids. Mutat. Res. **78:** 67–77.
27. BABUDRI, N., B. PANI, S. VENTURINI, M. TAMARO, C. MONTI-BRAGADIN & F. BORDIN. 1981. Mutation induction and killing of V79 Chinese hamster cells by 8-methoxypsoralen

plus near-ultraviolet light: relative effects of monoadducts and crosslinks. Mutat. Res. **91:** 391–394.

28. DREVON, C., C. PICCOLI & R. MONTESANO. 1981. Mutagenicity assays of estrogenic hormones in mammalian cells. Mutat. Res. **89:** 83–90.

29. KNAAP, A. G. A. C., B. W. GLICKMAN & J. W. I. M. SIMONS. 1981. Effects of ethionine on the replicational fidelity in V79 Chinese hamster cells. Mutat. Res. **82:** 355–363.

30. MALLON, R. G. & T. G. ROSSMAN. 1981. Bisulfite (sulfur dioxide) is a comutagen in *E coli* and in Chinese hamster cells. Mutat. Res. **88:** 125–133.

31. NODA, K., M. UMEDA & Y. UENO. 1981. Cytotoxic and mutagenic effects of sterigmatocystin on cultured Chinese hamster cells. Carcinogenesis **2:** 945–949.

32. OCHI, T., M. UMEDA, H. MASUDA & H. ENDO. 1981. Induction of chromosomal aberrations and 8-azaguanine-resistant mutations by aryldialkyltriazenes in cultured mammalian cells. Mutat. Res. **88:** 197–209.

33. PANI, B., N. BABUDRI, S. VENTURINI, M. TAMARO, F. BORDIN & C. MONTI-BRAGADIN. 1981. Mutation induction and killing of prokaryotic and eukaryotic cells by 8-methoxypsoralen, 4,5'-dimethylangelicin, 5-methylangelicin, 4'-hydroxymethyl-4,5'-dimethylangelicin. Teratogen. Carcinogen. Mutagen. **1:** 407–415.

34. TURCHI, G., S. BONATTI, L. CITTI, P. G. GERVASI & A. ABBONDANDOLO. 1981. Alkylating properties and genetic activity of 4-vinylcyclohexene metabolites and structurally related epoxides. Mutat. Res. **83:** 419–430.

35. FUSCOE, J. C., J. P. O'NEILL, R. M. PECK & A. W. HSIE. 1979. Mutagenicity and cytotoxicity of nineteen heterocyclic mustards (ICR compounds) in cultured mammalian cells. Cancer Res. **39:** 4875–4881.

36. JOHNSON, N. P., J. D. HOESCHELE, R. O. RAHN, J. P. O'NEILL & A. W. HSIE. 1980. Mutagenicity, cytotoxicity, and DNA binding of platinum(11)chloroamines in Chinese hamster ovary cells. Cancer Res. **40:** 1463–1468.

37. THOMPSON, L. H., R. M. BAKER, A. V. CARRANO & K. W. BROCKMAN. 1980. Failure of the phorbol ester 12-O-tetradecanoylphorbol-13-acetate to enhance sister chromatid exchange, mitotic segregation, or expression of mutations in Chinese hamster cells. Cancer Res. **40:** 3245–3251.

38. SNEE, R. D. & J. D. IRR. 1981. Design of a statistical method for the analysis of mutagenesis at the hypoxanthine-guanine phosphoribosyltransferase locus of cultured Chinese hamster ovary cells. Mutat. Res. **85:** 77–93.

39. JONGEN, W. M. F., P. H. M. LEHMAN, M. J. KOTTENHAGEN, G. M. ALINK, F. BERENDS & J. H. KOEMAN. 1981. Mutagenicity testing of dichloromethane in short-term mammalian test systems. Mutat. Res. **81:** 203–213.

40. MACHANOFF, R., J. P. O'NEILL & A. W. HSIE. 1981. Quantitative analysis of cytotoxicity and mutagenicity of benzo[*a*]pyrene in mammalian cells (CHO/HGPRT system). Chem. Biol. Interact. **34:** 1–10.

41. O'NEILL, J. P., N. L. FORBES & A. W. HSIE. 1981. Cytotoxicity and mutagenicity of the fungicides Captan and Folpet in cultured mammalian cells (CHO/HGPRT system). Environ. Mutagen. **3:** 233–237.

42. TAN, E. L. & A. W. HSIE. 1981. Mutagenicity and cytotoxicity of haloethanes as studied in the CHO/HGPRT system. Mutat. Res. **90:** 183–191.

43. HEARTLEIN, M. W., J. P. O'NEILL, B. C. PAL & R. J. PRESTON. 1982. The induction of specific locus mutations and sister chromatid exchanges by 5-bromo- and 5-chlorodeoxyuridines. Mutat. Res. **92:** 411–416.

44. KRAHN, D. F., F. C. BARSKY & K. T. McCOOEY. 1982. CHO/HGPRT mutation assay: evaluation of gases and volatile liquids. *In* Genotoxic Effects of Airborne Agents. R. R. Tice, D. L. Costa & K. M. Schaich, Eds.: 91–103. Plenum Publishing Corporation. New York, N.Y.

45. NEWBOLD, R. F., C. B. WIGLEY, M. H. THOMPSON & P. BROOKES. 1977. Cell-mediated mutagenesis in cultured Chinese hamster cells by carcinogenic polycyclic hydrocarbons: nature and extent of the associated hydrocarbon-DNA reaction. Mutat. Res. **43:** 101–116.

46. SELKIRK, J. K. 1977. Benzo[*a*]pyrene carcinogenesis: a biochemical selection mechanism. J. Toxicol. Environ. Health **2:** 1245–1258.

47. BIGGER, C. A. H., J. E. TOMASZEWSKI & A. DIPPLE. 1978. Differences between products of

binding of 7,12-dimethyl-benz[a]anthracene to DNA in mouse skin and in a rat liver microsomal system. Biochem. Biophys. Res. Commun. **80:** 229–235.

48.   LANGENBACH, R., C. HIX, L. OGLESBY & J. ALLEN. 1983. Cell-mediated mutagenesis of Chinese hamster V79 cells and *Salmonella typhimurium*. Ann. N.Y. Acad. Sci. (This volume.)

49.   KRAHN, D. F. 1979. *In* Mammalian Cell Mutagenesis: The Maturation of Test Systems. Banbury Report 2. A. W. Hsie, J. P. O'Neill & V. K. McElheny, Eds.: 251–261. Cold Spring Harbor Laboratory. Cold Spring Harbor, N.Y.

50.   LATT, S. A., J. ALLEN, S. E. BLOOM, A. CARRANO, E. FALKE, D. KRAM, E. SCHNEIDER, R. SCHRECK, R. TICE, B. WHITFIELD & S. WOLFF. 1981. Sister-chromatid exchanges: a report of the Gene-Tox Program. Mutat. Res. **87:** 17–62.

51.   PRESTON, R. J., W. AU, M. A. BENDER, J. G. BREWEN, A. V. CARRANO, J. A. HEDDLE, A. F. MCFEE, S. WOLFF & J. S. WASSOM. 1981. Mammalian *in vivo* and *in vitro* cytogenetic assays: a report of the U.S. EPA's Gene-Tox Program. Mutat. Res. **87:** 143–188.

THE L5178Y/TK GENE MUTATION ASSAY FOR THE DETECTION OF CHEMICAL MUTAGENS

David E. Amacher and Gail N. Turner

Drug Safety Evaluation Department
Pfizer Central Research
Groton, Connecticut 06340

INTRODUCTION

One important application of *in vitro* toxicological systems is the use of cultured mammalian cells for the detection of mutagens as presumptive genotoxic carcinogens. With *in vitro* models, the genetic toxicologist can simulate the biotransformation of xenobiotics, then examine well-characterized target cells for irreversible DNA damage, unmistakable evidence that genotoxic chemicals or metabolites were present.

The L5178Y/TK gene mutation assay[1] detects the heritable loss of thymidine kinase (TK) activity in heterozygous mouse lymphoma cells following acute exposure to genotoxic agents. Frequencies of bromodeoxyuridine- or trifluorothymidine-resistant variants are substantially increased following exposure of L5178Y cells to direct-acting mutagens or those chemicals requiring the presence of exogenous microsomal enzyme fractions for activation to detectable mutagens.[2] This assay is also suitable for cocultivation experiments with metabolically competent cell monolayers.[3]

Our purposes here are to reexamine some basic characteristics of this gene mutation assay, then describe some specific applications of the L5178Y $TK^{+/-} \rightarrow TK^{-/-}$ assay for the detection of mammalian cell gene mutations with special emphasis on exogenous metabolic biotransformation procedures.

MATERIALS AND METHODS

Chemicals

Cytosine-1-β-D-arabinofuranoside (ara-C), 5-fluorodeoxyuridine (FUdR), trifluorothymine deoxyriboside (TFT), 6-thioguanine (6TG), colchicine, ethyl methanesulfonate (EMS), 3-methylcholanthrene (3MCA) and 5-bromo-2'-deoxyuridine (BUdR) were from Sigma Chemical Co. (St. Louis, Mo.). Methotrexate was from ICN Pharmaceuticals (Plainview, N.Y.). [2-^{14}C]-Thymidine—specific activity, 57 mCi/mmol—was obtained from Amersham (Arlington Heights, Ill.). RPMI-1640 medium and horse serum were purchased from M. A. Bioproducts (Walkersville, Md.). RPMI combined with 5% horse serum is subsequently referred to as R_5 in the text. Other sources were: dimethylsulfoxide (DMSO), Pierce Chemical (Rockford, Ill.); hycanthone methanesulfonate (Hyc), Sterling-Winthrop (Rensselaer, N.Y.); 2,7-dinitrofluorene, Aldrich (Milwaukee, Wis.); safrole, MCB (Cincinnati, Ohio); and ICR-191, Polyscience, Inc. (Warrington, Pa.).

Metabolic Activation Procedures

Male rats [Crl:COBS CD(SD)BR] weighing 200–250 g, male mice [Crl:COBS CD-1(ICR)BR] weighing 32–36 g, and adult Syrian hamsters [LaK:LVG(SYR)] were obtained from Charles River Breeding Labs.

239

0077-8923/83/0407-0239 $01.75/0 © 1983, NYAS

Separate S-9 (9,000 × g supernatant) batches were prepared from the livers of untreated male CD rats or CD-1 mice, and from male and female hamsters by established techniques.[4] All animals had been fasted for 24 hours. Freshly prepared S-9 was stored at −80°C. Final S-9 concentration during mutagen testing was 5% (v/v).[5]

Urine from safrole-treated mice was obtained by previously described procedures.[6] Mice had been dosed by oral gavage with either 1,000 mg/kg safrole prepared in saline + 0.02% Tween 80 or solvent only (controls). Urine was collected for 18 hours over dry ice and stored frozen. Prior to mutagenicity testing, urine was adjusted to pH 6.8, centrifuged, combined with β-glucuronidase, filter sterilized, and incubated at 37°C.[6]

Cells and Culture Conditions

L5178Y/TK$^{+/-}$ cells (3.7.2C subclone) were obtained from D. Clive, Burroughs Wellcome Co., and stored in liquid N_2. Cell stocks designated as T1L1-T1L5 originated as large-colony TFT^r variants when methyl methanesulfonate–treated 3.7.2C cells were cloned in soft-agar medium containing 4 μg/ml TFT. These are described elsewhere.[7] T1L4 had been grown continuously in nonselective R_5 medium for 52 weeks; the other four had been grown continuously for 26 weeks when used in these experiments. A known TK$^{-/-}$ cell line (3.7.2C.110) was obtained from Dr. Brian Myhr, Litton Bionetics.

Cloning and Selection Procedures

THMG medium used to counterselect for the TK$^{+/-}$ phenotype consisted of RPMI-1640 soft-agar medium supplemented with 9.0 μg/ml thymidine, 15 μg/ml hypoxanthine, 22.5 μg/ml glycine, and 0.3 μg/ml methotrexate. Nonselective medium (R_5) contained RPMI-1640, 5% prescreened horse serum, and 0.5 mg/ml Pluronic F-68 (BASF Wyandotte).

All cell stocks were screened for mycoplasma and were negative. In general, cells were maintained on a daily basis and cloned in soft-agar medium (no antibiotics) as previously described.[8] TFT concentration was 4 μg/ml unless otherwise indicated. All mutant frequency determinations were based upon predominantly large (>0.3 mm) TFT^r colony counts on days 7–10.

TK Enzyme Assay

With minor variations, the procedure of Ives et al.[9] was used for the determination of intracellular TK enzyme activity. Cells (8 × 10^7) from each cell stock were washed twice with Hank's balanced salt solution, pelleted in 0.5 ml tris buffer + 1.0 mM mercaptoethanol, and sonicated. One hundred microliters of this supernatant were added to 100 μl of a reaction mixture containing 6.19 mg ATP, 1.24 mg $MgCl_2$, and 9 μCi [2-^{14}C]-thymidine per ml tris buffer (pH 8.0). This combination was incubated 30 minutes at 37°C with gentle rotation, then the reaction was terminated by immersion for 2 minutes in 100°C water. The glass microvials were cooled and centrifuged at 5,000 rpm for 5 minutes. Fifty microliters of the supernatant were spotted on 2.4-cm DE-81 disks. Disks were dried, washed 15 minutes with 1 mM ammonium formate (20 ml/disk), and washed twice (15 minutes each time) with methanol (20 ml/disk) to remove unreacted thymidine. Additional disks were spotted with 25 μl reaction mixture, dried, and placed directly into counting vials. These samples were used to calculate the total amount of available substrate. A third set of disks were spotted with 25 μl of unreacted assay mixture and washed as before. These reaction blanks were later subtracted as part of the background. All disks were placed into sample vials with HCl/KCl (0.1/0.2 M), eluted for 15 minutes, then mixed with 10 ml aquasol for scintillation counting. TK activity was computed as described by Ives et al.[9]

RESULTS AND DISCUSSION

The Biological Basis for TFT Resistance in L5178Y Cells

Trifluorothymidine-resistant (TFT^r) cells derived from mutagen-treated TK$^{+/-}$ stock have a greatly diminished capacity for thymidine uptake[10] and the phosphorylation of thymidine measured by a direct enzyme assay.[11] Phosphorylated TFT is an irreversible inhibitor of thymidylate synthetase in mammalian cells.[12]

If mutagen-induced TFT resistance represents a heritable loss of TK activity, TFT^r-clone-derived cells should also be resistant to inhibition by similar deoxyribonucleoside analogues that can act as alternate substrates for TK enzyme. We have shown, for example, that presumptive TK$^{-/-}$ mutants selected on the basis of TFT resistance are also cross-resistant to BUdR,[7,8] another TK substrate.[13] Phosphorylated FUdR is a powerful, competitive inhibitor of thymidylate synthetase.[12] We compared the effect of FUdR on the cloning efficiencies of TK$^{+/-}$ and 3.7.2C.110 cells, the latter having been derived as a large-colony TFT^r mutant from an EMS-treated 3.7.2C TK$^{+/-}$ cell culture.[14] As shown in FIGURE 1, the 3.7.2C.110 cell line was resistant to the presence of up to 0.1 μg/ml FUdR, concentrations lethal to TK$^{+/-}$ cells.

Thymidine kinase and deoxycytidine (CdR) kinase may be closely associated and under a common regulatory control. For example, it has been suggested that mitochondrial TdR and CdR kinases are one and the same enzyme.[15] Cells deficient in CdR kinase are resistant to the antimetabolite ara-C.[16] To determine if TK loss in L5178Y cells coincided with CdR kinase loss, we compared the cloning efficiencies of TK$^{+/-}$ stock versus TFT^r 3.7.2.110 cells. The results are shown in FIGURE 1. The 3.7.2.110 TK$^{-/-}$ and 3.7.2C TK$^{+/-}$ cell lines were equally resistant to ara-C concentrations less than 0.015 μg/ml, suggesting that mutagen-induced loss of TK activity was not necessarily accompanied by CdR kinase loss as measured by ara-C resistance.

FUdR produces a strong positive mutagenic response in L5178Y cells after three hours of treatment, as does 9–30 μM ara-C (data not shown); however, the FUdR response is obviously due to mutant selection rather than mutant induction.

The Mutagenic Response Measured by Increased TFT Resistance

For the L5178Y/TK assay, the spontaneous forward mutation rate at the thymidine kinase locus in 3.7.2 TK$^{+/-}$ cells is 2 x 10^{-6} mutations/locus per generation.[17] Background mutant frequencies (accumulated mutants of spontaneous origin) as low as 16×10^{-6} or as high as 297×10^{-6} have been reported within a single extensive study.[2] All determinations of experimental mutagenicity are ultimately based upon comparisons with concurrent mutant frequencies in solvent-treated control cultures. Therefore, it is important to maintain stock cultures with a minimal TFT^r colony count for maximum assay sensitivity. With slight modifications of the THMG selection procedures described by Clive and Spector,[18] we have maintained TK$^{+/-}$

3.7.2C cell stocks with an average TFT^r mutant frequency of 30 × 10^{-6} (FIGURE 2). Spontaneous mutant frequencies >90 × 10^{-6} are unacceptably high in our judgment and indicate either the failure to eliminate sufficiently preexisting TFT^r variants in TK$^{+/-}$ cell stocks or the existence of experimental conditions that result in incomplete selection for the TK$^{-/-}$ phenotype. In particular, some lots of sera can produce a concentration-dependent enhancement of observed TFT^r colony frequencies in untreated control cultures (data not shown).

Unlike some spontaneous TFT^r colonies,[7] mutagen-induced cell clones (large) are phenotypically stable.[7,8] This is further demonstrated in TABLE 1 where 4 TFT^r clonal

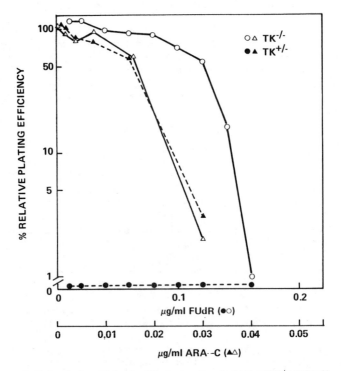

FIGURE 1. Relative plating efficiencies (% of solvent controls) for TK$^{+/-}$ (3.7.2C) and TK$^{-/-}$ (3.7.2C.110) cells seeded in soft-agar cloning medium (1 × 10^3 cells/dish) containing either FUdR or ara-C and incubated 7–9 days at 37°C.

lines grown in nonselective medium a total of 26 weeks and 1 TFT^r clonal cell line grown 52 weeks in nonselective medium were recloned in soft-agar medium containing either BUdR or THMG and assayed for intracellular TK enzyme activity. Normal generation times, complete cross-resistance to BUdR, total sensitivity to THMG medium, and trace levels of TK enzyme activity indicated the long-term retention of the TK$^{-/-}$ phenotype.

Our previous data suggest that the mutant frequencies based upon large TFT^r colony counts and obtained after the treatment of L5178Y cells with such mutagens as methyl methanesulfonate[7] or *trans*-diaminodichloroplatinum[19] are stable over varied

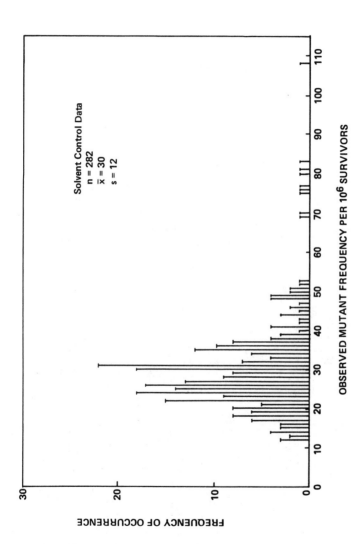

FIGURE 2. Cumulative mutant frequency data for TK$^{+/-}$ cells treated with either 100 μl DMSO or saline for 3 hours, washed, allowed 48 hours expression, and cloned in soft-agar medium.

TABLE 1

THE PHENOTYPIC TESTING OF PRESUMPTIVE TFT^R VARIANTS DERIVED FROM
MUTAGEN-TREATED L5178Y/TK$^{+/-}$ CELLS*

Cell Line	Generation Time (hours)	BUdR^r Colonies per 10^2 Survivors	THMG^r Colonies per 10^2 Survivors	TK Activity (%)
T1L1	11.0	93	0	0.15
T1L2	11.5	116	0	0.77
T1L3	11.5	85	0	0.00
T1L4	11.3	114	0	0.15
T1L5	10.8	131	0	0.02
TK$^{+/-}$	10.8	0.000105	63	100

*Clonal cell lines had originated as large TFT^R colonies following treatment with 213 μM methyl methanesulfonate (see Reference 7) and had been growing in nonselective R$_5$ medium for either 52 weeks (T1L4) or 26 weeks (all others). Resistance to 100 μg/ml bromodeoxyuridine (BUdR) and THMG medium (see text), and intracellular TK enzyme activity (percent of control) are shown.

expression times. This additional aspect of phenotypic stability is demonstrated in FIGURE 3. As shown, the frequencies of TFT^r variants in L5178Y cells treated with either 310 μg/ml EMS or 10 μg/ml hycanthone remain constant between two or four days expression or over a 1–12 μg/ml range of TFT.

With TK$^{+/-}$ cell stock relatively free of accumulated TFT^r variants, it is possible to

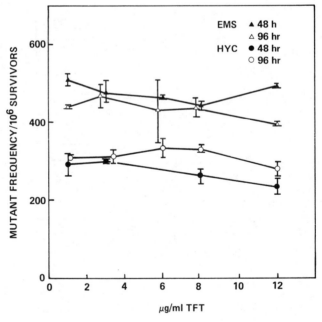

FIGURE 3. Mutant frequencies observed in L5178Y/TK$^{+/-}$cells treated for 3 hours with either 310 μg/ml EMS or 10 μg/ml hycanthone. Expression was either 48 or 96 hours. Cloning medium contained 1–12 μg/ml TFT.

measure chemically induced TFT^r colonies. Typical results for mutagens tested in the presence of liver postmitochondrial fraction from untreated rodents are shown in FIGURES 4 and 5. Tested in the presence of rat liver S-9, increasing concentrations of 2,7-dinitrofluorene cause a drop in both overall cell survival (relative total grown as calculated by Clive and Spector)[18] and absolute cloning efficiency (viable colony count) and cause a rise in mutant frequency (corrected for survival) as well as an increase in absolute mutant yield (TFT^r colony count) within an acceptable range of cell survival. Another example of a positive mutagenic response is shown in FIGURE 5 for 3MCA in the presence of hamster liver S-9. Total relative growth (not shown) ranged from 96–114% for male hamster S-9 and from 96–111% for the female hamster S-9. Insolubility of 3MCA in culture medium at >1 μg/ml accounts for the plateaued response shown in FIGURE 5. Cytotoxicity for a given 3MCA concentration varies with the rodent species or strain used as the S-9 source.[4]

All of our mutagenicity data are based upon predominantly large TFT^r colony counts. But in the presence of TFT, colony size distributions are always bimodal, regardless of mutagen treatment. Clive and coworkers have suggested that large (lambda) TFT^r colonies indicate gene mutations while small (sigma) TFT^r colonies are produced by specific chromosomal aberrations.[2,11,14] FIGURE 6 illustrates the typical association between large and small TFT^r colonies for ICR-191, perhaps the most potent mutagen we have tested in the L5178Y/TK assay. Shaded histobars represent mutant frequency data from two independent trials seven weeks apart. During the second trial, actual colony sizing was completed using a calibrated reticle. A dose-related nonlinear small-colony component similar to that previously described for mutagens such as methyl iodide[14] is apparent in FIGURE 6. But ICR-191 has been described as an agent that effectively produces single-gene mutagenesis without apparent chromosomal breakage.[20] The genetic significance of small colonies in this example is not obvious, but is representative of the ubiquitous association between sigma colonies and cytotoxicity.

The Biotransformation of Carcinogens to Mutagens

Pivotal in the testing of any chemical in an *in vitro* assay is the phenomenon of enzymatic conversion or biotransformation of some chemicals to primary, secondary, or tertiary metabolites that may differ considerably in toxic or mutagenic potential compared to the original substrate. This has been accomplished in the L5178Y/TK assay through the use of liver S-9 (9,000 × *g* supernatant),[2,4] the testing of urine from mutagen-dosed mice,[6] or the use of metabolically competent mammalian cell monolayers.[3]

Using these assay modifications, a comprehensive approach toward the mutagenicity testing of the carcinogen safrole can be demonstrated as follows. Safrole is not mutagenic in the conventional assay to several *Salmonella typhimurium* tester strains, even with microsomal activation.[21,22] In the rat and mouse, 1'-hydroxysafrole may be the proximate carcinogenic metabolite of safrole.[23] Male CD-1 mice can excrete 30% or more of a safrole dose as conjugated 1'-hydroxysafrole.[24] Jotz and Mitchell demonstrated that safrole was negative in the absence and positive in the presence of liver S-9 from aroclor-treated rats in the L5178Y/TK assay,[25] but the positive experiment was compromised by high solvent control mutant frequencies (98 × 10^{-6}). We tested safrole in the present study under a variety of metabolic activation schemes (FIGURES 7 and 8). Tested directly, a three-hour exposure of L5178Y cells to safrole concentrations up to 90 μg/ml caused moderate toxicity and no mutagenicity (FIGURE 7). In the presence of either rat or mouse S-9 (5% v/v), 4–32 μg/ml safrole were both

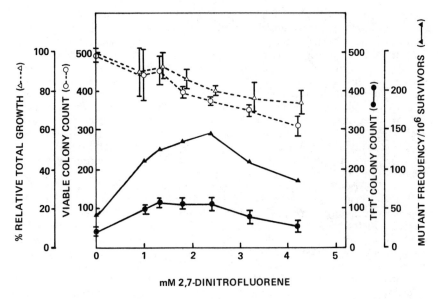

FIGURE 4. Actual colony counts and derived growth and mutant frequency data for L5178Y/ TK$^{+/-}$ cells treated with 2,7-dinitrofluorene for 3 hours followed by 48 hours expression.

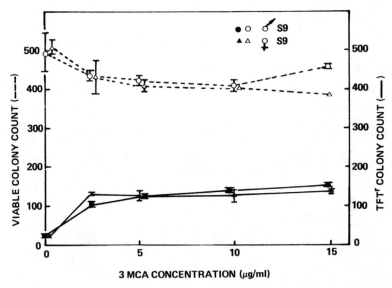

FIGURE 5. Actual colony counts for L5178Y/TK$^{+/-}$ cells treated for 3 hours with 3MCA in the presence of liver S-9 prepared from untreated male or female hamsters. Expression time was 48 hours.

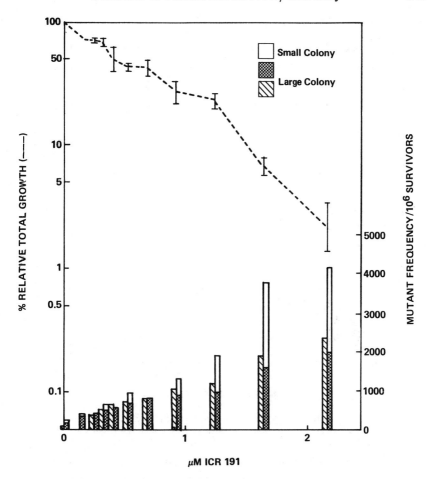

μM ICR 191

FIGURE 6. Mutagenicity and cytotoxicity in L5178Y/TK$^{+/-}$ cells treated for 3 hours with ICR-191. Shaded histobars show mutagenicity data based on large TFT^r colony counts from two independent trials. Combined total relative growth for the two trials is shown. Additional mutagenicity data based solely on small-colony counts are shown for one trial (clear histobars).

toxic and mutagenic compared to the nonactivation controls. These data show that both noninduced rat and mouse liver can activate safrole to a metabolite mutagenic to mammalian cells. We have previously questioned the necessity for aroclor pretreatment when preparing rodent liver S-9 for use in the L5178Y/TK assay.[5]

Using previously described techniques,[6] we compared urine from safrole and solvent-treated mice (FIGURE 8) or equivalent volumes of saline. Duplicate assays carried out on different days revealed the presence of mutagenic safrole metabolites in the urine of the treated mice, while even larger aliquots of urine from the solvent controls were neither mutagenic nor cytotoxic. Thus, the L5178Y cell is clearly sensitive to mutagenic metabolites of safrole produced by both *in vivo* and *in vitro* biotransformation techniques.

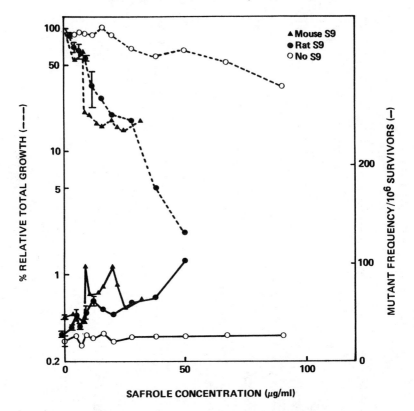

FIGURE 7. Mutagenicity and cytotoxicity of safrole for L5178Y/TK$^{+/-}$ cells in the presence or absence of liver S-9. Exposure time was 3 hours; expression time was 48 hours.

Specific Applications of the L5178Y/TK Assay for the Estimation of Mutagenic Potential

By analyzing published data where both carcinogens and noncarcinogens were tested,[5,8] the performance of the L5178Y/TK assay can be estimated as described by Cooper et al.[26] Sensitivity (number of carcinogens declared mutagenic/number of carcinogens tested) = 15/20 or 75%; specificity (number of noncarcinogens declared nonmutagenic/number of noncarcinogens tested) = 12/16 or 75%; and predictive value (number of carcinogens declared mutagenic/number of carcinogens or noncarcinogens declared mutagenic) = 15/19 or 79%. These are crude indicators based upon limited data (36 chemicals), but they do suggest acceptable reliability.

A notably wide range of chemical and physical agents have been tested for mutagenic potential in the L5178Y/TK assay.[1,2,5,8,19,27] Although there are obvious limitations regarding what should be tested for mutagenic potential (FUdR for example), there are specific applications where an assay of this nature can be used to supplement our knowledge of the overall toxicological properties of a substance. We have selected as examples a mitotic spindle inhibitor that can arrest cells in metaphase for chromosome analysis (colchicine)[28] and a purine analogue used as a selective agent for HGPRT enzyme–deficient mutants (6-thioguanine).[29]

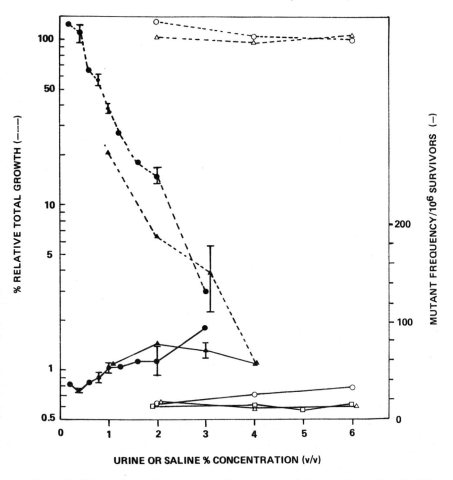

FIGURE 8. Mutagenicity and cytotoxicity of urine from safrole-treated mice for L5178Y/ TK$^{+/-}$ cells. Exposure time was 3 hours; expression time was 48 hours. Solid symbols indicate data for urine tested in two separate trials. Open squares indicate saline controls, and the testing of control urine is indicated by open circles and triangles. Sample means ± one-half range are shown where duplicate samples were available.

The results are shown in FIGURE 9. Hsu *et al.* have referred to nonmutagens or weak mutagens such as colchicine as mitotic poisons.[30] Such agents may cause mitotic arrest or spindle malfunction resulting in polyploidy or aneuploidy. Colchicine is the most toxic chemical we have so far tested in the L5178Y assay and provides an indication of the limits of assay sensitivity. Huberman and Heidelberger have previously measured the production of resistance to 8-azaguanine in Chinese hamster cells following four hours of exposure to the pyrimidine nucleoside analogues BUdR, TFT, FUdR, and ara-C,[31] but could not test purine analogues for obvious reasons. In this study, we measured resistance to TFT in L5178Y cells following three hours of exposure to 6-thioguanine. As shown in FIGURE 9, 6TG was quite mutagenic in the L5178Y/TK assay.

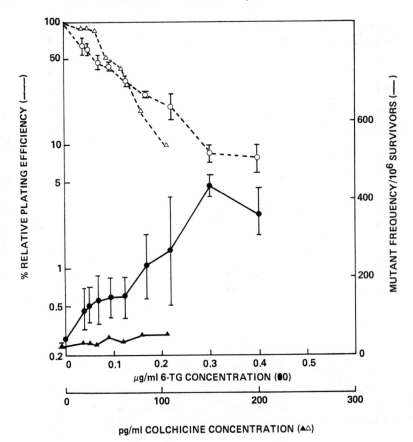

FIGURE 9. Mutagenicity and cytotoxicity of colchicine or 6-thioguanine in L5178Y/TK$^{+/-}$ cells following 3 hours exposure and 48 hours expression. Error bars indicate sample means ± one-half range.

CONCLUSIONS

Through analysis of the solvent control data collected in our laboratory over a recent interval, we have demonstrated that low background mutant frequencies are readily obtainable in the L5178Y/TK assay. Likewise, high cloning efficiencies are possible even after treatment with mutagens such as 3MCA or 2,7-dinitrofluorene. A positive mutagenic response is indicated by a dose-related increase in mutant yield within a range (~20–90% total relative growth) of cytotoxicity. Chemically induced TFT^r mutants (large colony) are phenotypically stable over varied expression times (two versus four days) and after removal of selective pressure for up to 52 weeks. Further, these stable TFT^r mutants possess little TK enzyme activity. The L5178Y/TK assay is readily coupled with a variety of exogenous metabolic activation systems or metabolite sources including rodent liver microsomal preparations, unpurified mouse urine, and cultured rodent cells. The L5178Y/TK assay is suitable for the testing of certain classes of agents, such as mitotic poisons, that cannot be adequately tested in prokaryotic mutagenesis assays. In summary, the L5178Y/TK assay provides an important function in a comprehensive genetic toxicology test battery.

ACKNOWLEDGMENTS

We thank Simone C. Paillet for technical assistance with the mutant stability and metabolic activation studies; John H. Ellis, Jr., Austin Joyce, and Martina French for their help during the urine metabolite studies; and Marge Kenyon for typing this manuscript.

REFERENCES

1. CLIVE, D., W. G. FLAMM, M. R. MACHESKO et al. 1972. A mutational assay system using the thymidine kinase locus in mouse lymphoma cells. Mutat. Res. 16: 77–87.
2. CLIVE, D., K. O. JOHNSON, J. F. S. SPECTOR et al. 1979. Validation and characterization of the L5178Y/TK$^{+/-}$ mouse lymphoma mutagen assay system. Mutat. Res. 59: 61–108.
3. AMACHER, D. E., S. C. PAILLET & I. ZELLJADT. 1982. Metabolic activation of 3-methylcholanthrene and benzo[a]pyrene to mutagens in the L5178Y/TK assay by cultured embryonic rodent cells. Environ. Mutagen. 4: 109–119.
4. AMACHER, D. E. & G. N. TURNER. 1980. Promutagen activation by rodent-liver postmitochondrial fractions in the L5178Y/TK cell mutation assay. Mutat. Res. 74: 485–501.
5. AMACHER, D. E. & G. N. TURNER. 1982. Mutagenic evaluation of carcinogens and non-carcinogens in the L5178Y/TK assay utilizing postmitochondrial fractions (S9) from normal rat liver. Mutat. Res. 97: 49–65.
6. AMACHER, D. E., G. N. TURNER & J. H. ELLIS, JR. 1981. Detection of mammalian cell mutagens in urine from carcinogen-dosed mice. Mutat. Res. 90: 79–90.
7. AMACHER, D. E. & S. C. PAILLET. 1981. Trifluorothymidine resistance and colony size in L5178Y/TK$^{+/-}$ cells treated with methyl methanesulfonate. J. Cell. Physiol. 106: 349–360.
8. AMACHER, D. E., S. PAILLET, G. N. TURNER et al. 1980. Point mutations at the thymidine kinase locus in L5178Y mouse lymphoma cells. II. Test validation and interpretation. Mutat. Res. 72: 447–474.
9. IVES, D. H., J. P. DURHAM & V. S. TUCKER. 1969. Rapid determination of nucleoside kinase and nucleotidase activities with tritium-labeled substrates. Anal. Biochem. 28: 192–205.
10. AMACHER, D. E., S. C. PAILLET & G. N. TURNER. 1979. Utility of the mouse lymphoma L5178Y/TK assay for the detection of chemical mutagens. In Mammalian Cell Mutagenesis: The Maturation of Test Systems. Banbury Report 2. A. W. Hsie, J. P. O'Neill & V. K. McElheny, Eds.: 277–289. Cold Spring Harbor Laboratory. Cold Spring Harbor, N.Y.
11. CLIVE, D. & M. M. MOORE-BROWN. 1979. The L5178Y/TK$^{+/-}$ mutagen assay system: mutant analysis. In Mammalian Cell Mutagenesis: The Maturation of Test Systems. Banbury Report 2. A. W. Hsie, J. P. O'Neill, & V. K. McElheny, Eds.: 421–429. Cold Spring Harbor Laboratory. Cold Spring Harbor, N.Y.
12. HEIDELBERGER, C. 1970. Chemical carcinogenesis, chemotherapy: cancer's continuing core challenges—G. H. A. Clowes memorial lecture. Cancer Res. 30: 1549–1569.
13. KRISS, J. P. & L. REVESZ. 1961. Quantitative studies of incorporation of exogenous thymidine and 5-bromodeoxyuridine into deoxyribonucleic acid of mammalian cells. Cancer Res. 21: 1141–1147.
14. MOORE, M. M. 1980. Analysis of the ability of the L5178Y/TK$^{+/-}$ mouse lymphoma assay to detect and quantitate mutagenic damage. Ph.D. Dissertation. University of North Carolina. Chapel Hill, N.C.
15. KIT, S. 1976. Thymidine kinase, DNA synthesis and cancer. Mol. Cell. Biochem. 11: 161-182.

16. ROGERS, A. M., R. HILL, A. R. LEHMANN *et al.* 1980. The induction and characterization of mouse lymphoma L5178Y cell lines resistant to 1-beta-D-arabinofuranosyl-cytosine. Mutat. Res. **69:** 139–148.
17. CLIVE, D. 1974. Mutagenicity of thioxanthenes (hycanthone, lucanthone and four indazole derivatives) at the TK locus in cultured mammalian cells. Mutat. Res. **26:** 307–318.
18. CLIVE, D. & J. F. S. SPECTOR. 1975. Laboratory procedure for assessing specific locus mutations at the TK locus in cultured L5178Y mouse lymphoma cells. Mutat. Res. **31:** 17–29.
19. AMACHER, D. E. & S. C. PAILLET. 1980. Induction of trifluorothymidine-resistant mutants by metal ions in L5178Y/TK$^{+/-}$ cells. Mutat. Res. **78:** 279–288.
20. KAO, F. T. & T. T. PUCK. 1969. Genetics of somatic mammalian cells. Quantitation of mutagenesis by physical and chemical agents. J. Cell. Physiol. **74:** 245–258.
21. McCANN, J., E. CHOI, E. YAMASAKI *et al.* 1975. Detection of carcinogens as mutagens in the Salmonella/microsome test: assay of 300 chemicals. Proc. Nat. Acad. Sci. USA **72:** 5135–5139.
22. WISLOCKI, P. G., E. C. MILLER, J. A. MILLER *et al.* 1977. Carcinogenic and mutagenic activity of safrole, 1'-hydroxysafrole, and some known or possible metabolites. Cancer Res. **37:** 1883–1891.
23. BORCHERT, P., J. A. MILLER, E. C. MILLER *et al.* 1973. 1'-Hydroxysafrole, a proximate carcinogen metabolite of safrole in the rat and mouse. Cancer Res. **33:** 590–600.
24. BORCHERT, P., P. G. WISLOCKI, J. A. MILLER *et al.* 1973. The metabolism of the naturally occurring hepatocarcinogen safrole to 1'-hydroxysafrole and the electrophilic reactivity of 1'-acetoxysafrole. Cancer Res. **33:** 575–589.
25. JOTZ, M. M. & A. D. MITCHELL. 1981. Effects of 20 coded chemicals on the forward mutation frequency at the thymidine kinase locus in L5178Y mouse lymphoma cells. *In* Evaluation of Short-Term Tests for Carcinogens. F. J. de Serres & J. Ashby, Eds.: 580–593. Elsevier/North-Holland. New York, N.Y.
26. COOPER, J. A., R. SARACCI & P. COLE. 1979. Describing the validity of carcinogen screening tests. Br. J. Cancer **39:** 87–89.
27. JACOBSON, E. D., K. KRELL, M. J. DEMPSEY *et al.* 1978. Toxicity and mutagenicity of radiation from fluorescent lamps and a sunlamp in L5178Y mouse lymphoma cells. Mutat. Res. **51:** 61–75.
28. EIGSTI, O. H. & P. DUSTIN, JR. 1955. Colchicine: In Agriculture, Medicine, Biology, and Chemistry. Iowa College Press. Ames, Iowa.
29. CHASIN, L. A. 1973. The effect of ploidy on chemical mutagenesis in cultured Chinese hamster cells. J. Cell. Physiol. **82:** 299–308.
30. HSU, T. C., C. J. COLLIE, A. F. LUSBY *et al.* 1977. Cytogenetic assays of chemical clastogens using mammalian cells in culture. Mutat. Res. **45:** 233–247.
31. HUBERMAN, E. & C. HEIDELBERGER. 1972. The mutagenicity to mammalian cells of pyrimidine nucleoside analogs. Mutat. Res. **14:** 130–132.

VIABLE CHROMOSOMAL MUTATIONS AFFECTING THE TK LOCUS IN L5178Y/TK$^{+/-}$ MOUSE LYMPHOMA CELLS: THE OTHER HALF OF THE ASSAY

D. Clive

Genetic Toxicology Laboratory
Burroughs Wellcome Co.
Research Triangle Park, North Carolina 27709

Human genetic diseases are of chromosomal as well as intragenic mutational origin. For genetic risk assessment of a test agent, both classes of events should be assayed. The potential of an agent to produce gene mutations is typically assessed in bacterial or cultured mammalian cell specific locus assays. Viability of these mutants is assured by the fact that the end point is almost always a growing mutant colony. Chromosomal aberrations are usually detected microscopically as morphological or numerical changes in the chromosomes of fixed, lysed mitotic cells. Chromosomal damage uncovered in such standard breakage studies is of unknown viability and, since dead "mutants" pose no genetic risk to either germinal or somatic cells, is of unknown relevance to human genetic risk assessment.

There are occasional exceptions to this rule that chromosomal damage is of unknown viability, such as the mutants obtained in the heritable translocation assay in mice, in a multiplicity of *Drosophila* tests, and, *in vitro,* in the L5178Y/TK$^{+/-}$ mouse lymphoma assay. The gene mutational aspect of this last system has been described in the present session by Amacher and Turner. I would like to discuss briefly the ability of this system to detect at least some classes of viable chromosome aberrations affecting the TK$^+$ chromosome [the distinctive chromosome 11 of these cells, which carries the single functional thymidine kinase (TK$^+$) gene]. Two poster papers (see Hozier *et al.* and Clive *et al.,* this volume) will expand upon some of what I am about to say.

The presence of a range of colony sizes in 5-bromodeoxyuridine (BUdR) or trifluorothymidine (TFT) selection plates has been known for about a decade. The significance of the small colonies (denoted as either σTFT^r, σTK$^{-/-}$, or simply σ mutants) to mutagenicity testing has been questioned, first of all by ourselves and then, after we were convinced that they were real mutants,[4–6,11] by Amacher.[1–3] Within the past three years, Hozier and Sawyer, in collaboration with Moore and myself, have provided firm evidence that σTFT^r colonies represent TK$^{-/-}$ mutants whose genetic fitness (measured as growth rate either during colony formation in soft-agar cloning medium or in nonselective liquid suspension medium) has been diminished as a result of aberrations to the TK$^+$ chromosome 11.[8,9,12,13]

Alternative interpretations of the nature of these σTFT^r colonies have been proposed. Amacher has espoused the view that they represent the effects of residual treatment-induced toxicity on possibly nonmutant cells.[1] However, such claims appear to result from the choice of culture medium (RPMI 1640), which, for some as-yet-unknown reason, does not allow as clean selection with TFT as does Fischer's medium. For example, 91 out of 95 spontaneous and all of 99 mutagen-induced σTFT^r variants isolated from Fischer's-based medium retained this phenotype following several generations of growth in nonselective suspension medium.[12] In a parallel experiment using RPMI medium, 27–57% (depending on TFT concentration during selection) of the spontaneous σTFT^r variants lost this phenotype following short-term culture in nonselective RPMI medium.[14]

253

0077–8923/83/0407–0253 $01.75/0 © 1983, NYAS

Other concerns have been expressed as to the origin of these $\sigma TK^{-/-}$ mutants. Accepting our evidence that they are real mutants with specific karyotypic abnormalities, Langenbach and Preston (personal communication) have raised the possibility that this damage results not from the mutagen treatment but rather from the exposure to the selective agent, TFT. At least four arguments can be brought against this possibility:

1. Different mutagen classes produce different proportions of small (σ) and large (λ) colony mutants. Ethylating agents (ethyl methanesulfonate, N-nitroso-N-ethylurea, and diethylnitrosamine) produce >75% $\lambda TK^{-/-}$ mutants, while methylating agents (methyl methanesulfonate, N-methyl-N'-nitro-N-nitrosoguanidine, and dimethylnitrosamine) give rise to >75% $\sigma TK^{-/-}$ mutants at comparable cytotoxicities and at the standard 48-hour expression time. The frameshift mutagens ICR-170 and hycanthone methanesulfonate produce mostly λ and mostly σ mutants, respectively. Most other mutagens/carcinogens give rise to predominantly σ mutants[5,11] (also Clive, unpublished results). It is difficult to see how TFT can exert this pattern of response by acting either alone or synergistically (two or more days after mutagen removal) with this wide a variety of mutagens.

2. The ratio of σ/λ mutants is dose dependent. $\lambda TK^{-/-}$ mutants are induced with linear kinetics, while for most mutagens, σ mutants occur with greater-than-linear kinetics with respect to either concentration or exposure time.[11] Again, it is difficult to explain this dose dependence in terms of a TFT effect acting two or more days after mutagen removal.

3. How does TFT seek out and cause visible damage to a specific chromosome (the distinctive chromosome 11 that carries the functional TK gene in our $TK^{+/-}$ heterozygote)? If the cell is already mutated to TK deficiency, the already implausible argument that TFT is interacting with the TK gene—and not its gene product—is lost. And if the cell is still TK competent, then one would have to explain how TFT simultaneously induces a chromosome 11 abnormality that *will* result in a $TK^{-/-}$ mutant following expression, and yet fails to arrest the growth of the still TK-competent cell.

4. It is too much of a coincidence to discover that "nonmutagenic" clastogens such as acyclovir[7] and caffeine[10] are mutagenic at the TK locus in this system and produce predominantly $\sigma TK^{-/-}$ mutants (FIGURES 1 and 2). Despite extensive efforts, no strict gene mutations have been observed for either compound with the possible exception of an early chemostat study done with *Escherichia coli*.[15]

Thus, the evidence is overwhelming that these $\sigma TK^{-/-}$ mutants are real and represent mutagen-induced chromosomal damage affecting the TK^+ chromosome 11 in these cells.

Greater uncertainty exists when we reverse our claim from "the induction of $\sigma TK^{-/-}$ mutants indicates that a clastogenic treatment has occurred" to "if compound X is clastogenic, then it will produce $\sigma TK^{-/-}$ mutants." At this point in our research we do not know all of the types of clastogenic damage capable of giving rise to this class of mutants. What we do know is that, 20–30 generations after cloning, banded karyotypes of these mutant cultures reveal trisomy 11, translocations to the telomeric end of chromosome 11, deletions of the terminal 2 bands of chromosome 11 (sometimes translocated elsewhere in the genome), 2-band insertions into the middle of chromosome 11, and in a minority of mutants, no visible karyotypic abnormality at this same 300-band level of resolution, at this late time in culture.[9,13]

Cytogenetic analysis of σ mutant colonies shortly after isolation suggests that the above chromosome 11 aberrations may not represent the initial chromosome lesion (see

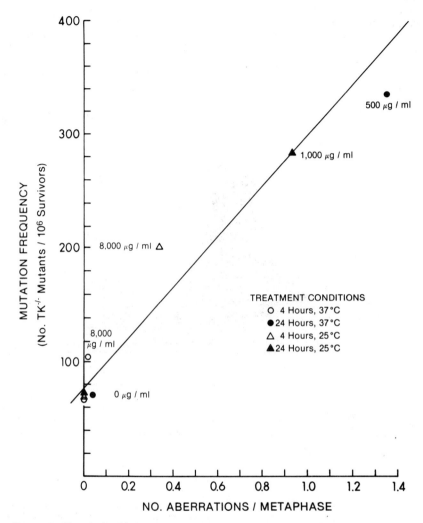

FIGURE 1. The relationship between chromosome aberration frequency (abscissa) and mutant frequency at the TK locus (ordinate) following various treatments with caffeine. Cells were exposed to various concentrations of caffeine for either 4 or 24 hours at either 25° or 37°C as indicated. Treated cultures yielding comparable survivals were cloned for mutant frequency 48 hours after start of treatment and sampled for cytogenetic analysis at the same time. Fifty metaphases were scored from each culture for both chromosome- and chromatid-type aberrations. Caffeine was neither mutagenic nor clastogenic under the standard assay conditions (4 hours, 37°C). All other treatment conditions induced both $TK^{-/-}$ mutants (a major proportion of which were small colonies) and chromosome breakage; the two end points increased proportionately to each other. Indicated on the graph are concentrations of caffeine analyzed at the corresponding treatment conditions.

Hozier *et al.*, this volume). Rather, a high frequency of telomere-to-telomere dicentrics, involving the TK$^+$ chromosome 11 as one of the two partners, is seen earlier in clonal history. Random breakage is seen to be occurring among daughter cells of a type that is capable of generating most of the lesions listed above for the later cultures. This may either indicate a propensity for these cells to form dicentrics or provide a model system for investigating the genesis and evolution of viable chromosome aberrations in mammalian systems. We are optimistic that future research with this system will shed significant insight into the origin and pathology of human chromosomal aberrations and genetic disease.

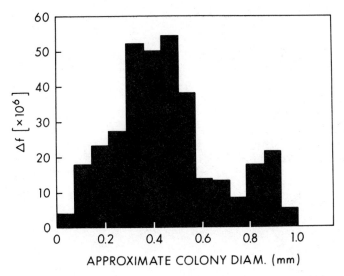

FIGURE 2. Distribution of TK$^{-/-}$ mutant colony sizes following treatment with acyclovir (4 hours exposure to 2,400 μg/ml at 37°C; 48 hours expression). Abscissa, diameter of TK$^{-/-}$ mutant colonies; ordinate, number of TK$^{-/-}$ mutants of indicated diameter per 10^6 survivors.

REFERENCES

1. AMACHER, D. E. & S. C. PAILLET. 1981. Trifluorothymidine resistance and colony size in L5178Y/TK$^{+/-}$ cells treated with methyl methanesulfonate. J. Cell. Physiol. **106:** 349–360.
2. AMACHER, D. E., S. PAILLET & V. A. RAY. 1979. Point mutations at the thymidine kinase locus in L5178Y mouse lymphoma cells. I. Application to genetic toxicological testing. Mutat. Res. **64:** 391–406.
3. AMACHER, D. E., S. C. PAILLET, G. N. TURNER, V. A. RAY & D. S. SALSBURG. 1980. Point mutations at the thymidine kinase locus in L5178Y mouse lymphoma cells. II. Test validation and interpretation. Mutat. Res. **72:** 447–474.
4. CLIVE, D., A. G. BATSON & N. T. TURNER. 1980. The ability of L5178Y/TK$^{+/-}$ mouse lymphoma cells to detect single gene and viable chromosome mutations: evaluation and relevance to mutagen and carcinogen screening. *In* The Predictive Value of Short-Term Screening Tests in Carcinogenicity Evaluation. G. M. Williams *et al.*, Eds.: 103–123. Elsevier/North-Holland Biomedical Press. Amsterdam, the Netherlands.
5. CLIVE, D., K. O. JOHNSON, J. F. S. SPECTOR, A. G. BATSON & M. M. M. BROWN. 1979.

Validation and characterization of the L5178Y/TK$^{+/-}$ mouse lymphoma mutagen assay system. Mutat. Res. **59**: 61–108.

6. CLIVE, D. & M. M. MOORE-BROWN. 1979. The L5178Y/TK$^{+/-}$ mutagen assay system: mutant analysis. In Mammalian Cell Mutagenesis: The Maturation of Test Systems. Banbury Report 2. A. Hsie, J. P. O'Neil & V. McElheny, Eds.: 421–430. Cold Spring Harbor Laboratory. Cold Spring Harbor, N.Y.

7. CLIVE, D., N. T. TURNER, J. C. HOZIER, A. G. BATSON & W. E. TUCKER, JR. The genetic toxicity of acyclovir. Fundamen. Appl. Toxicol. (In press.)

8. HOZIER, J., J. SAWYER, D. CLIVE & M. MOORE. 1982. Cytogenetic distinction between the TK$^+$ and TK$^-$ chromosomes in the L5178Y/TK$^{+/-}$-3.7.2C mouse lymphoma cell line. Mutat. Res. **105**: 451–456.

9. HOZIER, J., J. SAWYER, M. MOORE, B. HOWARD & D. CLIVE. 1981. Cytogenetic analysis of the L5178Y/TK$^{+/-}$ mouse lymphoma mutagenesis assay system. Mutat. Res. **84**: 169–181.

10. KIHLMAN, B. A. 1977. Caffeine and Chromosomes. Elsevier/North-Holland Biomedical Press. Amsterdam, the Netherlands.

11. MOORE-BROWN, M. M. & D. CLIVE. 1979. The L5178Y/TK$^{+/-}$ mutagen assay system: in situ results. In Mammalian Cell Mutagenesis: The Maturation of Test Systems. Banbury Report 2. A. Hsie, J. P. O'Neil & V. McElheny, Eds.: 71–88. Cold Spring Harbor Laboratory. Cold Spring Harbor, N.Y.

12. MOORE-BROWN, M. M., D. CLIVE, B. E. HOWARD, A. G. BATSON & K. O. JOHNSON. 1981. The utilization of trifluorothymidine (TFT) to select for thymidine kinase–deficient (TK$^{-/-}$) mutants from L5178Y/TK$^{+/-}$ mouse lymphoma cells. Mutat. Res. **85**: 363–378.

13. MOORE, M. M., D. CLIVE, J. HOZIER, B. E. HOWARD, A. G. BATSON, N. T. TURNER & J. SAWYER. 1982. Characterization of trifluorothymidine-resistant (TFT^r) mutants of L5178Y/TK$^{+/-}$ mouse lymphoma cells: mutant analysis. (In preparation.)

14. MOORE, M. M. & B. E. HOWARD. 1982. Quantitation of small colony trifluorothymidine-resistant mutants of L5178Y/TK$^{+/-}$ mouse lymphoma cells in RPMI-1640 medium. Mutat. Res. **104**: 287–294.

15. NOVICK, A. & L. SZILARD. 1951. Experiments on spontaneous and chemically induced mutations of bacteria growing in the chemostat. Cold Spring Harbor Symp. Quant. Biol. **16**: 337–343.

CELL-MEDIATED MUTAGENESIS OF CHINESE HAMSTER
V79 CELLS AND *SALMONELLA TYPHIMURIUM*

Robert Langenbach,*† Cathy Hix,†‡ Linda Oglesby,§
and James Allen†

INTRODUCTION

Most environmental chemicals exist in a promutagenic/procarcinogenic form and require metabolic activation to manifest biological activity.[1,2] *In vitro* assays to detect genotoxicity typically utilize well-characterized target cells whose growth and maintenance requirements are established in many laboratories. However, as these target cells are unable to activate metabolically many chemicals to which humans are exposed, some type of exogenous activation system is usually needed. The choice of activation system can significantly affect assay sensitivity as well as the number of different classes of chemicals detected. Similarly, the particular genotoxic end point assayed is of critical importance due to chemical specificities for certain types of genetic effects. Within this context, the use of intact cells (as compared with cell homogenates) as metabolic activation systems combined with the measurement of multiple genetic end points in Chinese hamster V79 cells or *Salmonella typhimurium* is discussed below. Applications of this approach for both bioassay system development and for studies of fundamental problems in carcinogenesis are presented. The use of intact cells as activation systems in genetic toxicology assays has recently been reviewed by Langenbach and Oglesby.[3]

MATERIALS AND METHODS

Primary rodent cells from liver, lung, kidney, and bladder tissues were prepared as previously described.[4-8] Methods for isolating bovine bladder and liver cells have also been described.[9-11] The medium used for all cell culture studies was Williams medium E plus 10% heat-inactivated fetal bovine serum supplemented with 2 mM L-glutamine, penicillin (100 units/ml), and streptomycin (100 μg/ml). Humidified incubators were used at 37°C with an atmosphere of 5% CO_2 in air. The V79 mutation protocols were previously published.[4,8] The cocultivation of metabolizing cells, V79 cells, and chemical was for 48 hours prior to reseeding of the V79 cells for ouabain resistance (two-day expression time) or 6-thioguanine resistance (six-day expression time).

The *S. typhimurium* assay was carried out according to the procedure of Ames *et al.*,[12] except that intact cells from the various tissues (usually 6×10^6/plate) were added in place of the S-9 mix.[9-11] The cells were not supplemented with cofactors.

Protocols used here for the simultaneous determination of sister chromatid exchanges (SCEs) and mutation in V79 cells have been reported by Allen *et al.*[13]

*Cellular and Genetic Toxicology Branch, National Toxicology Program, National Institute of Environmental Health Sciences, Research Triangle Park, N.C. 27709.
†Genetic Toxicology Division, U.S. Environmental Protection Agency, Health Effects Research Laboratory, Research Triangle Park, N.C. 27711.
‡National Research Council Resident Research Associate. Permanent affiliation: Tennessee Technological University, Cookeville, Tenn. 38501.
§Northrop Services, Inc., Research Triangle Park, N.C. 27709.

258

RESULTS AND DISCUSSION

Comparison of Intact Cells and Cell Homogenates

The most often used metabolic activation system in short-term assay systems has been liver homogenates (S-9) from rats pretreated with chemicals that induce drug-metabolizing enzymes. However, while S-9 preparations are convenient to prepare and use, they do not necessarily mimic *in vivo* metabolism.[3,14–17] We have felt that metabolically competent intact cells may better bridge this gap, and thus a primary objective of our laboratory has been the development of this alternative activation approach for use in genetic toxicity assays.[3–11] TABLE 1 shows a comparison of carcinogen concentration ranges for chemicals from four chemical classes when activated by intact cells or S-9 preparations, with mutagenesis of V79 cells scored as the end point. The rat hepatocyte–mediated V79 mutagenesis system was more sensitive than the rat liver S-9-mediated system for *N*-nitrosodimethylamine (DMN) and 7,12-dimethylbenz[*a*]anthracene (DMBA). The sensitivity for detecting DMBA was further increased when rat embryonic fibroblasts or rat lung epithelial cells (see

TABLE 1

CHEMICAL DOSE (μM) REQUIRED TO GIVE TENFOLD INCREASE OVER V79 CELL
BACKGROUND MUTATION FREQUENCY WITH INTACT RAT LIVER CELLS
OR RAT LIVER S-9 FOR METABOLIC ACTIVATION*

		Concentration (μM)	
Class	Chemical	S-9	Intact Cells
Nitrosamine	DMN	1000–5000	1–10
Hydrocarbon	DMBA	40–60	1–5
			0.1–1†
Aromatic amine	AAF	10–25	>50
Mycotoxin	AFB	0.5–1	0.1–1

*Ouabain at 1 mM was used as the selective agent with a two-day expression time. The spontaneous mutation frequency for ouabain varied from 1–2 mutants per 10^6 survivors.
†Rat embryonic fibroblast mediated.

TABLE 2) were used for metabolic activation. Aflatoxin B_1 (AFB) was about equally mutagenic when activated by intact hepatocytes or by S-9, while S-9 activation appeared slightly more effective with the aromatic amine 2-acetylaminofluorene (AAF). The conclusions drawn from these data are valid when V79 cell mutagenesis is the end point and, as discussed below, can change when other target cells or genetic end points are used.

Many considerations enter into the choice of using intact cells or S-9 preparations for exogenous metabolic activation systems in short-term assays. The profiles of carcinogen metabolites and the DNA adducts formed have been found to differ when studied with intact cells and tissue homogenates,[14–17] with intact cells better mimicking the *in vivo* situation. In intact cells, the combination of activation and detoxication enzymes presumably retain their normal relative activities, whereas conjugation processes in tissue homogenates probably do not occur because the appropriate cofactor levels are decreased. For screening purposes, lower levels of conjugation could be advantageous as reactive intermediate production would be maximized; however, these aberrant conditions may not be appropriate for mechanistic studies into *in vivo* events. With broken-cell preparations, there may be a differential loss of the enzymes that contribute to the activation of different classes of chemical carcinogens. Additionally, certain enzymes that participate in carcinogen activation in extrahepatic tissues may be present at reduced levels in liver S-9 preparations or the substrates necessary for their enzyme activity are not supplied in the standard cofactor mix. An example of such an enzyme is prostaglandin synthetase.[18,19] This enzyme should be present and functional in certain types of intact cells. Therefore, such phenomena as organ and species differences in carcinogen activation may be studied with greater relevance when intact cells are used. However, intact cells may not be able to interact with the target cells as closely as do the enzymes in S-9 preparations. A need for proximity between activating cell and target cell has been reported;[20] however, the need for proximity can vary with stability of the reactive intermediate(s) and ease of transport of the activated intermediate. Furthermore, intact cells generally retain xenobiotic metabolizing enzymes under assay conditions for longer times than do S-9 prepara-

TABLE 2

RELATIVE MUTAGENIC ACTIVITIES FOR V79 CELLS OF CHEMICALS IN VARIOUS RAT CELL–MEDIATED MUTAGENESIS SYSTEMS*

Chemical (μg/ml)	Activating Cell Type			
	Liver	Lung	Kidney	Bladder
Anthracene (3)	1	1	1	1
BaP (1)	2	56	5	6
DMBA (1)	14	154	12	21
DMN (100)	95	2	19†	11
AFB (3)	42	4	ND†	ND
AAF (50)	6	ND	ND	1

*Data are expressed as fold increases over background mutation frequency with ouabain used as the selective agent for the liver and lung cell–mediated systems and 6-thioguanine as the selective agent in the kidney and bladder cell–mediated systems. The background mutation frequency for ouabain was 1 per 10^6 survivors, and for 6-thioguanine the background frequency was 3 per 10^6 survivors. Fold increase over background has been shown to be a useful indicator of mutagenic potency for N-methyl-N'-nitro-N-nitrosoguanidine and benzo[a]pyrene although not for methyl methanesulfonate when different selective agents were compared with V79 cells.[36] Furthermore, BaP and DMBA have been studied in this laboratory, and both ouabain and 6-thioguanine resistance were measured simultaneously for each chemical, and the results from either marker agreed with those shown in the table.
†ND = not determined.

tions. Both S-9 preparations and intact cells can be frozen and stored, but current methodologies are not developed as well for freezing and subsequent utilization of primary intact epithelial cells. Further studies comparing intact cells and cell homogenates as activation systems are warranted in order to determine their relative merits and limitations for usage in short-term bioassays.

Cell-Mediated Mutagenesis of Mammalian Cells

The first cell-mediated mutagenesis system was developed by Huberman and Sachs.[21] Rodent embryonic fibroblasts were used for metabolic activation, and V79 cells as the mutable target. This system has subsequently been utilized to test the mutagenicity of several polycyclic aromatic hydrocarbons and to show a correlation between the carcinogenicity of the hydrocarbons and their mutagenicity in the cell-mediated assay.[21–26]

Because fibroblastic cells have limited xenobiotic metabolic activation potential, the cell-mediated approach was expanded to use primary liver cells for activation so that a broader spectrum of chemicals could be examined by this methodology. The liver has been reported to possess the broadest capability of any mammalian organ for carcinogen metabolism.[27] San and Williams and Langenbach *et al.* reported development of primary liver cell–mediated systems with a rat liver cell line and V79 cells, respectively, as the mutable targets.[28,4] The liver cell–mediated system has been widely used in many laboratories, and a large volume of data collected (for review, see Reference 3). The hepatocyte-mediated system has proven especially sensitive for detecting nitrosamines as mutagens. Nitrosamines are a class of environmentally important carcinogens ordinarily difficult to detect in many other *in vitro* systems. For a series of short-chain aliphatic nitrosamines, we have reported a correlation between Syrian hamster hepatocyte-mediated mutagenesis of V79 cells and carcinogenic activity in the Syrian hamster.[29–31] The correlation coefficient between carcinogenic activity and mutagenic activity was greater than 0.90 when tumors at all sites in the hamster, including liver, were used in the calculation. Thus, for nitrosamines, the hamster hepatocyte–mediated system appears to be a reliable predictor of multiorgan tumorigenicity. Also for a limited number of these nitrosamines, it has been demonstrated that the hamster hepatocyte–mediated mutagenesis of V79 cells correlates more closely with hamster carcinogenic activity than does mutagenesis with hamster S-9 activation coupled with *S. typhimurium* as the mutable target.[31] Structure-activity correlations of the nitrosamines when studied in the cell-mediated system have been analyzed,[30] and insight into possible mechanisms of activation obtained. Jones and colleagues have also reported that for a series of nitrosamines, mutagenic activity in the rat hepatocyte–mediated V79 cell mutagenesis system correlates with carcinogenic activity of the nitrosamines in the rat.[32,33] The hepatocyte-mediated mutagenesis system has been used to demonstrate a lack of mutagenic activity of certain organochlorine chemicals tested with hepatocyte activation, and thus a nongenetic mechanism for the carcinogenicity of these chemicals was suggested.[34]

While correlations between cell-mediated mutagenicity and *in vivo* carcinogenicity exist, it should be emphasized that cellular activation systems are providing only the metabolic activation component. Other factors such as pharmacokinetics, possible requirements for multiorgan activation, DNA repair, rate of cell turnover, nature of DNA adducts, promotional events, etc., all of which could be involved in *in vivo* carcinogenesis, are not accounted for by cell-mediated activation per se, or by S-9 activation. However, some of these factors may be amenable to study by variations in the intact cellular activation approach. For example, in cases where multiorgan activation is required and the initial metabolism occurs in one organ while the ultimate activation step occurs in the target organ, it would be possible to study activation by using cells from each organ separately and in combination when cocultivated with the appropriate target cell. With such an approach it may be possible to gain insight into the relative contribution of the two organs in the carcinogen activation process.

One interesting and useful application of the cell-mediated mutagenesis approach has been the investigation of cell type, organ, and species differences in the activation of chemical carcinogens. In our early studies employing cellular activation systems, a cell type specificity was observed for rat liver cells and rat embryonic fibroblasts in the activation of AFB and benzo[a]pyrene (BaP) to mutagens for V79 cells.[35] AFB, a potent liver carcinogen, was activated by hepatocytes but not by rat fibroblasts, whereas BaP, a potent skin and lung carcinogen that can produce fibrosarcomas at the site of injection, was activated to mutagens by fibroblasts but not by hepatocytes at the

BaP doses studied. More recent studies in our laboratory (unpublished data) and others[34] (see also Reference 3 and references therein) indicate that at higher concentrations of BaP, some mutagenic intermediates are produced and/or released by the hepatocytes.

We have extended the cell-mediated approach to investigate the organ specificity of carcinogen activation by intact epithelial cells from male adult rat liver, lung, bladder, and kidney tissues. The data in TABLE 2 show the relative mutagenic activity of chemicals from four different classes when primary cultures of intact cells from these organs served as the source of metabolic activation. Primary cultures of the various cell types are used because they presumably would contain the metabolic capability most closely resembling xenobiotic metabolism as it occurs in the respective cell type *in vivo*. Also, as most cancers originate from epithelial cells in the various organs, epithelial cells were the metabolic activating cell type used. Anthracene, the noncarcinogenic hydrocarbon, was not mutagenic to V79 cells in any of the cell-mediated systems. In the liver cell–mediated system, BaP appeared inactive at the concentration studied while DMBA, DMN, and AFB were mutagenic (TABLE 2). Somewhat surprisingly, the liver carcinogen AAF appeared only weakly mutagenic in the rat hepatocyte–mediated V79 cell system. By contrast to liver cell activation, rat lung cells extensively activated the hydrocarbons, but did not activate DMN. The kidney cells activated BaP, DMBA, and DMN to mutagenic intermediates. Primary rat bladder cells also activated BaP, DMBA, and DMN, but did not appear to activate AAF to intermediates mutagenic to V79 cells. As described below, the low level of AAF mutagenicity in V79 cells when activated by liver and bladder cells (both organs are target organs for this carcinogen) may not be due to the absence of genotoxic intermediate production by these activating cells. In general, the mutagenicity of the chemicals shown in TABLE 2 with activating cells from the various organs is in agreement with the relative carcinogenicity of the chemical to that organ.

In studies with rat and hamster hepatocytes for metabolic activation in the cell-mediated mutagenesis system, we have observed that hamster hepatocytes are about four times as active as rat hepatocytes in activating DMN to intermediates mutagenic to V79 cells. Thus, the approach can also be used to study species differences in the metabolic activation of carcinogens.

Genetic End Points Measured with Cellular Activation Systems

Mammalian cell mutation, cell transformation, cytogenetic alterations, cytotoxicity, and bacterial mutagenesis are all end points that have been measured in target cells using cell-mediated systems.[3] The data in TABLE 3 illustrate some results comparing mutagenesis and SCE induction in V79 cells and *S. typhimurium* mutagenesis using uninduced hamster hepatocytes for the metabolic activation of two carcinogens, DMN and AAF. These data indicate that while hepatocyte activation of AAF did not significantly induce mutagenesis of V79 cells to 6-thioguanine (6-TG) resistance, there was concomitantly a significant elevation of SCE frequencies in these cells. These findings are interesting because our initial interpretation of low V79 cell mutagenesis induced by AAF was that active intermediates from this carcinogen were not being formed at significant levels in the isolated hepatocytes or, because of instability, were not reaching the DNA of the V79 cells (see TABLE 2). However, the increase of SCEs suggests that metabolically activated intermediates of AAF are interacting with V79 cell DNA. Furthermore, when *S. typhimurium* (TA98) is used as the target cell, the reversion to histidine independence is also enhanced when the hepatocytes activate AAF. Therefore, while V79 cell mutagenesis is not a suitable indicator of genotoxic activity for hepatocyte activation of AAF, SCE induction in V79 cells and bacterial mutagenesis do provide evidence of AAF's genotoxic activity with hepatocyte activation. By contrast, as also shown in TABLE 3, different results are obtained for DMN with hamster hepatocyte activation. With this carcinogen, V79 cell mutagenesis and SCE induction are sensitive indicators of DMN activation to genotoxic intermediates, while S. typhimurium (TA100) was reverted in a DMN dose range that is cytotoxic to V79 cells. The different sensitivities of these end points for the two carcinogens emphasize the need for measuring multiple end points when assessing genotoxic activity of an unknown chemical and indicate how in vitro activity can vary with the end point measured.

Cell-Mediated Mutagenesis of S. typhimurium

The cell-mediated approach has been used to gain insight into the relative roles of the liver and bladder in the activation of aromatic amines, a class of chemicals known to cause bladder cancer. In general there have been two theories on the activation of

TABLE 3

COMPARISON OF V79 CELL MUTAGENESIS AND SCE INDUCTION AND S. TYPHIMURIUM MUTAGENESIS AS END POINTS WITH HAMSTER HEPATOCYTES FOR METABOLIC ACTIVATION

Chemical (μg/ml)	CE (%)	6-TG Resistant Mutants/ 10⁶ Survivors	SCE/ Cell*	S. typhimurium Revertants/Dish	
Control	72	5	6	15†	107‡
2-AAF					
(10)	69	8	19	51	
(25)	50	10	33	62	
(50)	8	11	37	94	
DMN					
(1)	78	14	33		87
(100)	12	281	61		84
(1000)	toxic	toxic	toxic		468

*Mean values determined from ≥40 cells/treatment.
†Data for S. typhimurium strain TA98.
‡Data for S. typhimurium strain TA100.

chemicals that cause bladder cancer. One theory has postulated that carcinogen metabolism, including conjugation, occurred in the liver with subsequent enzymatic or pH-dependent hydrolysis in the urine leading to the active species in the bladder. Alternatively, activation of the carcinogen, or a metabolite, could occur in the bladder epithelium itself. These theories are not necessarily mutually exclusive, and we initiated studies to investigate this phenomenon in vitro with the cell-mediated approach. As the aromatic amines are not readily detectable with V79 cell mutagenesis as the end point (see TABLE 3), we developed a bovine bladder and bovine liver cell-mediated S. typhimurium system to detect these genotoxic agents in a cell-mediated assay.[9–11] Bovine, rather than rat, tissues were used in these studies to insure adequate numbers of bladder cells. Previous reports have been published on rodent liver cell-mediated S. typhimurium systems (see Reference 3 for review).

The data in TABLE 4 demonstrate that bladder cells can activate aromatic amines to mutagenic intermediates and in fact, bovine bladder cells are more active than bovine liver cells. Several doses of each chemical were tested in both tester strains TA98 and TA100,[10,11] but only results from two doses with TA98 are shown in TABLE

4. On a per-cell basis, the bladder cells were about 10 and 4 times, respectively, more active than the liver cells in activating aminofluorene (AF) and AAF to mutagens for *S. typhimurium*. Additionally, while bovine bladder cells could activate 4-aminobiphenyl (ABP), benzidine (BZ), and 2-naphthylamine (2NA) to mutagenic intermediates, bovine liver cells did not appreciably activate these chemicals to mutagens under the conditions employed. The noncarcinogenic aromatic amine 1-naphthylamine (1NA) was not mutagenic with either bladder or liver cell activation and illustrates the ability of the bovine bladder cell-mediated bacterial assay to discriminate between carcinogenic and noncarcinogenic aromatic amines. It should be emphasized that the observed differences in mutagenic activities when bladder cells and liver cells are used for metabolic activation may not reflect total metabolism by the two cell types, as extent of conjugation, retention of active intermediates, etc., could also influence the observed mutagenic response. However, the data do indicate the capability of bladder urothelial cells to activate metabolically chemicals that are carcinogenic to the bladder. Further mechanistic studies into the interaction of these two organs in the activation of bladder carcinogens are possible with the cell-mediated approach.

TABLE 4

MUTAGENICITY TO *S. TYPHIMURIUM* OF AROMATIC AMINES ACTIVATED
BY INTACT BOVINE BLADDER OR LIVER CELLS

Chemical (μg/plate)		TA98 Revertants/Plate*	
		Bladder	Liver
AF	10	3098	305
	20	3546	390
AAF	20	2376	534
	40	2739	709
4ABP	10	195	8
	20	409	0
BZ	20	100	21
	40	222	4
2NA	20	40	0
	80	100	0
1NA	20	0	0
	80	4	0

*Data are expressed per 6×10^6 bladder or liver cells/plate with backgrounds subtracted.

SUMMARY

In the cell-mediated approach, intact cells metabolically activate the chemical and the genetic end points are measured in cocultivated or coincubated target cells. Cell-mediated systems have been used to study fundamental problems in carcinogenesis, such as organ and species specificity of carcinogen activation, and in screening for carcinogenic chemicals. In the studies discussed here, cells from various rat, hamster, or bovine tissues are used to metabolically activate the chemical, and mutation and/or SCE induction in V79 cells and mutation of *S. typhimurium* are measured as genetic end points. The detection of genetic activity of a chemical depends both on the cell (organ, species, type, etc.) used for metabolic activation and on the genetic end point measured. Hydrocarbons and nitrosamines are two classes of environmentally significant chemicals that are sensitively detected with cell-mediated systems. The cell-mediated approach provides a valuable metabolic activation component for short-term *in vitro* systems, and further studies are needed to utilize and evaluate its full potential.

REFERENCES

1. MILLER, J. A. 1970. Carcinogenesis by chemicals: an overview. Cancer Res. **30:** 559–576.
2. HEIDELBERGER, C. 1975. Chemical carcinogenesis. Annu. Rev. Biochem. **44:** 79–121.
3. LANGENBACH, R. & L. OGLESBY. 1983. The use of intact cellular activation systems in genetic toxicology assays. *In* Chemical Mutagens. F. de Serres, Ed. **8.** Plenum Press. New York, N.Y. (In press.)
4. LANGENBACH, R., H. J. FREED & E. HUBERMAN. 1978. Liver cell-mediated mutagenesis of mammalian cells by liver carcinogens. Proc. Nat. Acad. Sci. USA **75:** 2864–2867.
5. LANGENBACH, R., S. NESNOW, L. MALICK, R. GINGELL, A. TOMPA, C. KUSZYNSKI, S. LEAVITT, K. SASSEVILLE, B. HYATT, C. CUDAK & L. MONTGOMERY. 1981. Organ specific activation of carcinogenic polynuclear aromatic hydrocarbons in cell culture. *In* Polynuclear Aromatic Hydrocarbons. M. Cooke & A. J. Dennis, Eds.: 75–84. Battelle Press. Columbus, Ohio.
6. TOMPA, A. & R. LANGENBACH. 1979. Culture of adult rat lung cells: benzo[*a*]-pyrene metabolism and mutagenesis. In Vitro **15:** 569–578.
7. LANGENBACH, R., S. NESNOW, A. TOMPA, R. GINGELL & C. KUSZYNSKI. 1981. Lung and liver cell-mediated mutagenesis sytems: specificities in the activation of chemical carcinogens. Carcinogenesis **2:** 851–858.
8. LANGENBACH, R., L. MALICK & S. NESNOW. 1981. Rat bladder cell-mediated mutagenesis of Chinese hamster V79 cells and metabolism of benzo[*a*]pyrene. J. Nat. Cancer Inst. **66:** 913–917.
9. OGLESBY, L., C. HIX, P. MACNAIR, M. SIEG, L. SNOW & R. LANGENBACH. 1983. Metabolic activation of aromatic amines to mutagens by bovine bladder and liver cells. Environ. Health Perspect. **49:** 147–154.
10. OGLESBY, L., C. HIX, P. MACNAIR, M. SIEG, L. SNOW & R. LANGENBACH. Development of a bovine bladder cell mediated–*Salmonella typhimurium* mutagenesis system. (Submitted.)
11. HIX, C., L. OGLESBY, P. MACNAIR, M. SIEG & R. LANGENBACH. Metabolic activation of aromatic amines to mutagens by bovine bladder and liver cells. (In preparation.)
12. AMES, B. N., J. MCCANN & E. YAMASAKI. 1975. Methods for detecting carcinogens and mutagens with the *Salmonella*/mammalian-microsome mutagenicity test. Mutat. Res. **31:** 347–364.
13. ALLEN, J. W., R. LANGENBACH, S. NESNOW, K. SASSEVILLE, S. LEAVITT, J. CAMPBELL, K. BROCK & Y. SARIEF. 1982. Comparative genotoxicity studies of ethyl carbamate and related chemicals: further support for vinyl carbamate as a proximate carcinogenic metabolite. Carcinogenesis **3:** 1437–1441.
14. BIGGER, C. A. H., J. E. TOMASZEWSKI, A. DIPPLE & R. S. LAKE. 1980. Limitations of metabolic activation systems used with *in vitro* tests for carcinogens. Science **209:** 503–505.
15. SCHMELTZ, S., J. TOSK & G. M. WILLIAMS. 1978. Comparison of the metabolic profiles of benzo[*a*]pyrene obtained from primary cell cultures and subcellular fractions derived from normal and methylcholanthrene induced rat liver. Cancer Lett. **5:** 81–89.
16. BARTSCH, H., C. MALAVEILLE, A. M. CAMUS, G. MARTEL-PLANCHE, G. BRUN, A. HAUTEFEUILLE, N. SABADIE, A. BARBEN, T. KUROKI, C. DREVON, C. PICCOLI & R. MONTESANO. 1980. Validation and comparative studies on 180 chemicals with *S. typhimurium* strains and V79 Chinese hamster cells in the presence of various metabolizing systems. Mutat. Res. **76:** 1–50.
17. DECAD, G. M., D. P. HSIEH & J. L. BYARD. 1977. Maintenance of cytochrome P-450 and metabolism of aflatoxin B_1 in primary hepatocyte cultures. Biochem. Biophys. Res. Commun. **78:** 279–287.
18. ZENSER, T. V., M. B. MATTAMAL & B. B. DAVIS. 1980. Metabolism of *N*-[4-(5-nitro-2-furyl)-2-thiazolyl]formamide by prostaglandin endoperoxide synthetase. Cancer Res. **40:** 114–118.

19. ZENSER, T. V., M. B. ARMBRECHT & B. B. DAVIS. 1980. Benzidine binding to nucleic acids mediated by the peroxidative activity of prostaglandin endoperoxide synthetase. Cancer Res. **40:** 2839–2845.

20. KUROKI, T. & C. DREVON. 1978. Direct or proximate contact between cells and metabolic activation systems is required for mutagenesis. Nature **271:** 368–370.

21. HUBERMAN, E. & L. SACHS. 1974. Cell-mediated mutagenesis of mammalian cells with chemical carcinogens. Int. J. Cancer **13:** 326–333.

22. HUBERMAN, E. & L. SACHS. 1976. Mutability of different genetic loci in mammalian cells by metabolically activated carcinogenic polycyclic hydrocarbons. Proc. Nat. Acad. Sci. USA **73:** 188–192.

23. HUBERMAN, E., L. SACHS, S. K. YANG & H. V. GELBOIN. 1976. Identification of mutagenic metabolites of benzo[a]pyrene in mammalian cells. Proc. Nat. Acad. Sci. USA **73:** 607–611.

24. HASS, B. S., C. K. McKEOWN, D. J. SARDELLA, E. BOGER, P. K. GHOSHAL & E. HUBERMAN. 1982. Cell-mediated mutagenicity in Chinese hamster V79 cells of dibenzopyrenes and their bay-region fluorine-substituted derivatives. Cancer Res. **42:** 1646–1649.

25. HUBERMAN, E. & T. J. SLAGA. 1979. Mutagenicity and tumor-initiating activity of fluorinated derivations of 7,12-dimethylbenz[a]anthracene. Cancer Res. **39:** 411–414.

26. SLAGA, T. J., E. HUBERMAN, J. DiGIOVANNI, G. GLEASON & R. G. HARVEY. 1979. The importance of the "Bay region" diol-epoxide in 7,12-dimethyl-benz[a]anthracene skin tumor initiation and mutagenesis. Cancer Lett. **6:** 213-220.

27. WILLIAMS, G. M. 1980. The detection of chemical mutagens/carcinogens by DNA repair and mutagenesis in liver cultures. In Chemical Mutagens: Principles and Methods for Their Detection. F. J. de Serres & A. Hollaender, Eds. **6:** 61–79. Plenum Press. New York, N.Y.

28. SAN, R. H. C. & G. M. WILLIAMS. 1977. Rat hepatocyte primary cell culture–mediated mutagenesis of adult rat liver epithelial cells by procarcinogens. Proc. Soc. Exp. Biol. Med. **156:** 534–538.

29. LANGENBACH, R., C. KUSZYNSKI, D. NAGEL, T. LAWSON & P. POUR. Mutagenic activities of short chain aliphatic nitrosamines in the hamster hepatocyte-mediated V79 cell system. (In preparation.)

30. LANGENBACH, R., C. KUSZYNSKI, R. GINGELL, T. LAWSON, D. NAGEL, P. POUR & S. NESNOW. 1982. Structure activity analysis of propyl and related nitrosamines in the hepatocyte-mediated V79 cell mutagenesis system. In Structure Activity as a Predictive Tool in Toxicology. L. Golberg, Ed. Hemisphere Publishing Co. New York, N.Y. (In Press.)

31. LANGENBACH, R., R. GINGELL, C. KUSZYNSKI, B. WALKER, D. NAGEL & P. POUR. 1980. Mutagenic activities of oxidized derivatives of N-nitrosodipropylamine in the liver cell-mediated and Salmonella typhimurium assays. Cancer Res. **40:** 3463–3467.

32. JONES, C. A. & E. HUBERMAN. 1980. A sensitive hepatocyte-mediated assay for the metabolism of nitrosamines to mutagens for mammalian cells. Cancer Res. **40:** 406–411.

33. JONES, C. A., P. J. MARLINO, W. LIJINSKY & E. HUBERMAN. 1981. The relationship between the carcinogenicity and mutagenicity of nitrosamines in a hepatocyte-mediated mutagenicity assay. Carcinogenesis 2: 1075–1077.

34. TONG, C., M. FAZIO & G. M. WILLIAMS. 1981. Rat hepatocyte-mediated mutagenesis of human cells by carcinogenic polycyclic aromatic hydrocarbons but not organochlorine pesticides. Proc. Soc. Exp. Biol. Med. **167:** 572–575.

35. LANGENBACH, R., H. J. FREED, D. RAVEH & E. HUBERMAN. 1978. Cell specificity in metabolic activation of aflatoxin B_1 and benzo[a]pyrene to mutagens for mammalian cells. Nature **276:** 277–280.

36. BRADLEY, M. O., B. BHUYAN, M. C. FRANCIS, R. LANGENBACH, A. PETERSON & E. HUBERMAN. 1981. Mutagenesis by chemical agents in V79 Chinese hamster cells: a review and analysis of the literature. A report of the Gene-Tox Program. Mutat. Res. **87:** 81–142.

APPLICATION OF TRANSFORMATION SYSTEMS

R. J. Pienta

The MITRE Corporation
Metrek Division
McLean, Virginia 22102

J. A. Poiley and R. Raineri

Litton Bionetics, Inc.
Kensington, Maryland 21795

INTRODUCTION

In a recent comprehensive review, Heidelberger discussed cellular transformation as a basic tool for studying various aspects of chemical carcinogenesis *in vitro*.[1] In keeping with the theme of this conference on toxicity testing, this report is limited primarily to the application of selected cell-transformation systems for identifying chemicals that are potentially carcinogenic or that act to enhance carcinogenicity.

CELL-TRANSFORMATION SYSTEMS AVAILABLE FOR STUDY

In spite of the apparent rapid proliferation in the number of cell-transformation systems reported in recent years, the field of *in vitro* carcinogenesis had difficult origins. About four decades ago, Earle and Nettleship first described the induction of *in vitro* malignant transformation of cells from C3H and other established mouse cell lines following treatment with 3-methylcholanthrene.[2] Unfortunately, because of the spontaneous appearance of transformation in serially passaged cells that were not treated with the carcinogen, they concluded that cultured mouse cells were not suitable for studying chemical carcinogenesis. It was to take another two decades before an interest in the use of cell cultures for carcinogenicity studies was renewed. Berwald and Sachs showed that early passage (primary and secondary) Syrian hamster embryo (SHE) cells could be morphologically transformed by a number of carcinogenic hydrocarbons, whereas transformation was not obtained with untreated cells or those treated with noncarcinogenic hydrocarbons.[3,4] They grew as progressive tumors when inoculated subcutaneously into adult hamsters. Subsequent studies with a number of diverse transformed cells and carcinogens showed that morphological criteria for transformation correlated with malignancy and could be used as valid end points for detecting the carcinogenicity of chemicals.[5-9] These studies laid the foundation for the field of *in vitro* carcinogenesis and have resulted in the use of a broad spectrum of cell-culture systems for identifying carcinogenic chemicals and for studying mechanisms of carcinogenesis at the cellular level. TABLE 1 summarizes the systems employing fibroblastic cells. Any omissions are unintentional.

Initially, successful transformation of cells was limited to rodent cells, namely, hamster, mouse, rat, and guinea pig. More recently, the number of systems has expanded to include cell cultures derived from dog, cat, and human cells. To detect transformation, cells are treated with the candidate chemical either as mass cultures or as small numbers of cells. The treated cells are then cultured further and may or may not be passaged before they are examined for evidence of transformation. End points for transformation include the formation of foci of morphologically altered cells on a

267

0077–8923/83/0407–0267 $01.75/0 © 1983, NYAS

TABLE 1

TRANSFORMATION OF FIBROBLASTIC CELLS BY CHEMICAL CARCINOGENS

Species	Cell Culture	Comments	Principal Investigators
Hamster (Syrian)	early-passage embryo cells	altered colony morphology	Berwald & Sachs, 1963, 1965[3,4] DiPaolo et al., 1969, 1971[6,8]
		altered colony morphology, cryopreserved cells	Pienta et al., 1977[10]; Pienta, 1980[11] DiPaolo et al., 1980[12]
		focus formation	Casto et al., 1977[13]
		focus formation, SA7	Casto et al., 1973, 1974[14,15]
	virus-infected cells		Casto, 1980[16]
	BHK21, clone 13	growth in soft agar	DiMayorca et al., 1973[17] Mishra & DiMayorca, 1974[18] Ishii et al., 1977[19] Styles, 1977, 1980[20,21]
(Chinese)	(CH/L) lung cells	growth in soft agar	Borenfreund et al., 1966[22]
		altered colony morphology	Sanders & Burford, 1967[23]
	(CHO) ovary cells	altered colony morphology	Sanders & Burford, 1968[24]
	embryo cells	focus formation, enhancement of viral transformation	Diamond et al., 1974[25]
Mouse	BALB/3T3 embryo cell line	focus formation	DiPaolo et al., 1972[26] Kakunaga, 1973[27]
	C3H/10T1/2 embryo cell line	focus formation	Reznikoff et al., 1973[28]
	C3H/G23 prostate cell line	focus formation	Chen & Heidelberger, 1969[29]
	C3H/M2 prostate cell line	focus formation	Marquardt et al., 1976[30]
	NIH Swiss embryo cell line	focus formation, AKR-virus-infected cells	Rhim et al., 1971, 1974[31,32]
Rat	Fischer, embryo cell line	focus formation	Rhim & Huebner, 1973[33] Gordon et al., 1973[34]
		focus formation, RLV-virus-infected cells	Freeman et al., 1970, 1973, 1975[35-37] Price et al., 1971[38] Traul et al., 1981[39]
Guinea pig	strain 2, fetal cells	altered colony morphology	Evans & DiPaolo, 1976[40]
Dog	beagle embryo cells	focus formation	Rhim et al., 1978[41]
Cat	feline cell line 713—strain 2978	mass culture, transformed cells grow in soft agar or form aggregates in liquid overlay	Rhim et al., 1979[42]

TABLE 1 (*continued*)

Species	Cell Culture	Comments	Principal Investigators
Human	newborn foreskin cells	altered morphology, growth in soft agar	Milo & DiPaolo, 1978[43]
		growth in soft agar	McCormick *et al.*, 1980[44]
			Silinskas *et al.*, 1981[45]
	skin biopsy cells	focus formation, growth in soft agar	Kakunaga, 1978[46]
	osteosarcoma cells	focus formation	Rhim *et al.*, 1975[47]
	neurofibrosarcoma cells	focus formation	Igel *et al.*, 1975[48]
	skin fibroblasts from a patient with hereditary adenomatosis coli	transformed cells form aggregates in liquid culture or grow in soft agar	Rhim *et al.*, 1981[49] Miyaki *et al.*, 1982[50]

background monolayer of normal nontransformed cells, when mass cultures are treated. Alternatively, when small numbers of cells are exposed to a carcinogen, morphologically altered, discrete colonies are formed. In addition to these morphological end points, malignant transformation may also be detected by the growth of cells in semisolid media, such as 0.3% agar, an indication that the cells have become anchorage independent, a property not shared by normal cells, which must adhere to the substratum in order to propagate.

In a number of the cell systems mentioned in TABLE 1, transformation was reported for only a single or, at most, several carcinogens. Although of academic interest, their utility for detecting carcinogenicity by diverse chemicals awaits further development.

This report, therefore, will be limited to the most commonly used cell-transformation assay systems. These can be classified into three broad categories: first, those employing diploid cell strains whose cells have a limited life span; second, those employing heteroploid cell lines whose cells exhibit an unlimited life span, i.e., can be continually passaged; and third, those systems employing cells concomitantly infected with or expressing an oncogenic virus (TABLE 2).

DESCRIPTION OF ASSAY SYSTEMS

Cell Strains

Systems employing diploid cell strains are limited to those using Syrian hamster embryo cells and include procedures for both clonal and focus assays.

For the clonal-transformation assay, target cells derived from either freshly prepared primary or reconstituted cryopreserved cells are used. To obtain individual colonies, approximately 300 to 500 secondary passage cells are seeded into petri dishes previously seeded with 60,000 to 80,000 irradiated feeder cells.[6,8–11] On the following day, these cell cultures are exposed to graded doses of test chemical. The cultures are then incubated for seven to nine days at 37°C. The medium is then removed, and the cultures are washed, fixed, and stained. Stained colonies are then examined for morphological transformation as characterized by the random orientation, crisscrossing, and piling up of cells, primarily at the periphery of transformed colonies. These morphological changes are not observed in colonies derived from untreated cells.

The focus assay system employs similar secondary hamster embryo cell cultures, except that the target cells are seeded at a higher density, usually 50,000 cells per dish, in the absence of x-irradiated feeder cells. The cells are exposed to the test chemical for a total of 20 to 25 days. The end point is the formation of a focus of morphologically transformed cells on a monolayer of normal cells.

Cell Lines

The systems employing the most extensively used cell lines are comprised of either BALB/c-3T3 and C3H/10T1/2 mouse embryo cells, C3H M2 mouse ventral prostate cells, or the BHK21 strain of Syrian baby hamster kidney cells. The mouse cell lines rely upon the formation of a focus as an end point for activity. For the BHK21 cell line, a positive response is indicated by an increase in the rate of formation of anchorage-independent colonies in soft agar.

With minor variations regarding the number of cells seeded, time of exposure to chemical, refeeding interval, and duration of incubation, the procedures for mouse focus assays are essentially the same. Depending upon the particular cell line, either 10^3 or 10^4 cells from frozen stock cultures are seeded. On the following day, the cultures are exposed to the test chemical for one to three days and incubated further for a total of four to six weeks. Upon termination of the experiment, the cultures are washed, fixed, stained, and examined for the presence of transformed foci. Three types of foci have been described and are designated as types I, II, and III.[28] Usually both types II and III are scored with the C3H/10T1/2 system since cells derived from these foci grow as tumors when injected into susceptible animals. In the BALB/c-3T3 and mouse prostate systems, only foci comparable to type III described for C3H/10T1/2 are scored.

In addition to the mouse systems, a system using Syrian baby hamster kidney cells

TABLE 2

CELL TRANSFORMATION SYSTEMS IN WHICH MOST CHEMICALS HAVE BEEN TESTED

1. Cell strains—diploid cells with a limited life span
   a. Syrian hamster embryo cells—clonal assay
   b. Syrian hamster embryo cells—focus assay
2. Cell lines—heteroploid with an unlimited life span
   a. BALB/c-3T3 mouse embryo cells—focus assay
   b. C3H 10T1/2 mouse embryo cells—focus assay
   c. C3H M2 mouse ventral prostate cells—focus assay
   d. BHK 21 clone 12 Syrian baby hamster kidney cells—
      anchorage-independent growth in soft agar
3. Chemical-viral interactions
   a. RNA-virus-infected cells—retrovirus enhancement of
      chemical transformation
      • Fischer rat embryo cells infected with RLV
      • National Institutes of Health Swiss mouse embryo cells infected with AKR
   b. DNA-virus-infected cells—chemical enhancement of
      DNA-viral transformation
      • Syrian hamster embryo cells infected with adenovirus SA7
      • Fischer rat embryo cells infected with adenovirus SA7

TABLE 3

OVERALL NUMBER OF CHEMICALS TESTED IN CELL-TRANSFORMATION SYSTEMS

	Number of Chemicals	
System*	Tested	On EPA List (%)
Cell strains	167	56 (34)
SHE (colony)	156	48 (31)
SHE (focus)	11	8 (73)
Cell lines	114	30 (26)
BALB/c-3T3	51	23 (45)
C3H/10T1/2	33	10 (30)
Mouse prostate	46	4 (9)
Viral-chemical interactions	224	68 (30)
RLV/RE	119	47 (39)
AKR/ME	30	9 (30)
SA7/SHE	120	36 (30)
SA7/RE	6	4 (67)

*RE is rat embryo; ME is mouse embryo.

has been described.[17-21] Instead of assessing transformation by morphological criteria, the end point for activity in the BHK21 system is a two- to fivefold increase in the ability of the cells to grow in soft agar after exposure to a carcinogen. This system differs from the others in that BHK21 cells are not considered to be "normal" since they exhibit a background of anchorage-independent colonies in soft agar and are tumorigenic in hamsters before treatment with a carcinogen.

Viral-Chemical Interactions

In systems employing viral-chemical interactions, there is the enhancement of chemical transformation by retroviruses on the one hand and the enhancement of DNA-virus-induced cell transformation on the other.

In the retrovirus enhancement assay, a mass culture of rat embryo cells concomitantly infected with or expressing an endogenous leukemia virus is treated with the test chemical. After exposure for one week, the culture is split into both a vertical and a stationary set of cultures. The stationary set is incubated without further splitting and examined periodically for the appearance of transformed foci. The vertical set is split every two weeks to form other sets of stationary and vertical cultures. This procedure is repeated for about six population doublings. The stationary cultures are held for as long as 60 days.

In the assay system employing the enhancement of adenovirus-induced transformation, primary Syrian hamster embryo cells are treated with test chemical for either 2 or 18 hours. The cultures are then infected with approximately 200 focus-forming units of SA7 adenovirus. After allowing for the adsorption of the virus to the cells, the cultures are trypsinized and replated. The frequency of SA7-transformed foci is determined 21 to 25 days later. The enhancement of viral transformation is expressed as the increase in foci per 10^6 surviving cells and is determined by dividing the frequency of transformation in a culture that was treated with chemical and SA7 virus by the frequency observed in a culture treated with virus alone.

RESULTS OF TESTS WITH DIVERSE CHEMICALS

Results described here were obtained primarily from a review prepared for the Environmental Protection Agency Gene-Tox Program of mammalian cell transformation by chemical agents.[51] Among the systems critically reviewed for the Gene-Tox Program, 167 chemicals were tested in cell-transformation systems employing cell strains, 114 were tested in systems employing cell lines, and 224 were tested in systems involving viral-chemical interactions (TABLE 3). A number of compounds were tested in more than one test system and, hence, may be duplicated in the totals, giving rise to apparent discrepancies between the total number of chemicals tested in each category of test system and the sum of the number of chemicals tested in individual systems. Authenticity for the carcinogenicity or noncarcinogenicity of the chemicals tested was limited to information provided by the EPA merged carcinogen list of April 1980. A large number of the chemicals tested were not on the EPA list, therefore precluding definitive comparison of results among the test systems. In several instances, as specifically noted in the tables, authenticity of carcinogenicity of chemicals tested in a particular test system was not critically evaluated but taken at face value from the open literature.

The compounds tested in the various cell-transformation systems represent about 50 diverse chemical classes, which are summarized in TABLE 4. The complete list of chemicals is summarized in the review.[51] In many instances, an individual chemical was tested in only one test system. Of the total number of chemicals tested, with

TABLE 4

CLASSES OF CHEMICALS TESTED IN MAMMALIAN CELL-TRANSFORMATION SYSTEMS

Acridines	Hydrazines
Alcohols	Hydroxylamine
Aldehydes	Lactones
Aliphatic amines	Metals and derivatives
Alkaloids	Mycotoxins
Alkyl epoxides	Nitro compounds
Alkyl halides	Nitriles
Alkyl sulfates	Nitrogen mustards
Alkyl sulfoxides	Nitrofurans
Alkyl sulfones	Nitrosamines
Alkyl sulfonates	Nitrosamides
Amides	Nitrosoguanidines
Amine-N-oxides	Nitrosoureas
Amino acids and derivatives	Organic sulfur compounds
Anhydrides	Phenols
Antibiotics	Phenothiazines
Aromatic amines and amides	Polycyclic aromatic hydrocar-
Aromatic azo compounds	bons, dihydrodiols, haloge-
Aryl oxides and epoxides	nated derivatives
Azides	Purine derivatives
Aziridines	Pyrimidine derivatives
Azoxy compounds	Steroids and related hormones
Benzene-ring derivatives	Thioureas
Carbamates	Ureas
Carboxyl acids and esters	Zanthenes
Heterocyclic ring compounds	Unclassified

TABLE 5

CHEMICALS TESTED IN TWO OR MORE CELL-TRANSFORMATION SYSTEMS

Chemicals tested in nine systems	
Benzo[*a*]pyrene	
Dibenz[*a,h*]anthracene	
Chemicals tested in eight systems	
N-Methyl-*N'*-nitro-*N*-nitrosoguanidine	
Chemicals tested in five systems	
2-Naphthylamine	
Chemicals tested in four systems	
Dimethoxy-DDT	
Methylazoxymethanol acetate	
Thiourea	
Chemicals tested in three systems	
Acetamide	Ethyl carbamate
Aflatoxin B$_1$	Ethinylestradiol
4-Aminobiphenyl	4,4'-Methylenebis(*o*-chloroaniline)
p-Aminoazobenzene	1-Naphthylamine
Benzidine	4-Nitro-*o*-phenylendiamine
Benzo[*e*]pyrene	Progesterone
Cyclophosphamide	Propane sultone
Diethylstilbestrol	Propyleneimine
4-(Dimethylamino)azobenzene	Tetrachloroethylene
Dimethylnitrosamine	Thioacetamide
Diphenylnitrosamine	1,8,9-Trihydroxyanthracene
Chemicals tested in two systems	
Acrylonitrile	Glycidaldehyde
3-Amino-1,2,4-triazole	Methyl methanesulfonate
Diethylnitrosamine	4-Nitrobiphenyl
Ethyl methanesulfonate	β-Propriolactone
N-Ethylnitrosourea	Uracil mustard
5-Fluorouracil	Saccharin

authenticated carcinogenic and noncarcinogenic activities, 41 were tested in more than two transformation systems. Of these, two chemicals, benzo[*a*]pyrene (BaP) and dibenz[*a,h*]anthracene, were tested in all nine systems (TABLE 5). Dimethoxy-DDT and 4-nitro-*o*-phenylenediamine were the only noncarcinogens on the EPA merged list.

An attempt was made to evaluate the validity and usefulness of the cell-transformation systems for detecting carcinogens. Various terminologies have been used to describe the validity of a test system.[52] For purposes of this evaluation, *sensitivity* is defined as the ability of a test system to give a positive result for a carcinogen. It is the ratio of the number of carcinogens giving a positive response to the total number of carcinogens tested. *Specificity* is defined as the ability of a test system to give negative results for noncarcinogens and is the ratio of the number of noncarcinogens identified correctly to the total number of noncarcinogens tested. *Accuracy* refers to the overall response of the test and is the ratio of the total number of correct responses, for both carcinogens and noncarcinogens, to the total number of substances tested. *Predictive value* refers to the portion of positive results for carcinogens that are actually carcinogens. It is the ratio of the true-positive responses to the true-positive responses plus the false-positive responses. A reliable test system requires both high sensitivity and specificity. The predictive value of a test system is

TABLE 6

SUMMARY OF RESPONSES TO CHEMICALS TESTED IN SYRIAN HAMSTER
EMBRYO OR KIDNEY CELLS

	Test Systems					
	SHE Clonal One Lab	SHE Clonal Multiple Labs	SHE Clonal Combined	SHE Focus	SHE Clonal FCRC*	BHK 21, CL13 Styles*
Chemicals tested						
Carcinogens	37	8	45	8	74	58
Noncarcinogens	2	0	2	0	31	62
Carcinogenicity undetermined (UD)	93	15	108	3	11	0
Responses						
Carcinogens positive	24	8	32	8	73, 67	54
Carcinogens negative (false negative)	4	0	4	0	1, 7†	4
Noncarcinogens negative	1	—	1	—	31	61
Noncarcinogens positive (false positive)	1	—	1	—	0	1
Transformation questionable	9	0	9	0	0	
UD positive	53	11	64	3	8	—
UD negative	30	2	32	0	3	—
UD transformation questionable	10	2	12	0	0	—
Evaluation						
Sensitivity	0.65	1.0	0.89	1.0	0.97, 0.91	0.93
Specificity	0.50	?	0.50	?	1.0	0.98
Accuracy	0.68	1.0	0.70	1.0	0.99, 0.93	0.96
Predictive Value	0.96	?	0.97	?	1.0	0.98

*Carcinogenicity as reported in the open literature.
†Positive with exogenous metabolic activation by hamster liver S-9 or cultured hamster hepatocytes.

dependent upon the proportion of carcinogens among the total number of chemicals tested as well as on the sensitivity and specificity of the test. The higher the proportion of carcinogens, the higher the predictive value of the test provided that sensitivity and specificity are also high.

In general there is good correlation between *in vitro* responses and *in vivo* carcinogenicity with the known carcinogens and noncarcinogens. A large number of chemicals whose carcinogenicity has not been authenticated were also tested. Although they are not used in the validation of the short-term tests, results with these chemicals may be used to provide supporting evidence when marginal results are obtained in animal bioassays. Responses to chemicals tested in transformation systems employing Syrian hamster cells are summarized in TABLE 6. It is immediately apparent that in the presence of the large number of chemicals whose carcinogenicity is considered undetermined and the small numbers, or in some instances the complete absence, of noncarcinogens tested, a definitive evaluation is not possible. Nevertheless, some useful information is presented.

Combined results from chemicals tested either in a single laboratory or in more than one laboratory indicate a high sensitivity (89%) and predictive value (97%) for the SHE clonal cell-transformation system. The values for the specificity (50%) and accuracy (70%) of the system should be taken with caution since they are influenced by the small number of noncarcinogens tested. Since no noncarcinogens were tested in the SHE focus system, no judgment can be made about the specificity of this test although the sensitivity with this small number of carcinogens was 100%.

When over 100 chemicals, whose reported carcinogenic activity was obtained from the open literature at the time, were tested at the National Cancer Institute Frederick Cancer Research In Vitro Carcinogenesis Laboratory, excellent correlations with reported *in vivo* activity for carcinogens and noncarcinogens were obtained.[11] Both sensitivity and predictive value were over 90%. Specificity and accuracy were 100%.

Summary of data obtained with chemicals tested in BHK21 clone 13 cells in one laboratory also gave values well over 90% for sensitivity, specificity, accuracy, and predictive value.

Results of the responses to chemicals tested on transformation systems employing the three mouse cell lines are summarized in TABLE 7. With the limited number of authenticated known carcinogens and noncarcinogens, the sensitivities of the test systems ranged from 53 to 80%. The accuracy ranged from 50 to 83%, and the predictive value ranged from 83 to 90%. A more definitive evaluation is hampered by

TABLE 7

SUMMARY OF RESPONSES TO CHEMICALS TESTED IN MOUSE CELL LINES

	Test Systems				
	BALB/c-3T3 One Lab	C3H/10T1/2 One Lab	Mouse Prostate One Lab	Cell Lines Multiple Labs	Cell Lines Combined
Chemicals tested					
Carcinogens	17	5	1	5	27
Noncarcinogens	1	1	0	0	2
Carcinogenicity undetermined (UD)	20	18	37	9	84
Responses					
Carcinogens positive	9	5	0	4	17
Carcinogens negative (false negative)	8	0	1	1	10
Noncarcinogens negative	0	0	—	—	0
Noncarcinogens positive (false positive)	1	1	—	—	2
Transformation questionable	0	0	0	0	0
UD positive	7	7	23	5	42
UD negative	13	11	13	3	40
UD transformation questionable	0	0	1	1	2
Evaluation					
Sensitivity	0.53	1.0	0	0.80	0.63
Specificity	0	0	?	?	0
Accuracy	0.50	0.83	—	0.80	0.59
Predictive value	0.90	0.83	?	?	0.90

TABLE 8

SUMMARY OF RESPONSES TO CHEMICALS TESTED IN SYSTEMS INVOLVING
VIRAL-CHEMICAL INTERACTIONS

	SA7/ SHE	SA7/ RE	SA7/ SHE-RE Combined	AKR/ ME	RLV/ RE*	RLV/ REt	RLV/RE Combined
Chemicals tested							
Carcinogens	33	4	37	6	27	19	41
Noncarcinogens	2	0	2	0	3	1	3
Carcinogenicity undetermined (UD)	84	2	86	18	32	48	72
Responses							
Carcinogens positive	27	3	30	5	27	16	39
Carcinogens negative (false negative)	6	1	7	1	0	3	1
Noncarcinogens negative	2	—	2	—	2	1	2
Noncarcinogens positive (false positive)	0	—	0	—	1	0	1
Transformation questionable	0	0	0	0	0	0	1
UD positive	55	2	57	10	26	25	41
UD negative	28	0	28	8	6	21	25
UD transformation questionable	1	0	1	0	0	2	6
Evaluation							
Sensitivity	0.82	0.75	0.81	0.83	1.0	0.84	0.95
Specificity	1.0	?	1.0	?	0.67	1.0	0.67
Accuracy	0.83	0.75	0.82	0.83	0.97	0.85	0.93
Predictive value	1.0	?	1.0	?	0.96	1.0	0.98

*From the laboratory of Traul et al.[39]
†From the laboratory of Freeman et al.[35–38]

the preponderance of chemicals for which authentic carcinogenic activities have not yet been assigned and the extremely small number of noncarcinogens tested.

Similar results were obtained with test systems employing viral-chemical interactions (TABLE 8). Two hundred and twenty-four individual chemicals representing about 36 classes of compounds have been tested in the four transformation systems. Of these chemicals, only 68 were on the EPA merged list. The sensitivity of the individual test systems ranged from 75–95% depending upon whether results were from single or combined tests. The specificity ranged from 67 to 100%, although no more than three noncarcinogens were tested in any one system.

METABOLIC ACTIVATION

In general, the carcinogenicity of most chemicals can be detected in the various cell-transformation systems. However, some compounds fail to transform cells, presumably due to a lack of cellular enzymes required for the metabolic conversion of these compounds to active forms. In bacterial and mammalian mutagenesis assay systems, these enzymes can be provided by the addition of rodent liver homogenates. The use of exogenous metabolic activation systems has been limited almost exclusively to the hamster embryo cell clonal assay system.[52,53] Several carcinogens (auramine, diethylnitrosamine, 3-methoxy-4-aminozobenzene, procarbazine) could be activated by the addition of hamster liver S-9 homogenate.[8,54] Negative results were obtained with 4-aminoazobenzene and saccharin. When intact hamster hepatocytes were used as a source of enzymes, diethylnitrosamine, 2-nitrofluorene, and 4-aminoazobenzene were activated sufficiently to induce transformation of hamster embryo cells.

The hamster embryo cell system has the capacity to use a variety of activation systems. For example, the system has been used to compare the potential of cells from a variety of organs such as liver, kidney, lung, bladder, small intestine, or early passage embryo fibroblasts to activate acetylaminofluorene (AAF) or to detoxify BaP (TABLE 9).

Although certain selected batches of hamster embryo cells are not transformed by AAF, they are transformed by the metabolic products aminofluorene (AF) or *N*-hydroxy-AAF (TABLE 9). Poiley *et al.* have demonstrated that cells from the various organs possess varying levels of AAF deacetylase and *N*-hydroxylase activities and are able to activate and bind covalently to AAF.[55] They reported that kidney cells from rats and hamsters, lung cells from rats, small intestine cells from hamsters, and hepatocytes from rats and, to a lesser extent, from hamsters activate AAF to a transforming agent. Because hamster embryo cells respond to more than one of the metabolites of AAF, a clear correlation between metabolism, binding, and transformation cannot be made. The system, however, provides a means for investigating further

TABLE 9

ACTIVATION OR INACTIVATION OF CARCINOGENS BY CELLS DERIVED
FROM VARIOUS RAT OR HAMSTER ORGANS

Chemical	Cells	Transformation	
Activation			
None	none	−	
AAF	none	−	
N-Hydroxy-	none	+	
AAF	none	+	
AF			
		Rat*	Hamster*
AAF	liver	+	
AAF	kidney	+	+
AAF	lung	+	−
AAF	small intestine	−	+
AAF	bladder	−	−
Inactivation			
BaP	none	+	
		Rat	Hamster
BaP	liver	−	−
BaP	kidney	+	+
BaP	lung	+	+
BaP	small intestine	+	+
BaP	trachea	+	+

*Source of cells.

the metabolic potential of a variety of cell types that cannot be studied in the intact animal.

Hamster embryo cells are transformed by BaP without the need for an exogenous metabolic activation system. However, in the presence of hepatocytes from the rat or hamster, detoxification or inactivation may occur as reflected by a decrease in transformation frequency. On the other hand, in the presence of intact cells derived from kidney, small intestine, lung, or trachea, no change in transformation frequency was observed. Differences in metabolic properties of the cells may be responsible for the divergent results. For example, hepatocytes were shown to produce large amounts of water-soluble conjugated metabolites, whereas the other cell types did not. Similarly, intact liver cells have a higher capacity for the conjugation of BaP than do disrupted liver cells.[56] The inhibition of BaP-induced transformation by intact cells more closely mimics the events occurring in the intact animal, where detoxification is a primary function of the liver. In general S-9 homogenates tend to activate while intact cells may also inactivate carcinogens.

PROMOTION AND ENHANCEMENT

Various agents have been shown to increase the carcinogenic response. Operationally, a chemical is considered to be a promoter if an increase in response occurs when it is applied sequentially to a system that has been treated previously with an initiating agent. The chemical is considered to be an enhancer if an increase in response is elicited when the second chemical is applied before or concomitantly with the initiating carcinogen.

Heidelberger and his associates first demonstrated the initiation and promotion stages of chemical carcinogenesis in cell cultures, employing C3H/10T1/2 cells.[57,58] In this system, target cells were initiated with carcinogenic polycyclic hydrocarbons, such as BaP, 7,12-dimethylbenzanthracene, or 3-methylcholanthrene (3-MC), given at very low doses capable of inducing few or no transformed foci. When the initiated cells were then treated repeatedly with a dose of 12-O-tetradecanoylphorbol-13-acetate (TPA) that by itself did not transform the cells, a large number of transformed foci were observed. Subsequently, they showed that saccharin acted as a promoter in this system when cells were initiated with 3-MC.[59] Sivak and Tu were unable to confirm the promoter activity of saccharin when BALB/c-3T3 mouse cells were initiated with 3-MC and subsequently treated with saccharin.[60] Donovan and DiPaolo reported that caffeine increased the transformation rate of Syrian hamster embryo cells induced by BaP, N-acetoxy-2-AAF, or N-methyl-N'-nitro-N-nitrosoguanidine (MNNG). Caffeine was added to the cultures usually one hour after treatment with carcinogen. The effect depended upon the concentration of caffeine, time of addition of caffeine, length of exposure, and carcinogen used. DiPaolo et al. also reported that pretreatment of Syrian hamster embryo cells for one hour with the alkylating agents methyl methanesulfonate or ethyl methanesulfonate enhanced transformation mediated by the potent carcinogens dimethylbenz[a]anthracene, 3-MC, N-acetoxy-2-AAF, and MNNG.[62]

In the C3H/10T1/2 cell-culture system, transformation mediated by 3-methylcholanthrene could be increased by pretreating the cell cultures with modifiers of arylhydrocarbon hydroxylase (AHH) and 3-methylcholanthrene-11,12-oxide hydrase (EH) activities.[63] Pretreatment with benz[a]anthracene (BA) (AHH and EH inducer), styrene oxide, cyclohexene oxide, and 1,2,3,4-tetrahydronaphthalene (precursor of 1,2,3,4-tetrahydronaphthalene-1,2-oxide) resulted in two- to threefold increases in transformation. When the cells were pretreated with a combination of BA and cyclohexene, 3-MC-mediated transformation was increased almost eightfold.[68]

Poiley *et al.* demonstrated the enhancement of dibenz[a,c]anthracene-induced transformation when Syrian hamster embryo cells were treated with TPA 30 days after treatment with subtransforming doses of carcinogen.[64] Earlier application of TPA inhibited the transformation response.

Rivedal and Sanner described the potentiating effect of cigarette smoke extract on the morphological transformation of hamster embryo cells by BaP.[65] Hamster embryo cells were transformed by either BaP or smoke extract alone. A combined treatment of BaP and smoke extract resulted in higher transformation rates than were observed with each of the compounds tested separately. In experiments to determine promoter activity, where cells were treated sequentially with BaP for four days followed by smoke extract for four days, the transformation frequency was significantly higher than expected on the basis of the compounds tested separately. These experiments demonstrated a synergistic effect between BaP and cigarette smoke extract and a promoterlike effect of smoke extract on BaP-initiated cell transformation. Further studies by these investigators demonstrated a synergistic effect on the transformation of Syrian hamster embryo cells by nickel sulfate and smoke extract.[66,67] In this study, they also showed the promotion of BaP-induced transformation by either TPA or smoke extract. In experiments with combinations of nickel sulfate and BaP or 3-MC, a synergistic effect could be detected between nickel sulfate and BaP but not between nickel sulfate and 3-MC.

Rivedal and Sanner examined a number of metal salts for their synergistic and promoter properties.[68] A synergistic enhancement of the transformation frequency was found for the combined treatment with the organic carcinogens BaP, N-hydroxy-2-AAF, and 4-nitroquinoline 1-oxide and the metal salts nickel sulfate, cadmium acetate, or potassium chromate. Chromic chloride and zinc chloride did not induce transformation by themselves, and they had no effect on the transformation frequency when tested in combination with BaP. When the cells were first exposed to BaP, both nickel sulfate and cadmium acetate showed a promotionlike effect similar to that observed with TPA.

CONCLUSIONS

Fibroblastic cell systems derived from mouse, rat, hamster, guinea pig, cat, dog, and human tissues and transformable by chemical agents have been identified. A review of those transformation systems in which the largest number of chemicals have been tested shows good agreement between *in vitro* responses and *in vivo* carcinogenicity. Valid short-term tests should have high specificity as well as sensitivity. Further testing designed to validate these systems as screens for predicting carcinogenicity should include a reasonable number of known carcinogens to assure specificity of the particular test system so that false-positive responses are minimized.

In addition to detecting chemicals that are potentially carcinogenic, the cell-transformation systems may have utility for identifying carcinogen promoters and enhancers.

REFERENCES

1. HEIDELBERGER, C. 1980. Cellular transformation as a basic tool for chemical carcinogenesis. *In* Advances in Modern Environmental Toxicology. Mammalian Cell Transformation by Chemical Carcinogens. N. Mishra, V. Dunkel & M. Mehlman, Eds. 1: 1–28. Senate Press, Inc. Princeton Junction, N.J.
2. EARLE, W. R. & A. NETTLESHIP. 1943. Production of malignancy in vitro. V. Results of injections of cultures into mice. J. Nat. Cancer Inst. 4: 213–227.

3. BERWALD, Y. & L. SACHS. 1963. In vitro transformation with chemical carcinogens. Nature **200:** 1182–1184.
4. BERWALD, Y. & L. SACHS. 1965. In vitro transformation of normal cells to tumor cells by carcinogenic hydrocarbons. J. Nat. Cancer Inst. **35:** 641–661.
5. HUBERMAN, E. & L. SACHS. 1966. Cell susceptibility to transformation and cytotoxicity by the carcinogenic hydrocarbon benzo[a]pyrene. Proc. Nat. Acad. Sci. USA **56:** 1123–1129.
6. DiPAOLO, J. A., P. J. DONOVAN & R. L. NELSON. 1969. Quantitative studies of *in vitro* transformation by chemical carcinogens. J. Nat. Cancer Inst. **42:** 867–876.
7. DiPAOLO, J. A., P. J. DONOVAN & R. L. NELSON. 1971. *In vitro* transformation of hamster cells by polycyclic hydrocarbons: factors influencing the number of cells transformed. Nature **230:** 240–242.
8. DiPIAOLO, J. A., R. L. NELSON & P. J. DONOVAN. 1971. Morphological, oncogenic, and karyological characteristics of Syrian hamster embryo cells transformed in vitro by carcinogenic polycyclic hydrocarbons. Cancer Res. **31:** 1118–1127.
9. DiPAOLO, J. A., R. L. NELSON & P. J. DONOVAN. 1972. In vitro transformation of Syrian hamster embryo cells by diverse chemical carcinogens. Nature **235:** 278–280.
10. PIENTA, R. J., J. A. POILEY & W. B. LEBHERZ III. 1977. Morphological transformation of early passage golden Syrian hamster embryo cells derived from cryopreserved primary cultures as a reliable in vitro bioassay for identifying diverse carcinogens. Int. J. Cancer **19:** 642–655.
11. PIENTA, R. J. 1980. A transformation bioassay system employing cryopreserved hamster embryo cells. *In* Advances in Modern Environmental Toxicology. Mammalian Cell Transformation by Chemical Carcinogens. N. Mishra, V. Dunkel & M. Mehlman, Eds. **1:** 47–83. Senate Press Inc. Princeton Junction, N.J.
12. DiPAOLO, J. A. 1980. Quantitative *in vitro* transformation of Syrian golden hamster embryo cells with the use of frozen stored cells. J. Nat. Cancer Inst. **64:** 1485–1489.
13. CASTO, B. C., N. JANOSKO & J. A. DiPAOLO. 1977. Development of a focus assay model for transformation of hamster cells in vitro by chemical carcinogens. Cancer Res. **37:** 3508–3515.
14. CASTO, B., W. J. PIECZYNSKI & J. A. DiPAOLO. 1973. Enhancement of adenovirus transformation by pretreatment of hamster cells with carcinogenic polycyclic hydrocarbons. Cancer Res. **33:** 819–824.
15. CASTO, B. C., W. J. PIECZYNSKI & J. A. DiPAOLO. 1974. Enhancement of adenovirus transformation by treatment of hamster embryo cells with diverse chemical carcinogens. Cancer Res. **34:** 72–78.
16. CASTO, B. C. 1980. Detection of chemical carcinogens and mutagenesis in hamster cells by enhancement of adenovirus transformation. *In* Advances in Modern Environmental Toxicology. Mammalian Cell Transformation by Chemical Carcinogens. N. Mishra, V. Dunkel & M. Mehlman, Eds. **1:** 241–271. Senate Press Inc. Princeton Junction, N.J.
17. DiMAYORCA, G., M. GREENBLATT, T. TRAUTHEN, A. SOLLER & R. GIRDANO. 1973. Malignant transformation of BHK_{21} clone 13 cells *in vitro* by nitrosamines—a conditioned state. Proc. Nat. Acad. Sci. USA **70:** 46–49.
18. MISHRA, N. K. & G. DiMAYORCA. 1974. *In vitro* malignant transformation of cells by chemical carcinogens. Biochem. Biophys. Acta **355:** 205–219.
19. ISHII, Y., J. A. ELLIOTT, N. K. MISHRA & M. W. LIEBERMAN. 1977. Quantitative studies of transformation by chemical carcinogens and ultraviolet radiation using a subclone of BHK_{21} clone 13 Syrian hamster cells. Cancer Res. **37:** 2023–2039.
20. STYLES, J. A. 1977. A method for detecting carcinogenic organic chemicals using mammalian cells in culture. Br. J. Cancer **36:** 558–563.
21. STYLES, J. A. 1980. Use of $BHK_{21}/C113$ cells for chemical screening. *In* Advances in Modern Environmental Toxicology. Mammalian Cell Transformation by Chemical Carcinogens. N. Mishra, V. Dunkel & M. Mehlman, Eds. **1:** 85–131. Senate Press Inc. Princeton Junction, N.J.
22. BORENFREUND, E., M. KRIM, F. K. SANDERS, S. S. STERNBERG & A. BENDICH. 1966. Malignant conversion of cells in vitro by carcinogens and viruses. Proc. Nat. Acad. Sci. USA **56:** 671–679.

23. SANDERS, F. K. & B. O. BURFORD. 1967. Morphological conversion of cells *in vitro* by N-nitrosomethylurea. Nature **213:** 1171–1173.

24. SANDERS, F. K. & B. O. BURFORD. 1968. Morphological conversion, hyperconversion, and reversion of mammalian cells treated *in vitro* with N-nitrosomethylurea. Nature **220:** 448–453.

25. DIAMOND, L., R. KNORR & Y. SHIMURU. 1974. Enhancement of simian virus 40–induced transformation of Chinese hamster embryo cells by 4-nitroquinolone 1-oxide. Cancer Res. **34:** 2599–2604.

26. DIPAOLO, J. A., K. TAKANO & N. C. POPESCU. 1972. Quantitation of chemically induced neoplastic transformation of BALB/3T3 cloned cell lines. Cancer Res. **32:** 2686–2695.

27. KAKUNAGA, T. 1973. A quantitative system for assay of malignant transformation by chemical carcinogens using a clone derived from BALB/3T3. Int. J. Cancer **12:** 463–473.

28. REZNIKOFF, C. A., J. S. BERTRAM, D. W. BRANKOW & C. HEIDELBERGER. 1973. Quantitative and qualitative studies of chemical transformation of cloned C3H mouse embryo cells sensitive to post confluence inhibition of cell division. Cancer Res. **33:** 3239–3249.

29. CHEN, T. T. & C. HEIDELBERGER. 1969. Quantitative studies on the malignant transformation of mouse prostate cells by carcinogenic hydrocarbons *in vitro*. Int. J. Cancer **4:** 166–178.

30. MARQUARDT, H. 1976. Malignant transformation *in vitro*. A model system to study mechanisms of action of chemical carcinogens and to evaluate the oncogenic potential of environmental chemicals. IARC Sci. Publ. No. 12: 389–410.

31. RHIM, J. S., B. CREASY & R. J. HUEBNER. 1971. Production of altered cell foci by 3-methylcholanthrene in mouse cells infected with AKR leukemia virus. Proc. Nat. Acad. Sci. USA **68:** 2212–2216.

32. RHIM, J. S., D. K. PARK, E. K. WEISBURGER & J. H. WEISBURGER. 1974. Evaluation of an *in vitro* assay system based on prior infection of rodent cells with nontransforming RNA tumor virus. J. Nat. Cancer Inst. **52:** 1167–1173.

33. RHIM, J. S. & R. J. HUEBNER. 1973. Transformation of rat embryo cells *in vitro* by chemical carcinogens. Cancer Res. **33:** 695–700.

34. GORDON, R. J., R. J. BRYAN, J. S. RHIM, C. DEMOISE, R. G. WOLFORD, A. E. FREEMAN & R. J. HUEBNER. 1973. Transformation of rat and mouse embryo cells by a new class of carcinogenic compounds isolated from particles in city air. Int. J. Cancer **12:** 223–232.

35. FREEMAN, A. E., P. J. PRICE, H. J. IGEL, J. C. YOUNG, J. M. MARYSK & R. J. HEUBNER. 1970. Morphological transformation of rat embryo cells induced by diethylnitrosamine and murine leukemia viruses. J. Nat. Cancer Inst. **44:** 65–78.

36. FREEMAN, A. E., E. K. WEISBURGER, J. H. WEISBURGER, R. G. WOLFORD, J. M. MARYAK & R. J. HUEBNER. 1973. Transformation of cell cultures as an indication of the carcinogenic potential of chemicals. J. Nat. Cancer Inst. **51:** 799–808.

37. FREEMAN, A. E., H. J. IGEL & P. J. PRICE. 1975. Carcinogenesis *in vitro*. I. *In vitro* transformation of rat embryo cells: correlations with the known tumorigenic activities of chemicals in rodents. In Vitro **11:** 107–116.

38. PRICE, P. J., A. E. FREEMAN, W. T. LANE & R. J. HUEBNER. 1971. Morphological transformation of rat embryo cells by the combined action of 3-methyl cholanthrene and Rauscher leukemia virus. Nature **230:** 144–146.

39. TRAUL, K. A., R. J. HINK, J. S. WOLFF & K. WLODYMR. 1981. Chemical carcinogens *in vitro*: an improved method for chemical transformation in Rauscher leukemia virus–infected rat embryo cells. J. Appl. Toxicol. **1:** 32–37.

40. EVANS, C. H. & J. A. DIPAOLO. 1975. Neoplastic transformation of guinea pig fetal cells in culture induced by chemical carcinogens. Cancer Res. **35:** 1035–1044.

41. RHIM, J. S., D. K. PARK, P. ARNSTEIN & W. A. NELSON-REES. 1978. Neoplastic transformation of canine embryo cells in vitro by N-methyl-N'-nitro-N-nitrosoguanidine. Int. J. Cancer **22:** 441–446.

42. RHIM, J. S., W. A. NELSON-REES & M. ESSEX. 1979. Transformation of feline cells in culture by a chemical carcinogen. Int. J. Cancer **24:** 336–340.

43. MILO, G. E., JR. & J. A. DIPAOLO. 1978. Neoplastic transformation of human diploid cells *in vitro* after chemical carcinogen treatment. Nature **275:** 130–132.

44. MCCORMICK, J. J., K. C. SILINSKAS & V. M. MAHER. 1980. Transformation of diploid

human fibroblasts by chemical carcinogens. *In* Carcinogenesis: Fundamental Mechanisms and Environmental Effects. B. Pullman, P. O. P. Ts'O & H. Gelbain, Eds.: 491–498. D. Reidel Publishing Co. Boston, Mass.

45. SILINSKAS, K. C., S. A. KATELEY, J. E. TOWER, V. M. MAHER & J. J. McCORMICK. 1981. Induction of anchorage-independent growth of human fibroblasts by propane sultone. Int. J. Cancer **26:** 565–569.

46. KAKUNAGA, T. 1978. Neoplastic transformation of human diploid fibroblast cells by chemical carcinogens. Proc. Nat. Acad. Sci. USA **75:** 1334–1338.

47. RHIM, J. S., C. M. KIM, P. ARNSTEIN, R. J. HUEBNER, E. K. WEISBURGER & W. A. NELSON-REES. 1975. Transformation of human osteosarcoma cells by a chemical carcinogen. J. Nat. Cancer Inst. **55:** 1291–1294.

48. IGEL, H. J., A. E. FREEMAN, J. E. SPIEWAK & K. KLEINFELD. 1975. Carcinogenesis *in vitro*. II. Chemical transformation of diploid human cell cultures: a rare event. In Vitro **11:** 117–129.

49. RHIM, J. S., R. J. HUEBNER, P. ARNSTEIN & L. KOPELOVICH. 1980. Chemical transformation of cultured human skin fibroblasts derived from individuals with hereditary adenomatosis of the colon and rectum. Int. J. Cancer **26:** 565–569.

50. MIYAKI, M., N. AKAMATSU, T. ONO, A. TONOMURA & J. UTSUNOMIYA. 1982. Morphologic transformation and chromosome changes induced by chemical carcinogens in skin fibroblasts from patients with familial adenomatosis coli. J. Nat. Cancer Inst. **68:** 563–571.

51. HEIDELBERGER, C., A. E. FREEMAN, R. J. PIENTA, A. SIVAK, J. A. BERTRAM, B. C. CASTO, V. C. DUNKEL, M. W. FRANCIS, T. KAKUNAGA, J. B. LITTLE & L. M. SCHECHTMAN. 1983. Cell transformation by chemical agents: a review and analysis of the literature. Mutat. Res. (In press.)

52. COOPER, J. A., R. SARACCI & P. COLE. 1979. Describing the validity of carcinogen screening tests. Br. J. Cancer **39:** 87–89.

53. POILEY, J. A., R. RAINERI & R. J. PIENTA. 1979. The use of hamster hepatocytes to metabolize carcinogens in an *in vitro* bioassay. J. Nat. Cancer Inst. **63:** 519–524.

54. PIENTA, R. J. 1979. A hamster embryo cell model system for identifying carcinogens. *In* Carcinogens: Identification and Mechanisms of Action. A. C. Griffin & C. R. Shaw, Eds.: 121. Raven Press. New York, N.Y.

55. POILEY, J. A., R. RAINERI, M. K. ERNST, A. W. ANDREWS & R. J. PIENTA. 1981. Activation of *N*-2-acetylaminofluorene by target and non-target Syrian golden hamster cells. In Vitro **17:** 222.

56. SELKIRK, J. K. 1977. Divergence of metabolic activation systems for short-term mutagenesis assays. Nature **270:** 604–607.

57. MONDAL, S., D. W. BRANKOW & C. HEIDELBERGER. 1976. Two-stage chemical oncogenesis in cultures of C3H-10T1/2 cells. Cancer Res. **36:** 2254–2260.

58. MONDAL, S. & C. HEIDELBERGER. 1976. Transformation of C3H-10T1/2 Cl 8 mouse embryo fibroblasts by ultraviolet irradiation and a phorbol ester. Nature **260:** 710–711.

59. MONDAL, S., D. W. BRANKOW & C. HEIDELBERGER. 1978. Enhancement of oncogenesis in C3H-10T1/2 mouse embryo cell cultures by saccharin. Science **201:** 1141–1142.

60. SIVAK, A. & A. S. TU. 1980. Cell culture tumor promotion experiments with saccharin, phorbol myristate acetate and several common food materials. Cancer Lett. **10:** 27–32.

61. DONOVAN, P. J. & J. A. DiPAOLO. 1974. Caffeine enhancement of chemical carcinogen–induced transformation of cultured Syrian hamster cells. Cancer Res. **34:** 2720–2727.

62. DiPAOLO, J. A., P. J. DONOVAN & B. C. CASTO. 1974. Enhancement by alkylating agents of chemical carcinogen transformation of hamster cells in culture. Chem. Biol. Interact. **9:** 351–364.

63. NESNOW, S. & C. HEIDELBERGER. 1976. The effect of modifiers of microsomal enzymes on chemical oncogenesis in cultures of C3H mouse cell lines. Cancer Res. **36:** 1801–1808.

64. POILEY, J. A., R. RAINERI & R. J. PIENTA. 1979. Two-stage malignant transformation of hamster embryo cells. Br. J. Cancer **39:** 8–14.

65. RIVEDAL, E. & T. SANNER. 1980. Potentiating effect of cigarette smoke extract on morphological transformation of hamster embryo cells by benzo[*a*]pyrene. Cancer Lett. **10:** 193–198.

66. RIVEDAL, E., J. HEMSTAD & T. SANNER. 1980. Synergistic effects of cigarette smoke extracts, benzo[*a*]pyrene and nickel sulphate on morphological transformation of hamster embryo cells. *In* Mechanisms of Toxicity and Hazard Evaluation. B. Homstedt, R. Lauwerys, M. Mercier & M. Roberfroid, Eds.: 259–263. Elsevier/North-Holland Biomedical Press. Amsterdam, the Netherlands.
67. RIVEDAL, E. & T. SANNER. 1980. Synergistic effect on morphological transformation of hamster embryo cells by nickel sulphate and benz[*a*]pyrene. Cancer Lett. **8:** 203–208.
68. RIVEDAL, E. & T. SANNER. 1981. Metal salts as promoters of *in vitro* morphological transformation of hamster embryo cells initiated by benzo[*a*]pyrene. Cancer Res. **41:** 2950–2953.

EPITHELIAL *IN VITRO* CELL SYSTEMS
IN CARCINOGENESIS STUDIES*

Carmia Borek

Radiological Research Laboratory
Departments of Radiology and Pathology
College of Physicians & Surgeons
Columbia University
New York, New York 10032

INTRODUCTION

The development of epithelial cells systems to study oncogenic transformation has presented a major challenge in the field of carcinogenesis.[1,2] Fibroblast cell systems have served well in qualitative and quantitative studies of radiogenic and chemically induced toxicity and transformation.[3,4]

In both diploid fibroblast cell strains and heteroploid fibroblast cell lines, transformation can be recognized phenotypically by morphological changes in the transformants that are not displayed by the normal cells. These changes appear within 10 days or 4–6 weeks, depending on the cell system used.[3–4] Thus, these cultures (e.g., hamster embryo, 3T3, or C3H 10T 1/2)[3,4] have been most effective tools for the pragmatic evaluation of the carcinogenic potential of environmental toxic agents[5] and for investigating underlying mechanisms in neoplastic development and its modulation by promoters and inhibitors.[6]

However, because there exists in man a preponderance of carcinomas over sarcomas, the importance of studying oncogenic transformation in epithelial cells is of great relevance to human disease. The difficulty lies in the fact that different tissues contain epithelial cells with singular differentiated characteristics, which must be defined to assert the differentiated nature of the cells being used.

These differentiated characteristics may disappear in time with progressive culture of the cells, thus rendering the cells still useful for transformation studies yet allowing the questioning of the mind as to how they represent the original tissue from which they were derived.[1,2]

Liver cells in culture are a case in point.[7–22] By careful maintenance, however, and optimal culture conditions, one can maintain many of the differentiated characteristics of the cells for prolonged periods of time[1,8,9] (FIGURE 1).

EPITHELIAL CELL TRANSFORMATION

Cell transformation has been studied in liver cells,[7–22] in epidermal cells,[23,24] in cells from salivary glands,[25] and in tracheal cells,[26] bladder,[27] and mammary epithelium.[28]

*This investigation was supported by Grant No. CA 12536-11 to the Radiological Research Laboratory, Department of Radiology, awarded by the National Institutes of Health, Department of Health and Human Services.

Liver Cells

Epithelial cells derived from liver have the longest history in transformation studies,[1,7–22] while primary nonreplicating[19] cultures have been used to study metabolic and enzymatic processes that occur following cellular exposure to specific chemicals (for review, see Reference 1). The established replicating cell lines[19] have been used in a variety of qualitative and quantitative transformation studies.[7–22] The study of specific markers in liver cell transformation has indicated that markers differ from those in fibroblasts[3,21] and that a certain amount of inconsistency of findings exists which may vary from one laboratory to another.[1,8–20] The uniformity found in fibroblast cultures does not hold true for epithelial cell cultures, even when derived from the same tissue source. Strain, species, age, modes of culture, as well as conditions of culture all play a role in determining the quality of the cells on hand. Morphological changes and

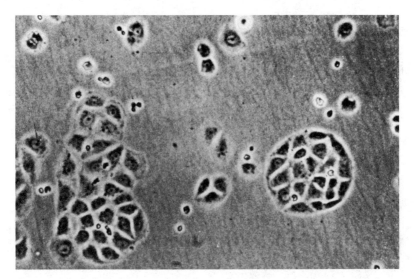

FIGURE 1. Normal liver cells in clonal culture.[11]

appearance of foci in culture are sometimes observed,[11] but in other systems cells remain flat even in the transformed state[12,20] (FIGURES 2 and 3 and TABLE 1).

Differences in the expression of phenotypic characteristics, especially those that are membrane associated (topography, enzymes),[20,21] may be cell-cycle related.

In applying liver cells to quantitative analysis of toxicity and transformation, it is important to assess markers that would most closely be associated with the malignant potential of the cells *in vivo* and thus reflect this potential when assayed *in vitro* following exposure to the carcinogen. Such markers have been studied in detail and reported by San *et al.*[22] A comparison of normal, transformed, and tumorigenic cells with bona fide liver carcinoma cells has indicated that of 11 cell and population markers, colony efficiency in agar was the most reliable in reflecting a progressive neoplastic state.[22] These studies illustrated the difficulty with epithelial cell transformation markers, since it was evident that there exists an independent acquisition of transformation properties that had different temporal expression.[9,22]

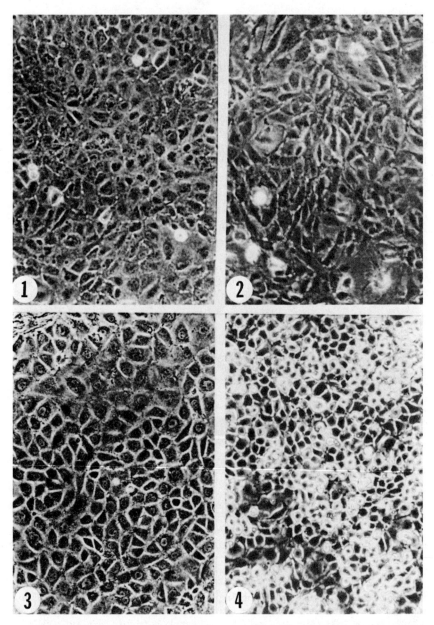

FIGURE 2. (1) Normal liver cells in confluent cultures.[12] (2) Transformed liver cells in culture. Note some pleomorphism, but cells are flat. (3) A normal liver cell line.[11] (4) The same line transformed *in vitro*.[11] Note piling up of cells.

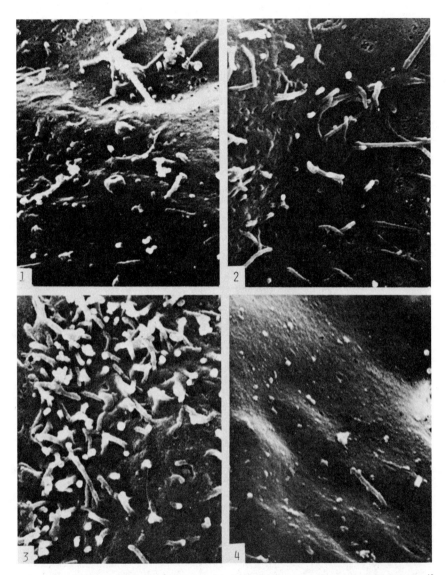

FIGURE 3. High resolution (70 Å) of normal and chemically transformed epithelial liver cells.[12] (1) Normal TRL cells in confluency. Note the sparse number of microvilli. Scanning election microscopy (SEM), × 10,080. (2) TRL cells transformed *in vitro* by 2 aminoanthroquinone. Note the moderate number of surface microvilli. SEM, × 11,340. (3) TRL cells transformed *in vitro* by nitrosomethyl urea. Note the large number of microvilli here, a marked difference from normal. SEM, × 10,080. (4) TRL cells transformed *in vitro* by aflatoxin B. Note the relatively smooth surface. SEM, × 11,340. All transformed cells induce carcinomas.[12,34] (Reproduced from Reference 20 with permission.)

Epidermal Cells

Besides liver cells, a relatively large amount of attention has focused on transformation of epidermal cells in culture (References 23, 24, and 29; reviewed in Reference 30). Tumor induction in skin is of historical importance in chemical carcinogenesis, and the development of cell cultures enabled an approach to epidermal cell transformation at a cellular level. Epidermal cells in culture and differentiated liver cells in culture present different problems, since epidermal cells are differentiating continuously and their proliferation and differentiation are highly dependent on calcium.[23,24,29] At low calcium (0.05–0.1 mM), epidermal cells stratify and have a limited growth potential.

TABLE 1

CHARACTERISTICS OF RODENT LIVER EPITHELIAL CELLS AND FIBROBLASTS MALIGNANTLY
TRANSFORMED *IN VITRO* AS COMPARED TO THEIR NORMAL PARENTAL CELLS*

Property	Epithelial Liver Cells
Morphology (light microscopy)	Often not dramatically different from normal, somewhat more pleomorphic in some cases.
Topography (scanning electron microscope)	Inconsistent changes. Sometimes an increase in microvilli.
Cell density	Inconsistent. Depending on the cell line and the tissue of origin. In some cell lines piling up of cells, in others maintenance of monolayer.
Serum requirement for growth	Low as in the normal (in liver cell). Has not been sufficiently studied in a variety of systems.
Growth in low calcium medium	Yes
Altered cell surface glycoproteins and glycolipids	Yes
Agglutinability by low concentrations of lectins	Yes
Altered Na/K-ATPase	In some lines.
Increased protease production	Inconsistent
Changes in cytoskeleton	Inconsistent
Growth in agar	Yes
Tumorigenicity	Yes

*For additional markers see References 9 and 22.

Work by Yuspa *et al.* has indicated that exposure of epidermal cells to chemical carcinogens results in an alteration in the regulation of maturation of a basal cell and its progeny.[31] The initiated cells proliferate under conditions where normal cells cease to proliferate, namely, in the presence of a calcium level of 1.4 mM. However, the initiated cells retain their capability to differentiate.[31] Altered calcium level has been reported following retroviral transformation[32] and can serve as a selective method to identify transformed epidermal cells, but the system has not been developed sufficiently for reproducible quantitative assessments of chemically induced transformation.

HUMAN CELLS

Epidermal Cells

The field of transformation of human epithelial cells is limited. So far, only epidermal cells have been transformed *in vitro* by a variety of chemical carcinogens and by ultraviolet light.[33] The cells were derived from foreskin and exhibited a loss of anchorage independence following progressive growth *in vitro* after treatment.[33] Their neoplastic potential was assayed in organ cultures of chick embryo skin, rather than in nude mice. The system is promising, though quantitative assays in human cell transformation have proved to be complex even in systems of human fibroblasts.[3]

Parenchymal Cells

Recently an *in vitro* organ culture has been developed of human pancreas.[5] The organ cultures, which exhibited differentiated qualities, responded to a variety of nitroso compounds with altered growth kinetics and subsequent tumorigenicity in animals. The events were rapid, with hyperplasia showing within three weeks and tumorigenicity within one month.[35]

This system may prove useful in some aspects of qualitative carcinogenesis. The system must be further developed for quantitative studies since the complex nature of an organ culture system renders it unsuitable to assess quantitatively at a cellular level dose-response relationships in carcinogenesis.

CONCLUSIONS

In recent years there has been an effort to develop epithelial cell systems for studying transformation. Of these, cells derived from liver have been the most explored and the most useful for quantitative studies on the toxicology and carcinogenicity of specific chemicals.[1,8–22] It will take time and effort to develop other epithelial systems for studying carcinogenesis. The differences among tissues in their patterns of differentiation call for various and specific kinds of defined culture media that would enable optimal conditions for cellular growth and the differentiated nature of the cells. Because of the importance in assessing toxicology and carcinogenesis in particular target tissues that correspond to those affected *in vivo,* as observed in animals and in epidemiological studies, such efforts are well worthwhile.

REFERENCES

1. BOREK, C. & G. M. WILLIAMS, Eds. 1980. Differentiation and Carcinogenesis in Liver Cell Cultures. Ann. N.Y. Acad. Sci. **349.**
2. HARRIS, C. C. & H. ANTRUP, Eds. Human Carcinogenesis. Academic Press. New York, N.Y. (In press.)
3. BOREK, C. 1982. Adv. Cancer Res. **37:** 159–232.
4. HEIDELBERGER, C. 1973. Adv. Cancer Res. **18:** 317–366.
5. MISHRA, N., V. DUNKEL & M. M. MEHLMAN, Eds. 1980. Advances in Modern Environmental Toxicology I. Mammalian Cell Transformation by Chemical Carcinogens. Senate Press, Inc. Princeton, N.J.

290   Annals New York Academy of Sciences

6. BOREK, C. 1982. *In* M. D. Anderson Symposium on Molecular Interrelationships of Nutrition and Cancer. M. Arnott *et al.*, Eds.: 337–350. Raven Press. New York, N.Y.
7. WILLIAMS, G. M. 1981. *In* Chemical Indices and Mechanisms. S. Brown & D. Davies, Eds.: 131–145. Pergamon Press. New York, N.Y.
8. BOREK, C. 1980. *In* Advances in Modern Environmental Toxicology I. N. Mishra, V. Dunkel & M. M. Mehlman, Eds.: 297–319. Senate Press, Inc. Princeton, N.J.
9. WILLIAMS, G. M. 1981. Cancer Forum **6:** 120–127. (Australian Cancer Society.)
10. KATSUTA, H. & T. TAKAOKA. 1968. *In* Cancer Cells in Culture. H. Katsuta, Ed.: 321–334. Tokyo Press. Tokyo, Japan.
11. BOREK, C. 1972. Proc. Nat. Acad. Sci. USA **69:** 956–959.
12. WILLIAMS, G. M., J. M. ELLIOTT & J. J. WEISBURGER. 1973. Cancer Res. **33:** 606–612.
13. MONTESANO, R., L. SAINT VINCENTE & L. TOMATIS. 1973. Br. J. Cancer **28:** 214–220.
14. DIAMOND, L., R. MCFALL, Y. TASHIRO & D. SABATINI. 1973. Cancer Res. **33:** 2627–2636.
15. BORENFREUND, E., P. J. HIGGINS, M. STEINGLASS & A. BENDICH. 1975. J. Nat. Cancer Inst. **55:** 375–384.
16. SCHAEFFER, W. C. & N. H. HEINTZ. 1978. In Vitro **19:** 418–427.
17. IYPE, P. I. & T. D. ALLEN. 1979. Cancer Lett. **6:** 27–32.
18. WEINSTEIN, I. B., J. M. ORENSTEIN, R. GEBERT, M. G. KAIGHN & U. C. STADLER. 1980. Cancer Res. **35:** 253–263.
19. WILLIAMS, G. M. 1976. Methods Cell Biol. **14:** 295–298.
20. BOREK, C. 1979. J. Supramol. Struct. Cell Biochem. **12:** 295–298.
21. BOREK, C. & D. L. GUERNSEY. 1981. Stud. Biophys. **84:** 53–54.
22. SAN, R. G. H., T. SHIMADA, C. T. MASLANSKY, D. N. M. KREISER, M. F. LAPSIA, J. J. RICE & G. M. WILLIAMS. 1979. Cancer Res. **39:** 4441–4448.
23. ELIAS, P. M., S. H. YUSPA, M. GUILLIO, D. L. MORGAN, R. R. BATES & M. A. LUTAZER. 1974. J. Invest. Dermatol. **62:** 569–581.
24. FUSENIG, N. E. & P. K. M. WORST. 1975. Exp. Cell Res. **93:** 443–457.
25. KNOWLES, M. A. & L. M. FRANKS. 1977. Cancer Res. **37:** 3917–3928.
26. STEELE, V., A. MARCHOK & P. NETTESHEIM. 1979. Cancer Res. **39:** 3805–3811.
27. KAHAN, B. D., L. P. RUTZKY, R. OYASU, F. WEISMAN & S. LEGRUE. 1977. Cancer Res. **37:** 2866–2871.
28. RICHARDS, J. & S. NANDI. 1978. Proc. Nat. Acad. Sci. USA **75:** 3836–3840.
29. COLBURN, N. H., W. F. VORDEN BUEGGE, J. K. BATES, R. H. GRAY, J. D. ROSSEN, W. H. KELSEY, & T. SHIMADA. 1978. Cancer Res. **38:** 624–634.
30. YUSPA, S. H., U. LICHTI, D. MORGAN & H. HENNINGS. 1980. *In* Biochemistry of Normal and Abnormal Epidermal Differentiation. Bernstein & Mimideiji, Eds.: 172–188. Tokyo University Press. Tokyo, Japan.
31. YUSPA, S. H., H. H. HENNINGS, & U. LICHT. 1981. J. Supramol. Struct. Cell. Biochem. **17:** 245–247.
32. WEISSMAN, B. E. & S. A. AARONSON. 1983. Cell: 599–606.
33. MILO, G. E., I. NOYES, J. DONAHOE & S. WEISBRODE. 1981. Cancer Res. **41:** 5096–5012.
34. WEISBURGER, E. K. Personal communication.

ROLE OF GENE AND CHROMOSOMAL MUTATIONS
IN CELL TRANSFORMATION

J. Carl Barrett, David G. Thomassen, and Thomas W. Hesterberg

Environmental Carcinogenesis Group
Laboratory of Pulmonary Function and Toxicology
National Institute of Environmental Health Sciences
Research Triangle Park, North Carolina 27709

INTRODUCTION

There is a considerable body of evidence that is consistent with a role of mutagenesis in carcinogenesis;[1,2] however, no definitive proof or disproof of this relationship exists (see References 1 and 2 for a complete discussion of these points). To improve our ability to assess the carcinogenic risk of chemicals and to gain an understanding of the mechanisms of carcinogenesis, it is important to determine the relationship between mutagenesis and carcinogenesis.

There are three major difficulties that must be addressed in attempting to define the role of mutagenesis in neoplastic transformation. (1) Carcinogens induce a wide variety of heritable genetic changes (for example, gene and chromosomal mutations) in cells, and it is necessary to identify the types of mutations critical in carcinogenesis. (2) The relationship between the mutagenic and carcinogenic potential of a chemical should be determined in a system in which both end points can be measured concomitantly. Otherwise, exceptions to the correlation between mutagenesis and carcinogenesis may arise from differences in metabolic activation, DNA binding, or DNA repair in different cells. (3) Neoplastic transformation of cells is not a one-step process; therefore a simple relationship between mutation and transformation is not likely. Each step of the multistep process of cell transformation must be examined for the role of mutagenesis in that particular step.

In this paper we will address these difficulties in order to provide an approach for studying the relationship between mutagenesis and carcinogenesis. In addition, we will discuss new information on the relationship itself based on our recent studies, which suggest that a chromosomal mutation, rather than a gene mutation, is involved in the induction of cell transformation by some, and perhaps many, carcinogens.

MUTATION ASSAYS AND CARCINOGEN TESTING

Most cellular systems for carcinogen testing are predicated on the somatic-mutation theory of carcinogenesis. Numerous mutagenesis assays, particularly those detecting gene mutations, have been developed over the past several years that have proved to be very useful in predicting the carcinogenic activity of chemicals.[3] However, despite their usefulness these assays fail to measure accurately, either qualitatively or quantitatively, the carcinogenic potential of certain chemicals.[1-3] In particular, these assays do not detect several known human carcinogens. A better understanding of the role of mutagenesis in carcinogenesis may help to explain the failures of gene-mutation assays for identifying carcinogens and may facilitate the development of new assays.

One of the complexities in understanding the relationship between mutagenesis and carcinogenesis is the diversity of genetic changes that can be induced in a cell treated with a carcinogen. These changes include mutations at the gene as well as the chromosomal level. Theoretically, any one of these induced genetic changes may contribute to the heritable conversion of a normal cell to the malignant state. Even if

291

0077–8923/83/0407–0291 $01.75/0 © 1983, NYAS

one assumes that the mutational theory of carcinogenesis is correct, the nature of the critical mutational event for any carcinogen is unknown and it is possible that different carcinogens act via different types of genetic changes. Since gene-mutation assays usually detect only a single or limited spectrum of all genetic changes and the nature of the cellular changes responsible for carcinogenesis is unknown, the lack of activity of certain carcinogens in these assays may not be surprising. Therefore, it is important to identify different types of genetic changes and to distinguish those changes from epigenetic ones.

For the purpose of this chapter we will use the following definitions. A *genetic change* is any alteration in primary DNA structure that results in a change in phenotype.[4,5] This change may be a single base change, a deletion or an insertion of one or more bases, a rearrangement in DNA structure, or a change in the number of genes or entire chromosomes.[4,5] An *epigenetic change* is any change in a phenotype that does not result from an alteration in primary DNA structure.[4,5]

We will define a *mutation* as a heritable genetic change and use these two terms interchangeably. It is important to note that epigenetic changes may also be stable and heritable (for example, the differentiated phenotype of cells).

Genetic changes can be classified into four major types (TABLE 1) including gene mutations, gene duplication or amplification, chromosome aberrations, and aneuploidy. Most mutation assays with prokaryotic or eukaryotic organisms use phenotypic

TABLE 1

TYPES OF GENETIC CHANGES

1. Gene mutations (point mutations, frameshift mutations, deletion and insertion mutations)
2. Gene duplication or amplification (increased copies of a gene)
3. Chromosome aberration (translocations, dicentrics, inversions, and terminal deletions)
4. Aneuploidy (abnormal numbers of chromosomes)

markers that detect primarily gene mutations. Therefore, most comparisons between the mutagenicity and carcinogenicity of chemicals provide information only on the relationship between carcinogenicity and the induction of gene mutations. These systems fail to detect chemicals that induce changes in the number of copies of a gene (or genes), either by gene amplification or by aneuploidy.[2] The importance of gene-dosage mutations in carcinogenesis will be discussed in a later section.

Chromosomal aberrations are detected by some mutation assay systems.[2] However, since chromosomal mutations compared to gene mutations are more likely to result in the loss or inactivation of essential genes,[6] the efficiency of detection of chromosomal mutations relative to gene mutations is not known. A greater understanding of the induction and detection of chromosomal mutations is important for carcinogenesis studies since the findings of nonrandom chromosome rearrangements or deletions in certain tumors[7,32] provide evidence for a role of chromosomal mutations in carcinogenesis.[8,9]

USE OF CELL-TRANSFORMATION SYSTEMS FOR THE CONCOMITANT STUDY
OF MUTAGENESIS AND CARCINOGENESIS

The correlation between mutagenicity and carcinogenicity of a number of chemicals, which provides the major experimental support for the somatic-mutation theory, has been demonstrated primarily using bacterial mutagenicity assays.[14] While this correlation can be nearly 90% in some studies,[13,14] chemicals have been reported that are carcinogenic but not mutagenic, or mutagenic but not carcinogenic.[1,2] These exceptions to this correlation may provide insight into the mechanisms of neoplastic transformation and the relationship of this process to somatic mutation. On the other hand, these exceptions may not be real but simply result from the use of different assay systems to measure each process (i.e., mutagenicity in bacteria versus carcinogenicity in animals). A correlation between mutagenicity and carcinogenicity might be observed with all chemicals if both processes were measured in the same target cell. The use of mammalian cells in culture allows the concomitant study of somatic mutation and neoplastic transformation measured in the same target cell.[1,13] During the past five years several laboratories have completed such studies, and we have discussed these studies in detail in a recent review.[13]

Our laboratory has studied the relationship between carcinogenesis and mutagenesis in Syrian hamster embryo fibroblasts for several years.[15–17] Recently, we have also developed a quantitative cell-transformation system with rat tracheal epithelial cells, which is also amenable to such studies.[18]

The first unequivocal demonstration of chemical carcinogen-induced neoplastic transformation of cells in culture was by Berwald and Sachs.[19] They demonstrated that Syrian hamster embryo cells treated with carcinogens give rise to morphologically transformed cells. These cells are not neoplastic but have an increased propensity to become neoplastic following further growth in culture, and hence are termed preneoplastic cells.[11,12] Berwald and Sachs were able to quantitate the number of morphologically transformed colonies that resulted one to eight days following exposure of cells to a carcinogen. Studies by DiPaolo and colleagues and Pienta and coworkers demonstrated that morphological transformation of Syrian hamster embryo (SHE) cells occurs following treatment with a variety of chemical carcinogens, but not after treatment with structurally related noncarcinogens. In fact, this is now one of the most sensitive and selective short-term tests for carcinogens.[1,13]

In addition to transformation, mutagenesis studies can be performed with SHE cells[1,13,15] at two genetic loci. Dominant mutations of the Na^+/K^+-ATPase gene are measured by resistance to ouabain (Oua^R) and X-linked, recessive mutations of the hypoxanthine phosphoribosyl transferase (HPRT) locus are measured by resistance to the purine analogue 6-thioguanine ($6TG^r$). Thus the frequency of induction of mutations at these two genetic loci can be directly compared with the frequency of induction of morphological transformation with the same target cells.[13,15]

CARCINOGENS INDUCE EARLY, PRENEOPLASTIC CHANGES IN CELLS IN CULTURE AT A HIGH FREQUENCY

When Syrian hamster embryo fibroblasts are treated with N-methyl-N'-nitro-N-nitrosoguanidine (MNNG), a direct-acting mutagen and carcinogen, a high frequency (>1%) of morphologically transformed cells is observed seven days later.[15] The frequency of this preneoplastic change is ~100 times greater than the frequency of gene mutations at either the HPRT or the Na^+/K^+-ATPase locus (TABLE 2). Similar results are observed following treatment with other carcinogens, such as benzo[a]pyrene (BaP).[13,15] We have also determined that early, preneoplastic changes of rat tracheal epithelial cells occur with a high frequency (>1%) following treatment with MNNG or other carcinogens.[18,20]

The frequency of cell transformation in these studies is significantly higher than frequencies of gene mutations commonly observed in mammalian cells in culture.

These comparisons, although theoretically very attractive, can be misleading. Different loci can have quite different frequencies of induced mutation depending on the fraction of all changes within those genes that actually result in detectable phenotypic changes. For example, in human fibroblasts the maximum reported frequency of gene mutation at the adenine phosphoribosyl transferase locus (APRT)[26] is 2×10^{-2}. This value approximates the high frequencies of carcinogen-induced preneoplastic transformation of Syrian hamster embryo and rat tracheal epithelial cells reported above.

The relationship between early preneoplastic changes in cells in culture and early carcinogen-induced events *in vivo* is unknown, because quantitation of the frequency of these events *in vivo* is lacking. However, recent experiments by Nomura have demonstrated a high frequency (up to 30%) of tumors in offspring following parental exposure to a variety of carcinogens.[21] Nomura's germ-cell "mutations," which predispose progeny to tumorigenesis, are reportedly similar to other accepted germ-line mutations in terms of inducibility by mutagens, heritability, pattern of stage-dependent sensitivity of the germ cells, and doubling dose following x-ray treatment. In

TABLE 2

COMPARISON OF ABILITY OF CHEMICALS TO INDUCE CELL TRANSFORMATION,
GENE MUTATION, AND CHROMOSOME MUTATION

Compound	Cell Trans- formation	Gene Muta- tion	Chromosome Aberrations or Aneuploidy	Cell- Transformation Frequency/ Gene-Mutation Frequency (TG^r or Oua^r)	Cell- Transformation Frequency/ Chromosomal Mutation Frequency
Benzo[a]- pyrene	+	+	+	$\sim 10^2$	1/10 to 1/20
2-Aminopurine	weak	weak	weak	insufficient data	insufficient data
6-N-hydroxyl aminopurine	+	+	+	3–12.5	1/10 to 1/20
Diethylstilbestrol	+	−	+	$>10^4$	1/10 to 1/20
Asbestos	+	−	+	$>10^4$	1/10 to 1/20

addition, germ-line "tumor" mutations occurred much more frequently than other germ-line mutations in the same animals—for example, about 10 times more frequently than dominant skeletal mutations. The apparently high frequency of induction of these germ-line "tumor" mutations *in vivo* suggests a possible relationship between the high frequency of carcinogen-induced changes in cells in culture and the high frequency of neoplasia-related changes induced *in vivo*.[21]

At least four explanations can be offered for the high frequency of inducible preneoplastic transformation of fibroblasts and epithelial cells in culture, of germ-line "tumor" mutations *in vivo*, and of some gene mutations in cells in culture.[2,18] The high frequency may be due to (1) a large number or size of the target for these phenotypes, (2) mutational "hot spots" in the gene (or genes) controlling these phenotypes, (3) a mutational basis other than gene mutation for these phenotypic alterations, for example, chromosomal mutations, or (4) a nonmutational, epigenetic alteration in gene expression. Measurement of only the frequency of cell transformation and mutation cannot distinguish between these possibilities.

EXCEPTIONS TO THE CORRELATION BETWEEN MUTAGENESIS AND CARCINOGENESIS

Exceptions to the correlation between the mutagenicity and carcinogenicity of some chemicals have been reported.[2] Further study of these exceptions might provide insight into the role of mutagenesis in carcinogenesis. However, as previously noted, these exceptions may not be real but simply result from the use of different assay systems (bacterial versus mammalian) to measure each end point. We have studied representative examples of possible mutagens that are not carcinogens and carcinogens that are not mutagens for their mutagenic and carcinogenic potential in Syrian hamster embryo fibroblasts in culture to examine critically the relationship between mutagenesis and carcinogenesis. The compounds studied were 2-aminopurine,[16] a nucleic acid base analogue, which is a classical mutagen in prokaryotic systems but not a carcinogen in animal studies; diethylstilbestrol,[17] a synthetic estrogen; and asbestos,[22] a mineral fiber. The latter two are known human carcinogens but are inactive as mutagens in most systems.

2-AMINOPURINE AND OTHER MODIFIED PURINES

2-Aminopurine (2-AP) is a well-defined mutagen in bacteriophages and bacteria, inducing base-pair changes in DNA. However, 2-AP is inactive as a carcinogen in rats following weekly subcutaneous injections and in mice following skin painting plus promotion with croton oil. Although 2-AP is a very active mutagen in prokaryotic systems, it was only a weak mutagen in Syrian hamster embryo cells. 2-AP is a mutagen in other eukaryotic organisms, although not potent, and is not effective in some organisms.[16]

Another modified purine, 6-*N*-hydroxylaminopurine (6-HAP), is like 2-AP in that it induces base-pair changes in prokaryotic and eukaryotic cells.[16] Although 6-HAP is less potent than 2-AP in inducing rII mutants of phage T_4, 6-HAP is much more potent than 2-AP in inducing ad-3 mutants in *Neurospora crassa*. Also, 6-HAP is a weak carcinogen in rats under conditions where 2-AP is not a carcinogen. These findings prompted us to examine the effects of 6-HAP in Syrian hamster embryo cells. 6-HAP is a potent mutagen in these cells, and the relative mutagenicity of 6-HAP and 2-AP in Syrian hamster embryo cells is consistent with their mutagenicity in *Neurospora*.

Since 2-AP is mutagenic in bacteria but not carcinogenic in animals, it is of interest to compare the ability of 2-AP as well as 6-HAP to induce somatic mutation and transformation in hamster embryo cells. 2-AP failed to induce significant neoplastic transformation of Syrian hamster embryo cells in our studies. The results of this study indicate that 2-AP is, at best, a weak transforming agent. The inability to induce tumors with 2-AP in two animal species is consistent with these observations. In contrast to the results with 2-AP, 6-HAP induced morphological transformation of cells in a dose-dependent manner. This correlates with the greater mutagenic activity of 6-HAP than 2-AP in these cells. These experiments indicate that the mutagenicity of modified purines correlates with their ability to induce transformation when these end points are measured in the same cellular system. Thus, this class of compounds may not be an exception to the correlation of mutagenesis and carcinogenesis.[16]

There are, however, quantitative differences in the relative ability of 6-HAP to induce transformation and mutation when compared to BaP or MNNG.[16] In contrast to the ratio of approximately 100 in the transformation-to-mutation frequencies observed with the latter carcinogens, the ratio following 6-HAP treatment was only 3–12 (TABLE 2). Although 6-HAP can induce both transformation and mutation, it is much less potent than other carcinogens in inducing transformation when compared at equal levels of mutagenicity. This suggests a difference in the two processes.[16]

Diethylstilbestrol (DES) is known to be a carcinogen in humans and rodents. However, DES and its metabolites are inactive as mutagens in the Ames test, even with a variety of metabolic activating systems. In collaboration with John McLachlan and Annette Wong of our institute, we demonstrated that DES has the ability to induce neoplastic transformation of Syrian hamster embryo fibroblasts in culture.[17,23] While the proliferation of estrogen-sensitive tissue may play an important part in the promotional phase of DES carcinogenesis, our results indicate that DES has the ability to induce cell transformation directly in culture in the absence of any measurable induction of cell proliferation.[23] Furthermore, the ability of DES and related analogues to transform cells in culture does not correlate with the reported estrogenic potency of the compounds *in vivo*, but does correlate with the ability of these compounds to be metabolized by peroxidase-mediated oxidation.[23] These results indicate that DES has a cell-transforming activity that is distinct from its hormonal activity.

Under the same conditions that result in cell transformation, DES failed to induce measurable gene mutation at either of two loci (HPRT or Na^+/K^+-ATPase) (TABLE 2). This is the first example of a definitive dissociation of gene mutation and cell transformation measured in the same cellular system.[17] Our results with DES may suggest an epigenetic basis for morphological transformation; however, we have proposed an alternative mechanism based on the ability of DES to induce numerical chromosomal mutations. In collaboration with T. Tsutsui and H. Maizumi of Nippon Dental University, Tokyo, we found that concentrations of DES that induce morphological transformation also induce a high level of aneuploidy (up to 20%), but not chromosome aberrations, in the same cells[24] (TABLE 2). More specifically, DES increased the number of cells with an aneuploid chromosome number that was ±2 chromosomes from the diploid chromosome number. Moreover, the neoplastic cell lines induced by DES also have a near diploid, aneuploid modal chromosome number.[24] An aneuploid DNA content is also observed in premalignant vaginal lesions of women exposed to DES prenatally.[25] When synchronized cell cultures are treated with DES during different phases of the cell cycle, cells treated during mitosis are the most sensitive to both cell transformation and aneuploidy induction.[24] Thus, we have postulated that DES induces nondisjunction due to its colchicinelike properties and that this property is important in its ability to induce cell transformation and cancer.[2,24]

Asbestos is another important human carcinogen that has been assumed to be "epigenetic" in its action because it has not been found to be mutagenic in most of the systems in which it has been tested.[3] Consequently, some have postulated that asbestos acts only by promoting the action of initiating agents such as benzo[a]pyrene and other components of cigarette smoke. Although epidemiological studies indicate that asbestos acts synergistically with cigarette smoke in the induction of bronchogenic carcinoma, asbestos workers who do not smoke also have an increased incidence of this form of cancer.[27] Furthermore, asbestos and other mineral fibers can induce tumors in animals.[28] Thus, it is important to determine whether or not asbestos can directly induce the neoplastic transformation of cells. Our recent experiments indicate that exposure of cultures of Syrian hamster embryo cells to asbestos results in a dose-dependent induction of neoplastic transformation,[22] which confirms a preliminary study by DiPaolo *et al.*[33]

The SHE cell-transformation system appears to model *in vivo* induction of mesotheliomas, since fiber dimension rather than chemical composition is the most important determinant of their neoplastic potency.[28] For instance, we have shown in the SHE cell system that long, thin fibers are more effective than short, thick fibers in transforming cells. Similarly, Stanton's group has reported that long, thin fibers, regardless of chemical composition, are more effective than short, thick fibers in the induction of mesotheliomas when injected into the pleural cavity of rats.[28]

Under conditions that resulted in morphological transformation of SHE cells, we have shown that asbestos failed to induce any detectable gene mutations at the HPRT or Na^+/K^+-ATPase locus (TABLE 2). Since asbestos is not mutagenic in SHE cells, we undertook studies to elucidate the mechanism by which asbestos induces neoplastic transformation in these cells. We have shown that asbestos is taken up by SHE cells 2 hours after exposure and that by 24 hours, many of the asbestos fibers are located in the perinuclear region of the cytoplasm, lined up tangentially around the nucleus, with few or no fibers being found in the rest of the cytoplasm or in the nucleus. Furthermore, cytogenetic studies have revealed that by 24 hours after asbestos treatment, some metaphases of SHE cells are aneuploid or polyploid and exhibit chromosomal aberrations (TABLE 2). Thus, although asbestos does not appear to cause gene mutations, it does indeed induce chromosomal mutations. Based on these findings we propose that asbestos, which can gain access to the perinuclear region of cells, causes transformation by binding to microtubules or other cytoskeletal proteins that are important in the disjunction of chromosomes during mitosis, thereby resulting in aneuploidy and polyploidy. We have previously discussed the proposed role of numerical chromosome changes in cell transformation and the development of cancer.[2,24]

It appears, then, that asbestos and other mineral fibers can be direct-acting carcinogens that do not induce gene mutations. This does not mean, however, that their action is "epigenetic." On the contrary, we suggest that they accumulate about the nucleus and cause changes in chromosome morphology and number, i.e., chromosomal mutations. How these changes relate to neoplastic transformation is a question to be addressed in future studies.

ROLE OF CHROMOSOMAL MUTATIONS IN CARCINOGENESIS

From the examples discussed above, our studies indicate that mutagenesis, measured by gene mutations, in some cases can be dissociated from carcinogenesis, measured by a cell-transformation assay, both qualitatively (e.g., DES and asbestos) and quantitatively (e.g., MNNG versus modified purines). On the other hand, a good correlation is observed between the induction of chromosomal changes and cell transformation (TABLE 2). All the chemicals discussed above induce chromosomal mutations, including the modified purines; and at equal transforming doses, quantitatively similar levels of chromosomal mutations are induced. DES induces only aneuploidy, while the other compounds induce both aneuploidy and chromosomal aberrations. It is not possible at this time to determine if one type of chromosomal mutation is more important than another in the induction of cell transformation.

Our hypothesis that a chromosomal-type mutation is involved in the morphological transformation of Syrian hamster embryo cells by some and perhaps most carcinogens provides an explanation for several observations. First, it explains the ability of DES to induce neoplastic transformation without measurable somatic mutations at two loci. The Oua^R and TG^r systems do not detect changes in chromosome number. Therefore, the lack of mutagenicity of DES with these markers indicates that DES fails to induce gene mutations under the conditions that induce neoplastic transformation; but such results are still consistent with morphological transformation resulting from a chromosomal mutation. Second, the frequency of chromosomal lesions is high in carcinogen-treated cells,[2] which might explain the high frequency of morphological transformation. Third, this hypothesis provides an explanation for the variation in the ratio of transformation to gene mutation induced by different classes of carcinogens and mutagens (TABLE 2). Chemicals that induce aneuploidy readily, but not gene mutations, would have a high ratio of transformation to gene mutation (e.g., DES and asbestos). Chemicals that are effective inducers of point mutations would have a low ratio of transformation to gene mutation (e.g., 6-HAP and possibly 2-AP). Interestingly, the ratio of cell-transformation frequency to chromosome-mutation frequency is relatively constant for different classes of chemicals (TABLE 2).

Nonrandom chromosomal changes, both numerical and structural, are observed in a variety of human and rodent tumors. This observation further strengthens a role for these changes in carcinogenesis. The origin of these changes is unknown, but the fact that chemicals produce both chromosomal changes and cancers suggests that these alterations may be the consequence of exposure to a chemical carcinogen. Further studies emphasizing the induction of these changes are needed.

ROLE OF MUTATION IN A MULTISTEP MODEL OF CARCINOGENESIS

There is now considerable evidence that neoplastic development occurs as a progressive process through qualitatively different stages.[1,2] This chapter has discussed the role of gene and chromosomal mutations in the induction of early, preneoplastic morphological changes in Syrian hamster embryo fibroblasts in culture. However, morphologically transformed SHE cells are not tumorigenic.[1,11,12] Rather these cells undergo a series of progressive changes ultimately resulting in neoplastic conversion of the cells. A multistep process such as tumor progression might be comprised of a number of mutational steps, a number of nonmutational steps, or a combination of both mutational and nonmutational steps. Quantitative studies of the induction of changes expressed late in the progression to neoplasia have been made.[29,30] These kinds of studies need to be done using agents that induce different types of mutations to assess the role of gene, chromosomal, and gene-number mutations in the induction of changes expressed late in neoplastic progression.

Permanent cell lines (e.g., C3H 10T 1/2, Balb/c 3T3, BHK) are frequently used for studies on the role of mutation in carcinogenesis. We have previously discussed the idea that permanent cell lines may already have undergone changes making them preneoplastic.[10] While such lines can play an important role in studies of neoplastic transformation and mutation, results obtained with these lines may be providing information on late, rather than early, changes involved in neoplastic transformation and thus the results should be viewed accordingly. For example, DES induces morphological transformation in diploid Syrian hamster embryo cells[17] but not in C3H 10T 1/2 cells.[31] If our hypothesis that the transformation of Syrian hamster embryo cells by DES is due to aneuploidy induction is correct, then it is not surprising that DES fails to induce transformation of an already aneuploid cell line. Either the morphological transformation of C3H 10T 1/2 cells is not due to aneuploidy induction or the degree of numerical chromosome changes that are needed in tetraploid cells for transformation to occur is greater than that induced by DES.

SUMMARY

To understand the role of mutagenesis in carcinogenesis fully, we must consider all types of mutations including gene, chromosomal, and gene-number mutations and all changes involved in the progressive development of neoplastic cells. We have found that certain known human carcinogens (i.e., DES and asbestos), which were classified as epigenetic carcinogens based on gene-mutation assays, have mutational activity at the chromosomal level that correlates with their ability to induce cell transformation. This should caution against classification of chemicals as genotoxic or epigenetic[3] without a complete understanding of their mechanism of action. Furthermore, our studies indicate that more than gene-mutation assays is needed for carcinogen testing. In particular, chromosomal changes induced by chemicals, both aberrations and aneuploidy, need to be carefully assessed. In addition, the role of all types of mutation in the overall process of neoplastic transformation needs to be determined. This can only be examined by studying each individual change involved in neoplastic progression. Thus, any attempt to assess a chemical's carcinogenic potential should consider not only all types of mutational changes but both early and late changes involved in neoplastic transformation.

REFERENCES

1. BARRETT, J. C., B. D. CRAWFORD & P. O. P. TS'O. 1981. The role of somatic mutation in a multistage model of carcinogenesis. *In* Mammalian Cell Transformation by Chemical Carcinogens. V. C. Dunkel & R. A. Mishra, Eds.: 467–501. Pathtox Publishing Co. Princeton, N.J.
2. BARRETT, J. C. 1981. Cell transformation, mutation, and cancer. Gann Monogr. Cancer Res. **27**: 195–206.
3. WEISBURGER, J. H. & G. M. WILLIAMS. 1981. Carcinogen testing: current problems and new approaches. Science **214**: 401–407.
4. DEMARS, R. 1974. Resistance of cultured human fibroblasts and other cells to purine and pyrimidine analogues in relation to mutagenesis detection. Mutat. Res. **24**: 335–364.
5. SIMINOVITCH, L. 1976. On the nature of hereditable variation in cultured somatic cells. Cell **7**: 1–11.
6. MOORE, M. M. & D. CLIVE. 1982. The quantitation of $TK^{-/-}$ mutants of L5178Y/$TK^{+/-}$ mouse lymphoma cells at varying times posttreatment. Environ. Mutagen. **4**: 499–519.
7. SANDBERG, A. 1980. The Chromosomes in Human Cancer and Leukemia. Elsevier. New York, N.Y.
8. BOVERI, T. H. 1929. The Origin of Malignant Tumors. Williams & Wilkins. Baltimore, Md.
9. WOLF, U. 1974. Theodor Boveri and his book, "On the Problem of the Origin of Malignant Tumors." *In* Chromosomes and Cancer. J. German, Ed.: 3–20. John Wiley. New York, N.Y.
10. BARRETT, J. C. & D. G. THOMASSEN. 1983. Use of quantitative cell transformation assays in risk estimation. *In* Proceedings of Workshop on Quantitative Estimation of Risk to Human Health from Chemicals. John Wiley. New York, N.Y. (In press.)
11. BARRETT, J. C. & P. O. P. TS'O. 1978. Evidence for the progressive nature of neoplastic transformation *in vitro*. Proc. Nat. Acad. Sci. USA **75**: 3761–3765.
12. BARRETT, J. C. 1980. A preneoplastic stage in the spontaneous neoplastic transformation of Syrian hamster embryo cells in culture. Cancer Res. **40**: 91–94.
13. BARRETT, J. C. & E. ELMORE. 1983. Comparison of carcinogenesis and mutagenesis of mammalian cells in culture. *In* Handbook of Experimental Pharmacology. Mutagenesis and Carcinogenesis. L. S. Andrews, R. J. Lorentzen & W. G. Flamm, Eds. Springer-Verlag. Berlin, Federal Republic of Germany. (In press.)
14. MCCANN, J. & B. N. AMES. 1976. Detection of carcinogens as mutagens in the Salmonella/microsome test assay of 300 chemicals. Discussion. Proc. Nat. Acad. Sci. USA **73**: 950–954.

15. BARRETT, J. C. & P. O. P. TS'O. 1978. The relationship between somatic mutation and neoplastic transformation. Proc. Nat. Acad. Sci. USA **75:** 3297–3301.

16. BARRETT, J. C. 1981. Gene mutation and cell transformation of mammalian cells induced by two modified purines, 2-aminopurine and 6-N-hydroxylaminopurine. Proc. Nat. Acad. Sci. USA **78:** 5685–5689.

17. BARRETT, J. C., A. WONG & J. A. MCLACHLAN. 1981. Diethylstilbestrol induces neoplastic transformation of cells in culture without measurable somatic mutation at two loci. Science **212:** 1402–1404.

18. BARRETT, J. C., T. E. GRAY, M. J. MASS & D. G. THOMASSEN. 1983. A quantitative clonal assay for carcinogen-induced alterations of respiratory epithelial cells in culture. *In* Application of Short-Term Bioassays in the Analysis of Complex Environmental Mixtures. M. Walters & S. Sandhu, Eds. Plenum Press. New York, N.Y. (In press.)

19. BERWALD, Y. & L. SACHS. 1965. *In vitro* transformation of normal cells to tumor cells by carcinogenic hydrocarbons. J. Nat. Cancer Inst. **34:** 641–661.

20. PAI, S. B., V. E. STEELE & P. NETTESHEIM. 1983. Quantitation of early cellular events during neoplastic transformation of tracheal epithelial cell cultures. Carcinogenesis. (In press.)

21. NOMURA, T. 1982. Parental exposure to X-rays and chemicals induces heritable tumors and anomalies in mice. Nature **296:** 575–577.

22. HESTERBERG, T. W., T. CUMMINGS, A. R. BRODY & J. C. BARRETT. 1982. Asbestos induces morphological transformation in Syrian hamster embryo cells in culture. J. Cell Biol. **95:** 449.

23. MCLACHLAN, J. A., A. WONG, G. DEGEN & J. C. BARRETT. 1982. Morphological and neoplastic transformation of Syrian hamster embryo fibroblasts by diethylstilbestrol and its analogs. Cancer Res. **42:** 3040–3045.

24. TSUTSUI, T., H. MAIZUMI, J. A. MCLACHLAN & J. C. BARRETT. Aneuploidy induction and cell transformation by diethylstilbestrol: a possible chromosomal mechanism in carcinogenesis. (Submitted.)

25. FU, Y. S., J. W. REAGAN, R. M. RICHART & D. E. TOWNSEND. 1979. Nuclear DNA and histologic studies of genital lesions in diethylstilbestrol-exposed progeny. I. Intraepithelial squamous abnormalities. Am. J. Clin. Pathol. **72:** 503–514.

26. STEGLICH, C. S. & D. DEMARS. 1982. Mutations causing deficiency of APRT in fibroblasts cultured from humans heterozygous for mutant APRT alleles. Somatic Cell Genet. **8:** 115–141.

27. SELIKOFF, I. J. 1979. Asbestos and smoking. J. Am. Med. Assoc. **242:** 458–459.

28. STANTON, M. F., M. LAYARD, A. TEGERIS, E. MILLER, M. MAY, E. MORGAN & A. SMITH. 1981. Relation of particle dimension to carcinogenicity of amphibole asbestos and other fibrous minerals. J. Nat. Cancer Inst. **67:** 965–975.

29. BOUCK, N. & G. DIMAYORCA. 1982. Chemical carcinogens transform BHK cells by inducing a recessive mutation. Mol. Cell. Biol. **2:** 97–105.

30. THOMASSEN, D. G. & R. DEMARS. 1982. Clonal analysis of the stepwise appearance of anchorage-independence and tumorigenicity in a permanent line (CAK) of mouse cells. Cancer Res. **42:** 4054–4063.

31. LILLEHAUG, J. R. & R. DJURHUUS. 1982. Effect of diethylstilbestrol on the transformable mouse embryo fibroblast C3H/10T 1/2 C18 cells. Tumor promotion, cell growth, DNA synthesis, and ornithine decarboxylase. Carcinogenesis **7:** 797–799.

32. KLEIN, G. 1981. The role of gene dosage and genetic transpositions in carcinogenesis. Nature. **294:** 313–318.

33. DIPAOLO, J., A. J. DEMARINIS & J. DONIGER. 1982. Asbestos and benzo[a]pyrene synergism in the transformation of Syrian hamster embryo cells. J. Environ. Pathol. Toxicol. **5:** 535–542.

EFFECTS OF TUMOR PROMOTERS

Walter Troll

Department of Environmental Medicine
New York University Medical Center
New York, New York 10016

Cocarcinogens, tumor promoters, and hormones contribute to cancers at specific sites. Among human cancers, the promoters for bladder, breast, colon, and prostate cancers have been proposed. The importance of these cancer-enhancing agents cannot be evaluated until we have test systems that measure them. This requires a knowledge of their mechanism of action. This is a difficult task because many observations of enzyme induction and cell responses (the pleiotropic action of tumor promoters) compete for this doubtful honor. The mechanism of tumor promotion is of additional interest so that we can devise measures to counteract or prevent tumor promotion. Our studies dealing with preventive agents have led to the idea that free oxygen radicals are the proximal agent of the tumor promoter 12-*O*-tetradecanoyl-phorbol-13-acetate (TPA).

We noted that in the classical two-stage mouse skin promotion model, antiinflammatory hormones and protease inhibitors prevent the appearance of polymorphonuclear leukocytes (PMNs). In PMNs, protease inhibitors can also prevent the production of superoxide anions (O_2^-) induced by TPA. This work has led to a reinvestigation of the role of oxygen radicals in tumor promotion. A variety of tumor promoters, including the indole alkaloids (teleocidin, lyngbyatoxin, and aplysiatoxin, recently isolated by Fujiki and Sugimura), are active in causing O_2^- formation in PMNs. Known inhibitors of tumor promotion (e.g., protease inhibitors and retinol) block the O_2^- formation of all promoters. Further support for the role of free oxygen radicals in tumor promotion has come from the work of Slaga, who showed that organic peroxides are tumor promoters and that antioxidants inhibit tumor promotion. Moreover, Birnboim has shown that TPA causes single-strand breaks in DNA of PMNs through the production of free oxygen radicals, which he postulated to be the key event in the promotion phase of cancer.

Tumor promoters combine with specific sites (receptors), which in turn are responsible for the multiplicity of the biological responses. We are indebted to the work of Peter Blumberg, who used a less lyophillic derivative of TPA, the ^{3}H-phorbol 12,13 dibutyrate (BUP), to identify and measure a phorbol ester receptor in mouse skin. One of the indole alkaloid promoters, dihydroteleocidin, competes with the binding of BUP, indicating that this chemically unrelated promoter combines with the identical receptor. The receptor binding studies may identify a group of promoters, but there is apparently a larger group of cocarcinogens and promoters that does not use the phorbol receptor. Among these materials are benzoyl peroxide and phenobarbital. Benzoyl peroxide is a promoter for skin carcinogenesis, and phenobarbital is a promoter of acetylaminofluoride-initiated liver cancer. Benzoyl peroxide presents a promoter proximal to TPA. Phenobarbital may specifically give rise to free oxygen radicals in liver. James Trosko has developed an ingenious, all-inclusive test system, which depends on quantitating cellular metabolic cooperation in V79 hamster cells. Tumor promoters inhibit metabolic cooperation, which perhaps permits the initiated cells to grow on their own without restraint.

The steps leading from the tumor promoter to the unsocial behavior of cells may involve specific combination with a cell receptor (as identified by Dr. Blumberg),

301

formation of toxic materials including free oxygen radicals, and damage to metabolic cell cooperation (as shown by Dr. Trosko). Some of the materials that are positive in this test are not recognized by the phorbol receptor assay and may present the toxic materials directly responsible for tumor promotion (e.g., benzoyl peroxide) or the materials that use another receptor system, leading to free oxygen radicals (e.g., phenobarbital).

The development of test systems specific for tumor promoters that are distinct from initiating carcinogens offers opportunities for unraveling multistage carcinogenesis.

PHORBOL ESTER RECEPTORS AND THE *IN VITRO* EFFECTS OF TUMOR PROMOTERS

Peter M. Blumberg, K. Barry Delclos, Joseph A. Dunn,
Susan Jaken,* Karen L. Leach,† and Emily Yeh

Molecular Mechanisms of Tumor Promotion Section
Laboratory of Cellular Carcinogenesis and Tumor Promotion
National Cancer Institute
Bethesda, Maryland 20205

Although the relative contributions of carcinogens and tumor promoters to the frequency of cancer in man is uncertain, a significant role for tumor promoters has been suggested.[1,2] This symposium portrays the impressive progress made in developing *in vitro* systems for detecting carcinogens. As yet, satisfactory, general assays for tumor promoters have not been documented. Development of such assays may need to await a clearer understanding of the mechanisms by which tumor promoters act.

This article will focus on studies on the mechanism of action of the phorbol esters, potent tumor promoters for mouse skin. The current evidence that the actions of the phorbol esters are mediated through phorbol ester receptors will be reviewed. Data at the whole animal, cellular, and biochemical levels of analysis that argue for functional subclasses of receptors will be highlighted. The implication of functional receptor subclasses is that the phorbol esters may possess multiple mechanisms of action and, in consequence, that *in vitro* assays predicated upon a single mechanism of action may be inadequate.

The classic properties of tumor promoters as elucidated in the mouse skin system by Berenblum, Boutwell, and others (see References 3 and 4 for reviews) are summarized in FIGURE 1. By themselves, tumor promoters are not carcinogenic or genotoxic (line 4). However, if applied chronically after a limited dose of carcinogen, the so-called initiator, they cause the appearance of tumors.‡ Tumor promoters are distinguished from cocarcinogens in that the tumor promoters can function if applied long after the initiating agent, e.g., a year in the case of mouse skin promotion (line 5), whereas cocarcinogens enhance tumor yield only if applied at or close to the same time as the carcinogen.

An important feature of tumor promoters is that, unlike initiating agents, their action is relatively reversible. Thus, chronic treatment with the tumor promoter before addition of the initiating agent does not lead to tumors (line 6). Likewise, if sufficient intervals are permitted between applications of the promoter, then tumors do not arise (line 7). Postulated mechanisms of tumor promoter action should be consistent with the *in vivo* behavior of the tumor promoters. As discussed previously,[5] most variants of

*The Johns Hopkins Oncology Center, Baltimore, Md. 21205.
†Supported by Public Health Service Grant CA 07054 awarded by the National Cancer Institute, Department of Health and Human Services.
‡The actual experimental result is that tumor promoters by themselves cause a low level of tumors. Thus, in whole-animal experiments they may be detected as weak carcinogens. The generally accepted interpretation is that the promoters are acting on spontaneously initiated cells present at a low level. Since extrapolation of risk from high experimental doses to low doses depends on the postulated mechanism of action, whether an apparent weak carcinogen is actually acting as a carcinogen or as a promoter will lead to very different conclusions.

0077–8923/83/0407–0303 $01.75/0 © 1983, NYAS

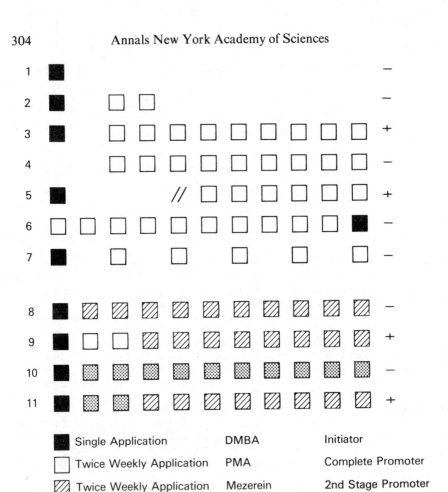

FIGURE 1. Characteristics of tumor promotion in mouse skin. Tumor response is indicated by $+/-$. DMBA, dimethylbenz[a]anthracene.

promoter mechanisms involving activation of a latent mutation, e.g., by induction of aneuploidy, do not predict reversibility of action.

Among tumor promoters for mouse skin, the most extensively studied have been the phorbol esters, the active promoters initially isolated from the seed oil of *Croton tiglium*.[6,7] The parent alcohol phorbol, which lacks biological activity, is a tetracyclic diterpene (FIGURE 2). The active derivatives of phorbol are esterified in the 12 and 13 positions. The most potent of these is phorbol 12-myristate 13-acetate (PMA). The phorbol esters may pose an environmental hazard to people under special circumstances.[8,9] Their primary importance for study, however, lies in their very high biological potency relative to most other promoters in skin (see Reference 10 for comparison). Their mechanism of action should therefore be less likely to be obscured by nonspecific secondary effects.

One of the important recent *in vivo* findings in skin promotion is that, just as skin carcinogenesis may be divisible into the stages of initiation and promotion, so promotion itself may be divisible into stages showing distinct structure-activity requirements. The observations of Slaga and colleagues demonstrating such stages[11,12] are as follows (FIGURE 1). The phorbol-related derivative mezerein is itself only of low potency as a tumor promoter (perhaps 1/50 the potency of PMA)[13] (line 8). However, if the mouse is treated for a short period of time (one to four weeks) with the complete promoter PMA, mezerein shows a potency comparable to PMA for completion of the promotion process (line 9). Conversely, 4-*O*-methyl PMA is of very low potency as a complete promoter[14] (line 10). However, treatment with 4-*O*-methyl PMA for two weeks followed by mezerein treatment leads to at least a partial response (line 11). These findings have been confirmed in principle although not in detail in a different strain of mice and with different combinations of phorbol derivatives by Marks and colleagues.[15] Further support for the subdivision of promotion into stages is the demonstration by Slaga and colleagues that the different stages of promotion show a differential spectrum of sensitivity to inhibitors.[16] The findings of multiple stages of promotion suggest the existence of multiple subclasses of phorbol ester receptors showing distinct structure-activity requirements. The results emphasize, moreover, the necessity of understanding mechanisms. A combination of compounds ineffective both as carcinogens and as complete promoters may have promoting activity, provided they are applied in the right sequence!

Early efforts to demonstrate receptors for the phorbol esters were frustrated by the lipophilicity of PMA, which resulted in a high level of nonspecific binding.[17] A solution for this problem, introduced by our laboratory, was to use as a radioactive ligand phorbol 12,13-dibutyrate (PDBu), a derivative that we had predicted to have an optimal ratio of binding affinity to lipophilicity. Using this derivative, we and subsequently others were able to demonstrate specific, saturable binding both to intact cells and to particulate preparations of both cells and tissues.[18-25] Specific binding represents the difference between total and nonspecific binding. Nonspecific binding is measured in the presence of a vast excess of nonradioactive PDBu (usually 30 μM final concentration). Typical data, obtained for binding of ^{3}H-PDBu to particulate preparations of mouse 3T3 cells,[26] are illustrated in FIGURE 3. Depending on the system, the dissociation constant for binding ranges from 7–50 nM. The number of binding sites ranges from 1.5–28 pmol/mg protein or 0.5–10 $\times$ 10^5 sites/cell. In contrast to the behavior of carcinogens, the binding of the phorbol ester is reversible.[18] The bound radioactivity can be shown to be the unaltered ligand, thus indicating that the phorbol esters are the "ultimate promoters." In most but not all systems examined (see below),

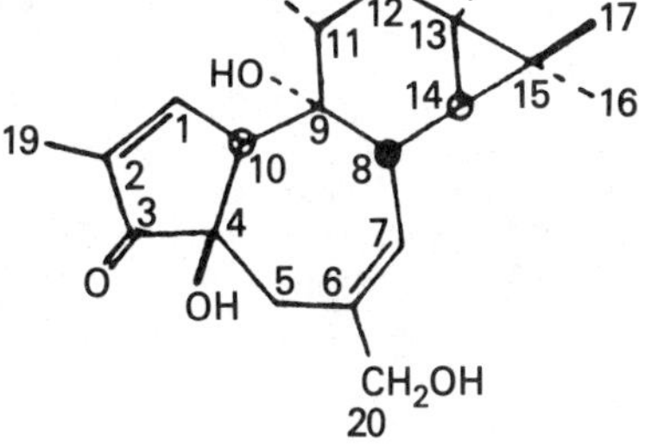

FIGURE 2. Structure of phorbol.

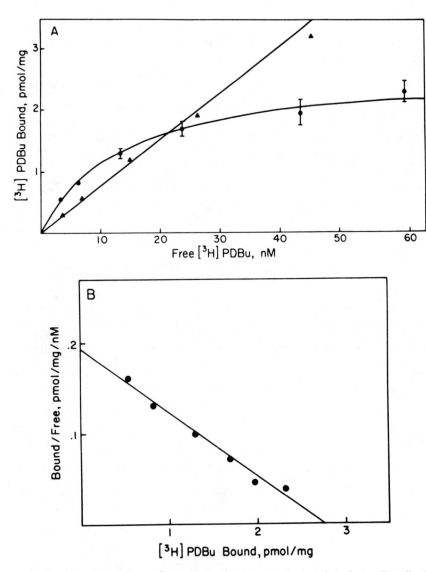

FIGURE 3. Specific binding of ^{3}H-PDBu to the particulate fraction from Swiss 3T3 cells. **A.** Saturation curve. Circles: specific binding, points are the average of triplicate determinations in a single experiment. Triangles: nonspecific binding, measured in the presence of 30 μM nonradioactive PDBu, points represent single determinations. **B.** Scatchard plot of data in A. (Reproduced from Reference 26 with the permission of the publisher.)

Scatchard analysis of the binding is consistent with a homogeneous class of binding sites.

Although the studies on phorbol ester promotion have been carried out predominantly with the mouse, the phorbol esters have been reported to have biological effects on a variety of other vertebrates (see References 5 and 27 for reviews) as well as on sea urchins[28] and hydra.[29] Consistent with such biological observations, PDBu receptors were found not only for multiple vertebrate species but also for the nematode *Caenorhabditis elegans*,[30] the fruit fly *Drosophila melanogaster*,[31] and the sea urchin *Lytechinus pictus*.[31] Receptor levels and binding affinities for these invertebrates were within the range found for vertebrate systems. In the case of the nematode, structure-activity relations were very similar to those for the mouse. Invertebrate systems may thus be appropriate for studying certain aspects of the mechanism of action of the phorbol esters.

Phorbol ester promotion has been demonstrated not only for mouse skin but also for other tissues such as vagina,[32] forestomach,[33] and trachea.[34] *In vitro* effects have been observed, moreover, for cultured cells derived from numerous tissues.[5,27] As expected from these results, we and others have been able to demonstrate phorbol ester receptors in virtually all tissues examined with the exception of erythrocytes.[31,35] A particularly important finding from these studies was that the level of binding in brain is markedly higher than that in other tissues (e.g., 7.5-fold that in skin). The absolute level of binding in brain, approximately 30 pmol/mg protein, is much greater than that for neurotransmitter receptors (<2 pmol/mg) or almost all hormone receptors. It is similar to the levels reported for the Na^+/K^+-ATPase or the Ca^{2+}-ATPase in brain. We had therefore suggested that the phorbol ester receptor is not analogous to a hormone receptor but rather represents a modulatory binding site on an enzyme, transport, or structural protein[20] (see below for further discussion).

A critical issue in the analysis of a binding activity for a ligand is the demonstration that the binding in fact mediates the biological responses to the ligand. The two usual approaches are genetic and pharmacological. As yet, although a number of laboratories have isolated variants unresponsive to the phorbol esters, in no instances have the variants been shown to be lacking binding activity.[26,36-39] In contrast, the pharmacological evidence that the binding detected with ³H-PDBu mediates phorbol ester responses is quite strong. First, excellent quantitative agreement was obtained between the potency of phorbol esters to induce fibronectin loss in chicken embryo fibroblasts and their ability to inhibit ³H-PDBu binding. Eight derivatives were examined, spanning a range of 6×10^4 in binding affinity. All were within a factor of 3.5 of the binding affinity predicted from the biological potency data.[18] Good structure-activity correlations have been reported for a variety of other biological responses in various systems.[25] Often, however, the data are based on a relatively limited series of compounds.

A second strong pharmacological argument that binding mediates responsiveness is based on results with the compounds lyngbyatoxin and dihydroteleocidin B (FIGURE 4). These compounds, produced by the blue-green alga *Lyngbya majuscula* and by *Streptomyces mediocidicus* respectively, have been shown to be promoters *in vivo* and to induce the same biological effects *in vitro* as do the phorbol esters.[40] Likewise, although structurally dissimilar from the phorbol esters, they inhibit ³H-PDBu binding with nanomolar affinity.[41,42]

A third argument that the receptors mediate responsiveness, weaker than the first two, is provided by studies on receptor down modulation. Collins and Rozengurt demonstrated that incubation of Swiss 3T3 cells with PDBu for 24 hours led to complete loss of PDBu receptors.[43] Concomitantly, cells became unresponsive to PDBu-induced mitogenesis and to inhibition by PDBu of epidermal growth factor (EGF) binding. The interpretation is weakened because the experiments cannot distinguish between causal and parallel changes in binding and responsiveness. In addition, phorbol ester receptor down modulation in most systems is only partial and the generality of Rozengurt's observations is uncertain.[25,44]

Two features of the structure-activity relations for inhibition of ^{3}H-PDBu binding are noteworthy. First, in all systems examined PMA has higher binding affinity than does PDBu. Thus, although PDBu is used as the ligand of convenience for measuring binding, the receptor being measured is more properly viewed as a PMA receptor. Second, in chick embryo fibroblasts,[18] rat pituitary cells,[21] and most other systems examined,[20,30,43] mezerein shows markedly lower affinity (20- to 136-fold) than does PMA. This relation reflects that expected for mezerein's effects as a complete promoter. It implies the existence of an additional target showing the structure-activity relations observed for inflammation, hyperplasia, and second-stage promotion in skin, viz., similar potencies for PMA and mezerein.

The focus of this symposium is on *in vitro* assays for toxicity testing. We and others

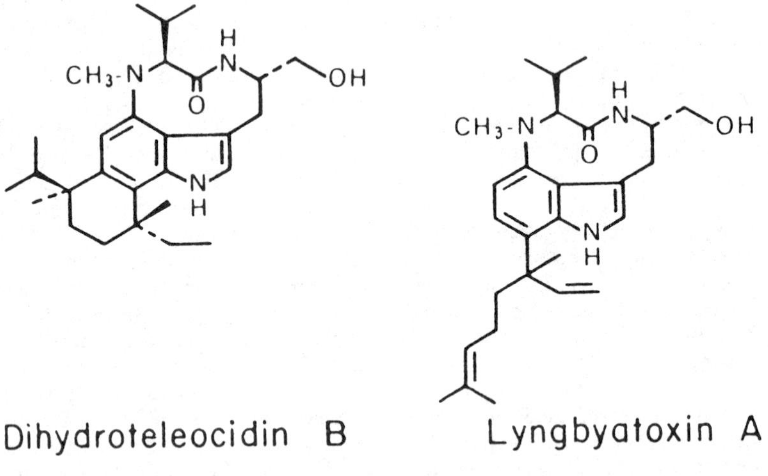

Dihydroteleocidin　B　　　　Lyngbyatoxin　A

FIGURE 4. Structures of dihydroteleocidin B and lyngbyatoxin.

had previously observed that tumor promoters structurally unrelated to the phorbol esters (the exceptions of lyngbyatoxin and dihydroteleocidin B are discussed above) failed to induce the same *in vitro* biological responses as did the phorbol esters.[10,45] Such agents would therefore not be predicted to interact directly at the phorbol ester receptor. Direct assay confirmed this expectation; the skin tumor promoters anthralin, cantharidin, iodoacetic acid, and phenol all failed to inhibit PDBu binding.[18,19] Inhibition of phorbol ester binding thus does *not* provide a general assay for skin tumor promoters. The studies on phorbol ester receptors may, however, provide insight into the molecular mechanism of this highly potent class of promoters. Since phorbol ester receptors are present in people, the normal function of these receptors and factors that modulate their activity may prove of potential importance in human cancer causation.

The *in vivo* studies with mezerein are most easily explained by two classes of receptors. The relative potencies of PMA and mezerein for three responses—induction of inflammation,[14] hyperplasia,[13] and ornithine decarboxylase[13]—correspond with

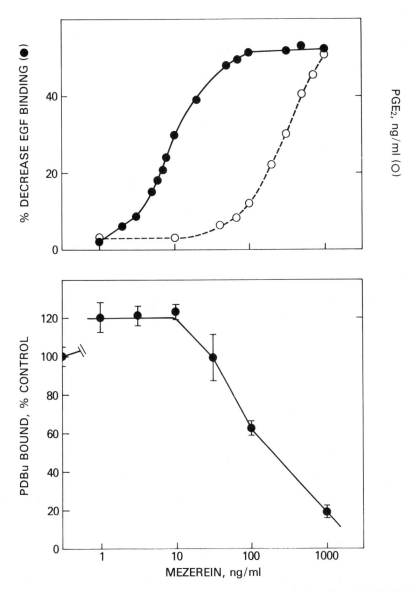

FIGURE 5. Relationship of biological responses to mezerein in G-292 cells and inhibition of ^{3}H-PDBu binding. Top panel: medium containing the indicated concentrations of mezerein was added to G-292 cells and specific ^{125}I-EGF binding (filled circles) and PGE$_2$ production (open circles) were measured following a 90-minute incubation at 37°C. Bottom panel: inhibition of ^{3}H-PDBu binding by mezerein was determined using a ^{3}H-PDBu concentration of 2 nM.

their relative potencies for second-stage promotion,[11] whereas their difference in potencies as complete promoters is much greater.[14] The apparent differences in potency might be an artifact, however, arising from differences in the time courses for the different responses measured together with differences in the pharmacokinetics of PMA and mezerein in mouse skin. The same criticism holds, albeit to a lesser degree, for the demonstration *in vitro* that the potency of mezerein for stimulating 2-deoxyglucose uptake in chicken embryo fibroblasts was 50-fold greater than that for inducing fibronectin loss.[46] Although drug absorption would not be a problem *in vitro,* serum concentrations and times of incubation still differed in the two assays. These sources of variability were eliminated in studies with the G-292 human osteosarcoma cell line, in which mezerein yielded dose-response curves for inhibition of EGF binding and for induction of prostaglandin E_2 (PGE_2) production that differed by 25-fold in their ED_{50} (half maximally effective dose) values under identical culture conditions and incubation times (FIGURE 5, top).[47] Competition for ^{3}H-PDBu binding indicated a relatively low affinity, which corresponded with the ED_{50} for induction of PGE_2 production (FIGURE 5, bottom). This cell line may therefore prove of particular utility in clarifying the mode of action of mezerein.

Another compound, 12-deoxyphorbol 13-isobutyrate (DPB), has provided direct evidence for phorbol ester receptor heterogeneity. The 12-deoxyphorbol derivatives were of interest because they had been initially reported to be highly inflammatory but only very weakly promoting, leading Hecker to conclude that "the definite irritant activity of 12-deoxi-phorbol-esters does not necessarily mean cocarcinogenic activity, and thus these biological activities may be considered as virtually independent biological properties of these diterpene esters."[48] Later reports suggested that these derivatives were somewhat less irritant than initially reported,[49,50] although this probably cannot account for their lack of promoting activity. The 12-deoxyphorbol esters induce a somewhat more transient erythema than does PMA.[51] In addition, these derivatives appear to differ from PMA in inducing higher mortality during promotion assays.[48]

We analyzed the binding of ^{3}H-DPB (FIGURE 6) to particulate preparations from mouse skin,[52] since skin was the tissue in which the structure-activity relations for the 12-deoxyphorbol derivatives had been most extensively characterized. A biphasic binding curve was obtained, consistent with two phorbol ester binding sites (designated PBS-1 and PBS-2) present at 0.14 and 1.55 pmol/mg protein and possessing dissociation constants (K_d) for ^{3}H-DPB of 6.9 and 86 nM, respectively. The structure-activity relations at PBS-2 were measured by competition of ^{3}H-DPB binding at 47 nM ^{3}H-DPB. Under these conditions, 80% of the total occupied receptors corresponded to PBS-2 and errors induced by neglecting the contribution to binding by PBS-1 led to

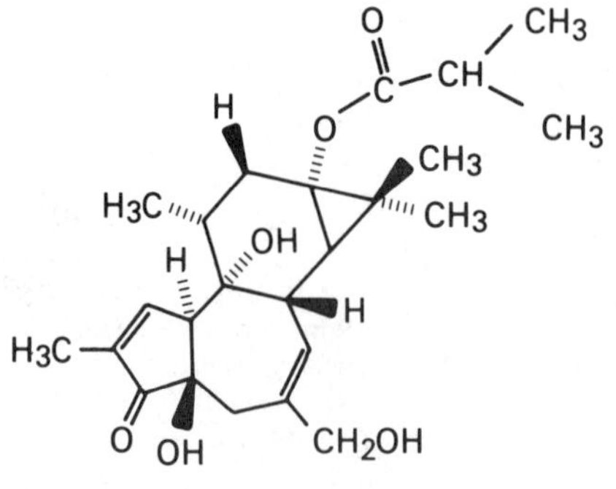

FIGURE 6. Structure of 12-deoxyphorbol 13-isobutyrate.

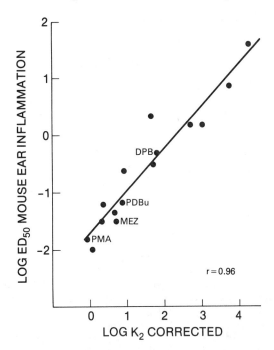

FIGURE 7. Correlation of binding affinity at PBS-2 and mouse ear inflammatory potency. Binding affinities of PBS-2 were determined from competition of ^{3}H-DPB binding and are corrected for changes in the concentrations of free ligand due to nonspecific binding.

only small errors ($\leq 30\%$) in the measured affinity constants. For 15 derivatives examined, very good agreement was obtained between binding affinity and inflammatory activity (FIGURE 7) (5×10^4 range in binding affinity, $r = 0.96$). Of particular note, mezerein fitted on the correlation line with the other derivatives. Because of the low amount of radioactivity incorporated in PBS-1, detailed structure-activity analysis has not yet been carried out for this site.

Analysis of ^{3}H-PDBu binding to the mouse skin preparations yielded a curved Scatchard plot, again indicative of heterogeneous binding. The level of specific binding at saturation was higher than that for ^{3}H-DPB. Since PDBu could block all of the ^{3}H-DPB binding, PBS-1 and PBS-2 appeared to represent a subset of the total PDBu-sensitive sites. The PDBu binding was therefore analyzed for three binding components after fixing the amounts of sites PBS-1 and PBS-2 from the ^{3}H-DPB-binding data. A value for PBS-3 of 1.87 pmol/mg was obtained. The binding affinities for PDBu at PBS-1, 2, and 3 were 0.7 nM, 10.3 nM, and 53 nM, respectively. As expected, the affinity of DPB at PBS-3 was very low, 5,300 nM.

Analysis of the structure-activity relations at PBS-3 by competition of ^{3}H-PDBu binding was complicated by the appreciable fraction of total binding (64%) attributable to PBS-1 and PBS-2 at 33 nM ^{3}H-PDBu. Errors in the measured affinities of the nonradioactive ligands at these sites would therefore be magnified in the determination of the affinities at PBS-3. Nonetheless, the data clearly indicate different structure-activity relations at the two sites. PMA and phorbol 12,13-didecanoate (PDD) yielded similar 50% inhibitory concentrations with either radioactive ligand, whereas DPB yielded a 10-fold difference in values. Although the problems in analysis make detailed conclusions uncertain, the structure-activity relations for PBS-3 better fit those for complete promoters than do those for PBS-2. Thus, mezerein was 100-fold less potent than PMA at this site. Likewise, for a series of symmetrically substituted phorbol esters, binding potency as a function of chain length decreased more rapidly at PBS-3 than at PBS-2. A similar differential relationship has been reported for promoting as compared to inflammatory activity.[53] More precise analysis of the structure-activity relations will need to await the development of ligands with greater selectivity for PBS-3. Nonetheless, the present binding data strongly support the biological evidence for heterogeneity in the structure-activity relations.

The basis for receptor heterogeneity could be multiple genes that code for receptor variants and that had arisen through gene duplication and divergence. Alternatively, heterogeneity could reflect a single gene product that has been differentially modified. Such modification could be either covalent, e.g., by proteolysis or phosphorylation, or noncovalent, e.g., by different microenvironments arising from different lipid domains.

Inhibition of binding by heating (100°C, 5 minutes) or by treatment either with protease or with phospholipase A_2 suggested that the receptor is a protein dependent on its lipid environment.[20] Results with the photoaffinity probe phorbol 12-p-azidobenzoate 13-benzoate (PaBB) support this interpretation. In the dark, PaBB reversibly bound to the phorbol ester receptor of brain membranes with high affinity, 0.8 nM.[54,55] Ultraviolet (UV) irradiation yielded specific, irreversible labeling with good efficiency (35–45%). The specific, irreversible binding could be inhibited either by heating or by prior treatment of the membranes with protease. Nonetheless, virtually all of the specific labeling was extractable with chloroform-methanol. Fractionation yielded two adducts on thin-layer chromatography. The lower-mobility product was identified as a PaBB-phosphatidylserine adduct on the basis of its susceptibility to the specific enzyme phosphatidylserine decarboxylase. The product of the reaction, a phorbol ester–phosphatidylethanolamine adduct, cochromatographed with the higher-mobility peak.

The failure of PaBB also to label the protein moiety of the receptor may be because the nitrene generated on the ligand is inadequately reactive, resulting in selective insertion into the double bonds of the phospholipid. Such behavior has been observed for some but not other analogous systems.[56,57] Alternatively, since the nitrene is located at the end of the side chain, it may be buried in the lipid bilayer and not in adequate proximity to the protein portion of the binding site. Although the photoaffinity labeling experiments with PaBB indicate that phosphatidylserine and phosphatidylethanolamine are preferentially associated with the receptor, they cannot indicate whether specific phospholipids are essential for binding activity.

Evidence that they may be so. is provided by another system. We had previously found that the phorbol ester receptors in brain are entirely membrane bound. We have recently demonstrated, however, that addition of phosphatidylserine to brain cytosol reconstituted specific, high-affinity binding activity in that fraction.[58] The binding affinity of ^{3}H-PDBu was 3.14 ± 0.09 nM. At saturation, 23 ± 2 pmol/mg protein were bound. The structure-activity relations for binding generally resembled those for the membrane receptor, although affinities were sometimes slightly higher. PMA, for example, had an affinity of 0.041 nM, as compared to 0.066 nM for membranes. Analysis of phospholipid dependence revealed that phosphatidylserine was most effective at yielding reconstitution. The acidic phospholipids phosphatidic acid and phosphatidylinositol gave a partial response, as did sphingomyelin, whereas phosphatidylcholine and phosphatidylethanolamine were relatively ineffective. The relationship between the membrane receptor and the cytosolic apo-receptor remains to be determined.

The mechanism of transformation by Rous sarcoma virus and several other of the acute transforming viruses has been shown to be by introduction into the cell of a tyrosine kinase, apparently analogous to a normal cellular enzyme but expressed at an enhanced level.[59] Based on the similarity in phenotypes induced in chick embryo fibroblasts by Rous sarcoma virus and PMA,[60,61] we had suggested several years ago that "the immediate response of the phorbol ester receptor interaction would be predicted to be also a change in kinase or phosphatase activity or specificity."[5] In fact, the properties of the phorbol ester receptors as characterized by us and others show marked similarities to those of the Ca^{2+}-phospholipid-dependent protein kinase described by Nishizuka and colleagues.[62] It is therefore of substantial interest that this latter group has reported that the phorbol esters can stimulate this enzyme directly.[63] The membrane-bound form of the kinase may thus correspond to at least a subclass of phorbol ester receptors, with the cytosolic phorbol ester apo-receptor corresponding to the cytosolic form of the kinase. Since the data of Nishizuka and colleagues suggest that PMA functions similarly to diacylglycerol to sensitize the kinase to activation by calcium, other factors that modulate lipid composition or structure or internal calcium concentrations may be of interest for further study as possible modulators in promotion.

SUMMARY

The evidence for the multistage nature of tumor promotion *in vivo* and for multiple subclasses of phorbol ester receptors *in vitro* argues that multiple mechanisms of tumor promotion exist. The existence of multiple mechanisms suggests that brute force assay for tumor promoters *in vivo* may be inadequate and that understanding of mechanisms may be essential. The interest in the phorbol esters is not primarily that they are environmental hazards for man, but rather that they provide a probe for phorbol ester receptors. These receptors are found in people, and modulation of their activity may play a role in tumor promotion in man.

REFERENCES

1. HIGGINSON, J. & C. S. MUIR. 1979. J. Nat. Cancer Inst. **63:** 1291–1298.
2. WEISBURGER, J. H., B. S. REDDY, L. A. COHEN, P. HILL & E. L. WYNDER. 1982. *In* Cocarcinogenesis and Biological Effects of Tumor Promoters. E. Hecker, N. E. Fusenig, W. Kunz, F. Marks & H. W. Thielmann, Eds.: 175–182. Raven Press. New York, N.Y.
3. SCRIBNER, J. D. & R. SÜSS. 1978. Int. Rev. Exp. Pathol. **18:** 137–198.
4. BOUTWELL, R. K. 1964. Prog. Exp. Tumor Res. **4:** 207–250.
5. BLUMBERG, P. M. 1980. Crit. Rev. Toxicol. **8:** 153–234.
6. HECKER, E. 1971. Methods Cancer Res. **6:** 439–484.
7. HECKER, E. 1968. Cancer Res. **28:** 2338–2349.
8. HECKER, E. 1981. J. Cancer Res. Clin. Oncol. **99:** 103–124.
9. HIRAYAMA, T. & Y. ITO. 1981. Prev. Med. **10:** 614–622.
10. DRIEDGER, P. E. & P. M. BLUMBERG. 1978. Int. J. Cancer **22:** 63–69.
11. SLAGA, T. J., S. M. FISCHER, K. NELSON & G. L. GLEASON. 1980. Proc. Nat. Acad. Sci. USA **77:** 3659–3663.
12. SLAGA, T. J., S. M. FISCHER, C. E. WEEKS, K. NELSON, M. MAMRACK & A. J. P. KLEIN-SZANTO. 1982. *In* Cocarcinogenesis and Biological Effects of Tumor Promoters. E. Hecker, N. E. Fusenig, W. Kunz, F. Marks & H. W. Thielman, Eds.: 19–34. Raven Press. New York, N.Y.
13. MUFSON, R. A., S. M. FISCHER, A. K. VERMA, G. L. GLEASON, T. J. SLAGA & R. K. BOUTWELL. 1979. Cancer Res. **39:** 4791–4795.

314   Annals New York Academy of Sciences

14. HECKER, E. 1978. In Mechanisms of Tumor Promotion and Cocarcinogenesis. T. J. Slaga, A. Sivak & R. K. Boutwell, Eds.: 11–48. Raven Press. New York, N.Y.
15. FÜRSTENBERGER, G., D. L. BERRY, B. SORG & F. MARKS. 1981. Proc. Nat. Acad. Sci. USA 78: 7722–7726.
16. SLAGA, T. J., A. J. P. KLEIN-SZANTO, S. M. FISCHER, C. E. WEEKS, K. NELSON & S. MAJOR. 1980. Proc. Nat. Acad. Sci. USA 77: 2251–2254.
17. LEE, L. S. & I. B. WEINSTEIN. 1978. J. Environ. Pathol. Toxicol. 1: 627–639.
18. DRIEDGER, P. E. & P. M. BLUMBERG. 1980. Proc. Nat. Acad. Sci. USA 77: 567–571.
19. DELCLOS, K. B., D. S. NAGLE & P. M. BLUMBERG. 1980. Cell 19: 1025–1032.
20. DUNPHY, W. G., K. B. DELCLOS & P. M. BLUMBERG. 1980. Cancer Res. 40: 3635–3641.
21. JAKEN, S., A. H. TASHJIAN, JR. & P. M. BLUMBERG. 1981. Cancer Res. 41: 2175–2181.
22. HOROWITZ, A. D., E. GREENEBAUM & I. B. WEINSTEIN. 1981. Proc. Nat. Acad. Sci. USA 78: 2315–2319.
23. SHOYAB, M. & G. J. TODARO. 1980. Nature 288: 451–455.
24. P. M. BLUMBERG, K. B. DELCLOS, W. G. DUNPHY & S. JAKEN. 1982. In Cocarcinogenesis and Biological Effects of Tumor Promoters. E. Hecker, N. E. Fusenig, W. Kunz, F. Marks & H. W. Thielman, Eds.: 519–535. Raven Press. New York, N.Y.
25. BLUMBERG, P. M., K. B. DELCLOS, J. A. DUNN, S. JAKEN, K. L. LEACH & E. YEH. In Tumor Promotion and Cocarcinogenesis In Vitro. T. J. Slaga, Ed. CRC Press. Boca Raton, Fla. (Submitted.)
26. BLUMBERG, P. M., E. BUTLER-GRALLA & H. R. HERSCHMAN. 1981. Biochem. Biophys. Res. Commun. 102: 818–823.
27. DIAMOND, L., T. G. O'BRIEN & W. M. BAIRD. 1980. Cancer Res. 32: 1–74.
28. BRESCH, H. & U. ARENDT. 1978. Naturwissenschaften 65: 660–662.
29. SHIBA, Y. 1982. Experientia 38: 498–499.
30. LEW, K. K., S. CHRITTON & P. M. BLUMBERG. 1982. Teratogen. Carcinogen. Mutagen. 2: 19–30.
31. BLUMBERG, P. M., K. B. DELCLOS & S. JAKEN. 1982. In Symposium on Organ and Species Specificity in Chemical Carcinogenesis. R. Langenbach, S. Nesnow & J. M. Rice, Eds.: 201–229. Plenum Publishing Co. New York, N.Y.
32. GOERTTLER, K., H. LÖHRKE & B. HESSE. 1980. Carcinogenesis 1: 707–713.
33. GOERTTLER, K., H. LOEHRKE, J. SCHWEIZER & B. HESSE. 1979. Cancer Res. 39: 1293–1297.
34. TOPPING, D. C. & P. NETTESHEIM. 1980. Cancer Res. 40: 4352–4355.
35. SHOYAB, M., T. C. WARREN & G. J. TODARO. 1981. Carcinogenesis 2: 1273–1276.
36. COLBURN, N. H., T. D. GINDHART, G. A. HEGAMYER, P. M. BLUMBERG, K. B. DELCLOS, B. E. MAGUN & J. LOCKYER. 1982. Cancer Res. 42: 3093–3097.
37. SANDO, J. J., M. L. HILFIKER, M. J. PIACENTINI & T. M. LAUFER. 1982. Cancer Res. 42: 1676–1680.
38. SOLANKI, V., T. J. SLAGA, M. CALLAHAM & E. HUBERMAN. 1981. Proc. Nat. Acad. Sci. USA 78: 1722–1725.
39. FISHER, P. B., U. COGAN, A. D. HOROWITZ, D. SCHACHTER & I. B. WEINSTEIN. 1981. Biochem. Biophys. Res. Commun. 100: 370–376.
40. FUJIKI, H., M. MORI, M. NAKAYASU, M. TERADA, T. SUGIMURA & R. E. MOORE. 1981. Proc. Nat. Acad. Sci. USA 78: 3872–3876.
41. UMEZAWA, K., I. B. WEINSTEIN, A. HOROWITZ, H. FUJIKI, T. MATSUSHIMA & T. SUGIMURA. 1981. Nature 290: 411–413.
42. COLLINS, M. & E. ROZENGURT. 1982. Biochem. Biophys. Res. Commun. 104: 1159–1166.
43. COLLINS, M. K. L. & E. ROZENGURT. 1982. J. Cell. Physiol. 112: 42–50.
44. JAKEN, S., A. H. TASHJIAN, JR. & P. M. BLUMBERG. 1981. Cancer Res. 41: 4956–4960.
45. WIGLER, M., D. DEFEO & I. B. WEINSTEIN. 1978. Cancer Res. 38: 1434–1437.
46. DRIEDGER, P. E. & P. M. BLUMBERG. 1980. Cancer Res. 40: 339–346.
47. JAKEN, S., M. A. SHUPNIK, P. M. BLUMBERG & A. H. TASHJIAN, JR. 1982. Cancer Res. 43: 11–14.
48. HECKER, E. 1971. In Proceedings, 10th International Cancer Congress: 213–224. Year Book Medical Publishers. Chicago, Ill.
49. SCHMIDT, R. J. & F. J. EVANS. 1980. Arch. Toxicol. 44: 279–289.
50. HERGENHAHN, M., S. KUSUMOTO & E. HECKER. 1974. Experientia 30: 1438–1440.

Blumberg *et al.*: Phorbol Ester Receptors   315

51. EVANS, F. J., A. D. KINGHORN & R. J. SCHMIDT. 1975. Acta Pharmacol. Toxicol. **37:** 250–256.
52. DUNN, J. A. & P. M. BLUMBERG. (Submitted.)
53. THIELMANN, H. W. & E. HECKER. 1969. *In* Deutscher Fortschritte der Krebsforschung. C. G. Schmidt & O. Wetter, Eds. **7:** 171–179. Schattauer-Verlag. Stuttgart & New York.
54. DELCLOS, K. B. 1982. Ph.D. Thesis. Harvard University. Cambridge, Mass.
55. DELCLOS, K. B. & P. M. BLUMBERG. 1983. Proc. Nat. Acad. Sci. USA. (In press.)
56. QUAY, S. C., R. RADHAKRISHNAN & H. G. KHORANA. 1981. J. Biol. Chem. **256:** 4444–4449.
57. BRUNNER, J. & F. M. RICHARDS. 1980. J. Biol. Chem. **255:** 3319–3329.
58. LEACH, K. L., M. L. JAMES & P. M. BLUMBERG. 1982. (abst.). J. Cell Biol. **95:** 431a.
59. ERIKSON, E., R. COOK, G. J. MILLER & R. K. ERIKSON. 1981. Mol. Cell. Biol. **1:** 43–50.
60. DRIEDGER, P. E. & P. M. BLUMBERG. 1977. Cancer Res. **37:** 3257–3265.
61. WEINSTEIN, I. B., L.-S. LEE, P. B. FISHER, A. MUFSON & H. YAMASAKI. 1979. J. Supramol. Struct. **12:** 195–208.
62. TAKAI, Y., A. KISHIMOTO, Y. KAWAHARA, R. MINAKUCHI, K. SANO, U. KIKKAWA, T. MORI, B. YU, K. KAIBUCHI & Y. NISHIZUKA. 1981. Adv. Cyclic Nucleotide Res. **14:** 301–313.
63. CASTAGNA, M., Y. TAKAI, K. KAIBUCHI, K. SANO, U. KIKKAWA & Y. NISHIZUKA. 1982. J. Biol. Chem. **257:** 7847–7851.

THE ROLE OF TUMOR PROMOTERS ON PHENOTYPIC ALTERATIONS AFFECTING INTERCELLULAR COMMUNICATION AND TUMORIGENESIS*

James E. Trosko, Cy Jone, and Chia-cheng Chang

Department of Pediatrics/Human Development
College of Human Medicine
Michigan State University
East Lansing, Michigan 48824

INTRODUCTION

Empirical observations of tumorigenesis in animal and human systems clearly indicate that it is a multistep process.[1] After the stable conversion of a single normal cell to a "premalignant" cell, there appears to be clonal amplification of this premalignant cell.[2,3] During the clonal expansion process, additional phenotypic changes occur within the clone, ultimately giving rise to a heterogeneous collection of phenotypes within the clone.[4] Among some of these new phenotypes appearing during the clonal amplification process, a few are able (1) to escape dependency on either endogenous or exogenous factors contributing to the selective clonal expansion of the premalignant cells; (2) to escape suppression of their premalignant phenotype by surrounding normal cells; and (3) to acquire the phenotypes of invasiveness and metastasis.

The concepts of initiation and promotion were created to explain experimental observations related to mouse skin tumorigenesis.[5] More recently, the applicability of this initiation/promotion concept of tumorigenesis has been made for experimental colon, breast, and bladder carcinogenesis in various animal systems, with evidence indicating that it might also be relevant to human carcinogenesis.[6]

The molecular mechanisms contributing to the complex evolution of phenotypic changes leading to a metastasizing cell are obviously not known. However, based on a variety of *in vivo* and *in vitro* studies designed to characterize "initiators" and "promoters," some conclusions can be drawn about the general nature of the initiation and promotion steps of carcinogenesis. The conversion of a normal cell to a single stably altered premalignant cell can be brought about by mutagens[7] and possibly, in some unique situations,[8] chemicals that are not mutagenic (see next section). On the other hand, the clonal amplification of the single premalignant cell seems to be brought about by conditions and chemicals that act as, at least, selective "mitogens" for the initiated cell (cell removal, cell death, growth factors, exogenous chemicals).[9] The process that brings about phenotypic diversification during the promotion phase has been termed "progression" by some.[6] Whether the molecular mechanisms that bring about this final step of tumorigenesis are qualitatively different from the genotypic changes occurring during the initiation or promotion phases is not known.[10]

*Research was sponsored by an Environmental Protection Agency grant (R8085870) and a National Cancer Institute grant (CA21104) to James E. Trosko.

0077-8923/83/0407-0316 $01.75/0 © 1983, NYAS

Definitions Related to the Initiation/Promotion Concept of Carcinogenesis

Recent conceptual, experimental, and technical developments in the fields of carcinogenesis and toxicology have led to a massive amount of confusion due to terminology. The origin of some of that confusion seems to be related to one of the major conceptual advances in cancer research, namely, the introduction of the "initiation-promotion" model to explain the multistaged nature of carcinogenesis. Although by no means is it clear that all cancers conform to the model, and to the implied mechanisms of the model, clearly the original mouse skin tumorigenic studies, along with the more recent examples of breast, colon, lung, and stomach, suggest that the initiation-promotion model can be used to explain carcinogenesis in many species and organ systems. In effect, initiation referred to the process by which a normal cell was converted to a "premalignant cell," and promotion was the process that then allowed the initiated or premalignant cell to attain a series of phenotypes, ultimately conferring the malignant phenotype to the cell.

It was only natural that the "success" of this model led to the development of the terms "initiators" and "promoters," which we suppose referred to any physical, chemical, or biological agent that would affect either one or both phases.

As additional studies related to the biochemical and molecular *mechanisms* of the intiation and promotion model of carcinogenesis became available, it appears that many investigators used the terms "initiators" and "promoters" not only to explain the *biological process* of the conversion of normal cells to premalignant cells (i.e., initiators) and of premalignant cells to malignant cells (i.e., promoters) but also to explain the mechanisms of intiation and promotion. This appears to have created several real problems.

First, there is the tendency to assume (1) that there is only one mechanism for either the initiation or the promotion process, (2) that agents (particular chemicals) can be classified as either initiators or promoters, and (3) that chemicals (or physical agents) that can affect the initiation (or promotion) process cannot affect the promotion (or initiation) process.

Considering the first assumption, since the conversion of a normal cell to a premalignant cell appears to be a *permanent* or irreversible process, it was easy to presume that "initiators" were *mutagens* (genotoxic—see later discussion). Since, at least in mouse skin tumorigenesis, the early portion of the promotion phase seems "reversible," it was also easy to presume that "promoters" were nonmutagens (nongenotoxic or "epigenetic"—see later discussion).

At this point, we must agree on a working definition of a mutagen. We feel that any agent that can irreversibly (in practical terms) alter the original genetic information (quality or quantity) is a mutagen. This can be achieved by agents that damage DNA bases and DNA strands or cause additions, deletions, translocations, or transpositions of chromosome structure. Some agents, which might not damage DNA, could cause karyotypical changes by arresting normal DNA synthesis, only affecting membrane or cytoskeleton structure to bring about these effects.[11] In addition, some non-DNA-damaging chemicals that can alter the DNA information by interacting with other nonmutagens (i.e., comutagens),[12] by inhibiting DNA-repair enzymes, by changing the fidelity of polymerase,[13] and by modulation of nucleotide pools[14] can be considered, in the broadest sense, "mutagens." The term "genotoxins," as originally defined[15] and commonly used,[16] seems much more restrictive in that it refers to agents that interact "with the genetic material" and "are highly specific for nucleic acids." Clearly, then, not all mutagens would be genotoxins if one accepts that definition of genotoxin. We feel if the term "genotoxin" is to be used, it should be equivalent to the broadest definition of a "mutagen," which would include any agent that can alter, by any mechanism, the quality and quantity of genes and chromosomes.

There seems to be no doubt that agents that can induce gene or chromosomal damage (*mutagens*) can act as initiators of carcinogenesis. Some could argue that the demonstration of known mutagens, acting as initiators, does not prove a mutation is responsible for initiation but that a nonmutagenic potential of the mutagen caused the initiation. In any case, mutations, therefore mutagens, have the characteristic of inducing permanent change in a normal cell, thereby satisfying a major criterion for the initiation phase.

However, under very special conditions, a stable and heritable alteration in the expression of a gene does *not* involve either a qualitative or quantitative nature of the genetic information (i.e., mutation), but rather in its expression [at the transcription to posttranslational levels (epigenetic or nongenotoxic)—see later discussion] could conceivably "initiate" a normal cell to a premalignant cell. Possibly, the example of the mouse teratocarcinomas could fit this category.[8] A normal blastocyst, without exposure to a known mutagen, can be placed in the testicular capsule of an adult male mouse. In this unusual environment for the blastocyst, growth factors and hormones (presence or absence) would lead to abnormal growth control leading to an "initiated" state. As long as the cell is in that environment, without additional genetic changes, it continues to have abnormal growth control. If under the influence of normal growth factors (cell mass dilutes growth factors), the cells can differentiate. In addition, if teratocarcinoma cells are transplanted to a normal blastocyst, they can contribute to "normal" development, suggesting that out of an abnormal environment, these cells have a normal genotype. One can then infer that the teratocarcinoma cell is altered in its gene expression ("epigenetic"), not in its gene information ("mutagenic"). If this is true, then, one might argue that this presumptive nonmutagenic process can "initiate" a normal cell to a premalignant cell. In addition, if during the proliferation ("promotion" phase) of the "epigenetically initiated" cell a series of mutations occurred, some of the initiated cells could become malignant. In these cases, the teratocarcinoma cells, if placed back into a normal blastocyst, *would not* contribute to normal development.

Whether there are other examples such as this special one is hard to ascertain at present. In addition, whether there are chemicals (or other agents) that could alter gene expression, without inducing a gene or chromosomal mutation, such that a normal cell could be converted to a premalignant cell (i.e., initiated) is not known (possibly, ethionine or azacytidine).[17] If so, then *nonmutagenic, epigenetic,* or *nongenotoxic chemicals could be "initiators."* Chemicals can have multiple biological functions. There could be different mechanisms for the biological phenomenon of initiation.

In recent years, because of pressure on chemical and drug industries to characterize chemicals, there has been a tendency to use a few molecular, chemical, biochemical, or biological end points to classify chemicals. Once classified (i.e., mutagen in the Ames assay, "carcinogen" in animal bioassay), it is sometimes assumed that the characterization encompasses a chemical's complete potential biological consequence.

Clearly, agents that can induce mutations, gene or chromosomal, not only produce viable cells with altered phenotypes, but also some mutations will kill cells (cytotoxic). The death of cells, itself, can have major repercussions on the organism, depending on the number and kind of cells killed as well as the point in development when the killing occurred. The cytotoxicity of cells can induce compensatory hyperplasia of the surviving stem cells. That is to say, the *mutagenic* consequence of a chemical could include a cytotoxic component, which in turn could induce, *indirectly,* altered gene expression ("epigenetic") in the surviving cells by removing the stem cell from "contact inhibition"[18] or growth control.[19,20] Therefore, these chemicals can have *mutagenic, cytotoxic,* and *epigenetic* properties (FIGURE 1). The ability of certain tissues in certain species to metabolize the chemical, as well as the effective biological concentration found in the tissue, could influence any and all of these potential consequences of a chemical. A chemical cannot be characterized biologically outside the biological context.

Now because some agents are not mutagenic in a few *in vitro* or *in vivo* gene or chromosomal assays, there is the tendency to assume that the chemical is either "safe" or possibly an "epigenetic" hazard, implying it can be handled and treated differently from mutagens. In addition, if this nonmutagenic chemical seems to influence carcinogenesis, it might be a tumor promoter, therefore, it could be used differently, under all conditions, from mutagens. While this might be the case generally, it does have some serious flaws.

First, this "epigenetic" agent might under certain conditions act to alter the gene expression stably (i.e., initiate cells) or during critical periods, altering differentiation (i.e., acting as a teratogen),[21] or during maturation periods of tissue (i.e., blocking sperm maturation leading to aspermia).[22] Therefore, calling an agent "epigenetic" does not exonerate the chemical from potential human health hazard. At high concentrations and in specific types of cells, such chemicals, as all chemicals, can be cytotoxic. In addition, these "epigenetic" chemicals affect membranes (and membrane-attached spindle fibers) and could induce chromosome mutations in certain kinds of cells at high concentrations, which has been observed with 12-*O*-tetradeca-noylphorbol-13-acetate (TPA).[23]

Lastly, chemicals can be dangerous, not by mutating cells or directly altering gene expression, but by killing them (inhibiting crucial enzymes, disrupting membrane permeability, etc.).[24] This in turn could, as in the case of mutagens that are cytotoxic, induce indirectly the surviving cell to activate its genes to enter "compensatory" hyperplasia. *Mutagens and cytotoxins, as well as cell removal or surgery, could be "indirect" promoters.* Several different mechanisms can lead to tumor promotion.

The third assumption that "initiators" or "promoters" cannot influence the promotion or initiation phase, respectively, has already been challenged with the previous discussion. Moreover, because of at least four assumptions (namely, "initiators" can be mutagens, mutagens induce permanent genetic changes, initiation is a permanent or irreversible process, and promotion is a reversible process up to a point), we normally assume that chemicals that are known mutagens-initiators do not act during the promotion phase. In fact, some confusion arises when the word *initiator* is used a second time *after* an animal is exposed to a chemical mutagen. Technically, once a normal cell is initiated, it is already converted from a normal cell to a premalignant cell. Therefore, to expose initiated tissue that has been promoted to more initiators can lead to a confusion of concepts and mechanisms on the biological level (initiation and promotion) with concepts and mechanisms on the molecular level (mutagenesis, gene modulation). For example, Potter recently introduced a very useful hypothesis for experimental testing, namely, the "IPI" (initiation-promotion-initiation) experiment.[25] We all know what is meant in the proposed experiments (i.e., expose the animal to a mutagen to initiate the animal; expose the animal to a nonmutagenic promoter for various times, then follow the promotion with additional mutagens). In this case to refer to the chemical mutagen in both cases as an initiator (in the context of the experiment) is technically not consistent, since the hypothesis is that the chemical mutates an allele in a single cell to initiate it to a premalignant state; exposure to the promoter clonally amplifies the premalignant cell; and finally exposure to a mutagen now will mutate the second allele in one of the initiated cells, converting it to the malignant phenotype. Now while initiators added to previously initiated tissue

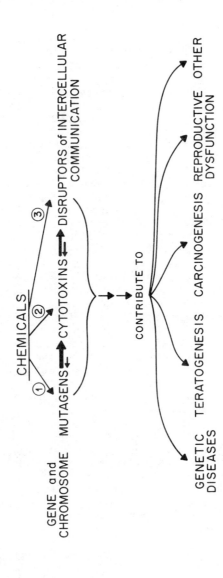

FIGURE 1. A heuristic scheme to classify chemicals on the basis of three biological end points, mutagenicity, cytotoxicity, and inhibition of intercellular communication. The arrows between the biological end points are designated to mean that chemicals can have multiple biological consequences in the direction indicated (i.e., mutagens can kill cells and the death of cells can cause the modulation of gene expression in some surviving cells). In addition, some chemicals that can inhibit intercellular communication at noncytotoxic levels could at much higher concentrations kill cells, possibly also inducing some chromosomal mutations at these high doses. It further depicts the speculated disease processes in which mutagenesis, cell death, and disrupted intercellular communication can play a role.

could initiate new normal cells to premalignant cells, the meaning implied by the term initiator in the IPI protocol is mutagen. It should also be pointed out that the term an "initiated cell" specifically implies that the cell has been converted by an initiator to a "premalignant" cell. Not all cells exposed to initiators are "initiated." This is an important point, which must be understood. If some initiators are mutagens, and if all genes of a cell are potentially mutable by initiators, only those cells in which specific genes being mutated can set the cell down the path of being malignant are "initiated" cells.

Yet another point of confusion that exists with regard to promoters, epigenetic agents, and "carcinogens" needs to be examined. To us, the word "carcinogens," in light of our understanding of the multistaged nature of carcinogenesis, is highly misleading since, by implication, it suggests that the "carcinogen" is *the* cause of cancer (i.e., a cancer-*causing* agent). If we accept carcinogenesis (the process of cancer formation) as involving a complex evolution of phenotypes from normal to premalignant to metastatic and if we assume the initiator and promotion processes are different (i.e., one involving a stable alteration of the normal genome to a premalignant phenotype—*initiation*—by either mutagenic or epigenetic mechanism; the other—promotion—involving the clonal expansion of cells to increase the chances of additional genetic or stable epigenetic events to occur), then it seems that few, if any, chemicals can "cause" cancer by *one* mechanism.[24]

To illustrate this point, it is well known experimentally that a given chemical, depending on the concentration or biological content, can act as a complete "carcinogen" (i.e., apparently elicit all the biological prerequisites to complete the carcinogenic process) or as only a portion of the carcinogenic process (incomplete carcinogen, i.e., initiation, promotion). When 2-acetylaminofluorene (AAF) is given to rats at a low, noncytotoxic level in a single or few exposures, physical or chemical promotion must follow in order that liver tumors appear (i.e., AAF acts as an "initiator"). If AAF is given to the rats at high and chronic levels, such that liver toxicity is visible, tumors will appear without additional intervention of physical or chemical promoters (i.e., AAF acts as both initiator and promoter—"complete carcinogen").

In conclusion, if the foregoing discussion is correct, then one of the major implications is that extreme care must be given to labeling chemicals on the basis of testing them in either a few *in vitro* or *in vivo* systems, without knowing the mechanisms by which they act and the specific biological context of their actions. To label a chemical an "epigenetic" agent because it was not mutagenic in the Ames assay does not exonerate the chemical as either an "initiator" under some special biological condition or as a potential health hazard on another biological level (i.e., teratogen, reproductive toxicant, etc.).

PROMOTION, PROMOTERS, AND CELL-CELL COMMUNICATION

The basic assumption of this manuscript is that genes that control intercellular communication play a significant (a necessary, but insufficient) role in the ultimate conversion of a normal cell to a premalignant and malignant cell. As in the case of the initiation process, where species, genetic, developmental, tissue, and cell-cycle factors can influence the potential of a chemical's initiation potential (i.e., drug-metabolizing enzyme, tissue distribution, repair enzymes, etc.), so also are there similar factors modulating the promotion process.[26] The demonstration that a given chemical can act as a promoter in one species and not another and one tissue or organ and not another[27] points to the complexity of the promotion process. In addition, evidence from mouse skin tumorigenesis suggests that promotion itself consists of many stages.[28] If one accepts the "progression" phase to be either the terminal stage of promotion or a distinct phase,[6,10] then one must account mechanistically for each of these phases. To complicate matters even more is the observation that the biological phenomenon called promotion can be accomplished by cytotoxic levels of an initiator (i.e., complete carcinogens), cytotoxic levels of noninitiators (i.e., viruses or chloroform), noncytotoxic levels of promoters, physical irritation, cell removal by surgery, and natural hyperplasia. The observation that the promotion process is often, but not always, accompanied by hyperplasia or that some hyperplasia is not associated with promotion or promoters suggests either that hyperplasia is not a prerequisite for promotion or that not all hyperplastic conditions are identical. For example, it has been shown that not all agents that induce hyperplasia are able to induce the same gene expression.[29] The implication is that in order for an agent or condition to act as a promoter of an initiated cell, it must be able to induce specific gene expression in that cell. If this is true, then certain promoters can do this without turning on genes associated with hyperplasia in the normal tissue (i.e., promoters without hyperplastic properties). Also, certain chemicals might be able to "turn on" genes associated with hyperplasia in normal cells without turning on critical genes in the initiated cell.

We must, however, remember that the tumor, a collection of cells from a clonal origin of a single altered cell,[2,3] is the result of at least the initiated cell having been allowed to proliferate and/or not to differentiate properly as its normal stem cell neighbors. This means that promotion, and promoters, induced "hyperplasia" of at least the initiated cell. The characteristics of the promotion process and of promoters have been reviewed.[30] In general, under the promoting conditions the "pure" promoters appear to be nonmutagenic, to be nongenotoxic, and to be "mitogenic" for certain cells under certain conditions.[26]

Several observations have contributed to the hypothesis that inhibition of intercellular communication is a critical process that occurs during tumor promotion, namely, cells that are tumorigenic do not communicate properly; "contact inhibition," a property of normal cells, appears to be absent in transformed and tumorigenic cells; cocultivation of certain transformed cells with normal cells suppresses the transformed phenotype; fusion of radiation-induced and chemically induced transformed cells with normal cells leads to nontransformed hybrids; coinjection of certain tumorigenic cells with normal cells suppresses the growth of the tumorigenic cells in an appropriate host animal; and many chemicals known to be tumor promoters of various organ systems in several species have been shown to inhibit intercellular communication (see reviews in References 9 and 26).

Putting the hypothesis in perspective, one must recognize that in a multicellular organism, the many different tissues have complex feedback systems, not only between the different tissues, but also between the pluripotent stem cells and the terminally differentiated cells of that cell lineage.[19,20] In addition, within the cells of like physiologies or differentiated states, there is a form of intercellular communication mediated by gap junctions.[31] This form of intercellular communication can help to keep cells of a differentiated state in concert or equilibrium by allowing a transfer of ions and small-molecular-weight metabolites.[32]

The regulation of intercellular communication has been linked to the regulation of proliferation and differentiation during development[33] and during organism homeostasis.[19] Interference with intercellular communication would in this perspective be dysfunctional in the sense that control of one tissue by another, control of one cell by another, and the orchestration of all cells within a tissue function would be hampered. In the context of cancer, if a cell has been altered such that it can proliferate more readily than a normal cell and/or that it does not differentiate properly, then it would tend to increase itself. However, if this cell can communicate via gap junctions with normal neighbors, any metabolic product it might lack that can be transferred via the gap junction can be shared. As a result, genetically "deficient" cells can be compensated by metabolic cooperation.[34,35] If these deficient cells are isolated from their normal neighbors by cell death, cell removal, or developmental or chemical suppression of gap junctions, then the deficient cell can manifest its premalignant phenotype. If the number of these deficient cells grows to a "critical mass," then even though they may "communicate" with each other, they are unable to make up for their metabolic deficiencies. As a consequence, they would still grow (possibly as benign tumors). If the genetically deficient cell is the "initiated" cell and the clone of initiated cells is a "benign" or premalignant tumor, then one could speculate that an additional genetic change must occur to give the premalignant cell the phenotypes of invasiveness and metastasis.

The chemical inhibition of metabolic cooperation might induce or help a cell to "mimic" some of the transformed or malignant phenotypes. Tumor promotion and, therefore, tumor promoters, by inhibiting metabolic cooperation and facilitating cell proliferation of the initiated cell, could enhance the chances of additional genetic changes (mutations or stable epigenetic events) that would permanently suppress gap junction communication.[26] In other words, promotion enhances the chances of a second and additional mutagenic events.[25,35]

The observation that some tumor promoters inhibited metabolic cooperation[36,37] has been used to develop a quantitative *in vitro* assay. Many known tumor promoters of several organ systems and several species that have been shown to inhibit metabolic cooperation were predicted and shown to be tumor promoters.[36-56] Although some tumor promoters have not been shown to inhibit metabolic cooperation *in vitro*,[57-59] it appears that the *in vitro* test conditions were not optimized to detect moderate or weak promoters.[60]

For example, a chemical might act as a tumor promoter *in vivo* by being cytotoxic, thereby allowing a surviving initiated cell to be freed from neighboring intercellular metabolic cooperation. That kind of chemical promoter would not be detected in the *in vivo* assay. If a promoter inhibited metabolic cooperation by reacting to specific cellular receptors, the absence of those receptors in the *in vitro* test cells would cause one to miss detection (i.e., hormones that would need specific receptors to act as promoters).[61]

To date, many chemicals that have either been shown to be "mutagenic" in bacterial systems but not "carcinogenic" in animals (i.e., dinitrofluorobenzene),[51] or to be nonmutagenic in bacterial and mammalian cells but to be "carcinogens" or tumor promoters in animals, have been shown to be inhibitors of metabolic cooperation (i.e., phorbol esters, saccharin, DDT, oleic and linoleic acids, phenobarbital, butylated hydroxytoluene, benzoyl peroxide, chlordane, dieldrin, aldrin, specific congeners of polybrominated biphenyls, polychorinated biphenyls, dilantin, kepone, teleocidin, anthralin). The fact that a few have inhibited metabolic cooperation but not acted as promoters *in vivo* might either be used as evidence that inhibition of metabolic cooperation has nothing to do with tumor promotion, or that the chemical was not tested properly *in vivo*. For example, valium was shown to inhibit metabolic cooperation *in vivo*,[62] but not shown to be a liver tumor promoter.[63] One might predict that if valium were to be a tumor promoter, an initiation-promotion experiment ought to be done on a brain tumor model system.

One implication of this *in vitro* system seems that *in vivo* chemical promoters that inhibit metabolic cooperation at noncytotoxic levels seem to exhibit a no-effect threshold. If the inhibition of intercellular communication by suppression of gap-junction-mediated transfer of critical metabolites is, in part, responsible for tumor promotion, then there is a theoretical basis for "no-effect" levels for single promoter exposure and for chronic exposures. In a three-dimensional space, a single initiated cell would be communicating with several normal cells. A single molecular "hit" by a chemical promoter on a cell would not be sufficient either to inactivate all the gap junctions directly or indirectly.[26,35]

Another generalization that seems to be emerging is that not all promoters are alike, although they might all act through the same mechanism (i.e., inhibit intercellular communication). For example, chronic exposure to a critical concentration of phorbol esters[64] and phenobarbital[65] seems to be required, because these chemicals are excreted and/or metabolized. Saccharin, however, has to be taken at high concentrations and regularly because, although not metabolized, it seems to be found in only certain tissues at high enough concentration before excretion (i.e., in the bladder).

Single application of nonexcreted and nonmetabolized chemicals, if given at high enough levels, appears to be sufficient to act as a promoter (i.e., polychlorinated biphenyls).[66] In these cases, this type of chemical binds to the fat cells and sets up an equilibrium in the body, thereby constantly "bathing" the system as a constant promoter.

Among the many unknown areas of the mechanism of tumor promotion is the potential synergistic effect of two or more no-effect levels of tumor promoters. Recognizing the obvious limitations of any, including this, *in vitro* assay, future studies to test the intercellular communication hypothesis of tumor promoter must be performed by validating the assay by comparative *in vitro* and *in vivo* studies, interspecies studies, and mechanism studies.

REFERENCES

1. FOULDS, L. 1975. Neoplastic Development. 2. Academic Press. London, England.
2. NOWELL, P. C. 1976. The clonal evolution of tumor cell populations. Science 194: 23–28.
3. FIALKOW, P. 1974. The origin and development of human tumors studied with cell markers. N. Engl. J. Med. 291: 26–35.
4. POSTE, G. & I. J. FIDLER. 1980. The pathogenesis of cancer metastasis. Nature 283: 139–146.
5. BERENBLUM, I. 1978. Historical perspective. In Carcinogenesis. T. J. Slaga, A. Sivak & R. K. Boutwell, Eds. 2: 1–10. Raven Press. New York, N.Y.
6. PITOT, H. C., T. GOLDSWORTHY & S. MORAN. 1981. The natural history of carcinogenesis: implications of experimental carcinogenesis in the genesis of human cancer. J. Supramol. Struct. Cell. Biochem. 17: 133–146.
7. TROSKO, J. E. & C.-C. CHANG. 1981. The role of radiation and chemicals in the induction of mutations and epigenetic changes during carcinogenesis. Adv. Radiat. Biol. 9: 1–36.
8. MINTZ, B. & K. ILLMENSEE. 1975. Normal genetically mosaic mice produced from malignant teratocarcinoma cells. Proc. Nat. Acad. Sci. USA 73: 3585–3589.
9. TROSKO, J. E. & C.-C. CHANG. Role of intercellular communication in tumor promotion. In Tumor Promotion and Cocarcinogenesis In Vitro. 3. CRC Press. Boca Raton, Fla. (In press.)
10. BARRETT, J. C. & D. G. THOMASSEN. Use of cell transformation assays for quantitative risk estimation. In Quantitative Estimation of Risk to Human Health from Chemicals. N. Nelson, Ed. John Wiley & Sons, Inc. New York, N.Y. (In press.)
11. HUANG, Y., C.-C. CHANG & J. E. TROSKO. Aphidicolin induces endoreduplication in Chinese hamster cells. Cancer Res. (In press.)
12. NAGAO, M., T. TAHAGI & T. SUGIMARA. 1978. Differences in effects of norharman with various classes of chemical mutagens and amounts of S-9. Biochem. Biophys. Res. Commun. 83: 373–378.
13. WEYMOUTH, L. A. & L. A. LOEB. 1978. Mutagenesis during *in vitro* DNA synthesis. Proc. Nat. Acad. Sci. USA 75: 1924–1928.

14. ANDERSON, D., C. R. RICHARDSON & P. J. DAVIES. 1981. The genotoxic potential of bases and nucleosides. Mutat. Res. **91:** 265–272.
15. RAMEL, C., Ed. 1973. AMBIO Special Report: Evaluation of Genetic Risks of Environmental Chemicals. Goteborgs Offsettryckeri: AB. Stockholm, Sweden.
16. BRUSICK, D., Ed. 1980. Principles of Genetic Toxicology. Plenum Press. New York, N.Y.
17. RAZIN, A. & A. D. RIGGS. 1980. DNA methylation and gene function. Science **210:** 604–610.
18. LEVINE, E. M., Y. BECK, C. W. BOONE & H. EAGLE. 1965. Contact inhibition, macromolecular synthesis and polyribosomes in cultured human diploid fibroblasts. Proc. Nat. Acad. Sci. USA **53:** 350–356.
19. POTTER, V. R. Cancer as a problem in intercellular communication: regulation by growth-inhibiting factors (chalones). *In* Progress in Nucleic Acid Research and Molecular Biology. W. E. Cohn, Ed. Academic Press. New York, N.Y. (In press.)
20. TUBIANA, M. & E. FRINDEL. 1982. Regulation of pluripotent stem cell proliferation and differentiation: the role of long-range humoral factors. J. Cell. Physiol. (Suppl. 1): 13–21.
21. TROSKO, J. E. & C.-C. CHANG & M. NETZLOFF. 1982. The role of inhibited cell-cell communication in teratogenesis. Teratogen. Carcinogen. Mutagen. **2:** 31–45.
22. TROSKO, J. E. & C.-C. CHANG. Implications of genotoxic and nongenotoxic mechanisms in carcinogenesis to risk assessment. *In* Quantitative Estimation of Risks to Human Health from Chemicals. N. Nelson, Ed. John Wiley & Sons, Inc. New York, N.Y. (In press.)
23. EMERIT, I. & P. A. CERUTTI. 1981. Tumor promoter phorbol-12-myristate-13-acetate induces chromosomal damage via indirect action. Nature **293:** 144–146.
24. TROSKO, J. E. & C.-C. CHANG. Potential role of intercellular communication in the rate-limiting step in carcinogenesis. J. College Toxicol. (In press.)
25. POTTER, V. R. 1982. A new protocol and its rationale for the study of initiation and promotion of carcinogenesis in rat liver. Carcinogenesis **3:** 1375–1379.
26. TROSKO, J. E., C.-C. CHANG & A. MEDCALF. Mechanisms of tumor promotion: potential role of intercellular communication. Cancer Invest. (In press.)
27. SHOYAB, M., T. C. WARREN & G. J. TODARO. 1981. Tissue and species distribution and developmental variation of specific receptors for biologically active phorbol and ingenol esters. Carcinogenesis **2:** 1273–1276.
28. SLAGA, T. J., S. M. FISCHER, K. NELSON & G. L. GLEASON. 1980. Studies on the mechanism of skin tumor promotion: evidence for several stages in promotion. Proc. Nat. Acad. Sci. USA **77:** 3679–3663.
29. LASKIN, J. D., R. A. MUFSON, L. PICCININI, D. L. ENGELHURDT & I. B. WEINSTEIN. 1981. Effects of the tumor promoter 12-*O*-tetradecanoylphorbol-13-acetate on newly synthesized proteins in mouse epidermis. Cell **25:** 441–449.
30. BERENBLUM, I. & V. ARMUTH. 1981. Two independent aspects of tumor promotion. Biochim. Biophys. Acta **651:** 51–63.
31. LOEWENSTEIN, W. R. 1979. Junctional intercellular communication and the control of growth. Biochim. Biophys. Acta **560:** 1–65.
32. FLAGG-NEWTON, J. L., I. SIMPSON & W. R. LOEWENSTEIN. 1979. Permeability of the cell-to-cell membrane channels in mammalian cell junction. Science **205:** 404–407.
33. GILULA, N. B. 1980. Cell-to-cell communication and development. *In* Cell Surface: Mediator of Developmental Processes. S. Subtelny & N. K. Wessells, Eds.: 23–42. Academic Press, Inc. New York, N.Y.
34. HOOPER, M. L. 1982. Metabolic cooperation between mammalian cells in culture. Biochim. Biophys. Acta **651:** 85–103.
35. TROSKO, J. E. & C.-C. CHANG. 1980. An integrative hypothesis linking cancer, diabetes and atherosclerosis: the role of mutations and epigenetic changes. Med. Hypoth. **6:** 455–468.
36. YOTTI, L. P., C.-C. CHANG & J. E. TROSKO. 1979. Elimination of metabolic cooperation in Chinese hamster cells by a tumor promoter. Science **206:** 1089–1091.
37. MURRAY, A. W. & D. J. FITZGERALD. 1979. Tumor promoters inhibit metabolic cooperation in cocultures of epidermal and 3T3 cells. Biochem. Biophys. Res. Commun. **91:** 395–401.
38. FITZGERALD, D. J. & A. W. MURRAY. 1979. Inhibition of intercellular communication by tumor-promoting phorbol esters. Cancer Res. **40:** 2935–2937.

39. UMEDA, M., K. NODA & T. ONO. 1980. Inhibition of metabolic cooperation in Chinese hamster cells by various chemicals including tumor promoters. Gann 71: 614–620.
40. TROSKO, J. E., B. DAWSON, L. P. YOTTI & C.-C. CHANG. 1980. Saccharin may act as a tumor promoter by inhibiting metabolic cooperation between cells. Nature 284: 109–110.
41. ENOMOTO, T., Y. SASASKI, Y. SHIBA, Y. KANNO & H. YAMASAKI. 1981. Tumor promoters cause a rapid and reversible inhibition of the formation and maintenance of electrical cell coupling in culture. Proc. Nat. Acad. Sci. USA 78: 5628–5632.
42. GUY, G. R., P. M. TAPLEY & A. W. MURRAY. 1981. Tumor promoter inhibition of intercellular communication between cultured mammalian cells. Carcinogenesis 2: 223–227.
43. SLAGA, T. J., A. S. P. KLEIN-SZANTO, L. C. TRIPLETT, L. P. YOTTI & J. E. TROSKO. 1981. Skin tumor promoting activity of benzoyl peroxide: a widely used free radical generating compound. Science 213: 1023–1025.
44. TROSKO, J. E., B. DAWSON & C.-C. CHANG. 1981. PBB inhibits metabolic cooperation in Chinese hamster cells in vitro: its potential as a tumor promoter. Environ. Health Perspect. 37: 179–182.
45. NODA, K., M. UMEDA & T. ONO. 1981. Effects of various chemicals including bile acids and chemical carcinogens on the inhibition of metabolic cooperation. Gann 72: 772–776.
46. WILLIAMS, G. M., S. TELANG & C. TONG. 1981. Inhibition of intercellular communication between liver cells by the liver tumor promoter 1,1,1-trichloro 2,2-bis(p-chloro phenyl)ethane. Cancer Lett. 11: 339–344.
47. NEWBOLD, R. F. & J. AMOS. 1981. Inhibition of metabolic cooperation between mammalian cells in culture by tumor promoters. Carcinogenesis 2: 243–249.
48. KINSELLA, A. R. 1981. Investigation of the effects of the phorbol ester TPA on carcinogen-induced forward mutagenesis to 6-thioguanine-resistance in V79 Chinese hamster cells. Carcinogenesis 2: 43–47.
49. TROSKO, J. E., L. P. YOTTI, S. T. WARREN, G. TSUSHIMOTO & C.-C. CHANG. 1982. Inhibition of cell-cell communication by tumor promoters. In Carcinogenesis. E. Hecker, N. E. Fusenig, W. Kunz, F. Marks & H. W. Thielmann, Eds. 7: 565–585. Raven Press. New York, N.Y.
50. TSUSHIMOTO, G., J. E. TROSKO, C.-C. CHANG & S. D. AUST. 1982. Inhibition of metabolic cooperation in Chinese hamster V79 cells in culture by various polybrominated biphenyl (PBB) congeners. Carcinogenesis 3: 181–185.
51. WARREN, S. T., D. J. DOOLITTLE, C.-C. CHANG, J. I. GOODMAN & J. E. TROSKO. 1982. Evaluation of the carcinogenic potential of 2,4-dinitrofluorobenzene and its implications regarding mutagenicity testing. Carcinogenesis 3: 139–145.
52. TSUSHIMOTO, G., J. E. TROSKO, C.-C. CHANG & F. MATSUMURA. 1982. Inhibition of intercellular communication by chlordecone (kepone) and Mirex in Chinese hamster V79 cells in vitro. Toxicol. Appl. Pharmacol. 64: 550–556.
53. TSUSHIMOTO, G., C.-C. CHANG, J. E. TROSKO & F. MATSUMURA. 1982. Cytotoxic, mutagenic and tumor-promoting properties of DDT, lindane and chlordane on Chinese hamster cells in vitro. J. Environ. Pathol. Toxicol. Oncol. 5: 397–410.
54. TSUSHIMOTO, G., S. ASANO, J. E. TROSKO & C.-C. CHANG. Inhibition of intercellular communication by various congeners of polybrominated biphenyl and polychlorinated biphenyl. In PCB's: Human and Environmental Hazards. J. Hook, Ed. Ann Arbor Science Publications. Ann Arbor, Mich. (In press.)
55. JONE, C. M., J. E. TROSKO, C.-C. CHANG, H. FUJIKI & T. SUGIMURA. 1982. Inhibition of intercellular communication in Chinese hamster V79 cells by teleocidin. Gann 73: 874–878.
56. YANCEY, S. B., J. E. EDENS, J. E. TROSKO, C.-C. CHANG & J.-P. REVEL. 1982. Decreased incidence of gap junctions between Chinese hamster V79 cells upon exposure to the tumor promoter 12-O-tetradecanoyl phorbol-13-acetate. Exp. Cell Res. 139: 329–340.
57. CHAMBERLAIN, M. 1982. The influence of mineral dusts on metabolic cooperation between mammalian cells in tissue culture. Carcinogenesis 3: 337–338.
58. KINSELLA, A. R. 1982. Elimination of metabolic cooperation and the induction of sister chromatid exchanges are not properties common to all promoting or co-carcinogenic agents. Carcinogenesis 3: 499–503.

59. BARRETT, J. C. & E. ELMORE. Inability of diethylstilbesterol to induce 6-thioguanine resistant mutants and to inhibit metabolic cooperation of V79 Chinese hamster cells. Mutat. Res. (In press.)

60. TROSKO, J. E., C. JONE, C. AYLSWORTH & G. TSUSHIMOTO. 1982. Elimination of metabolic cooperation is associated with the tumor promoters, oleic acid and authralin. Carcinogenesis **3:** 1101–1103.

61. YAGER, J. D. & R. YAGER. 1980. Oral contraceptive steroids as promoters of hepatocarcinogenesis in females. Sprague-Dawley rats. Cancer Res. **40:** 3680–3685.

62. TROSKO, J. E. & D. F. HORROBIN. 1980. The activity of diazepam in a Chinese hamster V79 lung cell assay for tumor promoters. IRCS **8:** 887.

63. HINO, O. & T. KITAGAWA. 1982. Effect of diazepam on hepatocarcinogenesis in the rat. Toxicol. Lett. **11:** 155–157.

64. VERMA, A. K. & R. K. BOUTWELL. 1980. Effects of dose and duration of treatment with the tumor promoting agent, 12-*O*-tetradecanoyl-phorbol-13-acetate on mouse skin carcinogenesis. Carcinogenesis **1:** 271–276.

65. PERAINO, C., E. F. STAFFELDT, D. A. HAUGEN, L. S. LOMBARD, F. J. STEVENS & R. J. M. FRY. 1980. Effects of varying the dietary concentration of phenobarbital on its enhancement of 2-acetylaminofluorene-induced hepatic tumorigenesis. Cancer Res. **40:** 3268–3273.

66. PEREIRA, M. A., S. L. HERREN, A. L. BRITT & M. M. KHOURY. 1982. Promotion by polychlorinated biphenyls of enzyme-altered foci in rat liver. Cancer Lett. **15:** 185–190.

GENOTOXIC AND EPIGENETIC CARCINOGENS: THEIR IDENTIFICATION AND SIGNIFICANCE

Gary M. Williams*

Naylor Dana Institute for Disease Prevention
American Health Foundation
Valhalla, New York 10595

INTRODUCTION

The induction of cancer by chemicals is a process that consists of a series of steps. These comprise two principal sequences, the conversion of the normal cell to a neoplastic cell and the progression of the neoplastic cell to formation of a tumor. Chemicals are involved in this process at several points in both of the sequences, primarily as "initiating" agents, which produce neoplastic conversion most likely as a result of reaction with DNA, and as promoting agents, which facilitate the development of altered cells into tumors. Effects by chemicals in both sequences can lead to an increase of cancer in animals, and consequently, chemicals can be carcinogenic by a variety of mechanisms.

In 1977, a proposal was made to categorize carcinogens into two types, genotoxic, which are capable of reacting with and damaging DNA, and epigenetic, which do not damage DNA but produce other biological effects that result in the production of tumors.[1,2] This basic dichotomy has been recognized by other investigators[3,4] and working groups.[5,6] Specific classes of carcinogens have been assigned to these two categories (TABLE 1) based upon the established capacity of representatives to produce DNA damage or other biological effects.[7]

IDENTIFICATION OF GENOTOXIC CARCINOGENS

A carcinogen can be determined to be genotoxic either by demonstrating through biochemical techniques that it damages DNA or by showing that it is active in tests that reliably measure DNA damage either directly or indirectly as mutagenesis or chromosomal effects.[8] Genotoxic carcinogens consist mainly of those that either in their parent form or after biotransformation by enzyme systems give rise to electrophilic reactants. In addition, some metal carcinogens have displayed activity in short-term tests indicative of genotoxicity.

The correlation between genotoxicity per se and carcinogenicity is very high,[9] and consequently, genotoxicity in several test systems may be taken as presumptive evidence of carcinogenicity.[9,10] For the detection of genotoxic carcinogens, a battery of short-term tests (TABLE 2) has been recommended.[9,10] Positive results for DNA damage, mutagenesis, and chromosomal effects are almost certainly indicative of carcinogenic potential and, regardless, demonstrate a definite genotoxic hazard of the chemical. Other combinations of test results may be equally significant. For example, positive results in our hepatocyte primary culture/DNA-repair test[11–13] and the Ames *Salmonella*/microsome test have always been associated with carcinogenic activity.[11,14]

*During the performance of most of the work reported in this paper, I was supported by Grant CA 17613 from the National Cancer Institute.

328

The correlation between genotoxicity and carcinogenicity is supported by the findings of a recent International Agency for Research on Cancer (IARC) working group.[15] Evaluation of the genotoxicity of chemicals tested for carcinogenicity revealed that for 41 chemicals classified as genotoxic, the results of animal studies on 88% were judged as showing sufficient or limited evidence of carcinogenicity (TABLE 3). Thus, a battery of short-term tests for genotoxicity is a reliable means for detecting one type of chemical carcinogen.

In addition to direct genotoxins, certain agents may be indirectly genotoxic, as in the case of metal carcinogens that alter the fidelity of DNA polymerases.[16] Also, toxic agents generate reactive oxygen species, which may damage DNA. Similarly, Shank and colleagues have shown that certain agents can produce aberrant methylation of DNA under toxic conditions.[17] In addition, production of other types of genetic effects, such as aneuploidy, that do not involve direct DNA damage may be important in

TABLE 1

CLASSIFICATION OF CARCINOGENIC CHEMICALS

Category and Class	Example
A. Genotoxic carcinogens	
1. Activation independent	alkylating agent
2. Activation dependent	polycyclic aromatic hydrocarbon, nitrosamine
3a Inorganic*	metal
B. Epigenetic carcinogens	
3b Inorganic*	metal
4. Solid state	plastics
5. Cytotoxic	nitrilotriacetic acid
6. Hormone modifying	estrogen
	amitrole
7. Immunosuppressor	purine analogue
8. Cocarcinogen	phorbol ester
	ethanol
9. Promoter	organochlorine pesticides
	saccharin

*Some are tentatively categorized as genotoxic because of evidence for damage of DNA; others may operate through epigenetic mechanisms such as alterations in fidelity of DNA polymerases.

carcinogenesis.[18] Thus, the mechanistic classification of carcinogens will undoubtedly require refinement.

IDENTIFICATION OF EPIGENETIC CARCINOGENS

Epigenetic carcinogens are defined as being nongenotoxic and producing a biological effect that could account for their carcinogenicity. Thus, the first step in identifying an epigenetic carcinogen is to establish that it does not damage DNA. This may be done using biochemical studies or short-term tests for genotoxicity. Then, evidence must be developed on other biological effects. Generally, this would be done in whole-animal studies, but *in vitro* systems are now becoming available for identifying one class of epigenetic agents, tumor promoters.[8,19,20]

The *in vivo* carcinogenic effects of a chemical provide indications of possible

TABLE 2

DECISION POINT FOR CARCINOGEN TESTING: *In Vitro* Battery*

1. Mammalian DNA repair
2. Bacterial mutagenesis
3. Mammalian mutagenesis
4. Sister chromatid exchange
(5) Cell transformation†

*From Reference 10.
†Optional.

epigenetic effects. For example, an agent that produces tumors only in hormonally responsive tissues may do so as a consequence of perturbation of the endocrine system. Another effect suggestive of an epigenetic action is the production of tumors only in a single organ. As an example of this, nitrilotriacetic acid has been negative in a variety of short-term tests and produces tumors only in the urinary tract. Extensive research by Anderson and colleagues has shown that this selective effect occurs only at doses that are toxic to the kidney.[21] Similarly, saccharin, which produces only bladder cancer in low yield, is nongenotoxic[22] and its administration results in a promoting action in bladder carcinogenesis.[23,24]

Of interest to this laboratory, several organochlorine pesticides have been found to produce tumors exclusively or predominantly in the livers of mice and rats.[20] These have been generally nongenotoxic in several different types of short-term tests[8] and, importantly, do not elicit DNA repair in hepatocytes from mice, rats, or hamsters.[25] One organochlorine pesticide, DDT, has been reported to be a liver tumor promoter,[26] and therefore, we have been investigating the possibility that these chemicals produce liver tumors through a promoting mechanism.[20]

An important *in vitro* approach to the study of tumor promotion has become available through the work of Yotti *et al.* and Murray and Fitzgerald, who described the action of tumor promoters in inhibiting intercellular communication.[27,28] In a completely liver-derived system, we confirmed this effect for the liver tumor promoter phenobarbital[8] and have now shown that the organochlorine pesticides DDT, chlordane, and heptachlor inhibit intercellular communication between cultured liver cells,[29,30] in support of studies by Trosko *et al.*[19]

Inhibition of intercellular communication *in vivo* by tumor promoters could serve

TABLE 3

CARCINOGENICITY SUMMARY OF IARC EVALUATIONS OF CHEMICALS WITH SUFFICIENT EVIDENCE OF GENOTOXICITY*

Evidence of Carcinogenicity	Chemicals
Sufficient	27
Limited	9
Inadequate	bleomycin
	chloroprene
	6-mercaptopurine
	methotrexate
	1-naphthylamine

*Data extracted from Reference 15.

to release dormant tumor cells from growth control by surrounding normal cells. This action would be part of the second sequence of steps in tumor development, the progression of neoplastic cells to formation of tumors. The production of liver tumors in mice and rats by nongenotoxic organochlorine pesticides may result from the promoting effect of these compounds on preexisting cells with an abnormal genotype.[31] Importantly, if it is established that a membrane effect is the basis for tumor promotion, this would represent a true epigenetic effect.

SIGNIFICANCE OF GENOTOXIC AND EPIGENETIC CARCINOGENS

The concept that carcinogens produce their effects by different modes of action has sound experimental support, but requires further elaboration. Nevertheless, the information already accrued has some definite implications.

In the evaluation of potential carcinogenicity of a chemical, reliable evidence of genotoxicity obviates the need for further testing since genotoxins are virtually certain to be carcinogenic under some conditions. However, the absence of activity in short-term tests does not preclude carcinogenicity through nongenotoxic mechanisms

TABLE 4

CHARACTERISTICS OF CARCINOGENICITY OF GENOTOXIC CARCINOGENS

1. Occasionally carcinogenic with single exposure.
2. Frequently carcinogenic at low (i.e., subtoxic) doses.
3. Can have additive or synergistic effects with one another.
4. Can be active transplacentally, and carcinogenicity often increased in neonates.
5. Subcarcinogenic effects can be made manifest by subsequent promoting action.
6. Effects can be enhanced by cocarcinogens.
7. Organotropism shifted by inhibitors of biotransformation.

in chronic studies. Thus, for a compound that is inactive in short-term tests, the decision to proceed to long-term studies must be based upon the type and extent of potential human exposure.

The lack of activity of epigenetic carcinogens, such as asbestos, hormones, and pesticides, in tests for genotoxicity has been interpreted by some as evidence of a lack of sensitivity of the tests. According to the concept of genotoxic and epigenetic carcinogens, however, such negative results are seen to be a true reflection of the biological activity of certain chemicals and to provide information on their mechanism of carcinogenicity. Thus, it is important that such negative results for nongenotoxic agents not be used to calculate misleading low percentages of correlation between short-term test results and carcinogenicity.

Another major implication of the concept that carcinogens operate through different mechanisms is that the hazard evaluation for each agent must take this into account.[32] Genotoxic carcinogens vary greatly in their potency, but as a group, their carcinogenic effects can pose extreme hazards (TABLE 4). In contrast, for at least one class of epigenetic agents, tumor promoters, the characteristics of their carcinogenic and promoting effects are quite different (TABLE 5). Therefore, these two types of agents represent different kinds of hazards to human health.[33,34]

In conclusion, the distinction between different types of carcinogens has a solid

TABLE 5

CHARACTERISTICS OF ACTIVITY OF PROMOTERS

1. Not demonstrated to be active with single exposure.
2. May be active at low dose, but require a level of exposure to produce relevant biological effect.
3. Additivity uncertain. Can inhibit one another.
4. Only diethylstilbestrol active transplacentally.
5. No evidence of enhanced susceptibility of neonates.
6. Shifts in organotropism not reported.

scientific base and leads to new approaches to the evaluation of data from short-term tests and carcinogenicity bioassays.

ACKNOWLEDGMENT

The constant collaboration of my colleague Dr. John H. Weisburger is gratefully acknowledged.

REFERENCES

1. WILLIAMS, G. M. 1979. A comparison of in vivo and in vitro metabolic activation systems. In Critical Reviews in Toxicology—Strategies for Short-term Testing for Mutagens/Carcinogens. B. Butterworth, Ed.: 96–97. CRC Press. West Palm Beach, Fla.
2. WILLIAMS, G. M. 1979. The status of in vitro test systems utilizing DNA damage and repair for the screening of chemical carcinogens. J. Assoc. Off. Anal. Chem. 62: 857–863.
3. KROES, R. 1979. Animal data, interpretation and consequences. In Environmental Carcinogenesis. P. Emmelot & E. Kriek, Eds.: 287–302. Elsevier/North-Holland Biomedical Press. Amsterdam, the Netherlands.
4. KOLBYE, A. C. 1980. Impact of short-term screening on regulatory action. In The Predictive Value of Short-Term Screening Tests in Carcinogenicity Evaluation. G. M. Williams, R. Kroes & H. W. Waaijers, Eds. 3: 311–326. Elsevier/North-Holland Biomedical Press. Amsterdam, the Netherlands.
5. Health Council of the Netherlands. 1980. The Evaluation of the Carcinogenicity of Chemical Substances. Government Publishing Office. The Hague, the Netherlands.
6. International Commission for Protection Against Environmental Mutagens and Carcinogens. 1983. Report of ICPEMC task group on the differentiation between genotoxic and nongenotoxic carcinogens. ICPEMC, Medical Biological Laboratory TNO. Rijswijk, the Netherlands.
7. WEISBURGER, J. H. & G. M. WILLIAMS. 1980. Chemical carcinogens. In Toxicology, The Basic Science of Poisons. J. Doull, C. D. Klaasen & M. O. Amdur, Eds. 2nd edit.: 84–138. Macmillan Publishing Co. Inc. New York, N.Y.
8. WILLIAMS, G. M. 1980. Classification of genotoxic and epigenetic hepatocarcinogens using liver culture assays. Ann. N.Y. Acad. Sci. 349: 273–282.
9. WILLIAMS, G. M. & J. H. WEISBURGER. 1981. Systematic carcinogen testing through the decision point approach. Annu. Rev. Pharmacol. Toxicol. 21: 393–416.
10. WEISBURGER, J. H. & G. M. WILLIAMS. 1981. Carcinogen testing. Current problems and new approaches. Science 214: 401–407.
11. WILLIAMS, G. M. 1980. The detection of chemical mutagens/carcinogens by DNA repair and mutagenesis in liver cultures. In Chemical Mutagens. F. J. de Serres & A. Hollaender, Eds. 6: 61–79. Plenum Press. New York, N.Y.
12. WILLIAMS, G. M. 1981. Liver culture indicators for the detection of chemical carcinogens.

In Short Term Tests for Chemical Carcinogens. R. H. C. San & H. F. Stich, Eds.: 581–609. Springer-Verlag. New York, N.Y.

13. WILLIAMS, G. M. 1981. The detection of genotoxic chemicals in the hepatocyte primary culture/DNA repair test. Gann Monogr. Cancer Res. **27**: 47–57.

14. WILLIAMS, G. M., M. F. LASPIA & V. C. DUNKEL. 1982. Reliability of the hepatocyte primary culture/DNA repair test in testing of coded carcinogens and noncarcinogens. Mutat. Res. **97**: 359–370.

15. International Agency for Research on Cancer. 1982. IARC Monographs on the Evaluation of the Carcinogenic Risk of Chemicals to Humans. Supplement 4. IARC. Lyon, France.

16. SIROVER, M. A. & L. A. LOEB. 1976. Metal-induced infidelity during DNA synthesis. Proc. Nat. Acad. Sci. USA **73**: 2331–2335.

17. BARROWS, L. R. & R. C. SHANK. 1981. Aberrant methylation of liver DNA in rats during hepatotoxicity. Toxicol. Appl. Pharmacol. **60**: 334–345.

18. PARRY, J. M., E. M. PARRY & J. C. BARRETT. 1981. Tumour promoters induce mitotic aneuploidy in yeast. Nature **294**: 263–265.

19. TROSKO, J. E., L. P. YOTTI, B. DAWSON & C.-C. CHANG. 1981. In vitro assay for tumor promoters. *In* Short-Term Tests for Chemical Carcinogens. H. F. Stich & R. H. C. San, Eds.: 420–427. Springer-Verlag. New York, N.Y.

20. WILLIAMS, G. M. 1981. Liver carcinogenesis: the role for some chemicals of an epigenetic mechanism of liver tumor promotion involving modification of the cell membrane. Food Cosmet. Toxicol. **19**: 577–583.

21. ANDERSON, R. L., C. L. ALDEN & J. A. MERSKI. 1982. The effects of nitrilotriacetate on cation disposition and urinary tract toxicity. Food Cosmet. Toxicol. **20**: 105–122.

22. ASHBY, J., J. A. STYLES, D. ANDERSON & D. PATON. 1978. Saccharin: a possible example of an epigenetic carcinogen/mutagen. Food Cosmet. Toxicol. **16**: 95–103.

23. HOOSON, J., R. M. HICKS, P. GRASSO & J. CHOWANIEC. 1980. Ortho-toluene sulphonamide and saccharin in the promotion of bladder cancer in the rat. Br. J. Cancer **42**: 129–147.

24. COHEN, S. M., M. ARAI, J. B. JACOBS & G. H. FRIEDELL. 1979. Promoting effect of saccharin and DL-tryptophan in urinary bladder carcinogenesis. Cancer Res. **39**: 1207–1217.

25. MASLANSKY, C. J. & G. M. WILLIAMS. 1981. Evidence for an epigenetic mode of action in organochlorine pesticide hepatocarcinogenicity: a lack of genotoxicity in rat, mouse and hamster hepatocytes. J. Toxicol. Environ. Health **8**: 121–130.

26. PERAINO, C., R. J. M. FRY & D. D. GRUBE. 1978. Drug-induced enhancement of hepatic tumorigenesis. *In* Carcinogenesis, Mechanisms of Tumor Promotion and Cocarcinogenesis. T. J. Slaga, A. Sivak & R. K. Boutwell, Eds.: 421–432. Raven Press. New York, N.Y.

27. YOTTI, L. P., C.-C. CHANG & J. E. TROSKO. 1979. Elimination of metabolic cooperation in Chinese hamster cells by a tumor promoter. Science **206**: 1089–1091.

28. MURRAY, A. W. & D. J. FITZGERALD. 1979. Tumor promoters inhibit metabolic cooperation in cocultures of epidermal and 3T3 cells. Biochem. Biophys. Res. Commun. **91**: 395.

29. WILLIAMS, G. M., S. TELANG & C. TONG. 1981. Inhibition of intercellular communication between liver cells by the liver tumor promoter 1,1,1-trichloro-2,2-bis (*P*-chlorophenyl) ethane (DDT). Cancer Lett. **11**: 339–344.

30. TELANG, S., C. TONG & G. M. WILLIAMS. 1982. Epigenetic membrane effects of a possible tumor promoting type on cultured liver cells by the nongenotoxic organochlorine pesticides chlordane and heptachlor. Carcinogenesis **3**: 1175–1178.

31. WILLIAMS, G. M. 1980. The pathogenesis of rat liver cancer caused by chemical carcinogens. Biochim. Biophys. Acta **605**: 167–189.

32. WEISBURGER, J. H. & G. M. WILLIAMS. 1981. Basic requirement for health risk analysis: the decision point approach for systematic carcinogen testing. *In* Proceedings of the Third Life Sciences Symposium on Health Risk Analysis. C. R. Richmond, P. J. Walsh & E. D. Copenhauer, Eds.: 249–271. Franklin Press. Philadelphia, Pa.

33. WILLIAMS, G. M. 1983. Epigenetic effects of liver tumor promoters and implications for health effects. Environ. Health Perspect. (In press.)

34. WILLIAMS, G. M. & J. H. WEISBURGER. 1983. Risk assessment of dietary carcinogens and tumor promoters. *In* Diet and Cancer: From Basic Research to Policy Implications. T. C. Campbell & D. Schottenfeld, Eds. A. R. Liss, Inc. New York, N.Y. (In press.)

IDENTIFICATION OF GENOTOXINS: A CORRELATION OF BACTERIAL MUTATION WITH HEPATOCYTE DNA REPAIR

G. S. Probst, C. Z. Thompson, L. E. Hill, J. K. Epp, S. B. Neal,
T. J. Oberly, and B. J. Bewsey

Toxicology Division
Lilly Research Laboratories
Division of Eli Lilly and Company
Greenfield, Indiana 46140

INTRODUCTION

Since the mid-1970s a variety of short-term tests have been developed for the identification of mutagens and carcinogens.[1] In addition to the *Salmonella*/mammalian microsome test of Ames,[2-4] which has found widespread application, other tests that measure direct DNA damage and repair,[5-9] mutation in cultured mammalian cells,[10-12] chromosomal effects,[13-17] and *in vitro* cell transformation[18-19] have also been proposed as predictive for carcinogenic chemicals.

A common feature of these systems is that the end point measured reflects the interaction of a chemical with DNA and, therefore, only carcinogens that express their activity through alterations to the genetic material are detected. Accordingly, this new research area has been named genetic toxicology and agents that elicit a positive response in these tests are operationally defined as genotoxins.[20] Within this framework, substances for which there is evidence of carcinogenicity and that are believed to exert a carcinogenic effect through pathways not involving DNA, such as hormonal imbalance, immune suppression, chronic tissue injury, or solid-state effects, are operationally classified as epigenetic carcinogens.[20] Presently, no reliable test for epigenetic carcinogens has been developed; however, recent progress with assays for tumor promoters[21,22] may contribute to our understanding of epigenetic carcinogenesis.

Considerable progress has been made in the development of tests for genotoxic chemicals, and an important outcome of the efforts from many laboratories has been the recognition that no single system is a definitive predictor of genotoxicity. Consequently, many industrial laboratories have adopted a testing strategy incorporating a battery of assays, which provides a biological hierarchy of testing using systems having diverse genetic end points and dissimilar mechanisms for metabolic activation.

For the past several years this laboratory has been committed to the development of a genetic toxicology test battery that now serves to interface with a research and development program to predict the genotoxicity of candidate pharmaceuticals and agriculturals. The test battery is conducted in tier fashion using systems of increasing biological complexity and relevance, and includes tests for bacterial mutation, DNA repair, mammalian cell mutation, and chromosomal effects *in vivo*. The largest number of compounds have been tested in the bacterial mutation and hepatocyte DNA-repair assays, and results in these systems provide the basis for the following discussion. The relatively large data base developed in these assays has allowed the selection of more interesting compounds for evaluation in the latter two tests. Pertinent findings will also be discussed below. The experience with this battery has reinforced the belief that no single test can adequately predict mutagenic or carcinogenic potential and that the collective findings of a test battery are essential for the comprehensive assessment of genotoxicity.

334

0077-8923/83/0407-0334 $01.75/0 © 1983, NYAS

A GENETIC TOXICOLOGY TEST BATTERY

Point mutations in bacteria were identified using a modification of the Ames test[2-4] as described by McMahon.[23,24] Both base-pair substitution and frameshift mutations were detected in this system, which employs eight histidine auxotrophs of *Salmonella typhimurium* (G46, TA1535, TA100, C3076, TA1537, D3052, TA1538) and two tryptophan auxotrophs of *Escherichia coli* (WP2, WP2uvrA⁻), which were simultaneously exposed to a 10,000-fold continuous concentration gradient of the test compound in agar. Metabolic activation was achieved by the incorporation of a rat liver microsomal fraction (S-9), and mutation was scored qualitatively by estimating the concentration over which mutant colonies appeared along the chemical gradient.

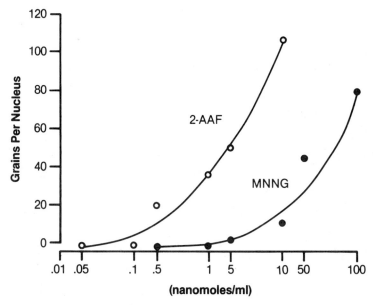

FIGURE 1. Dose-related positive responses for unscheduled DNA synthesis in cultured hepatocytes treated with N-methyl-N'-nitro-N-nitrosoguanidine (MNNG, filled circles) or 2-acetylaminofluorene (2-AAF, open circles).

Mutagenic potency was judged from the number of tester strains reverted as well as from the concentration range over which revertants were observed.

Chemically induced DNA damage was measured by the identification of unscheduled DNA synthesis (UDS) in primary cultures of adult rat hepatocytes as described by Williams[5,25,26] and Probst *et al.*[7] UDS was quantified by enumerating the number of silver grains over the cell nucleus in autoradiograms prepared from treated hepatocyte cultures. Primary cultures of adult rat hepatocytes are nondividing, yet retain the enzymes necessary for the metabolic activation of most procarcinogens, and consequently the system provided metabolic competency and a molecular target in the same cell wherein DNA-repair synthesis was not confounded by DNA-replicative synthesis. Dose-dependent positive responses were obtained for both direct-acting and activation-dependent genotoxins (FIGURE 1).

In mammalian cells, chemically induced mutation was measured in L5178Y TK$^{+/-}$ mouse lymphoma cells as described by Clive et al.[10,27] with modification by Amacher et al. and Oberly et al.[11,28,29] In this test, termed the mouse lymphoma assay, mutation was scored by the identification of mutant L5178Y TK$^{-/-}$ cells, which are deficient for thymidine kinase (TK) activity and were, therefore, resistant to the lethality of the DNA base analogue trifluorothymidine (TFT). Mutation was quantified by counting the number of colonies in soft-agar medium containing TFT, and dose-related positive responses for ultimate carcinogens and procarcinogens were obtained (FIGURE 2). The system required activation by rat liver microsomal enzymes and was redundant with the bacterial mutation test in terms of metabolic activation.

Chemically induced chromosomal alteration was measured in vivo by the quantification of sister chromatid exchange (SCE) in metaphase chromosomes from bone marrow of Chinese hamsters using an adaptation of the method of Allen et al.[30] as described by Neal and Probst.[31] The system was sensitive to direct-acting and activation-dependent genotoxins[31] (FIGURE 3) and has been reported to be more sensitive[32-34] and is more easily scored than conventional cytogenetic evaluations of chromosomal damage. The in vivo SCE assay is regarded as especially useful since it provides the advantage of various routes of chemical administration, multiple sites and pathways for metabolic activation and detoxification, and target cells that are removed from major sites of metabolism and requires the systemic distribution of genotoxic metabolites for substances that produce positive results.

CORRELATION OF BACTERIAL MUTATION AND HEPATOCYTE DNA REPAIR

Among the four tests in the battery, the largest number of compounds have been tested in the bacterial mutation and hepatocyte DNA-repair assays (1,942 and 555, respectively). Of these, 368 chemicals were tested in both systems; however, 116 were proprietary compounds and therefore are not discussed. Results for the remaining 252 compounds are summarized in TABLE 1 and have been reported elsewhere in greater detail.[7-9,35] The compounds have been arbitrarily classified by general structural or functional characteristics, and the distribution of positive and negative findings is further summarized in TABLE 2.

For the polycyclic aromatic hydrocarbons, agreement between the two tests occurred for 87% of the compounds. Two compounds, 1,2-benzanthracene and 9,10-dimethylanthracene, produced positive results only in hepatocytes.

Agreement for findings with aromatic amines was 86%, and two compounds, 3,5-diaminobenzoic acid and 9-aminofluorene, produced positive responses only in bacteria. In the case of 9-aminofluorene, the biological significance of the bacterial response was questionable since metabolic activation was not required.

Results with biphenyls showed an 83% agreement between tests. Three compounds produced positive responses only in bacteria. These were all ethynyl derivatives and did not require activation. This result was unusual since all other positive responses involved nitro- or amino-substituted biphenyls.

A large number of aniline derivatives were tested, and like the biphenyls, many of these compounds are also classifiable as nitroaromatics or aromatic amines. For the anilines listed in TABLE 1, agreement between the tests was obtained for 33 of 47 compounds (70%). Of the 14 positive responses, 11 occurred in bacteria and 3 in hepatocytes. Among the 11 bacterial positives were 5 nitro compounds that would not have been expected to be positive in hepatocytes.[7] The carcinogens o-toluidine, m-toluidine, and o-anisidine were not positive in either system.

Only 8 of 24 nitro compounds (33%) produced the same response in both tests.

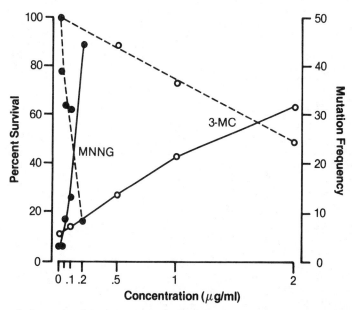

FIGURE 2. Dose-related responses for mutagenicity and toxicity in L5178Y cells treated *in vitro* with *N*-methyl-*N'*-nitro-*N*-nitrosoguanidine (MNNG, filled circles) or 3-methylcholanthrene (3-MC, open circles). Values for survival are indicated by the broken line, while values for mutation are indicated by the solid line.

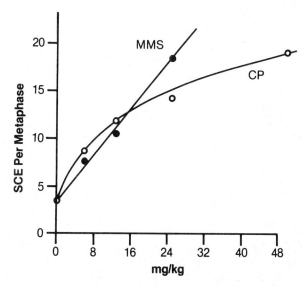

FIGURE 3. Dose-related positive responses for the *in vivo* induction of sister chromatid exchange in bone marrow of Chinese hamsters treated with methylmethanesulfonate (MMS, filled circles) or cyclophosphamide (CP, open circles).

TABLE 1

COMPARISON OF RESULTS FOR BACTERIAL MUTATION AND HEPATOCYTE DNA REPAIR*

Compound	Bacterial Mutation	Hepatocyte DNA Repair
Polycyclic aromatics		
Benzo[a]pyrene	+†	+
3-Methylcholanthrene	+†	+
2,3-Benzanthracene	+†	+
1,2,3,4-Dibenzanthracene	+†	+
7,12-Dimethylbenzanthracene	+†	+
1,2-Benzanthracene	−	+
9,10-Dimethylanthracene	−	+
Pyrene	−	−
Benzo[e]pyrene	−	−
Anthracene	−	−
Phenanthrene	−	−
2-Acetylphenanthrene	−	−
1,2,5,6-Dibenzanthracene	−	−
3,4,9,10-Dibenzpyrene	−	−
Fluorene	−	−
Aromatic amines		
2-Acetylaminofluorene	+†	+
2,7-Diacetamidofluorene	+†	+
2,7-Diaminofluorene	+†	+
2-Amino-9-fluorenone	+†	+
4-Amino-9-fluorenone	+†	+
2-Naphthylamine	+†	+
2-Aminoanthracene	+†	+
2-Aminofluorene	+†	+
2-Aminonaphthalene	+†	+
4-(4-Dimethylamino-3,5-dinitrophenyl)-maleimide	+†	+
3,5-Diaminobenzoic acid	+†	−
9-Aminofluorene	+	−
1-Naphthylamine	−	−
1-Aminoanthracene	−	−
Acridines		
Acridine yellow	+†	+
3,6-Diaminoacridine	+†	+
Acriflavin	+	+
ICR 191E	+	+
ICR 170F	+	+
3,6-Di-(dimethylamino)acridine	+	+
5,7-Dimethyl-1,2-benzacridine	+	+
9-Aminoacridine	+	−
Acridine	+	−
Dibenz[a,i]acridine	−	−
Anilines		
o-Phenylenediamine	+	+
m-Phenylenediamine	+	−
p-Phenylenediamine	+	−
4-Nitro-o-phenylenediamine	+	−
3-Nitro-o-phenylenediamine	+	−
2-nitro-p-phenylenediamine	+	−
m-Nitroaniline	+	−
p-Nitroaniline	+	−

TABLE 1   (*continued*)

Compound	Bacterial Mutation	Hepatocyte DNA Repair
Anilines (*continued*)		
o-Hydroxyaniline	+	−
m-Acetylaniline	+	−
m-Cyanoaniline	+	−
p-Trifluoromethylaniline	+ †	−
p-Toluidine	−	+
p-Ethylaniline	−	+
p-Bromoaniline	−	+
o-Toluidine	−	−
m-Toluidine	−	−
o-Ethylaniline	−	−
m-Ethylaniline	−	−
o-Nitroaniline	−	−
m-Hydroxyaniline	−	−
p-Hydroxyaniline	−	−
o-Anisidine	−	−
m-Anisidine	−	−
p-Anisidine	−	−
o-Phenetidine	−	−
m-Phenetidine	−	−
p-Phenetidine	−	−
o-Carboxyaniline	−	−
m-Carboxyaniline	−	−
p-Carboxyaniline	−	−
o-Acetylaniline	−	−
p-Acetylaniline	−	−
o-Cyanoaniline	−	−
p-Cyanoaniline	−	−
o-Chloroaniline	−	−
m-Chloraniline	−	−
p-Chloroaniline	−	−
o-Bromoaniline	−	−
m-Bromoaniline	−	−
o-Fluoroaniline	−	−
m-Fluoroaniline	−	−
p-Fluoroaniline	−	−
o-Trifluoromethylaniline	−	−
m-Trifluoromethylaniline	−	−
N,N-Dimethyl-4-nitrosoaniline	−	−
N,N-Diethyl-4-nitrosoaniline	−	−
Nitro compounds		
2,4,7-Trinitro-9-fluorenone	+	+
2,7-Dinitrofluorene	+	+
p-Nitrobenzylbromide	+	+
Furazolidone	+	+
2-Nitrofluorene	+	−
2,4-Dinitrofluorobenzene	+	−
p-Nitrobenzylacetate	+	−
p-Nitroanisole	+	−
2,4-Dinitrophenetol	+	−
Rondizole	+	−
Ipronidazole	+	−
Dimetridazole	+	−

(*continued*)

TABLE 1 (*continued*)

Compound	Bacterial Mutation	Hepatocyte DNA Repair
Nitro compounds (*continued*)		
Metronidazole	+	−
p-Dinitrobenzene	+	−
m-Dinitrobenzene	+	−
2,4-Dinitrophenylhydrazine	+	−
3-Nitro-1-vinyl-carboxypyrazole	+	−
4-Nitropyridine-1-oxide	+	−
1-(2-Hydroxyethyl)-2-methyl-5-nitroimidazole	+	−
2,4-Dichloronitrobenzene	+	−
p-Nitrophenol	−	−
2,4-Dinitrophenol	−	−
o-Dinitrobenzene	−	−
p-Nitrobenzoic acid	−	−
Biphenyls		
Benzidine	+†	+
3,3′-Diaminobenzidine	+†	+
3,3′-Dimethoxybenzidine	+†	+
4-Aminobiphenyl	+†	+
4-Hydroxylaminobiphenyl	+	+
4-Nitrobiphenyl	+	+
4-Nitrosobiphenyl	+	+
4-Ethynylbiphenyl	+	−
4-Ethynyl-4′-fluorobiphenyl	+	−
4-Ethynyl-2′-fluorobiphenyl	+	−
Biphenyl	−	−
2-Nitrobiphenyl	−	−
3-Nitrobiphenyl	−	−
2-Aminobiphenyl	−	−
3,3′,5,5′-Tetramethylbenzidine	−	−
4-Hydroxybiphenyl	−	−
2-Hydroxybiphenyl	−	−
Azo-bis-biphenyl	−	−
Nitrosamines		
N-Nitrosodimethylamine	+†	+
N-Nitrosodiethylamine	+†	+
N-Nitrosodipropylamine	+†	+
Nitrosopyrrolidine	+	+
N-Butyl-*N*-ethyl-*N*-nitrosamine	+	+
N-Allyl-*N*-methyl-*N*-nitrosamine	+	+
N-Ethyl-*N*-(2-methylallyl)-*N*-nitrosamine	+	−
N-Propyl-*N*-(2-methylallyl)-*N*-nitrosamine	+	−
N-Nitrosodiphenylamine	−	−
Nitrosoureas and amides		
N-Methyl-*N*′-nitro-*N*-nitrosoguanidine	+	+
N-Ethyl-*N*′-nitro-*N*-nitrosoguanidine	+	+
N-Propyl-*N*′-nitro-*N*-nitrosoguanidine	+	+
N-Nitrosomethylurea	+	+
Streptozotocin	+	+
N-Methyl-*N*-nitroso-*p*-toluene-sulfonamide	+	+
Esters and carbamates		
Methylmethanesulfonate	+	+
Ethylmethanesulfonate	+	+
Dimethylsulfate	+	+
Diethylsulfate	+	+

TABLE 1 (*continued*)

Compound	Bacterial Mutation	Hepatocyte DNA Repair
Esters and carbamates (*continued*)		
Diphenylpropynyl-*N*-cyclohexyl-carbamate	+	+
Di-(*p*-fluoropenyl)propynyl-*N*-cycloheptyl carbamate	+	+
Dimethylcarbamylchloride	−	−
Miscellaneous alkylating agents		
Cyclophosphamide	+†	+
Triethylenemelamine	+	+
2-Chloroethylamine	+	+
2-Chloroethylpiperidine	+	+
2-Chloroethyldiethylamine	+	+
2-Chloroethylmorpholine	+	+
2-Chloroethylpyrrolidine	+	+
Methyl-bis-2-chloroethylamine	+	+
N-Methyl-2-chloroallylamine	+	+
Nitrogen mustard	+	+
Hycanthone	+	+
Epichlorohydrin	+	−
Azo compounds		
4-Aminoazobenzene	+†	+
4-Dimethylamino-3′-methylazobenzene	+†	+
4-Dimethylamino-2-methylazobenzene	+†	+
4-Dimethylaminoazobenzene	+†	+
Methyl yellow	+†	−
Methyl orange	+†	−
Dimethylazobenzene	−	−
Trypan blue	−	−
Benzthiazoles and benzimidazoles		
2-Amino-6-methoxy-benzthiazole	+†	+
N-Methyl-2-amino-5-methoxy-benzthiazole	+†	+
2-Methylamino-6-methoxy-benzthiazole	+†	+
2-Aminobenzimidazole	+†	−
2-Amino-4-methoxy-benzthiazole	−	−
2-Amino-6-hydroxy-benzthiazole	−	−
2-Aminobenzthiazole	−	−
2-Amino-5-methoxybenzimidazole	−	−
Pesticides and polyhalogen compounds		
Ethidium bromide	+†	+
Diquat	+	−
Captan	+	−
Dimethoate	+	−
Lindane	−	−
Carbaryl	−	−
Chlordane	−	−
Heptachlor	−	−
DDT	−	−
Aldrin	−	−
Dieldrin	−	−
Mirex	−	−
Kepone	−	−
Methoxychlor	−	−
Endrin	−	−
Rotenone	−	−

(*continued*)

TABLE 1 (*continued*)

Compound	Bacterial Mutation	Hepatocyte DNA Repair
Pesticides and polyhalogen compounds (*continued*)		
Alachlor	−	−
Arochlor 1254	−	−
2,4-Dichlorophenoxyacetic acid	−	−
Quinolines		
4-Nitroquinoline-1-oxide	+	+
6-Nitroquinoline	+	+
8-Nitroquinoline	+	−
8-Hydroxyquinoline-1-oxide	−	−
Miscellaneous compounds		
N-Phenylsydnone	+	+
N-o-Chlorophenylsydnone	+	−
3-Homoveratrylsydnone	+	−
6-Thioguanine	+	−
Sodium azide	+	−
6-Bromo-4,4-dimethylcyclohexenone	+	−
Tilorone	+	−
Acetamidrazone	+	−
1-Naphthylthiourea	+†	−
1-Naphthylisothiocyanate	−	+
2-Fluorenecarboxaldehyde	−	+
8-Methoxypsoralen	−	+
Benzene	−	−
Diphenylamine	−	−
1-Acetylimidazole	−	−
Dibenz[*a,i*]carbazole	−	−
Phenazine	−	−
8-Azaguanine	−	−
6-Azauridine	−	−
Mycophenolic acid	−	−
Pergnenolon-16-α-carbonitrile	−	−
Diethylstilbestrol	−	−
Dimethylhydrazine	−	−
Brilliant blue G	−	−
D-Ethionine	−	−
L-Ethionine	−	−
m-Chloroperoxybenzoate	−	−
Safrole	−	−
Procarbazine	−	−
Reserpine	−	−
Isoniazid	−	−
Isonicotinamide	−	−
Isonicotinic acid	−	−
Isonicotinic acid-N-oxide	−	−
Anthihistamines		
Tripelennamine	−	+
Pyrilamine maleate	−	+
Methapyrilene	−	−
Chlorcyclizine	−	−
Diphenhydramine	−	−
Pyrrobutamine	−	−
Cyclopentamine	−	−
Chlorpheniramine	−	−

TABLE 1   (*continued*)

Compound	Bacterial Mutation	Hepatocyte DNA Repair
Phenols		
Naphthoresorcinol	+	−
2,4-Dichlorophenol	−	−
1-Naphthol	−	−
2-Naphthol	−	−
6-Benzoyl-2-naphthol	−	−
1-Nitroso-2-naphthol	−	−
Resorcinol	−	−
1,8-Dihydroxyanthraquinone	−	−
1-Naphthoflavone	−	−

*Detailed results for the compounds presented in this table have been published previously.[7-9,35]

†Metabolic activation with S-9 required.

Positive results noted only in bacteria occurred with 16 of 24 compounds (67%), and as is characteristic of the bacterial response to nitro compounds, metabolic activation was not required.[3,24] No nitro compound was uniquely positive in the hepatocyte UDS assay. The insensitivity of hepatocytes to nitro compounds has been described previously,[7] and this apparent shortcoming of the UDS system is thought to result from the inability of cultured hepatocytes to carry out aerobic nitro reduction.

For the quinolines and azo compounds, a 75% correlation in results was obtained. Among the four quinolines tested, a positive response was noted only in bacterial mutation with 8-nitroquinoline. The azo compounds, most of which were aminoazo derivatives, required metabolic activation to produce positive responses in bacteria, and it is uncertain whether the metabolism of the azo function or the aromatic amine was responsible for mutagenesis. Two aminoazo compounds, methyl red and methyl orange, induced bacterial mutation but not UDS.

With the exception of diphenylnitrosamine, which is of questionable carcinogenic potential, and two methylallyl nitrosamines, all other nitrosamines produced positive responses in both tests to yield a testing correlation of 78%. The positive bacterial response for the two methallyl nitrosamines was of questionable biological significance since these compounds were active without metabolic activation, contrary to the well-recognized fact that the genotoxicity of nitrosamines requires metabolic activation.

Nitrosoureas, amides, esters, and carbamates are reactive chemicals, and many behave as alkylating agents. It was, therefore, not surprising that the majority of these compounds were active in both systems and that metabolic activation was not required. For the nitrosoureas, amides, esters, and carbamates, a 100% testing correlation was obtained. Surprisingly, dimethylcarbamyl chloride was negative in both systems; however, the result may reflect instability of this chemical in aqueous solution.

A collection of miscellaneous alkylating agents, most of which were alkyl chlorides, produced activation-independent positive responses in both systems, and the agreement in results was 92%. The only compound for which test results differed was epichlorohydrin, which was active for bacterial mutation but not for UDS.

Acridines are generally regarded as DNA-intercalating agents and may be expected to be mutagenic in the absence of metabolic activation. Aromatic amine functions are characteristic of many acridines, and in the case of acridine yellow and 3,6-diaminoacridine, these amine functions may have been important for mutagenesis

TABLE 2

SUMMARY OF FINDINGS FOR CHEMICALLY INDUCED BACTERIAL MUTATION AND
HEPATOCYTE DNA REPAIR

Chemical Category	Total Tested	Positive Both Tests	Positive Bacteria Only	Positive UDS Only	Negative Both Tests	Test Agreement*
Polycyclic aromatics	15	5	0	2	8	87%
Aromatic amines	14	10	2	0	2	86%
Biphenyls	18	7	3	0	8	83%
Anilines	47	1	11	3	32	70%
Nitro compounds	24	4	16	0	4	33%
Quinolines	4	2	1	0	1	75%
Azo compounds	8	4	2	0	2	75%
Nitrosamines	9	6	2	0	1	78%
Nitrosoureas & amides	6	6	0	0	0	100%
Acridines	10	7	2	0	1	80%
Esters & carbamates	7	6	0	0	1	100%
Benzthiazoles & benzimidazoles	8	3	1	0	4	88%
Pesticides & polyhalogens	19	1	3	0	15	84%
Phenols	9	0	1	0	8	89%
Miscellaneous alkylating agents	12	11	1	0	0	92%
Antihistamines	8	0	0	2	6	75%
Miscellaneous	34	1	8	3	22	68%
Total	252	74	53	10	115	75%
Consolidated categories						
Hydrocarbons	62	6	10	4	42	77%
Amines	93	30	13	5	45	81%
Nitro compounds	39	8	21	0	10	46%
Alkylating agents	34	29	3	0	2	91%
Miscellaneous	24	1	6	1	16	71%
Total	252	74	53	10	115	75%
Total with nitro compounds omitted	213	66	32	10	105	80%

*Test agreement was the sum of compounds that were positive and negative in both tests divided by the number of compounds tested.

since bacterial mutation with these two compounds required metabolic activation. The overall agreement between results was 80% for the acridines. Two compounds, acridine and 9-aminoacridine, were positive only in bacteria.

All of the benzthiazoles and benzimidazoles tested contained aromatic amine functions, and it was expected that these were involved in the mutagenic response since all the compounds in this class that were bacterial mutagens required metabolic activation. An interesting observation was that the three benzthiazoles that were positive in both tests also contained a methoxy function. The methoxy function as well as the amine may have been necessary for genotoxicity since 2-aminobenzthiazole was negative in both tests. 2-Aminobenzimidazole produced an activation-dependent positive response only in bacteria and was the only compound for which the result between the two tests was at variance. Consequently, a testing correlation of 88% was obtained for this class.

A variety of pesticides and related polyhalogen compounds were tested and produced generally negative responses in both systems, and an 84% testing agreement was obtained. Exceptions included diquat, captan, and dimethoate, which were active only in bacteria without metabolic activation. Among the pesticides, several were regarded as rodent carcinogens, yet none produced positive responses in both tests. Furthermore, certain of these organochloride pesticides were reported as more active hepatocarcinogens in mice than in rats, yet UDS studies with mouse hepatocytes did not show a positive response with DDT, chlordane, or heptachlor.[20] Similarly, UDS

TABLE 3

BACTERIAL MUTATION AND HEPATOCYTE DNA REPAIR: RESPONSE TO NITRO COMPOUNDS

Compound	Bacterial Mutation	Hepatocyte DNA Repair
4-Nitroquinoline-1-oxide	+	+
6-Nitroquinoline	+	+
2,7-Dinitrofluorene	+	+
2,4,7-Trinitro-9-fluorenone	+	+
p-Nitrobenzylbromide	+	+
Furazolidone	+	+
4-Nitrobiphenyl	+	+
4-Nitrosobiphenyl	+	+
8-Nitroquinoline	+	−
2-Nitrofluorene	+	−
2,4-Dinitrofluorobenzene	+	−
p-Nitrobenzylacetate	+	−
p-Nitroanisole	+	−
2,4-Dinitrophenetol	+	−
Rondizole	+	−
Ipronidazole	+	−
Dimetridazole	+	−
p-Dinitrobenzene	+	−
m-Dinitrobenzene	+	−
2,4-Dinitrophenylhydrazine	+	−
3-Nitro-1-vinyl-carboxypyrazole	+	−
4-Nitropyridine-1-oxide	+	−
2,4-Dichloronitrobenzene	+	−
1-(2-Hydroxyethyl)-2-methyl-5-nitromidazole	+	−
4-Nitro-*o*-phenylenediamine	+	−
3-Nitro-*o*-phenylendiamine	+	−
2-Nitro-*p*-phenylenediamine	+	−
m-Nitroaniline	+	−
p-Nitroaniline	+	−
o-Nitroaniline	−	−
2-Nitrobiphenyl	−	−
3-Nitrobiphenyl	−	−
p-Nitrophenol	−	−
2,4-Dinitrophenol	−	−
o-Dinitrobenzene	−	−
p-Nitrobenzoic acid	−	−
1-Nitroso-2-naphthol	−	−
N,N-Dimethyl-4-nitrosoaniline	−	−
N,N-Diethyl-4-nitrosoaniline	−	−

was not obtained in hamster hepatocytes after treatment with mirex or endrin.[20] Another interesting observation was that ethidium bromide, a recognized intercalating agent, which is often employed as a stain for DNA in electrophoretic experiments, was active in bacteria only with metabolic activation.

Eight of nine phenolic compounds produced negative results in both tests (89% agreement). Naphthoresorcinol was the only exception and produced activation-independent mutation in *Salmonella*.

Considering all compounds tested, the incidence of positive results unique to the hepatocyte UDS assay was extremely low (4%), and it was therefore surprising that of the eight antihistaminic compounds tested, pyrilamine and tripelennamine were positive for hepatocyte UDS but not bacterial mutation. All others were negative in both tests. It is interesting to note that the potent rat hepatocarcinogen methapyrilene[36] was not active in either test.

A large number of compounds were combined in a miscellaneous category, and agreement among test results occurred for 68% of these compounds. Eight chemicals were active only for bacterial mutation, and three compounds were active only for hepatocyte UDS. Within this group were several putative carcinogens (e.g., benzene, D-ethamine, L-ethionine, safrole, procarbazine) and several mutagens (e.g., 8-azaguanine, 6-azauridine, and dimethylhydrazine), yet all were negative in both tests. The positive UDS response with 1-naphthylisothiocyanate and 8-methoxypsoralen (without ultraviolet exposure) was unexpected.

In order to facilitate further interpretation of the findings, the chemicals were reclassified in terms of functional activity into four broad groups including hydrocarbons, amines, nitro compounds, and alkylating agents. Results for these consolidated groups are included in TABLE 2, and it is apparent that the agreement between the bacterial mutation assay and the hepatocyte DNA-repair assay for alkylating agents, hydrocarbons, and amines exceeded 80%. In contrast, 21 of 39 nitro compounds (54%) were positive only in *Salmonella* (TABLE 3), and therefore, the correlation between the systems for this class of compounds was only 46%. Testing agreement for the total 252 compounds was 75%. The apparent insensitivity of the hepatocyte system to nitro compounds is thought to result from the inability of cultured hepatocytes to carry out nitro reduction *in vivo*[37-39] and indicates that *in vitro* hepatocyte UDS results may not be appropriate for evaluating the genotoxicity of nitro compounds. However, if the hepatocyte UDS assay is conducted by the *in vivo/in vitro* technique described by Mirsalis *et al.*,[40,41] nitro compounds are detected and the response requires metabolism of the compounds by the enteric bacteria.

In recognition of the insensitivity of the *in vitro* hepatocyte UDS assay to nitro compounds, a more realistic assessment of the correlation of this test with the *Salmonella* mutation assay can be made by deleting the nitro compounds from consideration, in which case the agreement between tests becomes 80%. A final observation from the overall findings was that the bacterial assay showed a relatively high frequency of unique positives (53 of 252 = 21%) while a similar response for hepatocyte UDS occurred with only 10 of 252 compounds (4%).

COMPARATIVE RESULTS IN A GENETIC TOXICOLOGY BATTERY

A variety of established or suspect carcinogens were represented by the 252 compounds tested, and positive responses for many of these carcinogens were obtained in both tests. Coincidentally, a number of carcinogens were detected in only one test (e.g., nitro compounds, some azo compounds) or were undetected by either test (e.g., organochloro pesticides, benzene, safrole, methapyrilene, ethionine, procarbazine, *o*-toluidine, and other anilines). For this reason a number of the compounds examined in this study were further tested in the mammalian cell mutation assay and sister chromatid exchange assay (TABLE 4). Consistent positive responses were obtained in each of the tests with many recognized direct-acting and activation-dependent carcinogens/mutagens, yet structurally related analogues of several of these com-

TABLE 4

FINDINGS IN A GENETIC TOXICOLOGY TEST BATTERY

Compound	Bacterial Mutation	DNA Repair	L5178Y Mutation	In Vivo SCE
Positives				
Methyl methanesulfonate	+	+	+	+
Ethyl methanesulfonate	+	+	+	+
4-Nitroquinoline-1-oxide	+	+	+	+
N-Methyl-N'-nitro-N-nitrosoguanidine	+	+	+	+
Hycanthone	+	+	+	+
Streptozotocin	+	+	+	+
Phenylsydnone	+	+	+	+
Nitrosomethylurea	+	+	+	+
2-Acetylaminofluorene	+*	+	+*	+
4-Aminobiphenyl	+*	+	+*	+
Benzidine	+*	+	+*	+
7,12-Dimethylbenz-anthracene	+*	+	+*	+
3-Methylcholanthrene	+*	+	+*	+
Negatives				
Fluorene	−	−	−	−
3,3',5,5'-Tetramethyl-benzidine	−	−	−	−
Diphenylnitrosamine	−	−	−	−
Cyclopentamine	−	−	−	−
Ambiguous				
Methapyrilene	−	−	−	−
Isoniazid	−	−	−	−
o-Toluidine	−	−	−	−
Safrole	−	−	−	−
Tripelennamine	−	+	−	−
p-Bromoaniline	−	+	−	−
Ethidium bromide	+*	+	−	−
2-Nitro-p-phenylendiamine	+	−	+	−†
3-Nitro-o-phenylenediamine	+	−	+	NT‡
4-Nitro-o-phenylenediamine	+	−	+	−†
2-Nitrofluorene	+	−	+	−/+§
9-Aminoacridine	+	−	+	−
Pyrilamine maleate	−	+	−	+
Procarbazine	−	−	+*	+
4-Nitrobiphenyl	+	+	−	+¶
Acridine orange	+*	+	+*	−
Nitrosopyrrolidine	+*	+	+*	−

*Metabolic activation with S-9 required.
†Administered by the oral route, all other compounds administered by the intraperitoneal route.
‡NT = not tested.
§Positive by the oral route, negative by the intraperitoneal route.
¶Positive by both the oral and intraperitoneal routes.

pounds that were generally believed to be nongenotoxic provided universally negative responses. Within the battery, the genotoxicity for several compounds remained ambiguous. For example, methapyrilene, o-toluidine, isoniazid, and safrole are all considered to be rodent carcinogens, yet each was negative in the test battery. Several compounds (tripelennamine and p-bromoaniline) were active in only one test; however, bioassay data were not available to delineate the significance of these findings. Ethidium bromide is known to interact with DNA, yet positive results were obtained for this chemical in only two of four tests and the compound was not active in the *in vivo* SCE assay. The nitrophenylenediamine compounds were considered as suspect carcinogens since bioassay results were inconclusive, yet the unique positive response in bacteria was corroborated by the positive results in the mammalian cell mutation assay. Similar results for mutation in bacteria and mutation in L5178Y cells were obtained for 9-aminoacridine and 2-nitrofluorene; however, these compounds were inactive for the induction of UDS and SCE. Pyrilamine maleate, which was positive in hepatocyte UDS, also induced SCE *in vivo* but not L5178Y TK$^{-/-}$ mutants *in vitro*. The carcinogen procarbazine, which had gone undetected in the *Salmonella* and UDS assays, was detected both in the L5178Y and *in vivo* SCE assays. The nitro compound 2-nitrofluorene gave the expected positive bacterial and negative UDS responses; however, the genotoxicity of this compound was confirmed in the L5178Y assay, and also in the *in vivo* SCE assay when the compound was administered by the oral route but not when administered by the intraperitoneal route. 4-Nitrobiphenyl was positive in all but the L5178Y assay.

Ambiguous results were obtained with seven nitro compounds (TABLE 4), and the inconsistencies in responses for these suggest dissimilar mechanisms for genotoxicity. 4-Nitrobiphenyl was the only nitro compound that was positive for UDS. In addition, the positive response for 4-nitrobiphenyl in the SCE assay was obtained by administering the compound by either the oral or the intraperitoneal route. These findings suggest some unique mechanism for the genotoxicity of 4-nitrobiphenyl, yet the compound was inactive for the induction of L5178Y mutation.

The response of the L5178Y assay to nitro compounds confirmed many results in the bacterial system. This was especially interesting since these compounds induced mutation in L5178Y cells without metabolic activation, thus suggesting either that L5178Y cells contain endogenous aerobic nitroreductase activity or that the compounds act directly in these cells. This observation, however, is complicated by the fact that the phenylenediamines were not active in the *in vivo* SCE assay when administered by either the oral or the intraperitoneal route.

Similar results were obtained with 2-nitrofluorene. This compound induced S-9-independent L5178Y mutation yet, contrary to the above findings, was positive only by the oral route in the *in vivo* SCE assay, thus suggesting the potential involvement of enteric bacteria in metabolic activation.

CONCLUSION

Over 250 compounds were tested for the induction of bacterial mutation and for unscheduled DNA synthesis in primary cultures of rat hepatocytes, and the collective findings showed an agreement in test results for 75% of the compounds. An important outcome of the study was the recognition that nitro compounds were good inducers of bacterial mutation (29 of 39 positive) but were poor inducers of hepatocyte UDS (8 of 39 positive). Consequently, if nitro compounds were deleted from consideration, then an agreement of 80% was obtained between tests.

In general, both tests detected a wide variety of genotoxic carcinogens (e.g., benzo[a]pyrene, benzidine, N-methyl-N'-nitro-N-nitrosoguanidine, etc.) and discriminated against nongenotoxic analogues of carcinogens (e.g., 3,3',5,5'-tetramethylbenzidine, fluorene, 2-aminobiphenyl, etc.). However, several carcinogens (o-toluidine, safrole, methapyrilene, etc.) were not detected in either system.

A comparison of results for bacterial mutation and hepatocyte DNA with results for L5178Y mutation and *in vivo* SCE induction revealed an excellent agreement between all tests for most potent genotoxins. It was further noted that the bacterial mutation and L5178Y assays tended to agree for nitro compounds, and it was interesting that the positive responses seen with the L5178Y system occurred without exogenous metabolic activation. Furthermore, the findings for SCE induction with nitro compounds could not be exclusively linked to the role of enteric bacteria for metabolic activation.

The collective results from the test battery illustrate the fact that the interaction of chemicals with biological systems is a complex and often unpredictable process, and that simple correlative findings between different tests may not be indicative of common pathways leading to genotoxicity. It was further concluded that the outcome of a battery of tests provided a stronger sense of confidence in delineating the genotoxic potential of chemicals in spite of the fact that some carcinogenic compounds remained undetected by any test in the battery.

REFERENCES

1. HOLLSTEIN, M., J. MCCANN, F. A. ANGELOSANTO & W. NICHOLS. 1979. Mutat. Res. 65: 133–226.
2. AMES, B. N., J. MCCANN & E. YAMASAKI. 1975. Mutat. Res. 31: 347–364.
3. MCCANN, J., E. CHOI, E. YAMASAKI & B. N. AMES. 1975. Proc. Nat. Acad. Sci. USA 72: 5135–5139.
4. MCCANN, J., N. E. SPINGARN, J. KOBORI & B. N. AMES. 1975. Proc. Nat. Acad. Sci. USA 72: 979–983.
5. WILLIAMS, G. M. 1977. Cancer Res. 37: 1845–1851.
6. WILLIAMS, G. M. 1978. Cancer Lett. 6: 199–206.
7. PROBST, G. S., L. E. HILL & B. J. BEWSEY. 1980. J. Toxicol. Environ. Health 6: 335–351.
8. PROBST, G. S. & S. B. NEAL. 1980. Cancer Lett. 10: 67–73.
9. PROBST, G. S., R. E. MCMAHON, L. E. HILL, C. Z. THOMPSON, J. K. EPP & S. B. NEAL. 1981. Environ. Mutagen. 3: 11–32.
10. CLIVE, D., K. O. JOHNSON, J. F. S. SPECTOR, A. G. BATSON & M. M. M. BROWN. 1979. Mutat. Res. 59: 61–108.
11. AMACHER, D. E., S. C. PAILLET, G. N. TURNER, V. A. RAY & D. S. SALSBURG. 1980. Mutat. Res. 72: 447–474.
12. KRAHN, D. F. & C. HEIDELBERGER. 1979. Mutat. Res. 46: 27–44.
13. ODASHIMA, S. & M. ISHIDATE. 1977. Mutat. Res. 48: 337–354.
14. PERRY, P. & H. J. EVANS. 1975. Nature 258: 121–125.
15. ABE, S. & M. SASAKI. 1977. J. Nat. Cancer Inst. 58: 1635–1641.
16. STETKA, D. G. & S. WOLF. 1976. Mutat. Res. 41: 333–342.
17. LATT, S. A., R. R. SCHRECK, K. S. LOVEDAY & C. F. SHULER. 1979. Pharmacol. Rev. 30: 501–535.
18. PIENTA, R. J., J. A. POILEY & M. B. LEBHERZ III. 1977. Int. J. Cancer 19: 642–655.
19. REZNIKOFF, C. A., J. S. BERTRAM, D. W. BRANDOW & C. HEIDELBERGER. 1973. Cancer Res. 33: 3239–3249.
20. WILLIAMS, G. M. 1980. Ann. N.Y. Acad. Sci. 349: 273–282.
21. YOHI, L. P., C.-C. CHANG & J. E. TROSKO. 1979. Science 206: 1089–1091.
22. TSUSHIMOTO, G., J. E. TROSKO, C.-C. CHANG & S. D. AUST. 1982. Carcinogenesis 3: 181–185.

23. CLINE, J. C. & R. E. MCMAHON. 1977. Res. Commun. Chem. Pathol. Pharmacol. 16: 523–533.
24. MCMAHON, R. E., J. C. CLINE & C. Z. THOMPSON. 1979. Cancer Res. 39: 682–693.
25. WILLIAMS, G. M. 1978. Cancer Lett. 4: 69–75.
26. WILLIAMS, G. M., E. BERMUDEZ & D. SCARAMUZZINO. 1977. In Vitro 13: 809–817.
27. CLIVE, D. & J. F. S. SPECTOR. 1975. Mutat. Res. 31: 17–29.
28. AMACHER, D. E., S. PAILLET & V. A. RAY. 1979. Mutat. Res. 64: 391–406.
29. OBERLY, T. J., C. E. PIPER & D. S. MCDONALD. 1982. J. Toxicol. Environ. Health 9: 367–376.
30. ALLEN, J. W., C. F. SHULER, R. W. MENDES & S. A. LATT. 1977. Cytogenet. Cell Genet. 18: 231–237.
31. NEAL, S. B. & G. S. PROBST. 1982. Mutat. Res. 113: 33–43.
32. ROSZINSKY-KÖCHER, G. & G. RÖHRBORN. 1979. Hum. Genet. 46: 51–55.
33. TAKEHISA, S. 1982. In Sister Chromatid Exchange. S. Wolff, Ed.: 87–147. John Wiley & Sons, Inc. New York, N.Y.
34. WOLFF, S. 1982. In Sister Chromatid Exchange. S. Wolff, Ed.: 41–57. John Wiley & Sons, Inc. New York, N.Y.
35. PROBST, G. S., C. Z. THOMPSON & L. E. HILL. Environ. Mutagen. (Submitted.)
36. LIJINSKY, W., M. D. REUBER & B. N. BLACKWELL. 1980. Science 209: 817–819.
37. POIRIER, L. A. & J. H. WEISBURGER. 1974. Biochem. Pharmacol. 23: 661–669.
38. SAZ, A. K. & R. B. SLIE. 1954. Arch. Biochem. Biophys. 51: 5–16.
39. SWAMINATHAN, S. & G. M. LOWER, JR. 1978. In Carcinogenesis, A Comprehensive Survey. Nitrofurans: Chemistry, Metabolism, Mutagenesis, and Carcinogenesis. G. T. Bryan, Ed.: 59–107. Raven Press. New York, N.Y.
40. MIRSALIS, J. C. & B. E. BUTTERWORTH. 1982. Carcinogenesis 1: 621–625.
41. MIRSALIS, J. C., T. E. HAMM, JR., M. J. SHERRILL & B. E. BUTTERWORTH. 1982. Nature 295: 322–323.

BACTERIAL-MAMMALIAN MUTAGENESIS CORRELATIONS: MECHANISTIC SIGNIFICANCE FOR CARCINOGENESIS*

Helmut Bartsch

Division of Environmental Carcinogenesis
International Agency for Research on Cancer
F-69372 Lyon Cedex 08, France

INTRODUCTION

Although the recognition of a substance as a human carcinogen has relied until now basically on epidemiological methods, this approach suffers from two main, unavoidable limitations: (1) it cannot, in most cases, control individual exposures, as it relies almost entirely on observations in human populations; and (2) since most people are exposed to such low levels that risk cannot be measured directly, this approach can only rarely, if ever, detect slight (<10%) increases in the incidence of commonly occurring tumors.[1]

The identification of chemicals as possibly carcinogenic to humans will therefore still rely largely on results from experimental systems. Of these, long-term animal tests are still the only ones capable, in the absence of adequate human data, of providing conclusive evidence of the carcinogenic effect of a chemical.

However, mutagenicity assays (in particular, prokaryotic test systems) offer a rapid and inexpensive means of acquiring information that is useful for selecting and ranking chemicals to be submitted to long-term carcinogen bioassays.

Whether mutagenicity and other short-term assays, singly or in combination, will eventually acquire the status of long-term animal tests in predicting human hazards depends entirely on further demonstrations of their consistency with results obtained in adequately conducted human epidemiological studies and animal tests.

This article presents an analysis of the qualitative relationship between carcinogenicity and mutagenicity in bacterial and mammalian cells *in vitro*, based on chemicals that are known to be or suspected of being carcinogenic to humans and/or to experimental animals.[2] In view of the increasing demand for quantitative data for the purpose of risk assessment, the question of whether or not mutagenicity data *in vitro* can be extrapolated to processes occurring in the intact mammalian body, i.e., the extent of covalent DNA binding or carcinogenic potency, is also briefly examined.

RESULTS AND DISCUSSION

Comparisons of Bacterial and Mammalian Mutagenicity Data Based on Chemicals Definitely or Probably Carcinogenic to Humans

Chemicals evaluated in volumes 1–29 of the *IARC Monographs on the Evaluation of the Carcinogenic Risk of Chemicals to Humans* as either definitely or probably carcinogenic to humans,[3] and which have been tested in various mutagenicity and other short-term assays, offer a basis for such an analysis.

*This research was supported in part by Contract No. 190-ENV-F with the Commission of the European Communities.

351

In 1982, an *ad hoc* working group updated Supplement 1 of the *IARC Monographs,*[4] i.e., all chemicals, groups of chemicals, industrial processes, and occupational exposures for which some data on carcinogenicity in humans were available were reevaluated, on the basis both of studies summarized previously in the monographs and of data published subsequently. Similar data from studies on experimental animals and from short-term tests were also summarized.[2] For the purpose of the following discussion, therefore, information and definitions are used that are available through this IARC program and that have been summarized in Supplement 4.[2]

Evaluation of Carcinogenic Risk to Humans

At present, no objective criteria exist to interpret data from studies in experimental animals or from short-term tests directly in terms of human risk. Thus, in the absence of *sufficient evidence* from human studies, evaluation of the carcinogenic risk to humans was based on consideration of both the epidemiological and experimental evidence. The chemicals, groups of chemicals, industrial processes, or occupational exposures were thus put into one of three groups:

Group 1. *The chemical, group of chemicals, industrial process, or occupational exposure is carcinogenic to humans.* This category was used only when there was *sufficient evidence* from epidemiological studies to support a causal association between the exposure and cancer.

Group 2 A and B. *The chemical, group of chemicals, industrial process, or occupational exposure is probably carcinogenic to humans.* This category included exposures for which, at one extreme, the evidence of human carcinogenicity was almost "sufficient" as well as exposures for which, at the other extreme, the evidence was inadequate. To reflect this range, the category was divided into higher (group A) and lower (group B) degrees of evidence. Usually, category 2A was reserved for exposures for which there was at least *limited evidence* of carcinogenicity to humans. Experimental data played an important role in assigning studies to category 2, and particularly those in group B; thus, the combination of *sufficient evidence* in animals and inadequate data in humans usually resulted in a classification of 2B.

Group 3. *The chemical, group of chemicals, industrial process, or occupational exposure cannot be classified as to its carcinogenicity to humans.*

Assessment of Data from Short-Term Tests

Because of the large number and wide variety of short-term tests that may be relevant for the prediction of potential carcinogens, the data relative to each compound have been tabulated, indicating both the end point of the test system and the degree of biological complexity (indicator organism). For the purpose of this discussion, results obtained in prokaryotes (i.e., bacteria) and mammalian cells *in vitro* (either rodent or human somatic cells or cell lines in culture) refer to tests in the presence or absence of an exogenous metabolic activation system. The end point "mutation" refers to heritable alterations in phenotype, including forward and reverse mutations.

The overall evidence in all the short-term tests considered and summarized in the tables was adjusted to fall into one of three categories, *sufficient, limited,* and *inadequate:*

1. *Sufficient evidence,* when there were at least three positive results in at least two of the three test systems measuring DNA damage, mutagenicity, or chromosomal effects. When two of the positive results were for the same genetic effect, they had to be derived from systems of different biological complexity.

2. *Limited evidence,* when there were at least two positive results, either for different end points or in systems representing two levels of biological complexity.

3. *Inadequate evidence,* when there were generally negative test results or only one positive result. Up to two positive test results were considered inadequate if they were accompanied by two or more negative test results.

The working group was unable to define criteria for "negative" evidence.

In TABLE 1 are listed chemicals either identified (group 1) or suspected to be carcinogenic in humans (group 2, A and B) and their mutagenic effect in prokaryotic and mammalian cellular test systems; only those compounds are included for which data in both systems were available. There was a high degree of parallel positive response for the 27 chemicals listed. This may indicate that, for 76% of all group 1 and 2 (A and B) carcinogens, DNA damage may be a critical event in the initiation of carcinogenesis. The results shown in TABLE 1 also provide retrospective evidence for the utility of these two types of mutagenicity tests, as they would have predicted the carcinogenicity of these chemicals. Accordingly, in some cases the working group[2] considered that positive results from short-term tests, providing sufficient evidence as well as the known chemical properties of the compounds, allowed their transfer from group 2B to group 2A; these included dimethylsulfate, nitrogen mustard, and procarbazine (TABLE 1).

Six of the established or probable human carcinogens (TABLE 2), i.e., arsenic and arsenic compounds, asbestos, benzene, chloroform, DDT, and diethylstilbestrol, showed a uniform lack of mutagenic effect in both prokaryotic and mammalian cell assays. Thus it would appear that these carcinogens do not act through DNA modification via electrophiles, but initiate or enhance carcinogenesis through mechanisms hitherto poorly understood. Thus, the two mutagenicity assays, even when applied in combination, would have failed to detect about 20% of the group 1 and 2 (A and B) carcinogens listed in TABLES 1 and 2. These results certainly imply a different mechanism of action for these carcinogens, but stress also the need for development of short-term tests that are able to detect carcinogens that possibly act at stages later than initiation. Reliable and validated tests for such assays are not available to date.

Five compounds (TABLE 2, numbers 8–12) were found to be mutagenic in bacterial, but not in mammalian cell, assays. It is difficult to conclude at present which of the two test systems produces false-negative results, as four out of the five compounds could not be classified with regard to their carcinogenicity in humans because of inadequate data; hydrazine, the only compound classified as a 2B carcinogen, was positively identified as a bacterial mutagen.

It is noteworthy that in this compilation (IARC Supplement 4),[2] no single compound was found to be mutagenic in mammalian cells but not in bacteria.

From this comparison of the mutagenic effect of chemicals in bacterial and mammalian cells *in vitro,* the following conclusions may be drawn: although the number of chemicals for which data were available for comparison was limited, the majority of established or possible human carcinogens showed a uniform positive response in both test systems (TABLE 1). Thus, results providing sufficient evidence from short-term tests appear useful to assist in interpretation of ambiguous epidemiological and experimental carcinogenicity data. However, the lack of detection of several human carcinogens emphasizes the need for development of tests for the

TABLE 1

COMPARISON OF CHEMICALS EITHER IDENTIFIED (GROUP 1) OR SUSPECTED TO BE
CARCINOGENIC (GROUP 2) IN HUMANS AND THEIR ACTIVITY IN TWO
MUTAGENICITY TEST SYSTEMS*

	Compound	Evidence for Mutagenicity in Prokaryotes[†]	Evidence for Mutagenicity in Mammalian Cells[†]	Evidence for Activity in Short-Term Tests[‡]	Summary Evaluation of Carcinogenic Risk to Humans: Groups[‡]
1	Adriamycin	+	+	sufficient	2B
2	Aflatoxins	+	+	sufficient	2A
3	Bischloroethyl nitrosourea	+	+	sufficient	2B
4	Benzyl chloride	+	+	sufficient	3
5	1-(2-Chloroethyl)-3-cyclohexyl-1-nitrosourea	+	+	sufficient	2B
6	Chromium and certain Cr (VI) compounds	+	+	sufficient	1
7	Cyclophosphamide	+	+	sufficient	1
8	1,2-Dichloroethane	+	+	sufficient	—§
9	Diethyl sulfate	+	+	sufficient	2A
10	Dimethylcarbamoyl chloride	+	+	sufficient	2B
11	Dimethyl sulfate	+	+	sufficient	2A¶
12	Epichlorohydrin	+	+	sufficient	2B
13	Ethylene dibromide	+	+	sufficient	2B
14	Ethylene oxide	+	+	sufficient	2B
15	Melphalan	+	+	sufficient	1
16	6-Mercaptopurine	+	+	sufficient	3
17	Methoxsalen + ultraviolet (A) light	+	+	sufficient	1
18	Myleran	+	+	sufficient	1
19	1-Naphthylamine	+	+	sufficient	3
20	2-Naphthylamine	+	+	sufficient	1
21	Nitrogen mustard	+	+	sufficient	2A¶
22	Procarbazine	+	+	sufficient	2A¶
23	Styrene oxide	+	+	sufficient	3
24	Triaziquinone	+	+	sufficient	2B
25	Thiotepa	+	+	sufficient	2B
26	Uracil mustard	+	+	sufficient	2B
27	Vinyl chloride	+	+	sufficient	1

*Chemicals assigned to groups 1, 2 (A and B), and 3 (from IARC Supplement 4 to *IARC Monographs* volumes 1–29)[2] for which positive data in both mutagenicity test systems were available are included.

†A plus sign indicates that the result was judged by the working group to be significantly positive in one or more assays.

‡For definitions, see text.

§No data in humans.

¶The working group considered that the known chemical properties of this compound and the results from short-term tests allowed its transfer from group 2B to 2A.

detection of late-stage carcinogens whose action in carcinogenesis does not apparently involve electrophiles that combine with the DNA of the target cells.

Comparison of the In Vitro *Mutagenicity and the* In Vivo *Covalent Binding Index of Some Chemicals Carcinogenic in Rodents*

In order to evaluate whether mutagenicity data *in vitro* can also predict quantitatively the processes that occur in intact mammalian organisms, the microsome-mediated mutagenicity in *Salmonella* was compared with the covalent binding index

TABLE 2

COMPARISON OF CHEMICALS EITHER IDENTIFIED (GROUP 1) OR SUSPECTED TO BE CARCINOGENIC (GROUP 2) AND THOSE THAT COULD NOT BE EVALUATED (GROUP 3) AND THEIR ACTIVITY IN TWO MUTAGENICITY TEST SYSTEMS*

	Compound	Evidence for Mutagenicity in Prokaryotes†	Mammalian Cells†	Evidence for Activity in Short-Term Tests‡	Summary Evaluation of Carcinogenic Risk to Humans: Groups‡
1	Arsenic and certain As compounds	−	−	limited	1
2	Asbestos	−	−	inadequate	1
3	Benzene	−	−	limited	1
4	Chlordane/heptachlor	−	−	inadequate	3
5	Chloroform	−	−	inadequate	2B
6	DDT	−	−	inadequate	2B
7	Diethylstilbestrol	−	−	inadequate	1
8	Chloroprene	+	−	sufficient	3
9	Dichloromethane	+	−	limited	3
10	Hydrazine	+	−	sufficient	2B
11	Styrene	+	−	sufficient	3
12	Vinylidene chloride	+	−	sufficient	3

*Only the chemicals assigned to groups 1, 2 (A, B), and 3 (from IARC Supplement 4)[2] for which positive or negative response data in both mutagenicity test systems were available are included.

†A plus sign indicates that the result was judged by the working group to be significantly positive in one or more assays. A minus sign indicates that it was judged to be negative from an evaluation of one or more assays.

‡For definitions, see text.

(CBI) of chemicals in rat liver *in vivo,* as summarized by Lutz.[5] The CBI in rat liver DNA was defined as μmol of carcinogen bound per mol nucleotides/mmol of carcinogen administered per kg body weight. CBIs and mutagenicity data were available for 44 compounds, about one-half of which were carcinogens with the liver as the principal target organ in the rat. As the data on mutagenicity were taken from the literature, possible limitations of this compilation are that different assay procedures were used to determine mutagenicity, i.e., plate, liquid incubation, or preincubation assays, and different enzyme inducers were used to pretreat rats before preparation of the liver homogenates utilized to activate the test chemicals. Similarly, different *S.*

typhimurium strains were tested; only those that gave the highest response were selected. The arithmetic mean and the extremes of the ranges for the CBI values for each compound are plotted in FIGURE 1. The CBIs for the 44 compounds varied over a range of five orders of magnitude; ethionine had the lowest and aflatoxin B_1 the highest values. For those compounds for which CBIs were available and that were mutagenic in *S. typhimurium*, mutagenic activities varied over a range of seven orders of magnitude, the lowest values being found for *N*-nitrosodiethylamine and the highest for aflatoxin B_1.

Eight chemicals that have been observed to bind covalently to rat liver DNA *in vivo* did not produce microsome-mediated mutagenicity in *Salmonella* strains, although their CBIs varied over four orders of magnitude. Six of these, diethylstilbestrol, benzene, carbon tetrachloride, 1'-hydroxysafrole, ethionine, and 1,2-dimethylhydrazine, have been shown to be carcinogenic in experimental animals. 1-Hydroxy-*N*-2-acetylaminofluorene, reported as being noncarcinogenic and also nonmutagenic,[6] is an exception in that it is the only compound for which a CBI value was found that has been reported to be noncarcinogenic. Apart from these 8 nonmutagenic compounds, all the remaining 36 chemicals showed a positive correlation between CBI and mutagenicity *in vitro* (r = 0.847; p < 0.001; n = 36) in a double logarithmic plot.

The 11 compounds that deviated from the approximately linear relationship can be grouped into two classes: (1) those for which the CBI values are too high or mutagenicity too low, or both; they include *N*-nitrosopyrrolidine, *N*-nitrosopiperidine, methane azoxymethanolacetate, methyl methanesulfonate, 1,2-bromoethane, and vinyl chloride; and (2) those for which the reported mutagenicity is either too high or the CBI too low, or both; they include 3-methylcholanthrene, benzo[*a*]pyrene, and 2-naphthylamine.

It is difficult to decide which of the two variables over- or underestimates the carcinogenicity of these 11 compounds, since the CBI in the liver and mutagenicity *in vitro* both have limitations as predictors of carcinogenic activity. It is reasonable that the CBI for the simple alkylating agents is too high, since both miscoding and nonmiscoding DNA adducts are measured, whereas only certain O-alkylated products are believed to have promutagenic potential.[7] Accordingly, compounds that have a CBI value that is too high were found to belong to this class of precarcinogens from which (monofunctional) alkylating agents are released.

By contrast, nitroso compounds have been reported to show too little mutagenic activity in the *Salmonella*/microsome assay, particularly in the plate assay.[8] They were also among the exceptions noted by Meselson and Russel,[9] which do not show the proportionality between carcinogenic potency and mutagenicity seen with certain other carcinogens/mutagens. The reasons for this low mutagenic activity may be related to metabolic parameters of the *in vitro* activation system used and to details of the assay procedure, as discussed in more detail elsewhere.[10,11]

On the other hand, it has been shown that, for some members of the class of compounds with too high mutagenicity or too low a CBI (2-naphthylamine, benzo[*a*]pyrene, and 3-methylcholanthrene), the mutagenic metabolites produced *in vitro* and their relative concentrations are not identical to those that are known to be relevant to carcinogenic processes. This discrepancy, as well as the fact that these three carcinogens do not have the liver as their principal target, may account for the deviation from linearity observed in FIGURE 1.

From FIGURE 1, the following conclusions may be drawn: providing that the comparison between CBI and *in vitro* mutagenicity data is confined to one organ from one animal species, i.e., rat liver *in vivo* and rat liver microsomal preparations *in vitro*, both variables (in a double logarithmic plot) were shown to be fairly well correlated in a linear fashion, as shown for 36 chemicals. Despite the known limitations of

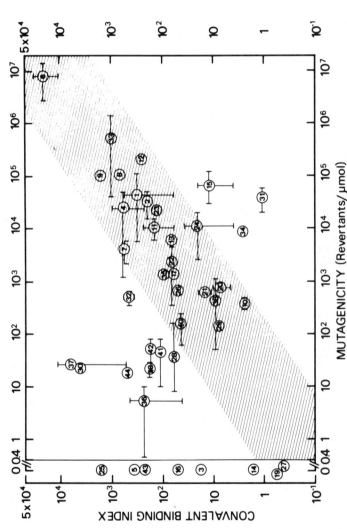

FIGURE 1. A double logarithmic plot of the covalent binding index (CBI) of 44 chemicals versus their mutagenicity in *Salmonella in vitro*. The mutagenicity data in *Salmonella typhimurium* his⁻ strains in the presence (for carcinogens requiring activation) or absence (for direct-acting agents) of rat liver 9,000 × g supernatant (S-9) are listed. The figures refer to the most sensitive strain tested following incubation with liver S-9 from rats treated with inducers (polychlorinated biphenyls, phenobarbitone, or 3-methylcholanthrene). The data were collected from the literature and have been summarized previously.[10] The numbers in the circles correspond to those of the chemicals listed in a summary table in Reference 10. When available, arithmetic means and ranges of CBI and mutagenicity values (indicated by bars) are indicated. Chemicals that did not fall within the shaded area are discussed in the text. (Reproduced from Reference 10 with permission from Academic Press, Inc.)

subcellular activation systems, *in vitro* mutagenicity data predicted *in vivo* DNA binding of chemicals with some accuracy, although several exceptions, i.e., chemicals that deviated from a linear relationship, were noted. These results identify species- and tissue-specific metabolic activation/detoxification processes as important variables for *in vitro/in vivo* extrapolation.

Quantitative Comparisons between Carcinogenicity, Electrophilicity, and Mutagenicity of Direct-Acting Alkylating Agents

Because the mutagenicity and carcinogenic activity of chemicals can vary over a range of one million,[12-14] it would be helpful for the estimation of the carcinogenic risk of chemicals in the human environment (in the absence of adequate animal data) if a quantitative relationship between carcinogenic and mutagenic activity were to be established. We have chosen to compare the biological and chemical activities of direct-acting alkylating agents, since differences between metabolic activation reactions leading to mutagenic metabolites *in vitro* and those giving rise to ultimate carcinogenic metabolites *in vivo* have become apparent for certain carcinogens. A brief summary of the results of these intercomparisons is reported below; details of this compilation will be reported elsewhere.

Ten monofunctional alkylating agents, *N*-nitroso-*N*-ethylurea (ENU), *N*-nitroso-*N*-methylurea (MNU), *N*-nitroso-*N*-methylurethane (MNUT), *N*-methyl-*N'*-nitro-*N*-nitrosoguanidine (MNNG), epichlorohydrin (ECH), glycidaldehyde (GA), ethyl methanesulfonate (EMS), methyl methanesulfonate (MMS), β-propiolactone (PL), and 1,3-propane sultone (PS), were assayed for mutagenicity in *S. typhimurium* TA1535 and TA100 strains, using plate incorporation or liquid incubation assays.[12]

In order to estimate TD_{50} values (TD_{50}, the total dose of carcinogen in mg/kg body weight required to reduce by one-half the probability of the animal being tumor free throughout a standard lifetime) for each of the 10 alkylating agents, carcinogenicity studies cited in the *IARC Monographs on the Evaluation of the Carcinogenic Risk of Chemicals to Humans*[15] were reviewed. Only those studies using mice, rats, and hamsters and that were adequately conducted and reported were considered. TD_{50} values were calculated according to the formula proposed by Hooper *et al.*:[16]

$$TD_{50} = Dt^3 \ln 2 / \ln[(1 - n_c/N_c)/(1 - n_e/N_e)] \qquad (1)$$

where D = total intake of carcinogen; t = experiment time/natural lifetime; n_c = number of tumor-bearing animals (TBA) among controls; N_c = total number of controls; n_e = number of TBA among experimental animals; N_e = total number of experimental animals.

Calculation of Substrate Constants and Ratios of N-7/O⁶-Alkylguanine Formation in DNA

Because Swain-Scott substrate constants (S) have not been reported in the literature for all the alkylating agents studied here, the Swain-Scott equation[7] was used to calculate the missing values by linear regression, as described previously.[17]

In a similar way, the Swain-Scott relation was used for the estimation of the ratio of N-7 to O⁶-alkylation of guanine (alkyl-G) in double-stranded DNA *in vitro* by five alkylating agents for which values have not been reported elsewhere (TABLE 1) (figures in parentheses in TABLE 3).

Quantitative Comparisons between Mutagenicity, Electrophilicity, and Carcinogenic Potency

A comparison of the mutagenic activities in TA100 strain (plate assays) versus the carcinogenic potency of the alkylating agents (expressed as TD_{50}) is shown in TABLE 3. The TD_{50} values for a given compound varied by as much as two orders of magnitude, according to route and schedule of administration, animal species, or strain.

Both the mutagenic and carcinogenic potency of the nine agents varied over a 10,000-fold range; there was no obvious proportionality between carcinogenicity in rodents and mutagenicity in *Salmonella* strain TA100 (TABLE 3) or TA1535 (data not shown). For example, ENU was the most potent carcinogen in rodents (low TD_{50}) when tested in adult or newborn animals or by different modes of administration, but it was only weakly active as a mutagen. Inversely, GA was among the most mutagenic compounds, but was one of the weakest carcinogens (high TD_{50}).

TABLE 3

COMPARATIVE MUTAGENICITY, ELECTROPHILICITY, AND CARCINOGENIC
ACTIVITY OF ALKYLATING AGENTS

Compound	Ratio of N-7/O^6-Alkylguanine in DNA*	Mutagenic Activity in TA100†	TD_{50} Values in Rodents‡ (range)
ENU	1.58	2790	1–35
MNU	9.09	550	7–155
MNNG	10.00	4	7–179
MNUT	(13.98)	9	7
PS	(76.44)	40	13–2167
PL	(124.20)	250	104–610
MMS	250.00	680	1082–1399
GA	(201.79)	15	1422–16865
ECH	(453.10)	1130	13718
EMS	33.34	14200	—§

*Values taken from Reference 17.

†Mutagenic activity is expressed as the concentration (μmol/l) required to produce 500 revertants of *S. typhimurium* strain TA100 per plate.

‡Calculated (in mg of compound/kg body weight) for rats, mice, and hamsters, using Equation 1 and data from *IARC Monographs*.[15]

§TD_{50} > 175, noncarcinogenic.

However, the ratio of N-7/O^6-alkylguanine formed when the agents react with double-stranded DNA *in vitro* (TABLE 3) showed a positive association with the TD_{50} values (rank correlation coefficient r_s for the median TD_{50} values: $r_s = 0.87$; $p < 0.005$; $n = 9$, EMS excluded); this single parameter reflected the high carcinogenicity of ENU and the low carcinogenic activity of MMS, GA, and ECH.

From this study on a limited number of carcinogens, it was concluded that a quantitative relation between carcinogenic and mutagenic activities in *Salmonella* strains could not be established from the results obtained with nine alkylating agents. Such mutagenic activities, therefore, cannot be used at present for the confident prediction of the carcinogenic potency of new, structurally related compounds. The positive correlation we have observed between carcinogenic activity and the initial ratio of N-7/O^6-alkylguanine formed in double-stranded DNA *in vitro* provides further circumstantial support to the hypothesis that oxygen atoms in DNA may be a critical target in carcinogenesis induced by alkylating agents.[18,7]

SUMMARY

Chemicals evaluated for their carcinogenic potential in the *IARC Monographs* (Supplement 4 to volumes 1–29)[2] are used to compare their response in bacterial and mammalian cell mutagenicity assays *in vitro*. Simultaneous positive and negative test results in both systems showed a high degree of parallelism. Several carcinogens active in animals/humans, however, were not detected in either assay.

The possibility of a quantitative extrapolation of bacterial mutagenesis data to processes occurring in intact mammals was further examined. Published covalent binding indices in rat liver DNA for 36 compounds were found to be correlated with their mutagenic effects in the *Salmonella*/liver-microsome test; several compounds deviated from this proportionality. The quantitative relationship between carcinogenicity in rodents (TD_{50}) and mutagenicity was examined, using 10 alkylating agents. Mutagenicity in *S. typhimurium* TA100 strain (plate and liquid assays) showed no correlation with carcinogenic potency. However, there was a positive relationship between TD_{50} values and the initial ratio of N-7-alkyl/O^6-alkyl guanine formed (predicted) after reaction with double-stranded DNA *in vitro*.

ACKNOWLEDGMENTS

The author wishes to acknowledge the invaluable contributions made to these studies by several members of the Division of Environmental Carcinogenesis, International Agency for Research on Cancer, and outside institutions, including L. Haroun, C. Malaveille, B. Terracini, L. Tomatis, J. Wahrendorf, and J. Wilbourn. The author is grateful to M. Courcier for assistance in preparing this manuscript.

REFERENCES

1. PETO, R. E. 1978. Carcinogenic effects of chronic exposure to very low levels of toxic substances. Environ. Health Perspect. **22:** 55–159.
2. IARC. 1982. IARC Monographs on the Evaluation of the Carcinogenic Risk of Chemicals to Humans. Supplement 4. Chemicals, Industrial Processes and Industries Associated with Cancer in Humans. International Agency for Research on Cancer. Lyon, France.
3. IARC. 1972–1982. IARC Monographs on the Evaluation of the Carcinogenic Risk of Chemicals to Humans. **1–29.** International Agency for Research on Cancer. Lyon, France.
4. IARC. 1979. IARC Monographs on the Evaluation of the Carcinogenic Risk of Chemicals to Humans. Supplement 1. Chemicals and Industrial Processes Associated with Cancer in Humans. International Agency for Research on Cancer. Lyon, France.
5. LUTZ, W. K. 1979. *In vivo* covalent binding of organic chemicals to DNA as a quantitative indicator in the process of chemical carcinogenesis. Mutat. Res. **65:** 289–356.
6. MCCANN, J., E. CHOI, E. YAMASAKI & B. N. AMES. 1975. Detection of carcinogens as mutagens in the *Salmonella*/microsome test: assay of 300 chemicals. Proc. Nat. Acad. Sci. USA **72:** 5135–5139.
7. LAWLEY, P. D. 1976. Carcinogenesis by alkylating agents. ACS Monogr. **173:** 83–244.
8. BARTSCH, H., A.-M. CAMUS & C. MALAVEILLE. 1976. Comparative mutagenicity of *N*-nitrosamines in a semisolid and liquid incubation system in the presence of rat or human tissue fractions. Mutat. Res. **37:** 149–162.

9. MESELSON, M. & K. RUSSEL. 1977. Comparisons of carcinogenic and mutagenic potency. *In* Origins of Human Cancer. H. H. Hiatt, J. D. Watson & J. A. Winsten, Eds.: 1473–1481. Cold Spring Harbor Laboratory. Cold Spring Harbor, N.Y.
10. BARTSCH, H., L. TOMATIS & C. MALAVEILLE. 1982. Qualitative and quantitative comparisons between mutagenic and carcinogenic activities of chemicals. *In* Mutagenicity: New Horizons in Genetic Toxicology. J. A. Heddle, Ed.: 35–72. Academic Press, Inc. New York, N.Y.
11. BARTSCH, H., T. KUROKI, M. ROBERFROID & C. MALAVEILLE. 1982. Metabolic activation systems *in vitro* for carcinogen/mutagen screening tests. *In* Chemical Mutagens—Principles and Methods for Their Detection. F. J. de Serres & A. Hollaender, Eds. 7: 95–161. Plenum Press. New York & London.
12. BARTSCH, H., C. MALAVEILLE, A.-M. CAMUS, G. MARTEL-PLANCHE, G. BRUN, A. HAUTEFEUILLE, N. SABADIE, A. BARBIN, T. KUROKI, C. DREVON, C. PICCOLI & R. MONTESANO. 1980. Validation and comparative studies on 180 chemicals with *S. typhimurium* strains and V79 hamster cells in the presence of various metabolizing systems. Mutat. Res. 76: 1–50.
13. MCCANN, J. & B. N. AMES. 1977. The Salmonella/microsome mutagenicity test: predictive value for animal carcinogenicity. *In* Origins of Human Cancer. H. H. Hiatt, J. D. Watson & J. A. Winsten, Eds.: 1431–1450. Cold Spring Harbor Laboratory. Cold Spring Harbor, N.Y.
14. NAGAO, M., T. SUGIMURA & T. MATSUSHIMA. 1978. Environmental mutagens and carcinogens. Annu. Rev. Genet. 12: 117–159.
15. IARC. 1972, 1974, 1976, 1978. IARC Monographs on the Evaluation of the Carcinogenic Risk of Chemicals to Humans. 1, 4, 7, 11, 17. International Agency for Research on Cancer. Lyon, France.
16. HOOPER, N. K., A. D. FRIEDMAN, C. B. SAWYER & B. N. AMES. 1977. Carcinogenic Potency: Analysis, Utility for Human risk Assessment, and Relation to Mutagenic Potency in Salmonella. International Agency for Research on Cancer. Lyon, France. (Progress Report for IARC/WHO meeting, Lyon, October 1977.)
17. BARTSCH, H., C. MALAVEILLE, L. TOMATIS, G. BRUN, B. DODET & B. TERRACINI. 1982. Quantitative comparisons between carcinogenicity, mutagenicity and electrophilicity of direct-acting *N*-nitroso compounds and other alkylating agents. IARC Sci. Publ. No. 41: 525–532.
18. LOVELESS, A. 1969. Possible relevance of O^6-alkylation of deoxyguanosine to the mutagenicity and carcinogenicity of nitrosamines and nitrosamides. Nature London 223: 206–207.

RELEVANCY ISSUES

V. A. Ray

Medical Research Laboratories
Pfizer Inc.
Groton, Connecticut 06340

The two most relevant or practical issues that the scientific community faces with mutagenicity testing for safety-evaluation purposes are: What testing models should we use routinely in defining a chemical as a mutagen? and What is the proper utility of the knowledge that a chemical is or is not a mutagen in one or more of these models? It is the purpose of this paper to examine several aspects of these two issues.

Recent contributions and commentary by scientists in many nations have resulted in the Organization for Economic Cooperation and Development (OECD) draft guidelines for the testing of chemicals.[13] Although these guidelines are still undergoing final comment, this document at least implies that there is a degree of concordance among a large sector of the genetic scientific community over which tests can be applied routinely as a minimum premarket data set. These guidelines also indicate that the tests identified are the result of the "state of the art" and that changes may be instituted as collaborative and review studies become available. Although a member country can make a request for a change in the guidelines to the Updating Panel of the Chemicals Program of OECD, there is no mechanism, to my knowledge, that provides for an automatic periodic review and updating process. In a recent article by Professor J. Z. Rubin, entitled "Caught by Choice," the author states that "ironically, entrapment arises from freedom of choice, the right to take one road over another, and that choices have consequences that we must learn to live with—those of the paths not taken as well as those that are."[15] It would indeed be unfortunate if segments of the scientific community were burdened by feeling the need to defend certain choices in the guidelines that experience and data accumulation identify as inappropriate or inadequate. An automatic updating mechanism would facilitate review and provide for change necessitated by scientific progress. Due to the rapid rate of development in this field, the time frame for such review probably should not be more than four years.

The second point I would like to explore is the positioning of mammalian cell and whole-animal test data to become the standard for safety-evaluation purposes. Due to the current availability of considerable data and experience in prokaryotic models and eukaryotic models other than mammalian cells, it is evident that such models will continue to play a substantive role in the testing of chemicals for some years. However, because a test is inexpensive or quick or has represented a means to explore multiple genetic end points does not establish that it is the most relevant to issues of safety evaluation for human beings. Although such attributes are attractive, they may not provide the toxicologist with sufficient consistently relevant data to decide on safety factors necessary in a regulatory context. Certainly, no single test has been able to provide this to date. Further, it is recognized that in defining safety, the totality of the information from all tests and sources is taken into consideration. However, let us not underestimate the potency of a single piece of information that a chemical produces a deleterious effect (mutation) in some single assay system.

There is also an issue of entrapment due to choices made previously that now create the perception of need for defense postures. All we have to do is recall the role of certain procedures in this field that have declined in value because of the evaluations produced by extensive use. It should be pointed out that this condition is not unique to genetic

362

0077-8923/83/0407-0362 $01.75/0 © 1983, NYAS

toxicology but applies to the whole field of toxicology and is common to any field of active scientific endeavor. In attempting to establish guidelines in a particular sector of toxicology, all of the factors just mentioned come into play, and I am using the draft OECD guidelines as a platform for comment, recognizing that much of this document is a constructive and necessary effort that forces analysis of test value and utility.

There are, currently, four OECD test guidelines proposed for final agreement:

1. *Salmonella typhimurium*—reverse-mutation assay.
2. *Escherichia coli*—WP2—reverse-mutation assay.
3. *In vitro* cytogenetics.
4. Micronucleus test.

In addition, there is a paper called "Principles for the Evaluation of the Mutagenic and Carcinogenic Potential of Chemicals," which is also undergoing comment. A test guideline on the sex-linked recessive lethal test in *Drosophila* is nearing final draft, and three other draft guidelines on *in vivo* cytogenetics, the dominant-lethal test, and mammalian cell gene mutation are being discussed currently for future test guidelines.

The principles paper has positioned the *S. typhimurium* or *E. coli* assay and *in vitro* cytogenetics as "tests for initial assessment of mutagenic potential" (minimum premarket data set). Two *in vivo* procedures, the micronucleus test and the *in vivo* bone marrow cell metaphase analysis, are also acceptable as substitutes for the *in vitro* cytogenetics assay.

This positioning of a prokaryotic organism for detecting point mutagens merits comment under the heading of a relevancy issue. In several other batteries of assays that have been proposed by individuals, committees, or governmental groups for safety-evaluation purposes, a mammalian cell gene-mutation assay has been included. Some of these batteries have been intended for detecting both mutagenic and carcinogenic potential of chemicals. Although there are many reasons for the inclusion of a mammalian cell model, those encountered most frequently are (1) to detect compounds that are negative in a bacterial assay; (2) to give perspective to results obtained in a bacterial assay; (3) to make allowance for different transport mechanisms and conditions between bacterial and mammalian cells, e.g., availability of the chemical to the nuclear apparatus; and (4) to provide for differences in repair capability.

TABLE 1 has several chemicals with discordant results between the Ames *Salmonella* and a mammalian cell assay. These are but a few examples of compounds showing uncertainty or discordance between microbial and mammalian cell models. Also, it has been pointed out time and time again that the result from a single assay is not adequate for predicting the carcinogenic potential of a chemical.

It is recognized that the extent of initial testing is determined by a number of factors, not all of which are scientific, and that different industries have different requirements and problems. The perception of testing requirements for foods and drugs can be quite different from that for bulk chemicals. However, the scientific objective is human safety and the assessment of a possible health hazard in the mammal called the human being. Since this is the objective, why is it not reasonable to place emphasis on mammalian cell and whole-animal models? In fact, is it unreasonable to establish such models as the standard in mutagenicity testing for safety-evaluation purposes? Obviously, the issue is of far greater magnitude in gene-mutation than in chromosomal level assays. All three of the assays indicated for chromosomal level effects in the initial assessment section of the OECD document would use mammalian cells or whole animals. If the scientific community were to proceed toward

TABLE 1

COMPARATIVE TEST RESULTS FOR SELECTED COMPOUNDS IN RODENT BIOASSAY (CAR) AND
AMES *SALMONELLA* (SAL) AND MOUSE LYMPHOMA (ML) L5178Y TK ASSAYS

Compound	CAR*	SAL†	ML†	Respective References
1. Acridine, 9 amino	unk	+	I	—, 9, 1
2. Aniline	+	−	+	11, 9, 1
3. Cadimum chloride	+	?	P	7, 5, 2
4. Fluorene, 4-acetylamino	−	+	−	9, 9, 4
5. Natulan	+	−	+,P	10, 5, 4
6. Nitrosamine, diphenyl	+	?	±	12, 9, 4
7. Phenol, *p*-amino	unk	−	+	—, 9, 3
8. Stilbestrol, diethyl	+	−	P	6, 5, 4

*Carcinogenicity assays: unk means unknown; +,− mean positive and negative respectively.
†Mutagenicity assays: +,− mean positive and negative respectively with metabolic activation.
P means positive without metabolic activation. I means indeterminate.

adopting a mammalian cell standard for the identification of gene or point mutagens in safety-evaluation work, what criteria would have to be met? Firstly, a series of validation steps would have to be satisfied. Standardized protocols have been established for the mouse lymphoma L5178Y, Chinese hamster ovary (CHO), and Chinese hamster V79 models by panels of the Environmental Protection Agency Gene-Tox program. Both the mouse lymphoma and V79 assays have over 100 compounds tested and reported in the open literature. It is unfortunate, for this purpose, that many of the almost 200 compounds reported for the V79 model were in a close chemical series, the polyaromatic hydrocarbons. The Gene-Tox panel reviews, which used the literature through early 1979, have some 48 compounds listed for mouse lymphoma, 189 for V79, and 16 for the CHO assay. An updating evaluation program is just now getting under way. Currently (April 1982), there were 14 testing laboratories offering the mouse lymphoma assay, 14 offering the CHO-HGPRT (hypoxanthine guanine phosphoribosyl transferase) assay, and 6 offering the V79-HGPRT assay in the United States. Reproducibility of results and intralaboratory variation are factors that need to be assessed, using an adequate number of laboratories.

Secondly, it would be helpful to compare results between several mammalian cell gene-mutation assays for the degree of concordance with a spectrum of chemicals. If results were sufficiently concordant, then one of these assays could be selected as a standard. Comparison to a group of carcinogens and noncarcinogens would be necessary to indicate the sensitivity and selectivity of the assays to predict carcinogenic potential. At this time we are proceeding well toward validation of mammalian cell gene-mutation assays. However, the number of chemicals tested in mammalian cell systems that have also been evaluated in rodent bioassays is still relatively small. This is especially true of the noncarcinogen group. In fact, the entire published genetic toxicology data base, including all assay models, is composed of predominantly positive compounds. This condition has certain consequences. Clearly, documentation of assertions that certain assays are good or bad indicators of carcinogenic potential is difficult. In fact, informed commentary would have to declare that many such assertions are outright premature. Knowledge of assay performance across chemical classes is still embryonic. Another consequence is the well-defined need to provide more genetic data on noncarcinogens as well as chemicals already found positive in rodent bioassays. I am proposing that we emphasize developing and using mammalian cell assays in safety evaluation of chemicals for use in the human mammal. Until the data base is more adequate, I believe it is essential to accumulate data side by side with *Salmonella* or other models to assist in making decisions on the mutagenic potential of chemicals.

The third issue I want to explore briefly is one aspect of the utility of genetic information in detecting and evaluating carcinogens. Currently, most toxicologists view carcinogenesis as a multistage process with mutation playing a role in the first stage, or initiation. The correlation between the mutagenic and carcinogenic activities of chemicals forms the basis for this view, which is reinforced by the knowledge of covalent binding to DNA and adduct formation by many of these chemicals. Other chemicals that have been designated as carcinogens by rodent studies do not produce mutation in models commonly used to detect such activity. Some of these substances have been called cocarcinogens, and others, promoters or modifiers of the carcinogenic process. This distinction, although poorly understood and vague in its boundaries, has tended to confine thought of the role of genetic activity of chemicals to the initiation stage. This tendency has continued despite knowledge of the high incidence of cancer associated with certain genetic disease syndromes and the knowledge that many tumor types have altered chromosome number. Recently, evidence has been accumulating that promoters can induce ploidy, or numerical, chromosome changes in mammalian cells[8] and in yeast.[14] TABLE 2 lists several promoters with associated genetic effects. These data, at the very least, suggest that involvement of genetic processes in carcinogenesis is not limited to mutation-initiation phenomena. Considering the diversity of effects associated with or induced by substances designated as promoters, it is scientifically imprudent to ascribe a single mechanism to such substances. In fact, attempting to model all promoters according to criteria based on phorbol ester activities is already scientifically untenable and distorts the concern and research effort that may be due such compounds in the carcinogenesis process. If current evidence is reinforced and expanded, solidifying the hypothesis that at least some promoters may be responsible for inducing genetic change, then hazard evaluation of such substances will have to be revised accordingly. There may be a spectrum of such activities that will be important to delineate in order to define the degree of concern. In any case, the role of the genetic toxicologist in detecting and evaluating carcinogens is an expanding one. Even the acquisition of data that some promoters can inhibit

TABLE 2

GENETIC EFFECTS OF SEVERAL TUMOR PROMOTERS*

Compound	Promoter or Cocarcinogen	Gene Mutation	Metabolic Cooperation Inhibitor	Aneuploidy	Mitotic Crossing Over
12-*O*-Tetradeca-noyl phorbol-13-acetate	+	−	+	+	−
Anthralin	+	−	−	+	−
Iodoacetic	+	−	−?	+	−
Oleic	+	−	−	+	−
Diethylstilbestrol	?	−?	−	+	−
Griseofulvin	+†	−	−?	+	NT‡

*Data from References 8 and 14.
†Cocarcinogen.
‡NT means not tested.

metabolic cooperation between cells is dependent on mutant recovery models. There-fore, in my view, it is a relevancy issue that in addition to concerns over exposure, pharmacokinetics, species differences, metabolism, and target-organ toxicity, proper attention be given to the role of genetic mechanisms in carcinogenesis that may underline several of the perceived stages in the process and not unduly limit our research and testing programs to the mutation-initiation stage.

REFERENCES

1. AMACHER, D. E., S. C. PAILLET, G. N. TURNER, V. A. RAY & D. S. SALSBURG. 1980. Point mutations at the thymidine kinase locus in L5178Y mouse lymphoma cells. II. Test validation and interpretation. Mutat. Res. 72: 447–474.
2. AMACHER, D. E. & S. C. PAILLET. 1980. Induction of trifluorothymidine-resistant mutants by metal ions in L5178Y/TK$^{+/-}$ cells. Mutat. Res. 78: 279–288.
3. AMACHER, D. E. & G. N. TURNER. 1982. Mutagenic evaluation of carcinogens and noncarcinogens in the L5178Y/TK assay utilizing postmitochondrial fractions (S9) from normal rat liver. Mutat. Res. 97: 49–65.
4. CLIVE, D., K. D. JOHNSON, J. F. S. SPECTOR, A. G. BATSON & M. M. M. BROWN. 1979. Validation and characterization of the L5178Y/TK$^{+/-}$ mouse lymphoma mutagen assay system. Mutat. Res. 59: 61–108.
5. HOLLSTEIN, M. & J. MCCANN. 1979. Short-term tests for carcinogens and mutagens. Mutat. Res. 65: 133–226.
6. IARC-WHO. 1974. IARC Monographs on the Evaluation of the Carcinogenic Risk of Chemicals to Man 6. Sex Hormones. International Agency for Research on Cancer. Lyon, France.
7. IARC-WHO. 1976. IARC Monographs on the Evaluation of the Carcinogenic Risk of Chemicals to Man 11. Cadmium, Nickel, Some Epoxides, Miscellaneous Industrial Chemicals and General Considerations on Volatile Anesthetics. International Agency for Research on Cancer. Lyon, France.
8. KINSELLA, A. R. 1982. Elimination of metabolic co-operation and the induction of sister chromatid exchanges as properties common to all promoting or co-carcinogenic agents. Carcinogenesis 3(5): 499–503.
9. MCCANN, J., E. CHOI, E. YAMASAKI & B. N. AMES. 1975. Detection of carcinogens as mutagens in the Salmonella/microsome test: assay of over 300 chemicals. Proc. Nat. Acad. Sci. USA 72(12): 5135–5139.
10. NCI bioassay of procarbazine (natulan) for possible carcinogenicity. Technical Report Series NCI-CG-TR-19. National Cancer Institute. Bethesda, Md.
11. NCI bioassay of aniline hydrochloride for possible carcinogenicity. Technical Report Series NCI-CG-TR-130. National Cancer Institute. Bethesda, Md.
12. NCI bioassay of N-nitrosodiphenylamine for possible carcinogenicity. Technical Report Series NCI-CG-TR-164. National Cancer Institute. Bethesda, Md.
13. Organization for Economic Co-operation and Development Chemicals Programme. Health Effects Testing Guidelines. OECD Publications Center, Suite 1207, 1750 Pennsylvania Ave. N.W., Washington, D.C. 20006.
14. PARRY, J. M., E. M. PARRY & J. C. BARRETT. 1981. Tumor promoters induce mitotic aneuploidy in yeast. Nature 294: 263–265.
15. RUBIN, J. Z. 1982. Caught by choice. The Sciences 22(7): 18–21.

APPLICATION OF SHORT-TERM TESTS TO SAFETY TESTING OF INDUSTRIAL CHEMICALS

C. A. Schreiner

Toxicology Division
Mobil Oil Corporation
Princeton, New Jersey 08540

According to a 1978 Federal Register entry, there are at present over 70,000 chemicals in commercial production.[1] New industrial chemicals are being introduced at a rate of 700–3,000 per year. Fishbein has reported that of the 6,000 chemicals tested for cancer-causing potential by lifetime studies in animals, only 1,000 have been tumorigenic.[2] With over 60,000 chemicals remaining, it becomes logistically impossible to test all of these compounds by classical toxicologic methods in animals. The lack of sufficient numbers of qualified scientists and research facilities, the time and cost expended to perform mammalian studies, and a growing public concern with the large numbers of animals used in biomedical research have led to the expanded development of "short-term" tests as predictors of mutagenesis and carcinogenesis.

Test materials evaluated in short-term tests cover a varied spectrum from wide usage, repeated exposure materials such as drugs, food and food additives, and pesticides to limited usage, minimal exposure industrial specialty chemicals. Because short-term *in vitro* tests require such relatively small quantities of test sample, they are useful in evaluating air or water pollutants and chemical residues in crops or in tissue of meat, fish, or poultry.

Properly performed and evaluated, short-term tests can be employed in selecting compounds for market development, identifying the potential of chemicals as work place hazards or environmental pollutants, monitoring corrective measures or altered environmental conditions, and prioritizing candidate compounds for longer, more extensive animal testing.

National and international government agencies (e.g., U.S. Food and Drug Administration, U.S. Environmental Protection Agency, Occupational Safety and Health Administration, Organization for Economic Cooperation and Development, European Economic Community) and trade associations have generated recommendations on approaches for short-term testing. Although preferences for specific assays vary, everyone seems to agree on a battery approach to identify (1) point mutations in microorganisms and mammalian cells, (2) chromosome damage, (3) DNA perturbation, which does not necessarily produce an observable mutation, and optionally, cell transformation *in vitro*. Some scientists and regulators believe an appropriate battery of short-term tests *in vitro* can be used for risk assessment or can be substituted for one of the two rodent species in animal bioassays for carcinogenicity; others believe that these tests should not be used in setting exposure limits for humans or in regulatory decision making but do have a role as biological indicators of potential hazard for carcinogenesis, mutagenesis, or reproductive malfunction.

TABLE 1 lists a general battery for genetic screening, combining *in vitro* and *in vivo* tests. All tests are not performed on every sample; the battery employed should be developed on a sample-by-sample basis depending on intended use, stage of development, status of material (potential product, process intermediate, or waste residues and effluent samples), and the likelihood and extent of human exposure.

The goal of the testing is twofold. *In vitro* test procedures are used to qualitatively

0077–8923/83/0407–0367 $01.75/0 © 1983, NYAS

TABLE 1

MUTAGENESIS BATTERY

Progress from *in vitro* tests for intrinsic activity to evaluation of risk in mammalian systems:

1. Point mutations *in vitro:* Ames test
   mouse lymphoma

2. Chromosome damage: *in vivo* cytogenetics
   Point mutation *in vivo: Drosophila*

3. DNA alteration: exposure *in vivo*/expression *in vitro*
   Unscheduled DNA synthesis
   Sister chromatid exchange
   Cell structure alteration: *in vitro* transformation
   　　　　　　　　　　　　　in vivo transplantation

4. Transmittable genetic and reproductive effects:
   dominant lethal assay

identify the intrinsic mutagenic activity of a material or component. These data are evaluated in conjunction with testing in animals to determine if a compound is activated, detoxified, sequestered, or excreted unaltered *in vivo* to contribute to a reliable quantitative assessment of risk to humans. Activity *in vivo* can be affected by species, sex, route and duration of administration, and the ability of a mutagenically active electrophile to reach the target cell at an appropriate time in the cell cycle.

In beginning with a discussion of *in vitro* procedures, I am going to skip the standard review of 90% correlations between microbial mutagenesis and animal carcinogenesis and concentrate on the difficulties encountered when the use of microbial and mammalian cell mutagenesis tests (e.g., Ames, mouse lymphoma) is extended from relatively pure, water-soluble chemicals to evaluating mutagenic activity in materials and products from a wide variety of industries. For instance, for food samples tested in the Ames test, an overgrowth of the histidine-deficient auxotrophs of *Salmonella* stimulated by endogenous histidine in the test sample may obscure a reliable assessment of the incidence of induced revertants. Many chemical products are a mixture of materials too complex to separate conveniently. When a significant lack of solubility in aqueous media is also a property of the complex mixture, you have an idea of the problem facing the genetic toxicologist in the petroleum industry. The standard solvents such as dimethylsulfoxide (DMSO) and acetone may solubilize the test material initially, but once introduced into the aqueous environment of the *Salmonella* assay or a mammalian cell system, the material may either precipitate to the bottom or form droplets on the surface, raising the question of how much of the original sample or what components of it are interacting with the cells. In the absence of toxicity as a crude end point of compound-cell interaction, a negative result in an *in vitro* system tells very little. TABLE 2 illustrates the frustration encountered by the American Petroleum Institute in testing fractions of crude petroleum in the Ames test.[3] Different grades of fuel and lubricants are distilled from crude petroleum in different boiling ranges. All of the fractions contain polynuclear aromatic hydrocarbons (PAHs) in varying concentrations and have been tested in mouse-skin-painting studies for induction of skin tumors. There is no correlation between bacterial mutagenesis using a standard protocol and carcinogenesis by this method. Mammalian cell systems are not much better. Projects are currently under way in a number of laboratories to determine if the solubility problem can be resolved to develop some predictive value from *in vitro* tests. In the Mobil laboratory, by simply changing the solvent for testing the D5 sample from DMSO to tetrahydrofuran, some mutagenesis with metabolic activation was produced. We are also exploring the use of emulsifiers and surfactants to increase the compound-cellular interaction without causing artificial mutagenesis or toxicity. Scientists at Chevron have reported preliminary success in altering the concentration of S-9 in the test system to improve the mutagenic response. Other groups are experimenting with microspheres or liposomes as carriers of the sample into the cell.

Chemical extraction procedures are effective methods of overcoming solubility problems, reducing the levels of biologically inert components in a complex mixture, and identifying fractions that have mutagenic activity. Epler and Guerin have published extensively on the mutagenic/carcinogenic activity of synthetic fuels in which chemical characterization and bioassay, principally the Ames *Salmonella* test, are combined to isolate and identify biologically hazardous agents in coal tars, synthetic oils, raw retorted shale oil, and aqueous wastes from these processes. Epler outlines a variety of extraction techniques for complex mixtures.[4] Using liquid-to-liquid extraction and column chromatography, it has been determined that for natural and synthetic crude oils, the mutagenic activity resides in neutral (principally PAH) and basic fractions (heterocyclic nitrogen compounds, aromatic amines, and aza-arenes). In the Mobil laboratory we have been studying those components of used motor oil responsible for mutagenic activity in the absence of metabolic activation, employing the Ames assay to follow mutagenic activity and to direct fractionation.

Identifying mutagenic activity in aqueous samples usually requires concentration of organics for accurate testing, as Loper *et al.* have done in their survey of drinking water[5] and as Yamasaki and Ames recommend for identifying mutagenic metabolites in urine.[6] The macroreticular polystyrene (XAD) resins are effective for this process.

The organic materials adsorbed on particulates from coal fly ash or diesel or gasoline engine exhaust emissions are eluted by Soxhlet extraction with methanol or methylene chloride prior to class fractionation and bioassay. Diesel-emission extracts are mutagenically active in the Ames test with and without metabolic activation, the

TABLE 2

CORRELATION OF AMES TEST RESULTS TO DERMAL CARCINOGENICITY*

Crude Sample	Distillation Range† (°F)	Carcinogenicity	Ames Test Results
C‡	OP-->1070	+	−
C-2	OP–350	+	−/+
C-3	350–550	+	−
C-4	550–700	+	−
C-5	700–1070	+	−
C-6	>1070	−	−
D‡	OP->1070	+	−
D-1	OP–120	−	−
D-2	120–350	+	−
D-3	350–550	+	−
D-4	550–700	+	−
D-5	700–1070	+	−
D-6	>1070	−	−

*Reproduced from Reference 3 with permission from the publisher.
†OP stands for initial boiling point.
‡Whole crude oil.

direct activity attributed largely to monosubstituted nitro-PAH compounds according to Pederson et al.[7] Other studies have shown that engulfment of diesel particles by mammalian cells results in gene mutation.[8] However, assay of diesel-emission particulates in different mutagenesis tests—mammalian cells in vitro,[9] in vivo exposure of Drosophila,[10] or mice for sister chromatid exchange (SCE) or chromosome breakage[11]—does not produce a mutagenic response.

This situation illustrates the necessity for caution in using in vitro test results on highly concentrated samples to make decisions for hazard estimates. Is the effect of the whole equal to the sum of the effects of the parts?

The detection or even generation of mutagenic activity may be a function of the fractionation scheme. The selection of resins or solvents may result in the loss or modification of specific compounds or eliminate synergism among components. A mixture may be regulated on the basis of mutagenic activity of highly concentrated materials that may exist in extremely dilute, insignificant quantities in the whole sample. For diesel emissions, the potential mutagens may be too tightly bound to the particulate to interact with the mammalian cell in vivo. Claxton and Barnes stated that the mutagenicity of extracts of diesel emissions in the Ames test can be altered by collection and extraction under exposure to nitrogen, oxygen, ultraviolet light (UV), NO_2, or SO_2.[12] Petrilli et al. reported that crude oil reduced the mutagenic activity of benzo[a]pyrene in a spiked sample by sequestering the material and making it unavailable for metabolic activation.[13] A similar phenomenon is demonstrated by petroleum materials in skin-painting studies where the potency of a sample in producing skin tumors is correlated with viscosity[14] and the likelihood of additional systemic carcinogenicity may relate to rate of percutaneous absorption.

It therefore is imperative to define the mutagenic activity observed in vitro in terms of in vivo response. It may be particularly useful to compare the same end point, e.g., chromosome breakage induced in Chinese hamster ovary (CHO) cells with chromosome breakage of bone marrow cells of treated animals, sister chromatid exchange, or unscheduled DNA synthesis (UDS) in vitro and in vivo. Since the available in vivo mammalian tests for point mutations are costly and somewhat complex, it is possible that SCE or UDS in vivo may provide reasonable confirmation of the point mutation events in vitro. Skinner et al. measuring UDS in peripheral lymphocytes and cytogenetic events in bone marrow cells from the same rats treated in vivo with cyclophosphamide, a clastogen, or 2-acetoamido fluorine, known to induce point mutations, demonstrated that both compounds caused UDS in vivo.[15]

Earlier in this presentation I referred to mutagenic activity and reproductive effects. Fifty percent of aborted human fetuses have chromosomal anomalies.[16] The increased fetal mortality that is the end point of the dominant lethal test may be caused

TABLE 3

TERATOLOGY VERSUS In Vivo CYTOGENETICS*

A. Positive teratology, positive cytogenetics	32
B. Negative teratology, negative cytogenetics	20
C. Positive teratology, negative cytogenetics	16
D. Negative teratology, positive cytogenetics	6
E. Inconclusive teratology or cytogenetics	12
F. Total compounds surveyed	86
Percent correlation (A + B/F × 100)	60%

*Reproduced from Reference 20 with permission from the publisher.

TABLE 4

TERATOLOGY AND *SALMONELLA* RESULTS*

A. Positive teratology, positive *Salmonella*	35
B. Negative teratology, negative *Salmonella*	9
C. Positive teratology, negative *Salmonella*	11
D. Negative teratology, positive *Salmonella*	3
E. Inconclusive teratology or *Salmonella*	2
F. Total compounds surveyed—conclusive results	58
G. Total compounds surveyed	60
Percent correlation (A + B/F × 100)	76%

*Reproduced from Reference 20 with permission from the publisher.

by chromosome damage or direct toxicity to developing sperm. It has been theorized that the electrophilic action of cytotoxic agents provides a common mechanism for mutagenicity and teratogenicity.[17,18] Schreiner and Holden surveyed chemicals tested for both teratogenesis and mutagenesis.[19] The predictive correlation between teratogenesis and *in vivo* or *in vitro* cytogenetic aberrations, including microtubule disruption of over 80 chemicals, was 65%. If one considers only whether a compound causing chromosome damage is also a teratogen, the correlation increases to 84% (TABLE 3). For the 60 carcinogens for which point mutation data in the Ames test and teratogenetic data were available, the correlation was 76% overall (TABLE 4). Again, if the ratio is limited only to whether a mutagenic carcinogen is a teratogen, the relationship increases to 92%. While I am not advocating mutagenic screens as definitive predictors of reproductive and teratogenic effects, the data suggest that a genotoxic compound may also pose a reproductive risk and should be tested accordingly.

Short-term tests for mutagenesis *in vivo* and *in vitro* have great potential as predictors of health hazards in humans. Given the large number and combinations of chemicals in the environment, these tests may be the only data available on some materials. Combined with definitive data on metabolism and pharmacokinetics, these tests could substantially reduce the number of lifetime carcinogenesis assays performed and the thousands of animals employed. They provide a means of monitoring how effectively we are cleaning up the environment for all living organisms. A significant concern at present is that we make grandiose leaps from the available data on observed mutagenic activity to human risk. For some materials such extrapolations are already appropriate—hycanthone causes point mutation *in vitro*, is a clastogen in animals and human cells,[20] and has been regulated internationally. AF-2 causes point mutations in *Salmonella* and stomach cancer in man.[21] Vinyl chloride is also mutagenic in *Salmonella*[22] and carcinogenic in man.[23] However, for complex mixtures it is inappropriate to draw such conclusions as yet. While discussions are under way on the role of short-term tests in regulating food additives or pesticides or drugs, many industries are still at the basic stage of determining how to perform the assays effectively with their compounds, identifying which tests are most appropriate, and confirming the reliability of results. Clearly the role of short-term tests in safety testing of industrial chemicals must be defined in the context of the types of materials being evaluated. The significance of test results should be determined for each material individually in relation to its general toxicology profile, its use, and its potential for human exposure.

REFERENCES

1. 1978. Fed. Regist. **43:** 50140–50147.
2. FISHBEIN, L. 1979. Potential Industrial Carcinogens and Mutagens. Studies in Environmental Science **4:** 2–4. Elsevier. New York, N.Y.
3. MACGREGOR, J. A., V. F. SIMMON, G. F. SHEPHERD, D. R. JAGANNATH, D. J. BRUSICK, L. M. SCHECHTMAN, A. S. PALMER, S. T. CRAGG & C. C. CONAWAY. 1982. An evaluation of the ability of Ames *Salmonella* assay to reflect the carcinogenic activities of complex petroleum hydrocarbon mixtures. Paper presented at the American Petroleum Institute Toxicology Symposium on Petroleum Hydrocarbons, Washington, D.C., May 11–13.
4. EPLER, J. L. 1980. The use of short-term tests in the isolation and identification of chemical mutagens in complex mixtures. *In* Chemical Mutagens, Principles and Methods for Their Detection. F. J. de Serres & A. Hollaender, Eds. **6:** 239–270. Plenum Press. New York, N.Y.
5. LOPER, J. C., D. R. LONG, R. S. SCHOENY, B. B. RICHMOND, P. M. GALLAGHER & C. C. SMITH. 1978. Residue organic mixtures from drinking water show *in vitro* mutagenic and transforming activity. J. Toxicol. Environ. Health **4:** 919–938.
6. YAMASAKI, E. & B. N. AMES. 1977. Concentration of mutagens from urine by absorption with the nonpolar resin XAD-2: cigarette smokers have mutagenic urine. Proc. Nat. Acad. Sci. USA **74:** 3555–3559.
7. PEDERSON, T. C. & J. S. SIAK. 1981. The role of nitroaromatic compounds in the direct-acting mutagenicity of diesel particle extracts. J. Appl. Toxicol. **1**(2): 54–60.
8. CHESCHEIR, G. M., N. E. GARRETT, J. D. SHELBURNE, J. L. HUISINGH & M. D. WATERS. 1981. Mutagenic effects of environmental particulates in the CHO/HGPRT system. Environ. Sci. Res. **22:** 337–350.
9. RUDD, C. J. 1980. Diesel particulate extracts in cultured mammalian cells. *In* Health Effects of Engine Diesel Emissions: Proceedings of an International Symposium **1:** 385–403. Publication No. EPA 600/9-80-057a. U.S. Government Printing Office. Washington, D.C.
10. SCHUBER, R. L. & R. W. NIEMEIER. 1980. A study of diesel emissions on *Drosophila*. *In* Health Effects of Engine Diesel Emissions **2:** 914–923. Publication No. EPA 600/9-80-057b. U.S. Government Printing Office. Washington, D.C.
11. PEREIRA, M. A., P. S. SABHARWAL, P. KAUR, C. B. ROSS, A. CHOI & T. DIXON. 1980. *In vivo* detection of mutagenic effects of diesel exhaust by short-term mammalian bioassays. *In* Health Effects of Engine Diesel Emissions **2:** 934–948. Publication No. EPA 600/9-80-057b. U.S. Government Printing Office. Washington, D.C.
12. CLAXTON, L. & H. M. BARNES. 1980. The mutagenicity of diesel exhaust exposed to smog chamber conditions as shown by *Salmonella typhimurium*. *In* Health Effects of Engine Diesel Emissions **1:** 309–326. Publication No. EPA-600/9-80-057a. U.S. Government Printing Office. Washington, D.C.
13. PETRILLI, F. L., G. P. DE RENZI & S. DE FLORA. 1980. Interaction between polycyclic aromatic hydrocarbons, crude oil and oil dispersants in the *Salmonella* mutagenesis assay. Carcinogenesis **1:** 51–56.
14. BINGHAM, E., R. P. TROSSET & D. WARSHAWSKY. 1980. Carcinogenic potential of petroleum hydrocarbons. J. Environ. Pathol. Toxicol. **3:** 483–563.
15. SKINNER, M. J., C. A. SCHREINER & F. T. DAVIS. 1982. Unscheduled DNA synthesis detection and metaphase analysis in a common test system *in vivo–in vitro*. J. Appl. Toxicol. **2**(3): 172–175.
16. CARTER, C. D. 1977. The relative contribution of mutant genes and chromosome abnormalities to genetic ill-health in man. *In* Progress in Genetic Toxicology. D. Scott, B. Bridges & F. Sobels, Eds.: 1–14. Elsevier/North-Holland Biomedical Press. Amsterdam, the Netherlands.
17. CONNER, T. A. 1975. Cytotoxic agents in teratogenic research. *In* Teratology Trends and Applications. D. E. Posivello & C. L. Berry, Eds.: 49–88. Springer-Verlag. Berlin, Federal Republic of Germany.
18. HARBISON, R. D. 1978. Chemical-biological reactions common to teratogenesis and mutagenesis. Environ. Health Perspect. **24:** 87–100.

19. SCHREINER, C. A. & H. E. HOLDEN. Mutagens and teratogens: a correlative approach. *In* Teratogenesis and Reproductive Toxicology. Handbook of Experimental Pharmacology. E. M. Johnson & D. Kochar, Eds. **65.** Springer-Verlag. Berlin, Federal Republic of Germany. (In press.)
20. RAY, V. A., H. E. HOLDEN, J. H. ELLIS, JR. & M. L. HYNECK. 1975. A comparative study on the genetic effects of hycanthone and oxamniquine. J. Toxicol. Environ. Health **1:** 211–227.
21. SUGIMURA, T., T. YAHAGI, M. NAGAO, M. TAKEUCHI, T. KAWACHI, K. HARA, E. YAMASAKI, T. MATSUSHIMA, Y. HASHIMOTO & M. OKADA. 1976. Validity of mutagenicity tests using microbes as a rapid screening method for environmental carcinogens. IARC Sci. Publ. **12:** 81–101.
22. RINKUS, S. J. & M. S. LEGATOR. 1979. Chemical characterization of 465 known or suspected carcinogens and their correlation with mutagenic activity in the *Salmonella typhimurium* system. Cancer Res. **39:** 3289–3318.
23. MILLER, E. C. & J. A. MILLER. 1981. Searches for the ultimate chemical carcinogens and their reactions with cellular macromolecules. Cancer **47:** 2327–2345.

LIMITED *IN VIVO* BIOASSAYS

G. Mazue, D. Gouy, B. Remandet, and J-M. Garbay

Clin-Midy Research Center
Sanofi Group
34082 Montpellier Cedex, France

INTRODUCTION

Limited *in vivo* bioassays are *in vivo* studies of a duration not exceeding one year and whose aim is to indicate the carcinogenic potential of a given substance and to determine its mechanism, either genotoxic or epigenetic, by means of an appropriate protocol. The moderate duration of this type of study requires a technique for rapid screening of the tumoral process, making use of anatomopathological or biochemical carcinogen markers.

These tests are in an intermediary position between classic long-term tests for carcinogenesis and short-term tests for mutagenesis.[1] In comparison with long-term studies, limited *in vivo* bioassays have the advantage of being less complex, less difficult to implement, and therefore less expensive, of supplying results more rapidly, and furthermore of detecting potential carcinogenic promoting effects. Moreover, these tests are performed on animals and take into account the metabolism of the chemical under test, the homeostatic environment, and the immune response and are therefore a good complement to the short-term tests which are often carried out *in vitro*.[2,3]

OVERVIEW

In the classic carcinogenicity studies, the compound under test is administered during the whole life span of the animals and the apparition of tumors is tested for in the whole organism. In limited *in vivo* bioassays where the aim is to shorten the study, methods that allow very early detection of the process or that ensure its more rapid development are needed. Consequently the experimental protocol must take these requirements into consideration, which results in a greater specificity of the type of tumors tested for and a need for carcinogen markers. It should be noted that this leads to a restriction on the range of organs explored. In this type of study, the protocol must also be adapted to the compound under test and the type of carcinogenesis being tested for. In particular, genotoxic activity will not be detected by the same method as promoting activity will and the protocol must be chosen according to the chemical structure of the molecule, the results obtained with short-term tests, and the other biological properties of the molecule.[4]

Principal Limited In Vivo *Bioassays*

As in the case of short-term tests, there exist a number of limited *in vivo* tests, but these being longer and more complicated to perform, they cannot be used as a battery. Consequently, one of these tests must be selected, for which a good knowledge of their respective points of interest and shortcomings is necessary. Moreover, this choice must be supported by a good knowledge of the molecule from both the chemical and

374

0077–8923/83/0407–0374 $01.75/0 © 1983, NYAS

biological points of view. This determines the place of these tests in the general strategy of development of the molecule. The method employed in these tests consists of using strains particularly predisposed for tumor development or applying a treatment that will indicate carcinogenic activity more rapidly.

Pulmonary Tumor Induction in Mice

This test is performed with mouse strains that are particularly sensitive to carcinogen agents and in which tumor development is very rapid (under 30 to 35 weeks).[5] The carcinogenic activity of the test compound can be assessed by the number of tumors occurring in the treated animals in comparison with the controls, but also by their type, which varies according to the substances under study. However, interpretation must be qualified: in general, a compound found to be active in this system gives rise to tumors in long-term tests, but a negative result does not exclude the risk of carcinogenesis, in particular in organs other than the lungs.

Breast Tumor Induction in Female Rats

The principle of breast tumor induction in female rats is identical to that of pulmonary tumor induction. It consists of choosing a strain sensitive to breast tumor development under the influence of certain carcinogenic agents. Development is rapid (less than nine months), and identification is straightforward. This test presents the same advantages as the preceding one and is subject to the same limitations.[6]

Skin Tumor Induction in Mice

It has been recognized for a long time that repeated application of certain products such as tar produces papillomas and carcinomas.[7] This principle can be used to reveal carcinogen initiating agents in a relatively short time by simultaneous association with a cocarcinogen or sequential application of a promoter, such as phorbol esters.[8] This accelerates the carcinogenesis process and also detects the potential initiating activity of an unknown product. Inversely, use of a known initiator serves as a test for potential promoting activity on the skin. This system is of particular interest as mouse skin contains the necessary enzymatic equipment for conversion to the metabolites responsible for initiating activity.

Altered Foci Induction in Rodent Liver

The earliest alteration that occurs during development of hepatic carcinogenesis is the altered foci,[9] which can be visualized by routine histologic and histochemical techniques. This is followed by the development of neoplastic nodules in the liver, and then by carcinomas. With known carcinogens, foci develop in 3 weeks and are numerous after 12 to 16 weeks of treatment. The principle of this test consists of testing for altered foci in the liver of rats receiving the test compound for several weeks, in comparison with control animals and with animals receiving a reference compound, such as *N*-2-fluorenylacetamide (FAA). After initiation of altered cell foci with a reference compound, the promoting activity of a substance may also be tested on this model[10] by the study of the evolution of these foci in comparison with control animals and also with a reference promoter such as phenobarbital. This model is particularly justified in studies of carcinogenic potential in drugs, as the great majority of these substances are transformed in the liver and give metabolites. This organ therefore undergoes extensive exposure to the test substance and also to all these metabolites. Potential activation of a procarcinogen is therefore carried out to perfection; it must however be noted that extrapolation to man requires knowledge of the comparative metabolisms of man and rodents. Furthermore, possible disactivation of the carcinogen is taken into account in this model. Altered cell foci induction in rodent liver is therefore an excellent model for the study of carcinogenic potential in drugs but requires a good knowledge of carcinogen markers.

Carcinogen Markers

A carcinogen marker must be easy to explore and representative of the cellular lesions occurring during the carcinogenic process. To facilitate exploration, these parameters must be measured on samples easy to take in laboratory animals, and the dosing technique must be easy, rapid, and reproducible. Furthermore, variations of this parameter in cells, tissues, and biologic liquids must have a good correlation with the histological lesions that define carcinogenesis.[11]

The ideal marker for carcinogenesis is specific to the cellular differentiation particularly of a given organ. Cellular carcinogen markers are, in general, associated with metabolic or membrane alterations. The metabolic alterations are increase of nucleic acid synthesis, responsible for cellular basophilia, and the reapparition of embryonic characters, such as production of α-fetoproteins or carcinoembryonic antigens. Enzyme or hormone secretion and excretion may also be subject to increase or decrease. Membrane alterations include alteration of surface glycoproteins and the apparition of surface neoantigens. We will leave aside these different markers and consider only those concerning rat liver, which are used routinely in limited *in vivo* bioassays.

Histologic Criteria

These can be divided into morphologic and histochemical criteria.

Morphologic Criteria. Altered cell foci, neoplastic nodules, and carcinomas occurring in rodent liver exposed to carcinogens have already been described.[12]

1. Altered foci or zones. The general structure of these foci is slightly altered in comparison with healthy tissue; they do not compress the adjacent parenchyma. They can be classified on hematoxylin eosin staining as clear cell foci; acidophilic cell foci; basophilic cell foci; or mixed cell foci composed of a combination of two or all three of these cell types.

2. Neoplastic nodules. These are composed of the above-mentioned cell types and are characterized by loss of internal structure and compression of the adjacent parenchyma.

3. Carcinoma. This is characterized by trabecular pattern and adenocarcinoma pattern, both of hepatic origin, and complex pattern, probably originating from transitional hepatic cells (oval cells).

4. Associated lesions. These lesions are consequent to the carcinogenic action. However, they are not specific and perturb interpretation of specific lesions: regenerative hepatocellular nodules resemble neoplastic nodules by their expansive centrifugal

nature, but lack the cellular characteristics of these, and their internal structure presents slight alterations only; hyperplasia of the biliary ducts, adenofibrosis or cholangiofibrosis, and biliary cysts; and hyperplasia of the oval cells, of transitional hepatocellular nature.[13] Such cells are probably of concern in the emergence of complex carcinomas with some promoters.

Histochemical Criteria. In addition to these distinctive morphologic aspects detectable by routine histology, the foci, nodules, and carcinomas present several abnormalities allowing their detection before morphology. They are the presence of gamma glutamyl transpeptidase (GGT),[14] an enzyme that is normally present in very small quantities in adult rat liver but appears in large quantities in fetal rat liver, which shows dedifferentiation of cells in neoplastic process—the foci are therefore GGT positive, as are neoplastic nodules and carcinomas; deficiency of glucose 6 phosphatase;[15] deficiency of adenosine triphosphatase;[16] and absence of iron accumulation after induction of hepatic siderosis.[17] This last parameter seems to be the most sensitive.

Biochemical Markers

Biochemical markers can be studied either in liver homogenate or in plasma.[18] Among these markers, GGT presents particular interest for the study of hepatic carcinogenesis in the rat. In fact, this enzyme is virtually undetectable in the liver of the adult rat. It is localized in the cells that are close to the bile ducts, and the plasmatic rate is almost nonexistent. But it appears in the liver homogenates, as in the plasma in the case of hepatic carcinogenesis. The plasmatic base rate being almost nonexistent, its increase is easily revealed. Therefore, this enzyme has an organ specificity as well as a dedifferentiation specificity and it presents the characteristics of the marker above described. Furthermore, this enzyme is quite stable and persists a long time in samples even at ambient temperature. Consequently sampling may be performed without necessitating precautions.

Thus the elements to be retained in routine practice for limited *in vivo* bioassays for altered foci induction in rodent liver are GGT plasmatic rate and anatomopathologic examination of the liver exposed to siderosis, in which foci, nodules, and carcinomas are counted. The tumoral process is easily identified by correlation between these parameters.

Application

The application of these assays presents the problem of why, when, and how they should be used in the global strategy of investigation of a molecule. The replies to these questions are interrelated.

The aim of these tests is to determine whether a molecule is carcinogenic and, if so, to indicate if it is genotoxic or a carcinogenic promoter. This aim conditions the strategy and implementation of the experiment. From the strategical point of view, application of these tests can take place after the short-term tests have been conducted to ascertain whether the molecule is epigenetic (promoter, cocarcinogen) or not in case of negativity of the short-term tests. But the type of limited *in vivo* bioassays must take into account knowledge concerning the metabolism and biologic properties of the compound, as well as its route of administration or exposure, in order to design a protocol well adapted to the case in question.

Long-term tests will be performed according to these results, which may avoid studies that are long, expensive, and of no worth in the event of the compound proving to be genotoxic. On the other hand, when limited *in vivo* bioassays yield negative results, long-term tests must still be envisaged.

In another way, limited *in vivo* bioassays can follow the long-term tests carried out after negative short-term tests, in the case of positive response to detail the nature of the compound—cocarcinogen, promoter, etc.

Protocols for limited *in vivo* bioassays differ, therefore, according to whether the activity tested is genotoxicity or promoting activity. In the case of pharmaceutical drugs, the limited test based on altered foci induction in iron-accumulation-resistant rodent liver is of particular interest. The liver is of course widely exposed to such substances and to their metabolites and has a high probability of exteriorizing lesions if the substance is carcinogenic, but a negative result does not exclude lack of potential tumoral effect on another organ. This test therefore does not exclude a long-term study but enables such a study to be undertaken with a much higher probability of obtaining a negative result.

In such tests development of tumors and obtention of results are much more rapid than with long-term tests. Handling is much easier, as reduced quantities of animals are used; examinations are simpler, as there is no need to search for tumors by palpation. Examinations are confined to a biochemical study, especially centered on GGT but also possibly on certain markers such as feritine or α-fetoprotein. As for anatomopathology, one organ only is tested, this representing a considerable gain of time in comparison with classic carcinogenicity studies. The kinetics of the lesions can also be examined during these tests.

EXAMPLES OF LIMITED *IN VIVO* BIOASSAYS IN RAT LIVER

Material and Methods

As mentioned previously, protocols for such limited *in vivo* bioassays differ according to whether the activity tested for is genotoxicity or epigenotoxicity, particularly promoting activity.

Genotoxic Activity

To test for genotoxic activity, the compound will be administered to male Fischer 344 rats for several weeks. This strain has been selected because of its almost systematic use in tests of this type described by various authors, who have found that animals of this sex and strain have a greater response sensitivity. Duration varies from 4 to 20 weeks, as does the number of dose levels in order to reveal a potential dose-, time-, or total-quantity-related effect. A comparison is made with negative controls and with positive controls receiving 0.02% of FAA in the diet. At termination of treatment, hepatic siderosis is induced by six injections of iron dextran solution (Inferon ND) at 100 mg iron/ml and 0.125 ml/100 g. The injections are administered at two-day intervals, the last injection on the day preceding autopsy. During the study, body-weight gain and food intake are measured and biochemical examinations with dosage of GGT plasmatic activity are carried out periodically. At autopsy, after weighing and exsanguination of the animal, the liver is weighed separately and fixed in 10% buffered formalin solution. After embedding of the four main lobes in paraplast, two serial sections 5 μ thick are taken from the following four lobes: left, medial left, medial right, and anterior right. The first section is stained with hematoxylin-eosin, and the second with prussian blue reaction in order to reveal hepatic siderosis. From the histological point of view, detection of potential initiating activity of the test compounds is based on counting the number of altered cell foci or zones and neoplastic nodules resistant to iron accumulation. This is calculated for each animal and for the four principal lobes. Liver fragments may also be frozen in order to dose GGT on the homogenates.

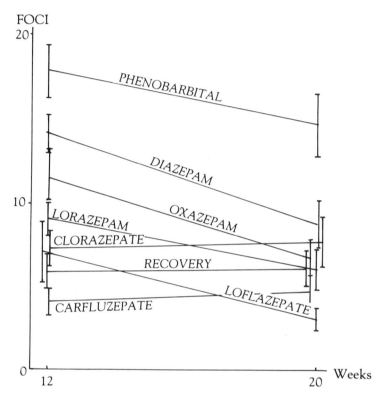

FIGURE 1. Promoting effect of some benzodiazepines (foci).

Promoting Activity

Testing for promoting effect is carried out on the same type of animal, with an initiating phase with FAA at 0.02% in the diet during 8 to 10 weeks. The duration of this phase may be modified according to variations in plasmatic GGT and the apparition of altered foci in control animals at the start of initiation. Following the initiation phase the animals are allowed a 2-week recovery period, following which promoting activity is tested in comparison with a group returning to basal diet (recovery) and with a positive control group receiving phenobarbital in the diet at 500 ppm. The same examinations are performed as for the study of genotoxic activity. Study length varies, sometimes reaching 20 to 40 weeks, but with a good promoter such as phenobarbital effects appear after 4 to 12 weeks. It is also interesting to carry out a kinetics study in order to follow the development of the potential promoting activity, to evaluate reversibility in the recovery group, and to modify the treatment period in the light of the preliminary anatomopathological and biochemical results. Moreover, the promoting phase should not be overprolonged, to avoid the arising of initiated lesions in rats fed basal diet (recovery), or high mortality rate by excess of the carcinogenic process, or the development of spontaneous tumors, which may occur as a result of age. Dose-related effect is also important to appreciate the promoting effect.

Study of the Initiating or Promoting Activity of Some Benzodiazepines

The initiating and promoting ability of some benzodiazepines (diazepam, chloraze-pate, oxazepam, lorazepam, loflazepate, and carfluzepate) was tested for in the model system using the following protocols.[19,20] In the test for initiating effect, treatment lasted 14 weeks; for the promoting effect an 8-week initiating phase with FAA was followed by a 4- or 12-week period of administration.

The anatomopathological study of the liver together with the biochemical study confirmed the initiating effect of FAA, the promoting effect of phenobarbital, and the lack of carcinogenic effect (either initiating—first study confirmed by the second study—or promoting—second study) of the benzodiazepines studied.

Foci (per liver) increased with FAA during the first 8 weeks, then abated during 8 to 20 weeks. With phenobarbital, foci decreased sharply between 8 and 20 weeks, and

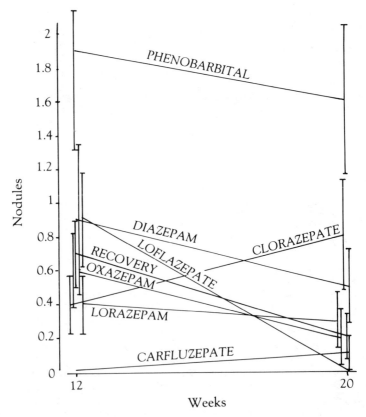

FIGURE 2. Promoting effect of some benzodiazepines (nodules).

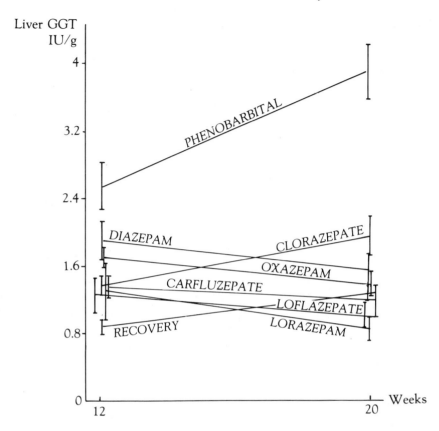

FIGURE 3. Promoting effect of some benzodiazepines (liver GGT).

nodules were maintained at the level reached at 8 weeks. All the benzodiazepine tranquilizers studied acted as recovery after FAA: decrease of foci and nodules in each case with statistical differences in comparison with phenobarbital (FIGURES 1 and 2).

Variations of hepatic or plasmatic GGT also discriminated between FAA, phenobarbital, and recovery as shown in FIGURES 3 and 4.

However, in this presentation one of the principal points is to see the correlations between the tumor marker GGT (hepatic or plasmatic) and the histological findings obtained in the promotion study.

These correlations (TABLE 1) were obtained with the individual results, and they show that plasmatic GGT and hepatic GGT are highly correlated, and also with foci and nodules. So plasmatic GGT appears as a good marker of liver carcinogenesis in such limited *in vivo* bioassays, and corresponds with most of the above-mentioned criteria.

Another study also showed these correlations with intervention of carcinomas.

Study of Polychlorinated Biphenyls and DDT Promoting Activity

Promoting activity is tested according to the previously mentioned general plan after FAA induction of altered foci during 10 weeks, followed by 2 weeks of recovery.[21]

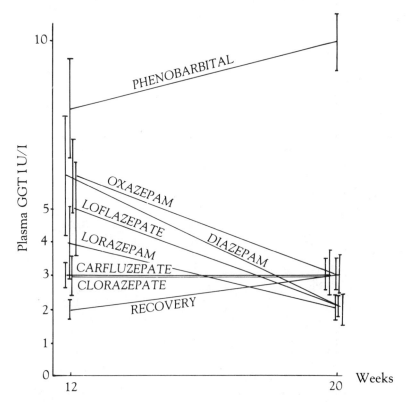

FIGURE 4. Promoting effect of some benzodiazepines (plasma GGT).

Following this initiating phase, promoting activity is tested during 25 and 48 weeks with administration of phenobarbital (PB) at 250 and 500 ppm, DDT at 250 ppm, polychlorinated biphenyls (PCB) at 250 and 500 ppm, and the associations of DDT and PCB at the dosages shown in FIGURE 5.

Pathology

During the initiating phase, lesions are consistent with those of the previous study.[22] At 37 weeks, with recovery foci and nodules regressed and no carcinomas were

TABLE 1

CORRELATIONS BETWEEN GGT, FOCI, AND NODULES

	Foci	Nodules	GGT of Plasma	GGT of Liver
Foci	1	—	—	—
Nodules	0.43*	1	—	—
GGT of plasma	0.57*	0.38*	1	—
GGT of liver	0.67*	0.48*	0.87*	1

*Significant at $p < 0.01$.

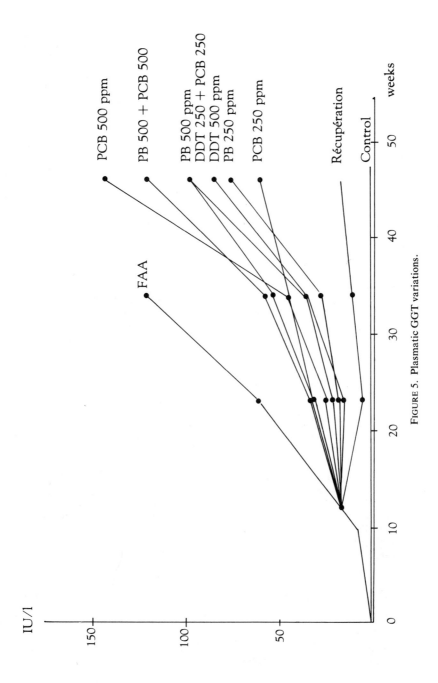

FIGURE 5. Plasmatic GGT variations.

observed. With DDT at 500 ppm and PB at 250 or 500 ppm, foci underwent regression similar to recovery. Nodules showed no regression but increased slightly with PB at 250 and 500 ppm. Carcinomas were noticed with PB at 250 ppm and DDT at 500 ppm. These results show the typical promoting activity of both compounds compared with FAA. With FAA, carcinomas are observed in all rats (two had abdominal metastasis) and they occupy 16% of the total liver surface examined. This point can explain the relative decrease of foci in comparison with PCB alone or in association, where carcinomas were also noticed in all rats at 500 ppm and more than 50% of rats at 250 ppm and occupied only 4% of the surface in the greatest cases.

The results obtained with FAA are also consistent with those recorded in literature by an increase of nodules when continuing FAA administration after initiation in comparison with recovery or a pure promoter (PB or DDT).

With PCB a large number of foci, nodules, and carcinomas were noticed. Such lesions seemed to be in an intermediate position between those obtained with the genotoxic carcinogen FAA and the pure promoters PB and DDT.

Therefore, whereas it is certain that PB and DDT are pure epigenetic promoters, such is not the case for PCB, for which direct hepatocarcinogenic action has been described when administered in the mouse[23] and, in certain cases, in the rat.[24] All this could explain the differences observed between the groups receiving PB or DDT and those receiving PCB. In fact, if administration of PB and DDT after FAA or FAA alone appears to have induced identical carcinomas, administration of PCB has caused in addition complex carcinomas with intervention of oval cells, suggesting a direct or indirect action of PCB on these cells.

At 60 weeks, all animals, even in the recovery group, presented carcinomas (about 16% of the surface in the recovery group, 50% with phenobarbital or DDT, and 100% with PCB alone or in association, and with FAA). Foci and nodules were inexploitable. Therefore, this type of test must not exceed 37 to 40 weeks in such analyses.

Biochemistry

Dosage of plasmatic GGT during the study yielded the following curves (FIGURE 5). Right after the end of the FAA initiating phase (results obtained at 10 weeks), the treated animals show a relatively high rate of plasmatic GGT in comparison with, on one hand, the control group and, on the other hand, GGT measured in the same animals prior to treatment. GGT continues to increase slightly until 12 weeks, at which time the various treatments begin. Afterwards, with recovery, GGT decreases in a rather marked but incomplete manner during the first weeks (results obtained at 23 weeks) and increases slightly during the following weeks (weeks 34, 46, and 58). With FAA, GGT decreases sharply, reaching more than 100 iu/l at 34 weeks, the date of the last sample before death of all animals in this group.

For the other groups treated with different promoters, GGT levels are in an intermediary position between the levels reached with FAA and those reached in the recovery group, with significant differences in comparison with both these groups at the different periods. At 23 weeks, the range of values of the different groups is relatively restricted, widening at weeks 34 and 46. At weeks 46 and 58 the results obtained in a limited number of animals (maximum five animals per group) must be considered with precaution. In fact, the slightest individual variation of GGT leads to important alterations of the mean value calculated for each group.

In brief, this assay in rat livers confirmed the promoting effect of DDT and phenobarbital and revealed a different activity of PCB, which appears to be a special promoter. Plasmatic GGT appears highly correlated with the liver lesions calculated with all individual results obtained at 37 weeks, especially with carcinomas. Without carcinomas, GGT is closely correlated with foci and to a lesser degree with nodules. These results show that the carcinogenic process can be evaluated before the emergence of carcinomas with foci and nodules only, the carcinomas inducing a decrease in correlations between GGT and early lesions.

REFERENCES

1. WILLIAMS, G. M. & J. H. WEISBURGER. 1981. Systematic carcinogen testing through the decision point approach. Annu. Rev. Pharmacol. Toxicol. **21:** 393–416.
2. WILLIAMS, G. M. 1976. The use of liver epithelial cultures for the study of chemical carcinogenesis. Am. J. Pathol. **85:** 739–753.
3. WILLIAMS, G. M. 1981. *In* Proceedings of a Symposium on Chemical Indices and Mechanisms of Organ Directed Toxicity. S. S. Brown & D. Davies, Eds.: 131–149. Pergamon Press. New York, N.Y.
4. WEISBURGER, J. H. & G. M. WILLIAMS. 1980. *In* Toxicology, the Basic Science of Poisons. J. Doull, C. D. Klaassen & M. O. Amdur, Eds. 2nd edit.: 84–138. Macmillan Publishing Co., Inc. New York, N.Y.
5. SHINKIM, M. B. & G. D. STONER. 1975. Lung tumors in mice: applications to carcinogenesis bioassays. Adv. Cancer Res. **21:** 2–58.
6. WEISBURGER, J. H. 1976. ACS Monogr.
7. VAN DUUREM, B. L. 1976. *In* The Physiopathology of Cancer. F. Homburger & S. Karger, Eds. Karger. Basel, New York & London.
8. SLAGA, T. J., A. SIVAK & R. K. BOUTWELL. 1978. Mechanisms of Tumor Promotion and Cocarcinogens. Raven Press. New York, N.Y.
9. FARBER, E. & M. B. SPORM. 1976. Early lesions and the development of epithelial cancer. Cancer Res. **36:** 419–438.
10. WATANABE, K. & G. M. WILLIAMS. 1978. Enhancement of rat hepatocellular altered foci by liver promoter phenobarbital: evidence that foci are precursors of neoplasms and that the promoter acts on carcinogen-induced lesions. J. Nat. Cancer Inst. **61(5):** 1311–1314.
11. MAZUE, G., D. GOUY & B. REMANDET. Tumor markers in experimental animals. *In* Safety Evaluation and Regulation of Chemicals. S. Karger. Basel, Switzerland. (In press.)
12. STEWART, H. L., G. M. WILLIAMS, C. H. KEYSER, L. S. LOMBARD & R. J. MONTALDI. 1980. Histologic typing of liver tumors of the rat. J. Nat. Cancer Inst. **64:** 177–207.
13. SELLS, M. A., S. L. KATYAL, H. SHIMOZURA, L. W. ESTER, S. SELLS & B. LOMBARD. 1981. Isolation of oval cells and transitional cells from the livers of rats fed the carcinogen DL ethonin. J. Nat. Cancer Inst. **66:** 355–362.
14. FIALA, S., A. MOHINDRU, W. G. KETTERING, A. E. FIALA & H. P. MORRIS. 1976. Glutathione and gamma glutamyl transpeptidase in rats during chemical carcinogenesis. J. Nat. Cancer Inst. **57:** 591–598.
15. SCHERER, E. & P. EMELOT. 1976. Kinetics of induction and growth of enzyme deficient islands involved in hepatocarcinogenesis. Cancer Res. **36:** 2544–2554.
16. PUGH, T. D. & S. GOLDBERG. 1978. Quantitative histochemical and autoradiographic studies of hepatocarcinogenesis in rats fed 2 acetylaminofluorene followed by phenobarbital. Cancer Res. **38:** 4450–4457.
17. WILLIAMS, G. M. 1976. Functional markers and growth behavior of preneoplastic hepatocytes. Cancer Res. **36:** 2540–2543.
18. GOUY, D., G. MAZUE & B. REMANDET. 1981. Hepatic and plasmatic gamma glutamyl transferase as initiation or promotion marker in rat carcinogenesis. J. Clin. Chem. Clin. Biochem. **19:** 682.
19. REMANDET, B. 1981. Les benzodiazepines sont-elles carcinogènes? Etude chez le rat par des tests de carcinogénèse à moyen terme. Veterinary Doctoral Thesis. Paul Sabatier University. Toulouse, France.
20. MAZUE, G., B. REMANDET, D. GOUY, J. BERTHE, R. RONCUCCI & G. M. WILLIAMS. 1981. Limited in vivo bioassays on some benzodiazepines: lack of experimental initiating or

386   Annals New York Academy of Sciences

promoting effect of the benzodiazepine tranquillizers diazepam, clorazepate, oxazepam
and lorazepam. Arch. Int. Pharmacodyn. Ther. **257:** 59–64.
21. GARBAY, J. M. 1982. Appréciation du rôle hépatocancérogène du phénobarbital, de
l'aroclor 1254 et du dichlorodiphényltrichloroéthane. Veterinary Doctoral Thesis. Paul
Sabatier University. Toulouse, France.
22. PERAINO, G., R. J. M. FRY & E. F. STAFFELOT. 1977. Effects of varying the onset and
duration of exposure to phenobarbital on its enhancement of 2 acetylaminofluorene
induced hepatic tumorigenesis. Cancer Res. **57:** 3623–3627.
23. NAGASAKA, H., S. TOMII, T. MEGA, M. MARUGAMI & N. ITO. 1972. Hepatocarcinogenicity
of polychlorinated biphenyls in mice. Gann **63:** 805.
24. KIMURA, N. T. & T. BABA. 1973. Neoplastic changes in the rat liver induced by
polychlorinated biphenyl. Gann **64:** 105–108.

AN OVERVIEW OF GENETIC TOXICITY TESTING IN THE NATIONAL TOXICOLOGY PROGRAM

Errol Zeiger

Cellular and Genetic Toxicology Branch
National Toxicology Program
National Institute of Environmental Health Sciences
Research Triangle Park, North Carolina 27709

The National Toxicology Program (NTP) was established by the Department of Health and Human Services (DHHS) in November 1978. The primary purposes of the NTP are threefold: to coordinate, within the DHHS, the testing of chemicals of public health concern; to develop and validate new and better test methods; and to broaden the toxicological characterization of those chemicals being tested. The NTP is comprised of elements of four DHHS agencies, the National Institute of Environmental Health Sciences, the National Cancer Institute (NCI), the Food and Drug Administration (National Center for Toxicological Research), and the Centers for Disease Control (National Institute of Occupational Safety and Health). The NTP coordinates toxicological testing and development within these agencies and tracks relevant toxicological programs in DHHS and other government agencies. Toxicological testing in the NTP can be divided into three major programs: cellular and genetic toxicology, general toxicology (which includes biochemical toxicology, immunotoxicology, pharmacokinetics, and reproductive toxicology), and carcinogenesis testing.[1]

The cellular and genetic toxicology information will be used in a number of phases within the NTP: (1) in the nomination and selection of chemicals for short-term and chronic testing; (2) in the evaluation and interpretation of carcinogenesis results; (3) to provide data in support of regulatory agency decisions; and (4) the construction of a data base that can be used to define the relationships of chemical structure to mutagenic and other genetic effects in various cell systems and animal species. Also, the information will be used to define the ability of various *in vitro* tests to predict genetic and carcinogenic effects in animals. In addition, the cellular and genetic toxicity program has the responsibility for developing and evaluating new test systems and new or improved protocols for established systems.

The tests being used for routine testing are listed in TABLE 1. Additional test systems scheduled for development, under development, or being evaluated are listed in TABLE 2. A number of the tests listed, such as transformation assays, the mouse specific-locus tests, the *in vivo* cytogenetics assay, and the DNA-repair tests, fit a dual role in that they are being evaluated and are also being used for limited testing.

The tests listed in TABLES 1 and 2 include systems in which effects are measured in somatic cells and those in which effects are measured in germ cells. The major difference between the two is that mutations arising in germ cells can be transmitted via the sperm or ova to the next generation, while somatic mutations can be expressed in the exposed individual. Mechanistically, induction of mutations in the two cell types is similar, although the cells may exhibit different sensitivities. Somatic cell systems are used to identify chemicals capable of producing genetic toxicity and to determine the spectrum of damage produced, and as predictors of carcinogenicity. Germ cell systems detect heritable genetic lesions, and have the capability of indicating whether the chemical exhibits cell-stage specificity in its activity. Data arising from heritable mutation studies in mice may be used in calculations for risk of induced genetic disease as has been applied in radiation biology.[2]

387

0077–8923/83/0407–0387 $01.75/0 © 1983, NYAS

TABLE 1

SYSTEMS USED FOR ROUTINE TESTING

Organism	Genetic End Point
1. *Salmonella typhimurium*	gene mutations
2. Mouse lymphoma (L5178Y TK$^{+/-}$) cells	gene mutations
3. Chinese hamster ovary (CHO) cells	chromosome aberrations
	sister chromatid exchanges
4. *Drosophila melanogaster*	sex-linked recessive lethal mutations
	reciprocal translocations

All systems being used for testing have undergone, or are undergoing, technical validation studies. Standardized protocols have been developed, and at least two laboratories (with the exception of the mouse heritable translocation and specific-locus tests) participate in a study in which the laboratories test the same, selected chemicals, under code. The results are compared at various stages during the study, and the chemicals are decoded at the completion of the exercise. If the laboratories show an acceptable level of agreement, the protocol is used for testing coded substances with unknown activities. All tests and evaluations of data are performed on coded samples; the codes are broken only after testing is completed and the data evaluated. During the course of the testing, chemicals with previously established responses are included among the samples sent for testing to serve as a quality control on the test system and the testing laboratory. The philosophy and organization of such a test-development program have been previously described.[3]

NOMINATION AND SELECTION OF CHEMICALS FOR TESTING

Chemicals are nominated for testing by agencies within the NTP, by other government organizations (both federal and state), and by private organizations and

TABLE 2

SYSTEMS UNDER DEVELOPMENT OR EVALUATION

End Point	Organism or Cell Line
1. Mammalian cell transformation	Balb/c 3T3
	Syrian hamster embryo
	Syrian hamster embryo, viral enhanced (SA-7)
	rat cells, virus infected
2. Unscheduled DNA synthesis	rat hepatocytes (*in vivo/in vitro*)
	rat hepatocytes *in vitro*
3. Chromosome aberrations and SCEs *in vivo*	mice
4. Aneuploidy	*Drosophila melanogaster*
	Saccharomyces cerevisiae
5. Induction of gene transposition	*Drosophila melanogaster*
	cultured mouse cells
6. Heritable translocation	mice
7. Specific-locus mutations (morphological and biochemical)	mice
8. Cell-to-cell communication	Chinese hamster lung (V79) cells

individuals. The majority of chemicals nominated for testing to date have been nominated for mutagenicity assays. Many of these chemicals were nominated because of human exposure. The majority of testing requested is in *Salmonella*, and occasionally *in vitro* cytogenetics or mammalian cell mutation to identify potential carcinogens. Based on their genotoxicity results, these chemicals may be nominated for carcinogenicity and other toxicity (either genetic or general) testing. Additionally, a large number of chemicals have been nominated for *Salmonella* testing based on their chemical structure. These data will be used to determine structure-activity relationships for future predictive efforts (see below). As a rule, all chemicals nominated for carcinogenicity or heritable mutagenicity testing will be tested in *Salmonella* and other *in vitro* genetic tests if these results are not already available. Chemicals will not be entered into lifetime or multigenerational tests without these data.

CARCINOGENICITY

A driving force behind the use of microbes or cultured mammalian cells for the testing of chemicals for mutagenic activity, preliminary to carcinogenicity testing, is the positive correlation observed between mutagenic activity (or cell-transforming activity) and carcinogenicity in rodents. Mutagenicity in the *Salmonella typhimurium* system of Ames,[4] the most widely studied *in vitro* assay, was originally reported to correlate with carcinogenicity in 90% of the chemicals tested.[5] Subsequent compilations by other authors produced correlations ranging from 76% to 92%;[6-10] many of the carcinogens used in these different correlations were the same. Similar high correlations with carcinogenicity have been reported with mammalian cell mutation and transformation systems.[6,8,11-13] One fact immediately obvious from these correlations was that certain classes of carcinogens, e.g., polycyclic aromatic hydrocarbons, *N*-nitrosoamines, alkylating agents, aromatic amines, and others, exhibited a high mutagenicity-carcinogenicity correlation—between 90% and 100%. However, other classes of carcinogens, best exemplified by metals, hormonally active chemicals, and chlorinated hydrocarbons, showed very low correlations, and the overall mutagenicity-carcinogenicity correlations obtained were dependent on the relative numbers of each class of chemical in the study. Such low correlations do not diminish the value of *Salmonella* and other tests for the detection of potential carcinogens but, instead, demand that the interpretation of results of mutagenicity tests with respect to potential carcinogenicity be tempered by knowledge of the chemical's structure or reactive properties (see below).

At the present time, because of the high degree of correlation between mutagenicity and carcinogenicity, chemicals nominated for the carcinogenesis bioassay are being tested in *Salmonella* and other test systems, and a positive result in any or all these tests adds to the justification for performing carcinogenesis testing. In addition, since the capacity of the NTP laboratories to perform carcinogenicity tests is limited, the results from *Salmonella* tests and other *in vitro* tests can provide information regarding the potential carcinogenicity of the selected chemicals and contribute to the setting of testing priorities. *Salmonella* mutagenicity data have not been the definitive test for setting priorities, but as our data base expands we can be more confident in applying these results. For example, as of August 1982, 214 chemicals, comprising a diversity of chemical classes, have been tested for both mutagenicity and carcinogenicity. The mutagenicity results were derived from the NTP Cellular and Genetic Toxicology program; carcinogenicity results for these chemicals were derived from NCI, NTP, and International Agency for Research on Cancer (IARC) summaries,[1,14,15] supplemented by selected positive results from the literature. This data base showed that of 129 chemicals that are carcinogens, 54% (69/129) were also mutagenic in *Salmonella*. If suspect carcinogens and chemicals that showed equivocal, or nonreproducible, mutagenic responses were included, the correlation increased to 58% (90/155). Earlier compilations, using smaller numbers of chemicals, produced similar correlations.[16] A number of these chemicals were also tested for mutagenicity in the L5178Y thymidine kinase (TK$^{+/-}$) system and/or for induction of chromosome aberrations and sister chromatid exchanges (SCEs) in Chinese hamster ovary (CHO) cells. When these results were combined with the *Salmonella* results so that a positive result at any of the four end points would be sufficient to label a chemical as a genotoxin, the correlations increased to 62% (80/129) for the positives and to 67% (104/155) when the suspect and equivocal responses were included.

These proportions are below the correlations reported in the literature (68% to 92%). However, this apparent discrepancy is a result of the distribution of chemical classes studied. Alkylating agents, polycyclic aromatic hydrocarbons, and *N*-nitrosamines, classes that show a high correlation, are present in a much smaller proportion in the NTP data base than in literature data bases. On the other hand, carcinogenic chlorinated hydrocarbons, both aliphatic and cyclic, for example, which tend to have a poor correlation with mutagenicity, comprise approximately 28% of this NTP data set.

The application of the *Salmonella* and other *in vitro* tests for identification of potential carcinogens reveals a problem not fully addressed by the above calculations. The concern of the NTP is not the number or proportion of carcinogens that are mutagens, as the above calculations are designed to show, but how predictive the mutagenic results are for carcinogenicity. To make this determination, the results used for the above correlations were recompiled. Of the 91 chemicals that were mutagens in *Salmonella*, 76% (69/91) were also carcinogens. Addition of results from the L5178Y and CHO cells resulted in the identification of additional carcinogens but did not significantly change the proportions (76%; 80/106). Incorporation of the suspect carcinogens and equivocal mutagens raised the predictivity to 84% (90/107) for *Salmonella* alone and 85% (104/122) for all three test systems. This indicates that, with the exception of a relatively small fraction of chemicals, mutagenicity in *Salmonella*, either with or without the other test systems, is predictive of carcinogenicity in the rodent bioassay. The lack of effect of the L5178Y and CHO results on these correlations is not significant at this time because of the relatively few chemicals used in these calculations.

There is, however, another aspect to these predictions: of the 107 chemicals that were not mutagenic in *Salmonella*, 61% (65/107) were either carcinogens or suspect carcinogens. When the L5178Y and CHO systems were included, the proportion dropped to 55% (51/92). The implication of these calculations is that nonmutagenicity in *Salmonella*, or in either of the two mammalian cell systems, is not predictive of noncarcinogenicity in rodents.

There were also 14 mutagens in *Salmonella* that were judged noncarcinogenic in the NCI carcinogenesis bioassay; three of these were also tested in the L5178Y system and were mutagenic. These chemicals are not concentrated in any specific chemical class, and the mutagenic responses range from weak to strong. These chemicals cannot be dismissed as "false positives" since they are capable of mutating *Salmonella*, and in some cases L5178Y cells as well, either directly or in the presence of S-9. They are considered chemicals of concern because of the uncertainty surrounding the evaluation of a chemical as noncarcinogenic based solely on rat and mouse feeding studies.

It is anticipated that as more data become available on the predictability of carcinogenic effects by genetic toxicity, the carcinogenesis protocol could be adjusted to maximize the information obtained. For example, the standard carcinogenicity assay could be bypassed in favor of a protocol designed to provide more information on dose-response relationships. Alternatively, if the chemical is of less importance it could be labeled as a probable carcinogen (unconfirmed) and the lifetime studies given a lower priority in favor of other chemicals of greater concern.

HERITABLE MUTAGENICITY

Short-term *in vitro* tests are also used to provide preliminary data prior to germ cell mutation studies. Here, as opposed to carcinogenesis, less extrapolation is needed because the end points being studied are identical to those seen in the *in vitro* and insect tests, i.e., gene mutations and translocations. As a rule, before a chemical is selected for either a heritable translocation or specific locus study it must be shown to induce gene mutations and/or chromosome aberrations in cultured mammalian cells and chromosome aberrations, with or without SCEs, in rodent cells *in vivo* (peripheral lymphocytes or bone marrow). In addition, chromosome aberrations, SCEs, other genetic damage, or the presence of the active form of the chemical should be demonstrable in mouse gonadal cells prior to the initiation of a germ cell study.

Results of *in vitro* genetic toxicity tests can also contribute to the design or selection of protocols for other *in vitro* or *in vivo* toxicological tests. For example, these results can provide information on whether or not the chemical is directly active or requires mammalian metabolic activation for its genetic activity, or induction of clastogenic effects in cultured cells will alert the pathologist to anticipate similar effects in the hematopoietic or gonadal cells of treated animals.

EVALUATION AND INTERPRETATION OF CARCINOGENICITY AND OTHER TOXICOLOGICAL RESULTS

When carcinogenesis testing is completed, results from the genetic toxicity tests will be included in the formal bioassay reports on each chemical. The evaluation of the carcinogenesis study will refer to these data, and the carcinogenicity test reports will incorporate the genetic toxicology results. A positive mutagenicity assay, especially in more than one test system, would probably strengthen the evaluation of a borderline or suspect carcinogen, whereas negative mutagenicity assays would not detract from carcinogenicity findings.

Positive *in vitro* genetic toxicity data will also be used to signal a potential "problem chemical." For example, if a chemical shows itself to have little or no overt toxicity in acute and subchronic tests, it would normally be given a low priority for lifetime studies. However, if that chemical is mutagenic in one or more *in vitro* assays, it might be evaluated in greater depth. If it is shown that the mutagenicity did not result from an impurity, the chemical would be considered for additional subchronic and, possibly, lifetime studies. On the other hand, negative results in the genetic toxicity assays would not be sufficient to remove a chemical from consideration for additional testing. Also, as mentioned previously, testing protocols for heritable mutagenicity will be designed around *in vitro* and short-term *in vivo* test results. The results from these germ cell tests will also be interpreted within the context of the short-term test data.

One use of genetic toxicity tests that is being pursued is the incorporation of genetic toxicity end points into acute and subchronic toxicity protocols. In the future it is anticipated that urine from animals on test will be tested for mutagenic activity during the course of testing and that *in vivo* cytogenetics assays will be performed on the animals at the time of sacrifice. A positive result in one of these *in vivo* tests would strengthen the identification of a chemical as a carcinogen.

REGULATORY AGENCY SUPPORT

Although one of the reasons for the creation of the NTP was to lend research, development, and testing support to the regulatory agencies, the NTP itself does not have regulatory authority, and the United States government agencies responsible for regulating hazardous chemicals (the Environmental Protection Agency, the Food and Drug Administration, the Consumer Product Safety Commission, and the Occupational Safety and Health Administration) are not empowered at this time to ban, or restrict the use of, chemicals solely on the basis of mutagenicity or genetic toxicity. However, a large number of chemicals are being tested, or being considered for testing, at the request of regulatory agencies. These chemicals were nominated for testing so that the regulatory agencies could obtain initial data on the chemicals in *Salmonella* and other *in vitro* test systems.

Another potential use for these data is to provide information for use in risk-assessment determinations by the regulatory agencies. For example, if a chemical is mutagenic, it implies that the carcinogenicity is a function of the genetic toxicity. However, if a carcinogen is not genotoxic, it may cause an increase in tumors through other mechanisms, e.g., "cocarcinogenesis" or "promotion." At the present time, a negative result in the tests listed in TABLE 1 is not sufficient for defining the chemical as nongenotoxic. Other indicators of genetic damage, such as chromosome nondisjunction (aneuploidy), DNA damage, repair, or adduct formation *in vitro* or *in vivo*, or the induction of genetic transpositions, would have to be examined before any definitive conclusions of "nongenotoxity" are reached. Additionally, the *in vitro* metabolic activation systems used do not always reflect *in vivo* metabolism and, therefore, may not detect chemicals that require intact cell or organ-specific metabolism. If the distinction of genotoxic versus nongenotoxic is valid, the estimates of risk would be different in each case. Since the regulatory agencies are required to make estimations of risk from exposure to the chemicals, knowledge of their mechanisms of action is important.

DATA BASE

Since the correlations between mutagenicity and carcinogenicity provide the rationale for using *Salmonella* and other *in vitro* tests in the process of deciding which chemicals should be tested for carcinogenicity, it is imperative that the validity of these correlations and their boundaries be established. The development of carcinogenicity-mutagenicity correlations for chemicals belonging to different structural classes will identify those classes where a positive *in vitro* mutagenic response is highly correlated with carcinogenicity in rodents, and a negative response could provide assurance that the substance would not be carcinogenic. The classes of chemicals for which there is a poor correlation would also be identified. In these classes, the presence or absence of a mutagenic response would not be a significant factor in the selection of these latter chemicals for chronic animal testing.

The *Salmonella* mutagenicity–rodent carcinogenicity data base currently contains over 200 chemicals. Additional chemicals are continually being added, including chemicals previously tested for carcinogenicity in the NCI Bioassay Program. (At the present time, there is insufficient overlap between *Drosophila*, L5178Y, and CHO results and carcinogenicity results to draw any conclusions regarding correlations, but the numbers of chemicals tested in these systems are increasing.) Because of the distribution and numbers of chemicals it is difficult, at this time, to make definitive statements about the correlations from the different chemical classes.

Because of the large number of short-term genetic toxicity tests available, there is a tendency to test chemicals in systems that measure differing end points, or similar end points but at different loci, regardless of the reliability of the tests used or the relation of the end points to events of interest. One goal of the NTP is the determination of which tests, either alone or in combination, are the most predictive for carcinogenesis or mutagenesis. Along with this goal is the selection of the most efficient tests, either individually or in a series (test battery), and the avoidance of uninformative or redundant test systems, that is, to avoid performing tests that only confirm the results of other tests rather than providing information or insight not obtained from the other tests. The sought-after test battery would be one that provides maximum information on the genetic toxicity of the chemical—types of mutations or other chromosome or DNA effects that can be induced, and the relative efficiency of the chemical in inducing the different types of damage; this battery would be the most predictive for carcinogenic or heritable mutagenic events.

REFERENCES

1. USPHS. 1981. National Toxicology Program Annual Plan for FY82. Publication NTP-81-94. Department of Health and Human Services, U.S. Public Health Service. Washington, D.C.
2. Committee 17. 1975. Environmental mutagenic hazards. Science 187: 503–514.
3. ZEIGER, E. & J. W. DRAKE. 1980. An environmental mutagenesis test development programme. IARC Sci. Publ. No. 27: 303–313.
4. AMES, B. N., J. McCANN & E. YAMASAKI. 1975. Methods for detecting carcinogens and mutagens with the Salmonella/mammalian microsome mutagenicity test. Mutat. Res. 31: 347–364.
5. McCANN, J., E. CHOI, E. YAMASAKI & B. N. AMES. 1975. Detection of carcinogens as mutagens in the Salmonella/microsome test: assay of 300 chemicals. Proc. Nat. Acad. Sci. USA 72: 5135–5139.
6. BARTSCH, H., C. MALAVEILLE, A. M. CAMUS, G. MARTEL-PLANCHE, G. BRUN, A. HANTEFENILLE, N. SABADIE, A. BARBIN, T. KUROKI, C. DREVON, C. PICCOLI & R. MONTESANO. 1980. Validation and comparative studies on 180 chemicals with S. typhimurium strains and V79 Chinese hamster cells in the presence of various metabolizing systems. Mutat. Res. 76: 1–50.
7. HEDDLE, J. A. & W. R. BRUCE. 1977. Comparison of tests for mutagenicity or carcinogenicity using assays for sperm abnormalities, formation of micronuclei and mutations in Salmonella. In Origins of Human Cancer. H. H. Hiatt, J. D. Watson & J. A. Winsten, Eds. C: 1549–1557. Cold Spring Harbor Laboratory. Cold Spring Harbor, N.Y.
8. PURCHASE, I. F. H., E. LONGSTAFF, J. ASHBY, J. A. STYLES, D. ANDERSON, P. A. LEFEVRE & F. R. WESTWOOD. 1978. An evaluation of 6 short-term tests for detecting organic chemical carcinogens. Br. J. Cancer 37: 873–959.
9. RINKUS, S. J. & M. S. LEGATOR. 1979. Chemical characterization of 465 known or suspected carcinogens and their correlation with mutagenic activity in the Salmonella typhimurium system. Cancer Res. 39: 3289–3318.
10. SUGIMURA, T., S. SATO, M. NAGAO, T. YAHAGI, T. MATSUSHIMA, Y. SEINO, M. TAKEUCHI & T. KAWACHI. 1976. Overlapping of carcinogens and mutagens. In Fundamentals in Cancer Prevention. P. N. Magee et al., Eds.: 191–215. University of Tokyo Press. Tokyo, Japan.
11. CLIVE, D., K. O. JOHNSON, J. F. S. SPECTOR, A. G. BATSON & M. M. M. BROWN. 1979.

Validation and characterization of the L5178Y/TK$^{+/-}$ mouse lymphoma mutagen assay
system. Mutat. Res. **59:** 61–108.

12. PIENTA, R. 1979. A hamster embryo system for identifying carcinogens. *In* Carcinogens:
Identification and Mechanisms of Action. A. C. Griffing & C. R. Shaw, Eds.: 121–141.
Raven Press. New York, N.Y.

13. STYLES, J. A. 1977. A method for detecting carcinogenic organic chemicals using
mammalian cells in culture. Br. J. Cancer **36:** 558–563.

14. GRIESEMER, R. A. & C. CUETO, JR. 1980. Toward a classification scheme for degrees of
experimental evidence for the carcinogenicity of chemicals for animals. IARC Sci. Publ.
No. 27: 259–281.

15. IARC. 1980. Annual Report. International Agency for Research on Cancer. Lyon,
France.

16. ZEIGER, E. 1982. Knowledge gained from the testing of large numbers of chemicals in a
multi-laboratory, multi-system mutagenicity testing program. *In* Environmental Mutagens and Carcinogens. T. Sugimura, S. Kondo & H. Takebe, Eds.: 337–344. University of
Tokyo Press. Tokyo, Japan. Alan R. Liss. New York, N.Y.

IMPACT OF SHORT-TERM TESTS ON REGULATORY ACTION

W. Gary Flamm and Virginia C. Dunkel

Bureau of Foods
Food and Drug Administration
Washington, D.C. 20204

Over the past several years, regulatory agencies such as the Food and Drug Administration have used short-term *in vitro* tests in three separate and distinct ways. These include (1) setting priorities for additional testing; (2) aiding in the evaluation of equivocal data from rodent bioassays, and (3) determining mechanisms of action.

The most widely recognized application of short-term *in vitro* assays is in the testing or screening of chemicals for setting priorities for additional testing, specifically, to determine the need for long-term carcinogenicity bioassays in rodents. Within the FDA, the program area in which short-term *in vitro* testing was applied on the widest scale was in the review of those food ingredients known as GRAS (generally recognized as safe) substances. When testing was initiated on GRAS substances in the early seventies, the main concern centered on the question of mutagenicity. The information obtained however could also be used as a means of assessing the need for carcinogenicity testing. The tier, or prescreening, approach as described by Flamm[1] was used because a large number of compounds, approximately 700 in all, needed testing. Indeed, when the number of compounds that may need chronic toxicity testing becomes that large, consideration has to be given to the use of a tier system approach, the characteristics of which must include a prescreen of short-term *in vitro* tests at the front end and a method capable of assessing, as accurately as possible, human risk at the conclusion of the process. The prescreen must, if the approach is to succeed, have a reasonably negligible false-negative-error rate. This is the major reason most approaches have included a battery of tests in the initial prescreen to avoid the possibility that compounds with carcinogenic activity will be missed by whatever detection system is used.

The review of GRAS substances is now essentially complete, and while very few substances that were on the GRAS list in 1974 were found to be genetically active in *in vitro* tests, this approach provided, on a timely basis, the information and assurances needed that these traditional food ingredients were not carcinogenic. Upon reflection, it seems that the overall program was highly successful.

Another example of how short-term tests are to be used as a screen for deciding which compounds should be tested for carcinogenicity is contained in the "Threshold Assessment Guideline,"[2] which the Food and Drug Administration has developed and published as a notice of availability in the *Federal Register* in March of 1982. This guideline is used by the Food and Drug Administration as a means of deciding which drugs administered to food-producing animals might have the potential to contaminate edible tissues with residues that pose a carcinogenic risk to humans and thus should be tested in a carcinogen bioassay. As described in the guideline, the decision is based on three factors: (1) the chemical structure of the compound; (2) toxicological data on the agent including the results from short-term *in vitro* tests; and (3) the use of the agent in food-producing animals and the residue level of the parent compound and/or its metabolites in edible tissue. The decision-making scheme is a sequential one in which the first consideration is the chemical structure of the drug and whether, based on structure-activity relationships, it is suspected of being carcinogenic or noncarcinogenic.

395

The second aspect of the process involves results from short-term *in vitro* and *in vivo* tests. The guideline does not require the use of specific *in vitro* tests but does recommend that "the battery should test the ability of the sponsored compound to induce mutation in two test systems that have been demonstrated to have a high correlation between detected mutagens and positive results in lifetime bioassays for carcinogenicity."[2] Mutation assays identified by the agency include those measuring point mutation in bacteria, point mutation in mammalian cells in culture, and the sex-linked recessive lethal test in *Drosophila*. In addition to the tests for mutation induction, an assay for DNA repair in mammalian cells completes the test battery. For those animal drugs whose structures appear suspect, carcinogenicity testing would be required unless the use of the drug in food-producing animals and the tissue residue indicate negligible human exposure. Even then, unless the drug is negative in a battery of short-term tests as well as negative in short-term animal tests, the drug will need to be subjected to carcinogenicity testing in long-term rodent bioassays before it can be marketed. If the drug is not of a suspect structure and if it proves negative in the tests described above, then residue tolerances can be set for the drug, based on toxicities other than neoplasia. In general, if there are reproducible positive results in the test battery, the carcinogenicity bioassay must be performed regardless of use or residue level. However, it must be recognized that this battery generates a complex set of data, hence rigid decision rules cannot be applied to each and every situation. In certain instances the type of additional testing required will necessitate consideration of the extent of positive data, the structure and residue level, the intended use, and any other relevant biological or chemical information available.

Mutagenicity and other short-term tests are used in other areas as well, and are likely to experience even greater use in the future in terms of making decisions about substances that are already on the market, as well as those for which approval is being sought for marketing. We believe it is fair and reasonable to predict that if the evidence supporting the validity of these applications strengthens, the reliance that is placed on their use for premarket clearance will increase proportionately. As mentioned earlier, there are two other major areas where *in vitro* tests have proved useful in scientific considerations that have influenced regulatory outcome. Indeed, more and more scientists, not just in the regulatory agencies, but in those national and international organizations that assess the results of carcinogenicity studies, are using short-term tests to help decide whether a compound that has been tested in the carcinogen bioassay is positive or negative. There are many occasions when the results in a rodent bioassay are neither positive nor negative, but instead are equivocal. The results from mutagenicity, DNA-repair, and neoplastic transformation tests can, along with other information, help in deciding whether the compound that has generated equivocal results in the animal bioassay is likely to be negative or positive.

The Board of Scientific Councilors within the National Toxicology Program, in its efforts to utilize all of the data available to it in making decisions about carcinogenicity, has on certain occasions used *in vitro* short-term tests in that manner. In fact, the FDA, in approving the permanent listing of FD&C Green No. 5, used as one of its bases the inactivity of this substance in mutagenicity tests. We would stress, however, that considerable caution be exercised in using ancillary information in coming to a conclusion about bioassay results. There are occasions and situations where such an application is highly defensible in scientific terms, but we are not at a point where we can support the universality of such an application. As a general rule, the more *in vitro* data available, the more different systems examined, and the stronger the evidence that the compound is negative, the greater the likelihood that this use of short-term tests is appropriate. But there are, of course, exceptions. This may be because the test substance belongs to a chemical class for which there is no or very little prior experience or because it is impossible to test the compound at a high enough level of sensitivity, or it could be due to a variety of causes.

The third and final major application of short-term tests is related to the question of mechanisms of action. Using saccharin as an example, it might well be argued that the negative results in *in vitro* tests establish that saccharin is not a genotoxic carcinogen, but is inducing or producing cancer by some, as yet unknown, mechanism. In recent years there have been a number of proposals that would, in essence, permit treating genotoxic carcinogens differently from nongenotoxic carcinogens. While many of those who have proposed such a scheme have been silent on exactly how they would deal with nongenotoxic carcinogens, it is assumed they would apply some sort of safety factor to the no-observed-effect level of the carcinogen. The position taken by the FDA has been that before a nongenotoxic substance can be treated differently from a classical carcinogen, there must be compelling evidence on its mechanism of action. An example of this is the endogenous estrogenic hormones, which, despite their carcinogenicity, are regulated based on their hormonal activity.

We can best summarize this area by referencing the conclusion reached by a panel of scientists at the winter 1982 meeting of the Toxicology Forum.[3] The consensus of the panel was that the scientific knowledge is insufficient to enable the establishment and application of a general approach that would be universally applied, but, on the other hand, scientific knowledge is believed sufficient to enable case-by-case application of these short-term tests to the question of mechanism in a manner that impacts on regulatory decisions. The next five years will be critical to the future of short-term *in vitro* tests and efforts to apply and utilize them for safety-evaluation purposes. Over this period it is likely that they will either advance considerably in terms of application and credibility, or fall by the wayside.

REFERENCES

1. FLAMM, W. G. 1974. A tier system approach to mutagen testing. Mutat. Res. **26:** 329–333.
2. Food and Drug Administration. 1982. Chemical compounds in food-producing animals; availability of criteria for guideline. Fed. Regis. **47:** 4972–4977.
3. Toxicology Forum. 1982. Transcript of annual winter meeting. Washington, D.C.

SHORT-TERM TESTS IN THE FRAMEWORK OF CARCINOGEN RISK ASSESSMENT TO MAN

Robert Kroes

Institute CIVO-Toxicology and Nutrition TNO
3700 AJ Zeist, the Netherlands

INTRODUCTION

During the last decades short-term test systems designed to detect possible carcinogenicity have been extensively refined. Especially the introduction of representative parts of the mammalian metabolism in the test protocol was a major improvement.[1] Presently many more or less simple and convenient systems are available to detect mutations, chromosome effects, DNA damage, and malignant transformation in a variety of species such as microbes, insects, and mammalian cells. Their relevance to carcinogenicity is reasonably good, but inconsistencies in the pattern of response indicate that their role as predictive indicators of carcinogenicity is still uncertain. The use of short-term tests in the framework of carcinogen risk assessment to man seems, however, very feasible, and they should be used as an important additive tool in this respect. In addition, other short-term tests are currently available that can be used to identify certain promoting properties of compounds, another important parameter in risk assessment.

In this presentation, consideration will be given to a systematic classification of properties of carcinogens with the ultimate goal to assess the risk to man. The importance of short-term tests in this systemic approach will be emphasized, and the need for further development of short-term tests *in vitro* and especially *in vivo* is stressed.

CHEMICAL BIOLOGICAL CONSIDERATIONS

Several authors have stressed that there may be a variety of mechanisms by which carcinogens exert their effect.[2-4] An important criterion in these different mechanisms is whether a carcinogen is able to interact with DNA, or whether it exerts its carcinogenic effect by a nongenotoxic mechanism. Since present knowledge of carcinogenesis mechanisms, however, is still insufficient, the proposed classification of carcinogens is simply based on two concepts: (1) the concept introduced by the Millers,[5,6] which emphasizes the relevance of metabolic activation of procarcinogens to form electrophilic reactants that interact with DNA (so-called genotoxicity), which subsequently may lead to partially transformed cells and/or neoplasia; and (2) the two-stage carcinogenesis model as proposed by Berenblum.[7] In this latter concept a promoter is an agent that permits tumor formation in altered cells that otherwise remain dormant. Tumor promoters are identified by their enhancement of tumor yield resulting from a previously administered carcinogen (initiating agent).

Thus in these concepts a genotoxic agent may initiate a normal cell to an altered cell, whereas a promoter may complete the conversion of a partially transformed cell to a neoplastic cell, or alternatively if a genotoxic carcinogen is necessary for the complete development of a neoplastic cell, a promoter may enable neoplastic cells to proliferate to a neoplasm[8] (SCHEME 1). These concepts were also the rationale for the success of genotoxicity testing in identifying carcinogens. Some carcinogens, however, show

398

0077–8923/83/0407–0398 $01.75/0 © 1983, NYAS

negative results in these genotoxicity tests, and it is reasonable to conclude that at least some of these carcinogens differ from those that do act on DNA. These so-called nongenotoxic carcinogens thus represent a different class of carcinogen. It seems too simple, however, to classify carcinogens only according to these properties since other parameters relevant to man may be of equal importance. Risk assessment for genotoxic and nongenotoxic carcinogens is therefore dependent on several other parameters as well, and these will be discussed later.

GENOTOXICITY

In general carcinogens that have genotoxic properties share several common characteristics:[2,4,5,9] they are electrophilic molecules or are capable of being metabolized into electrophilic molecules; they are able to alter isolated DNA from mammalian cells; they may induce DNA-repair synthesis; they may induce mutations in prokaryotic and/or eukaryotic systems; they may cause chromosomal aberrations; and they may cause malignant transformation—nongenotoxic agents, however, may also induce malignant transformation.

Detection of Genotoxicity

Many systems are available now to detect genotoxic properties. Most systems are highly dependent on the metabolic activation systems used in the assay. Therefore such tests usually contain two components: a microbe, cell, or insect in which the genetic change is expressed and a metabolizing system being an S-9 preparation, a mammalian (liver) cell, or an organism. Genetic end points are, among others, point mutations, deletions, chromosome breaks or transpositions, sister chromatid exchange, and chromosome aberrations.

Bacterial systems have been proved to be extremely useful in detecting genotoxic properties of potential carcinogens.[9–11] Their role in quantification of mutagenic potency in relation to carcinogenic potencies is still very uncertain. Their possible use, however, should be recognized and may be envisaged in several ways: (1) by direct extrapolation from *in vitro* tests; (2) by indirect extrapolation as a result of an *in vitro/in vivo* comparison of induced effects (the parallelogram method);[12,13] and (3) by using host-mediated assays to assess mutagenic potency of carcinogens in selected organs or mammals. It seems likely that a combination of the indirect extrapolation with host-mediated assays will become of value. In addition *mammalian cell mutation systems* are important in assessing genotoxicity.[11,14,15] Recent information also reveals substantial evidence that *chromosomal aberration tests in vitro* offer good possibilities.[16] Last but not least, *DNA-repair assays* are of considerable value,[9,14,17,18] especially those that make use of primary hepatocytes since this is an *in vitro* system in which the intact cell metabolism is used as a biotransformation system. Short-term *in vivo* assays such as host-mediated assays with prokaryotes,[19] the dominant lethal test,[20] and the micronucleus test[21] should be mentioned as well. Many of these tests, however, have a low sensitivity[11] and are therefore usually recommended as confirmatory tests rather than screening tests.

In vitro cell-transformation assays seem to be less relevant to genotoxicity as compared to the other systems discussed.[9,14,22] The main reason may be the fact that in this system, a phenotypic change is studied about which it is not certain that it refers to a genetic conversion of a normal to a tumor cell. Moreover, the lack of reproducible quantitative results among laboratories is another major difficulty. It becomes much

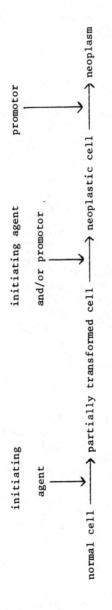

SCHEME 1. Scheme for the action of tumor promoters.[7]

more likely that *in vitro* transformation studies will be used as an important tool to detect promoting activities.[22]

The available genotoxicity tests provide the possibility for testing compounds qualitatively for genotoxicity. At least two or more of such tests with different genetic end points should be positive in order to prove genotoxicity. However, such tests should meet some minimal criteria:[11]

1. The protocol of the test should be supported by a formal validation study.
2. There should be a clear dose-response relationship, and results due to high toxicity should be excluded.
3. Positive controls should be used to check the reliability of the tests.
4. The results should be reproducible, preferably at different times and in separate laboratories.

If these requirements are met and genotoxicity has been found in different tests with different genetic end points, a compound should be treated as a genotoxic compound.

NONGENOTOXICITY OF CARCINOGENS

Modification of tumorigenic processes is a common feature in carcinogen bioassays. In fact an increase in tumor incidence of a "spontaneously" occurring tumor type is considered as a carcinogenic effect, although the compound provoking such a response may not be genotoxic. Such a nongenotoxic action may involve[2,3,4] nonspecific stimulation such as chronic irritation—cellular proliferation due to tissue damage; selective stimulation, for example, enzyme stimulation or enzyme inhibition; immune stimulation or immune suppression; hormonal imbalance; nutritional imbalance—deficiencies as well as hyperalimentations; and inhibition of DNA repair.

Detection of Nongenotoxicity

If a certain compound is found to be carcinogenic in a bioassay and a nongenotoxic mechanism is suspected, one should try to obtain more pertinent information concerning the mechanism of action. In a classical two-stage bioassay, an increased tumor incidence should be found when the compound is administered *following* an initiating (genotoxic) agent, and not when such a compound is given *before* the administration of an initiating agent. Until recently a classical two-stage promotion model has been the only available system to provide information on promotion.

With the increased interest in short-term tests, it has become clear that several tests may be of value in studying promotion or enhancement phenomena. *In vitro* short-term tests such as cell-transformation tests[22] and metabolic cooperation assays[8,23] with mammalian cells have been suggested as a tool to study these phenomena. These tests are of considerable importance as a qualitative tool for studying possible modifying properties, but most of them require validation.

For quantitative measurement, *in vivo* limited bioassays have become available. In such tests animals are first exposed to a limited, marginally carcinogenic dose of a genotoxic carcinogen. They are then exposed to the test agent to demonstrate supposed promotional activity as an enhanced tumor yield in the model. One should, however, realize that promotional effects may be tissue specific and therefore the use of an appropriate model is pertinent.

402 Annals New York Academy of Sciences

Nongenotoxicity versus Genotoxicity in Risk Assessment

Whereas genotoxic carcinogens possess both initiating and promoting properties (see SCHEME 1), nongenotoxic carcinogens show only promoting activity. In long-term bioassays it will often be impossible to discriminate between nongenotoxic and genotoxic carcinogens. Additional testing for genotoxicity, however, may in certain cases differentiate the two classes. As a rule nongenotoxic carcinogens may show a threshold in their dose-response curve, and such a threshold may be used in risk assessment. A no-effect level established in an animal experiment may then be the basis for setting acceptable levels for human exposure in the manner known from food toxicology.

Genotoxic carcinogens act on genetic material and therefore pose a clear qualitative hazard to man. They may be effective after one single exposure and act in a cumulative manner. For genotoxic carcinogens a threshold is usually not found and moreover cannot be expected to exist, as is likely from a theoretical point of view.[2] In risk-assessment procedures to man, this supposed nonexistence of a threshold will inevitably lead to extrapolation procedures other than those proposed above for nongenotoxic carcinogens.

Many mathematical models for extrapolation have been proposed.[24] In the use of such models for quantitative risk assessment, the notion of absolute safety is impossible to achieve and should be replaced by the concept of virtual safety. Mathematical models are either based on the assumption that dose-response curves arise from a distribution of individual thresholds within a population or on stochastic models that assume that a positive response is the result of the random occurrence of a number of biological events. Although such mathematical models may be appropriate for estimating low dose incidences in the animal species under investigation, they do not address the problem of the conversion of such data from animal to man. Nevertheless, mathematical models can be regarded as a useful tool in the process of quantitative risk assessment.

Calculations of virtual safe doses should be done at risk levels in the range of practical or societal concern. The choice of model is largely dependent on the experimental data. In the author's opinion a conservative procedure is best warranted, and therefore a linear extrapolation model should in general be preferred unless additional experimental data do justify otherwise.

It should be emphasized that in investigating the nongenotoxicity or genotoxicity of carcinogens the results of tests are not always as clear as one should like. If in a battery of tests only one test is found to be positive, this test should be repeated for confirmation. In case the result is confirmed it should be investigated to see whether particular reasons may be involved to explain the single positive result satisfactorily. If the result cannot be explained otherwise, a compound should be classified as "questionable genotoxic" and further risk assessment should be based on additional chemical and biological characteristics.

OTHER IMPORTANT PARAMETERS TO BE USED IN CARCINOGEN RISK ASSESSMENT

Chemical Structure

Similarities to known carcinogens may provide information on the possible action of a compound or on the possible pathway in metabolism. However, small changes in structure may affect the biological response of compounds considerably, even leading to complete noncarcinogenicity versus carcinogenicity.[25]

Biotransformation and Toxicokinetics

Most carcinogens undergo biotransformation[25] leading to detoxification or to metabolic activation.[5,6] Since biotransformation varies widely between species, this phenomenon deserves special attention in risk evaluation to man.

Functional and Morphological Effects in the Host

Different physiological effects may influence the carcinogenic response and indeed may be indirectly the reason for a carcinogenic response as well. Hormonal imbalance,[26,27] immune status changes,[28–30] nonspecific irritation and cytotoxicity with subsequent cell proliferation,[31,32] and stimulation of enzyme systems[33,37] have been identified as causal or contributing factors in carcinogenesis. Morphological events such as hyperplasia may also be of additional interest regarding the pathogenesis of the neoplasia involved. The histogenic sites at which a tumor occurs in the various species tested may also give additional information. Nongenotoxic carcinogens may affect only one or a few tissues, whereas the occurrence of malignancies at different sites may be a more specific characteristic for genetically active carcinogens. Furthermore, benign tumors provide less evidence for carcinogenic potential as compared to malignant tumors.

Results in the Bioassay, Species and Strain Differences

Long-term carcinogenicity studies offer a wide variety of possible results, which require a more definitive consideration before the carcinogenicity of a compound can be assessed properly. Important parameters to consider are among others latency period, relationship to background tumor incidence, and dose-response relationship. The evidence of carcinogenicity in a given study is more conclusive when positive results are found in a multidose study and they show a dose relation; primary rare malignant tumors do occur or a shift of benign to malignant tumors occurs within the same tissue; and latency period decreases and total incidence of spontaneous malignant tumor types increases.

The values of the effects described above in determining the nature of carcinogenic activity are not identical, and results must be scrutinized in each case. In case of induction of primary rare malignant tumors, the possibility that we are dealing with a genotoxic carcinogen is strong though never conclusive. If enhancement of incidences of tumors with a high "spontaneous" incidence occurs, or if the latency period of such a tumor type is shortened, it is uncertain whether we are dealing with a weak genotoxic or with a nongenotoxic carcinogen.

Latent period, usually difficult to determine, may, if determined in a proper way, give additional information: carcinogens producing a high yield of tumors after a short latency period are usually of the genotoxic type.

Species and strain specificity has been observed for some carcinogens.[25] Most carcinogens known to be active in man have been shown to be active in a variety of species, with the exception of benzene and arsenic.[38] Chemicals active in only one species may have mechanisms of action that do not occur widely in other species. Elucidation of such a mechanism may provide information on the risk or lack of risk to man.

Dose-response relationships provide information about the potency of carcinogens. Used in conjunction with other parameters (among others, short-term tests providing evidence for genotoxicity or nongenotoxicity, toxicokinetics, metabolism), dose-response relationships may help decide whether or not a given carcinogen poses a risk to man.

Results of Limited In Vivo Bioassays

Limited *in vivo* assays may be extremely useful as indicators of possible tumor-enhancement properties as well as determinants of initiating properties. Available systems are among others skin tumor induction in mice,[7] lung tumor induction in strain A mice,[39] breast cancer induction in rats,[40] liver foci or liver nodule formation in rodents,[41-44] and bladder tumor formation.[45,46] With the exception of the lung tumor model, these systems can be employed for the detection of initiating properties using a known promoter. If tumor-enhancement properties are investigated, the skin, breast, liver, or bladder tumor model may be used with a known initiator or a genotoxic carcinogen in a subcarcinogenic dose. As already stated, the development of other tissue- or organ-specific systems and of short-term *in vitro* systems should be pursued. For carcinogens that do not cause genetic damage, the use of these tests may provide insight into the carcinogenic mechanism.

Human Studies

Epidemiological information, metabolism, and toxicokinetics in man as well as results from health-surveillance studies will provide a basis for establishing the relevance of animal models. It should be realized that this type of information, if at all available, will be scarce. Where such information is available, however, risk assessment performed on the basis of animal data may be evaluated, thus giving a better understanding of their accuracy.

SHORT-TERM TESTS IN THE TOTAL FRAMEWORK OF RISK ASSESSMENT TO MAN

In assessing the possible risk to man of carcinogens, it may have become clear that a differentiation in genotoxic and nongenotoxic potential is rather important.[47,48] A flow diagram (SCHEME 2) is presented for a risk-assessment procedure for carcinogens. If carcinogens have to be evaluated one of the first steps will be to investigate the possible genotoxicity of the compound. If a certain carcinogen is genotoxic, additional information on structure, biotransformation, toxicokinetics, functional and morpho-logical effects, and results of animal data may enhance the likelihood of genotoxicity (and carcinogenicity) to man, or may provide information on which it is warranted to conclude that the compound does not constitute a carcinogenic risk to man.

In the case of likelihood of genotoxicity to man, a conservative approach in human risk assessment is indicated. Elimination of such a compound from the human environment will be the best solution. If, however, this is not possible or feasible for a variety of practical, societal, or economic reasons, linear extrapolation should be used in assessing the risk to man (procedure 1, SCHEME 2). Such a risk to man has to be compared to other known risks in order to be able to judge its significance, and in this way a maximum accepted average exposure can be permitted. A more "liberal" or "appropriate" extrapolation model may be applied when human studies so warrant (procedure 2).

Nongenotoxic carcinogens should be evaluated as well on the basis of available

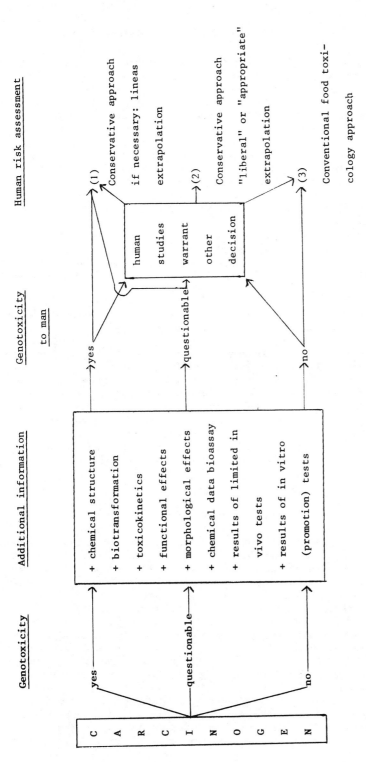

SCHEME 2. Flow diagram for risk-assessment procedures for carcinogens.

additional information. The likelihood that such information may lead to the conclusion that such a compound may be genotoxic to man is rather small but should be excluded wherever possible. If the additional body of information does not warrant another decision, a nongenotoxic carcinogen can be treated in risk assessment to man according to conventional toxicological procedures (procedure 3, SCHEME 2). Human studies, however, may warrant another decision, and this may lead to a more conservative approach using an appropriate extrapolation system.

Carcinogens for which the genotoxicity tests do not give a sufficiently definitive answer concerning their genotoxic nature should in general be treated as genotoxic carcinogens. The additional body of information may lead to the conclusion that a compound may be considered as a nongenotoxic compound to man, with the subsequent risk-assessment procedure (SCHEME 2), or may lead to the more conservative procedures 2 or 1.

The identification of genotoxic or nongenotoxic properties represents the two extremes of the total effect of carcinogens. With the consideration of the totality of effects of individual chemicals, and the possibility in future of quantitating these effects in a proper way, it seems likely that a range of carcinogens from pure initiators to pure promoters can be identified.

Risk assessment to man especially for carcinogens has been, and will be, a controversial issue. Whereas it is usually easier and safer for regulatory bodies to simply forbid the use of a known carcinogen irrespective of its mechanism, even this approach is not always possible. Many times man is exposed to carcinogens for which exposure cannot be eliminated. One of the best examples in this respect is the exposure of man to aflatoxin in many parts of the world.[49] Elimination of carcinogens from the human environment remains of course the primary goal, but valid reasons may exist to investigate in more detail whether an elimination is justified and whether it is at all possible. Therefore there is a need for procedures for evaluating the carcinogenic risk. This need is not only felt by regulators, but also society's concern has come to a stage where a procedure is urgently needed. The procedure proposed in this presentation is certainly not perfect, and hopefully can be improved considerably. However, in the author's opinion, it provides a tool to a better approach in the risk assessment for carcinogens to man.

SUMMARY

Short-term tests designed to detect possible carcinogenicity have been extensively refined during the last years. Presently, many more or less simple and convenient systems are available to detect mutations, chromosome effects, DNA damage, and malignant transformation. Although their relevance to carcinogenicity is often reasonably good, inconsistencies in the pattern of response indicate that their role as predictive indicators of carcinogenicity is still uncertain. The use of short-term tests in carcinogen risk assessment does seem feasible. These tests, however, should not be the only characteristic taken into consideration in such a risk assessment. Other characteristics such as chemical structure, biotransformation, and pharmacokinetics, qualitative and quantitative physiological and/or morphological effects, species, strain, and organ specificity, dose-response relation, and information on human studies, if available, are of importance too.

Current knowledge does not permit a rigid classification of carcinogens, but does warrant a subclassification into genotoxic and nongenotoxic compounds. Whereas for genotoxic compounds a real threshold cannot be expected on a theoretical basis, the existence of a threshold may well be expected for nongenotoxic compounds. In conjunction with other characteristics it may then be decided whether a genotoxic or nongenotoxic compound may be or may not be permitted in the human environment. In this evaluation process it is anticipated that for genotoxic compounds other extrapolation systems should be used, as compared to nongenotoxic compounds, where in fact a conventional food toxicology safety factor may be applied. Short-term tests are very important in the subclassification with respect to genotoxicity and seem to be of value for the detection of promoter activity as well.

ACKNOWLEDGMENT

The author gratefully acknowledges the administrative skill of Ms. S. Wildschut.

REFERENCES

1. AMES, B. N., W. E. DURSTON, E. YAMASAKA & F. D. LEE. 1973. Proc. Nat. Acad. Sci. USA 70: 2281–2285.
2. 1980. The Evaluation of the Carcinogenicity of Chemical Substances. VAR 5E. Ministerie van Volksgezondheid en Milieuhygiëne. Leidschendam, the Netherlands.
3. KROES, R. 1979. In Environmental Carcinogenesis. P. Emmelot & R. Kriek, Eds.: 298–302. Elsevier/North-Holland Biomedical Press. Amsterdam, the Netherlands.
4. WEISBURGER, J. H. & G. M. WILLIAMS. 1980. In Toxicology, the Basic Science of Poisons. J. Doull, C. D. Klaassen & M. O. Amdur, Eds. 2nd edit.: 84–138. McMillan Publishing Company, Inc. New York, N.Y.
5. MILLER, J. A. & E. C. MILLER. 1969. Prog. Exp. Tumor Res. 11: 273–301.
6. MILLER, J. A. & E. C. MILLER. 1976. ACS Monogr. 173: 737–762.
7. BERENBLUM, I. 1974. Carcinogenesis as a Biological Problem. Elsevier/North-Holland Inc. New York, N.Y.
8. WILLIAMS, G. M. 1981. Food Cosmet. Toxicol. 19: 577–583.
9. WILLIAMS, G. M., R. KROES, H. W. WAAIJERS & K. W. VAN DE POLL. 1980. The Predictive Value of Short Term Screening Tests in Carcinogenicity Evaluation. Applied Methods in Oncology 3. Elsevier/North-Holland Biomedical Press. Amsterdam, the Netherlands.
10. MOHN, G. R. 1981. Mutat. Res. 87: 191–210.
11. PURCHASE, I. F. H. 1982. Mutat. Res. 99: 53–71.
12. SOBELS, F. H. 1980. Arch. Toxicol. 46: 21–30.
13. 1982. Report on the mutagenicity of chemical substances of the Netherlands Health Council. The Hague, the Netherlands. (In press.)
14. KROES, R. 1981. Chemospheer 10: 605–619.
15. HSIE, A. W., D. A. CASCIANO, D. B. COUCH, D. F. KRAHN, J. P. O'NEILL & B. L. WHITFIELD. 1981. Mutat. Res. 86: 193–214.
16. ISHIDATE, M., T. SOFUNI & K. YOSHIKAWA. 1981. Gann Monogr. Cancer Res. 27: 95–108.
17. WILLIAMS, G. M. 1981. Gann Monogr. Cancer Res. 27: 45–55.
18. PROBST, G. S., R. M. MCMAHON, L. E. HILL, C. E. THOMPSON, J. K. EPP & S. B. NEAL. 1981. Environ. Mutagen. 3: 11–32.
19. MOHN, G. R. 1977. Arch. Toxicol. 38: 109–133.
20. BATEMAN, A. J. & S. S. EPSTEIN. 1971. In Chemical Mutagens: Principles and Methods for Their Detection. M. A. Hollaender, Ed.: 541–568. Plenum Press. New York, N.Y.
21. JENSSEN, D. & C. RAMEL. 1980. Mutat. Res. 75: 191–202.
22. BROOKES, P. 1981. Mutat. Res. 86: 233–242.
23. YOTTI, L. P., C. C. CHANG & J. E. TROSKO. 1979. Science 206: 1089–1091.
24. Interagency Regulatory Liaison Group. 1980. Annu. Rev. Public Health 1: 345–393.
25. WEISBURGER, J. H. & G. M. WILLIAMS. 1982. In Cancer, a Comprehensive Treatise. F. F. Becker, Ed. 2nd edit. 1: 241–334. Plenum Publishing Co. New York, N.Y.

26. IARC-WHO. 1974. IARC Monographs on the Evaluation of Carcinogenic Risk of Chemicals to Man 6. International Agency for Research on Cancer. Lyon, France.
27. IARC-WHO. 1979. IARC Monographs on the Evaluation of Carcinogenic Risk of Chemicals to Humans 21. International Agency for Research on Cancer. Lyon, France.
28. KROES, R., J. W. WEISS & J. H. WEISBURGER. 1975. Recent Results Cancer Res. 52: 65–75.
29. PENN, I. 1976. Cancer 37: 1024–1032.
30. STUTMAN, O. 1975. Adv. Cancer Res. 22: 261–422.
31. GRASSO, P. 1971. Food Cosmet. Toxicol. 9: 463–478.
32. FLAKS, A., J. M. HAMILTON & D. B. CLAYSON. 1973. J. Nat. Cancer Inst. 51: 2007–2008.
33. PERAINO, C., R. J. M. FRY & E. STAFFELDT. 1973. J. Nat. Cancer Inst. 51: 1349–1350.
34. PERAINO, C., R. J. M. FRY, E. STAFFELDT & J. P. CHRISTOFER. 1975. Cancer Res. 35: 2884–2890.
35. NISHIZUMI, M. 1976. Cancer Lett. 2: 11–15.
36. ROSSI, L., M. RAVERA, G. REPETTI & L. SANTI. 1977. Int. J. Cancer 19: 179–185.
37. MALANSKY, C. J. & G. M. WILLIAMS. 1981. J. Toxicol. Environ. Health 8: 121–130.
38. IARC-WHO. 1979. IARC Monographs on the Evaluation of the Carcinogenic Risk of Chemicals to Man. Suppl. 1. International Agency for Research on Cancer. Lyon, France.
39. SHIMKIN, M. B. & G. D. STONER. 1975. Adv. Cancer Res. 21: 2–58.
40. HUGGINS, C., L. C. GRAND & F. P. BRILLANTES. 1961. Nature 189: 204–207.
41. FARBER, E. & M. B. SPORN. 1976. Cancer Res. 36: 2476–2705.
42. WILLIAMS, G. M. & K. WATENABE. 1978. J. Nat. Cancer Inst. 61: 113–121.
43. WILLIAMS, G. M., S. KATAYAMA & T. OHMORI. 1981. Carcinogenesis 2: 1111–1117.
44. PITOT, H. C., T. GOLDSWORTHY, H. A. CAMPBELL & A. POLAND. 1980. Cancer Res. 40: 3616–3620.
45. HICKS, R. M., J. J. WAKEFIELD & J. CHOWANILC. 1975. Chem. Biol. Interact. 11: 225–233.
46. COHEN, S. M., M. ARAS, J. B. JACOBS & G. H. FRIEDELL. 1979. Cancer Res. 39: 1207–1217.
47. MUNRO, I. C. & D. R. KREWSKI. 1981. In Health Risk Analysis. C. R. Richmond, P. J. Walsh & E. D. Copenhader, Eds.: 443–459. Franklin Institute Press. Philadelphia, Pa.
48. WEISBURGER, J. H. & G. W. WILLIAMS. 1981. In Health Risk Analysis. C. R. Richmond, P. J. Walsh & E. D. Copenhader, Eds.: 249–270. Franklin Institute Press. Philadelphia, Pa.
49. IARC-WHO. 1972. IARC Monographs on the Evaluation of the Carcinogenic Risk of Chemicals to Man 1. International Agency for Research on Cancer. Lyon, France.

POSSIBILITY OF USING RAT MESOTHELIAL CELLS IN CULTURE TO TEST CYTOTOXICITY, CLASTOGENICITY, AND CARCINOGENICITY OF ASBESTOS FIBERS

M. C. Jaurand, I. Bastie-Sigeac, M. J. Paterour,
A. Renier, and J. Bignon

*Group for Research and Study of Respiratory
Infections and the Environment
INSERM Unit 139
CHU Henri Mondor
94010 Créteil Cedex, France*

Pleural diseases, namely, fibrosis (pleural plaques) and cancer (mesothelioma), may be related to the inhalation of asbestos fibers.[1] Recently, methods for long-term culturing of rat pleural mesothelial cells (PMCs) have been described.[2-4] In the present report, cultures of PMCs were used to determine some of the *in vitro* effects of UICC (International Union Against Cancer) chrysotile A fibers. The toxicity was assessed first by the determination of the growth rate of the cells treated with various concentrations of fibers; second by the measurement of the number of sister chromatid exchanges (SCEs) induced by the fibers; and third by observation of the transformed colonies when the cells were treated, in a two-stage model, with benzo[a]pyrene (BaP) as initiator and chrysotile as promoter.

UNTREATED PLEURAL MESOTHELIAL CELLS

PMCs were obtained from healthy rats, as described elsewhere.[4] In primary or secondary cultures, PMCs form a monolayer of epithelial-like contact-inhibited cells. The mean population doubling time (PDT) was about 30 hours. The cells were diploids ($n = 21$); their morphology and chromosome number were stable for about 40 population doublings. PMCs were able to metabolize BaP.[5]

CYTOTOXICITY TESTING

PMCs were cultured in NCTC 109 + 10% fetal bovine serum (FBS). Twenty-four hours after plating, 5 to 50 μg/ml of chrysotile fibers were added to the culture medium for 48 hours. The cells were counted in place every day (Nachet NS 1002). Chrysotile at 5–10 μg/ml induced a cytoplasmic vacuolation and the formation of binucleated cells. Moreover, the PDT was increased in a dose-dependent manner. With 20 μg/ml, there was no growth.

SISTER CHROMATID EXCHANGES

PMCs were cultured in Ham's F_{10}. Sixty hours after plating, the medium was replaced with medium containing 2 μg/ml chrysotile fibers and 10 μg/ml 5-bromo-

TABLE 1

SISTER CHROMATID EXCHANGES*

Cell Line (Passage Number):	I_8 (12)		F_{24} (4)	
Treatment:	None	Chrysotile	None	Chrysotile
n	42	40	40	38
m	7.6	11.4	4.9	6.0
p		<0.1%		<2.5%

*n, number of metaphases; m, mean number of SCEs per metaphase; p, from test F.

2'-deoxyuridine (BrdU) for 48 hours. PMCs were arrested in metaphase with 0.3 μg/ml colcemid. Slides were treated with Hoechst 33558, ultraviolet light, and stained with 4% Giemsa.

TABLE 1 shows that chrysotile induced a low but significant increase in the number of SCEs per metaphase. This does not confirm our preliminary data,[6] but the discrepancy could be due to the cell line. Livingston et al. found that other kinds of asbestos fibers increased SCEs,[7] but this was not observed by Price Jones et al.[8]

TRANSFORMATION ASSAY

The method was derived from the report of Lasne et al.[9] In our experiments, BaP (1 μg/ml) was used as initiator and added to the culture medium 24 hours after plating. Under these conditions, there was no inhibition of cell growth. PMCs occurred at the 12th passage. Chrysotile fibers (5 μg/ml) were then added at each passage under the same conditions used to test cytotoxicity. The treatment was repeated 11 times. The morphological transformation was determined in colonies grown from cells plated at low density (200 cells per 60-mm² dish) and seeded each 5 passages. Classification was as follows: I, sparse criss-crossing; II, overgrowth; and III, piling up. TABLE 2 shows that, under these conditions, chrysotile did not have a promoter effect.

CONCLUSION

The culture of rat PMCs provides a useful model to study asbestos-related modifications of cells. Studies on the interactions between asbestos and mesothelial cells are now just beginning; they have to be developed. The data reported here show

TABLE 2

NUMBER OF COLONIES ACCORDING TO MORPHOLOGICAL CLASS

Treatment	Number of Treatments							
BaP	0	0	0	0	1	1	1	1
Chrysotile	0	0	6	11	0	0	6	11
Passage Number:	18	23	17	22	17	22	17	22
Normal	69	98	35	87	60	81	68	92
I	26	2	52	11	27	14	24	7
II	5	0	13	2	13	5	7	1
III	0	0	0	0	0	0	0	0

some of the possible fields of investigation that can be used to explain the mechanisms of action of the mineral fibers.

REFERENCES

1. BECKLAKE, M. R. 1982. Asbestos-related diseases of the lungs and pleura. Am. Rev. Respir. Dis. **126:** 187–194.
2. THIOLLET, J., M. C. JAURAND, H. KAPLAN, J. BIGNON & E. HOLLANDE. 1978. Culture procedure of mesothelial cells from the rat parietal pleura. Biomed. Express Paris **29:** 69–73.
3. ARONSON, J. F. & V. J. CRISTOFALO. 1981. Culture of epithelial cells from the rat pleura. In Vitro **17:** 61–70.
4. JAURAND, M. C., J. F. BERNAUDIN, A. RENIER, H. KAPLAN & J. BIGNON. 1981. Rat pleural mesothelial cells in culture. In Vitro **17:** 98–105.
5. JAURAND, M. C., I. BASTIE SIGEAC, L. MAGNE, M. HUBERT HABART & J. BIGNON. Studies on in vitro chrysotile–pleural mesothelial cell interaction. Morphological aspects and metabolism of benzo-3-4-pyrene. Environ. Health Perspect. (In press.)
6. KAPLAN, H., A. RENIER, M. C. JAURAND & J. BIGNON. 1980. Sister chromatid exchanges in mesothelial cells cultured with chrysotile fibres. *In* The In Vitro Effects of Mineral Dusts. R. C. Brown, M. Chamberlain, R. Davies & I. P. Gormley, Eds. **1:** 251–254. Academic Press. London, England.
7. LIVINGSTON, G. K., W. N. ROM & M. V. MORRIS. 1980. Asbestos-induced sister chromatid exchanges in cultured Chinese hamster ovarian fibroblast cells. J. Environ. Pathol. Toxicol. **3:** 373–382.
8. PRICE JONES, M. J., G. GUBBINGS & M. CHAMBERLAIN. 1980. The genetic effects of crocidolite asbestos; comparison of chromosome abnormalities and sister-chromatid exchanges. Mutat. Res. **79:** 331–336.
9. LASNE, C., A. GENTIL & I. CHOUROULINKOV. 1974. Two stage transformation of rat fibroblasts in tissue culture. Nature **247:** 490–491.

ANTIONCOGENIC ACTIVITY OF DIVERSE CHEMICAL CARCINOGENS ON RETROVIRUS-INDUCED TRANSFORMATION OF HUMAN SKIN FIBROBLASTS*

J. R. Blakeslee,[†‡§] A. M. Elliot,[†] and L. J. Carter[†]

†Department of Veterinary Pathobiology
‡Department of Microbiology
§Comprehensive Cancer Center
Ohio State University
Columbus, Ohio 43210

INTRODUCTION

The utilization of a rapid *in vitro* assay to evaluate chemicals for carcinogenic activity would greatly reduce the time now required to determine carcinogenicity in animal test systems. The objectives of these studies were to determine whether chemicals altered ST-FeSV virus–directed transformation of human skin fibroblasts in a predictable manner and to correlate the alteration with carcinogenic or noncarcinogenic activity of the test chemical. The procedures used to study these interactions were described previously.[1,2]

The results, to date, showed that diverse classes of carcinogens inhibited virus transformation when virus-infected cells were exposed to test chemical two hours postinfection while noncarcinogens had no significant effect on transformation at two hours postinfection. Continued studies showed that the carcinogens inhibited a specific viral gene function, i.e., transformation, but not other viral gene products (feline oncornavirus–associated cell membrane antigen, group-specific antigen, and reverse transcriptase) that were shown to be cell mediated.[3] The studies reported here expanded the classes of carcinogens and noncarcinogens tested to include hydrazines, jet fuels, asbestos, formalin, methylazoxy methanol acetate,[4] and aromatic amines used as lubricant additives—alpha and beta naphthylamine and phenyl alpha and phenyl beta naphthylamine.

MATERIALS AND METHODS

Cells and Virus

Detroit 550, normal male human skin fibroblasts (HSFs) were grown under standard conditions as previously described.[1] The Snyder-Theilen strain of feline sarcoma virus was prepared from feline embryo cell culture, titrated in HSFs, and stored at −85°C.

Transformation Assays

Preconfluent log phase growth HSFs were seeded into 16-mm diameter wells at 4×10^4 cells/well in 1.5 ml complete medium and incubated 18 hours prior to

*Supported by Air Force Office of Scientific Research Contract F49629-80-C-0087.

412

Blakeslee *et al.*: Antioncogenic Activity 413

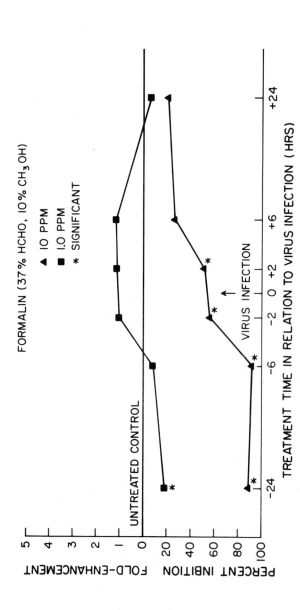

FIGURE 1. Commercial formalin (37% HCHO–10% CH₃OH) was diluted in complete medium at 10 ppm and 1 ppm final concentration. Cells were treated and infected as described in Materials and Methods. Significance was determined by Student's t-test.

treatment in 5% CO_2 at 37°C. Cells were infected with 20 focus-forming units/well. Cells chemically treated prior to virus infection were exposed to chemicals for 30 minutes, washed, and refed until virus infection. Virus-infected cells were treated 30 minutes with chemicals, washed, and refed with complete medium. Seven to 10 days post infection, cells were fixed and stained with Giemsa and foci enumerated with a 25–40× dissecting microscope. Data were evaluated statistically by the Student "t"-test.

TABLE 1

CORRELATION BETWEEN INHIBITION OF VIRUS TRANSFORMATION AND DIVERSE CHEMICALS INDUCING TUMORS IN RODENTS*

Chemical Group	Reported Activity†	Inhibition of ST-FeSV Transformation
I. Aromatic amines		
A. Naphthyl amines		
1. Two (2)	C	yes
2. Phenyl-alpha	C	yes
3. Phenyl-beta	C	yes
4. N-Acetoxy 2 fluorenylacetamide	C	yes
II. Polycyclic Hydrocarbons		
1. Benzo[a]pyrene	C	yes
2. Pyrene	NC	no
III. Hydrazines		
1. Hydrazine	C	yes
2. Mono-methyl hydrazine	C	yes
3. 1,1 Dimethyl hydrazine	C	yes
4. 1,2 Dimethyl hydrazine	C	yes
IV Other		
1. Aflatoxin B_1	C	yes
2. Amosite asbestos	C	yes
3. JP5 (shale)	NC	no
4. JP5 (petrol)	NC	no
5. RJ5	NC	no
6. Diesel fuel, marine	NC	no
7. Acetone	NC	no
8. Methyl azoxymethanol-acetate	C	yes
9. Formalin	C	yes
10. Triton X-100	?	no

*Data from References 5–7.
†C, carcinogen; NC, noncarcinogen.

RESULTS

The data displayed in FIGURE 1 are typical of the results for chemical-treated virus-infected cells. Predetermined nontoxic doses of commercial formalin (37% HCHO–10% CH_3OH) at 10 ppm and 1 ppm were used. Ten parts per million significantly inhibited virus transformation by values ranging from 50% to 90% of control values, while 1 ppm affected virus transformation to a lesser degree. Although transformation was inhibited at four time periods for formalin, the results of our studies with over 20 different chemicals (TABLE 1) show that the time periods *most affected* are two and six hours postinfection. The results shown in TABLE 1 are based on these time periods. The data compiled from this assay correlated well with the carcinogenic potential determined by the standard rodent assays.

REFERENCES

1. BLAKESLEE, J. R. & G. E. MILO. 1978. Chem. Biol. Interact. **23:** 1–11.
2. BLAKESLEE, J. R. 1981. AGARD Conf. Proc. **309:** B6, 1–6.
3. BLAKESLEE, J. R., A. M. ELLIOT & D. G. TURNER. 1980. *In* Advances in Comparative Leukemia Research 1979. D. S. Yohn, B. A. Lapin & J. R. Blakeslee, Eds.: 87–88. Elsevier/North-Holland. New York, N.Y.
4. FIALA, E. 1977. Cancer **40:** 2436–2445.
5. MISHRA, N., V. DUNKEL & M. MEHLMAN, Eds. 1980. Advances in Modern Environmental Toxicology. **1.** Senate Press. Princeton Junction, N.J.
6. SELIKOFF, I. J. 1977. *In* Origins of Human Cancer. H. Hiatt, J. D. Watson & J. A. Winsten, Eds.: 1765–1784. Cold Spring Harbor Laboratory. Cold Spring Harbor, N.Y.
7. Toxicology Branch, Aerospace Medical Research Laboratory, Wright Patterson Air Force Base. Unpublished data.

IN VITRO ALTERNATIVE IRRITANCY ASSAYS: COMPARISON OF CYTOTOXIC AND MEMBRANE TRANSPORT EFFECTS OF ALCOHOLS*

E. Borenfreund, C. Shopsis, O. Borrero, and S. Sathe

Laboratory Animal Research Center
The Rockefeller University
New York, New York 10021

The ever increasing public concern over the use of animals for experimental purposes emphasizes the need for the establishment of reliable alternative toxicologic assays. One area of concern affected by these considerations is the poorly reproducible Draize rabbit eye test, which is used for the assessment of the damage inflicted by irritants. In an effort to replace this assay, we set out to develop sensitive ànd reliable alternative in vitro tests. We report here two such assays in which the toxicity of a series of alcohols on cells in culture is examined, ranked according to severity of effect, and compared with the ratings obtained with the Draize test. Our experiments were designed to study the effects of alcohols on HepG2, an established epithelial human hepatoma cell line,[1] and Balb/c 3T3, a murine fibroblast cell line, by assaying for cytotoxicity and membrane transport effects.

For the cytoxicity assays, cells suspended in Dulbecco's minimum essential medium containing 10% fetal bovine serum and antibiotics were seeded to 24-well Falcon tissue-culture trays. Twenty-four or 48 hours later, fresh medium containing test alcohols in a wide range of concentrations was added to the cultures. At the time of addition of the test compounds, cells had reached a density of about 7×10^4 per cm^2. After 24 and 48 hours of incubation at 37°C in a water-saturated 5% CO_2 atmosphere, cells were scored by phase microscopy for morphological alteration (shape, vacuolization, and granularity), detachment from the substrate, and loss of viability, as judged by lysis and trypan blue dye exclusion. The highest tolerated dose (HTD) was the highest concentration at which no marked morphological alterations could be observed. Concentrations found to be nontoxic after 24 hours of incubation were usually also nontoxic after 48 hours.

For the membrane transport inhibition assays,[2] cells were plated at 5×10^4 (3T3) or 1×10^5 (HepG2) cells per 35-mm dish. After 48 hours, medium was removed and replaced with fresh medium (for control cultures) or fresh medium containing various concentrations of the alcohols to be tested. Four hours later, this medium was removed and the cells washed twice with 37°C phosphate-buffered saline (PBS) followed by the addition of 1.0 ml of PBS containing ^{3}H-uridine (33 μM, 1.67 μCi/ml). The cells were incubated at 37°C for 15 minutes, then washed three times with cold PBS and lysed in 0.5 N sodium hydroxide. One aliquot of each lysate was neutralized and counted in a Packard scintillation counter to determine ^{3}H-uridine uptake. A separate aliquot was analyzed for protein content to determine the extent of cell detachment during the treatment and assay. The concentration of alcohols required to induce a 50% decrease in uridine uptake (UI_{50}) was derived from a graphic representation of the data. A typical graph is shown in FIGURE 1.

Results from both assays are compared in TABLE 1 and FIGURE 2. As can be seen,

*This research was supported in part by a grant from Revlon Inc.

416

0077-8923/83/0407-0416 $01.75/0 © 1983, NYAS

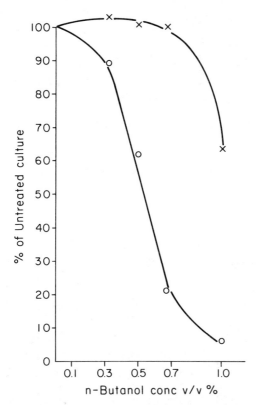

FIGURE 1. Effects of *n*-butanol on Balb/c 3T3 cells. Percent inhibition of uptake of 33 μM ³H-uridine, 1.67 μCi/ml (circles) and percent change in cell protein per dish (crosses) were measured as described in the text after four hours of treatment with the indicated concentrations of *n*-butanol.

TABLE 1

CYTOTOXIC AND MEMBRANE EFFECTS OF ALCOHOLS ON CELLS *IN VITRO*

	3T3		HepG2		
Alcohols	UI_{50} (v/v%)	HTD (v/v%)	UI_{50} (v/v%)	HTD (v/v%)	Ocular Irritancy*
1-Octanol	0.01	0.01	0.04	0.25	not available, other isomers severe
1-Pentanol	0.15	0.1	0.3	0.6	severe
Allyl	0.08	0.1	1.9	0.6	severe–moderate
1-Butanol	0.45	0.8	1.1	0.7	moderate
Tetrahydrofurfuryl	1.3	0.6	1.6	0.6	moderate
Isopropyl	1.1	0.8	3.1	0.7	moderate
Ethanol	2.1	2.5	3.5	2.5	moderate–mild
Propylene glycol	3.5	3.2	4.1	3.2	negative
Methanol	2.7	3.5	4.6	4.5	mild

*Rabbit eye irritancy as reported in the literature.

3T3 fibroblasts are more sensitive to the various alcohols than are human hepatoma cells, particularly with respect to 1-pentanol and allyl alcohol. There is for both cell lines a rather good rank correlation of the relative potencies of the various alcohols examined by the two assay systems. Thus alcohols that are more cytotoxic are also more potent inhibitors of ^{3}H-uridine uptake. Among the aliphatic alcohols tested, the severity of the reaction appears to be directly related to the carbon-chain length. The results obtained are in good agreement with the relative irritancies of these compounds reported in Draize rabbit eye test results.[3]

We are optimistic that these experiments, in conjunction with other approaches

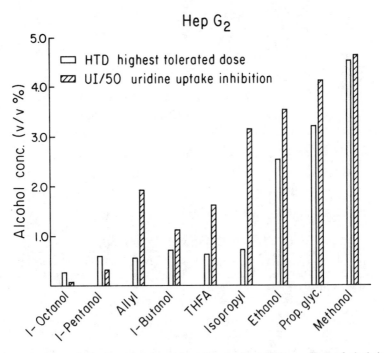

FIGURE 2. Comparison of cytotoxicity and uridine uptake inhibitory effects of alcohols on HepG2 cells. The highest tolerated dose for a 24-hour incubation period with various alcohols and the concentrations of the alcohols that induced a 50% inhibition of uridine uptake were determined as described in the text.

under investigation in our laboratories,[4] may provide the components of a battery of *in vitro* tests that can replace the Draize test.

ACKNOWLEDGMENT

We are indebted to Dr. Barbara Knowles, Wistar Institute, Philadelphia, Pa., for the human hepatoma cell culture.

REFERENCES

1. KNOWLES, B. B., C. C. HOWE & D. P. ADEN. 1980. Human hepatocellular carcinoma cell lines secrete the major plasma proteins and hepatitis B surface antigen. Science **209:** 497–499.
2. PRASAD, R., C. SHOPSIS & J. HOCHSTADT. 1981. Nutrient transport in a bovine epithelial cell line. J. Cell Physiol. **107:** 231–236.
3. CLAYTON, C. D. & F. E. CLAYTON, Eds. 1982. Patty's Industrial Hygiene and Toxicology. 3rd rev. edit. **2C.** John Wiley & Sons, Inc. New York, N.Y.
4. STARK, D. M., C. SHOPSIS, E. BORENFREUND & J. WALBERG. 1983. Alternative approaches to the Draize assay—chemotaxis, cytology, differentiation and membrane transport studies. *In* Product Safety Evaluation. A. Goldberg, Ed. Mary Ann Liebert, Inc. New York, N.Y. (In press.)

"SINGLE-GENE" AND VIABLE CHROMOSOME MUTATIONS AFFECTING THE TK LOCUS IN L5178Y MOUSE LYMPHOMA CELLS

D. Clive

Genetic Toxicology Laboratory
Burroughs Wellcome Co.
Research Triangle Park, North Carolina 27709

J. Hozier

Department of Biological Sciences
Florida Institute of Technology
Melbourne, Florida 32901

M. M. Moore

Mutagenesis and Cellular Toxicology Branch
Genetic Toxicology Division
U.S. Environmental Protection Agency
Research Triangle Park, North Carolina 27709

Small (σ) colony $TK^{-/-}$ mutants represent viable and heritable damage to the chromosome 11 bearing the single functional thymidine kinase (TK) gene (TK^+ chromosome) in the $TK^{+/-}$-3.7.2C heterozygote cell line. The decreased growth rate of these $\sigma TK^{-/-}$ mutants is attributed to simultaneous effects on postulated nearby σ genes needed for optimal growth (FIGURE 1). The type and extent of the TK^+ chromosome 11 damage detected at the 300-band level of resolution may be mutagen specific. Hycanthone and methyl methanesulfonate (MMS) produce predominantly large translocations onto this chromosome's terminal end. Ethyl methanesulfonate (EMS) and the "nonmutagenic" clastogens pyrimethamine, methotrexate, and acyclovir produce predominantly smaller scale damage (e.g., 1–2 band insertions or deletions), no visible damage, or even predominantly large (λ) colony $TK^{-/-}$ mutants.

Two mutants, each deleted of its TK^+ chromosome E1-E2 terminal bands (FIGURE 1), have been studied genetically. One of these originated as a deoxygalactose-resistant colony, which had lost the single functional galactokinase (GK) gene present in $TK^{+/-}$ $GK^{+/-}$-3.7.2C cells (mutant $GK^{-/-}$-C.692); the other originated as a $\sigma TK^{-/-}$ mutant, C.540F. Thus the TK^+ and GK^+ genes are on the same (i.e., TK^+) chromosome 11 in 3.7.2C cells.

Despite having similar TK^+ chromosome 11 deletions, each mutant has retained the other, nonselected gene: their genotypes are $TK^{+/-}$ $GK^{-/-}$-C.692 and $TK^{-/-}$ $GK^{+/-}$-C.540F. In C.540F the deleted chromosome 11 bands were found translocated to chromosome 6b. This, coupled with the extreme rarity of double ($TK^{-/-}$ $GK^{-/-}$) mutants and the overall lower mutability of the GK^+ gene (TABLE 1), argues for a lethal gene closely associated with the GK^+ gene (FIGURE 1). It further suggests that the GK^+ gene is distal to the TK^+ gene and that, in the formation of the $TK^{-/-}$ $GK^{+/-}$ mutant, the break occurred at the TK gene. The break in mutant $TK^{+/-}$ $GK^{-/-}$-C.692 is presumed to have occurred distally to the TK^+ gene.

The reconstituted karyotype of C.540F is accompanied by a much reduced spontaneous mutability of the presumably translocated GK^+ locus; this is not

420

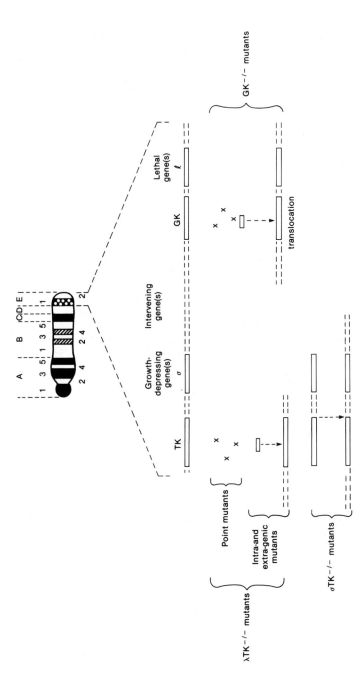

FIGURE 1. Tentative map of the two terminal bands (E1-E2) of the TK⁺ chromosome 11 of the L5178Y/TK⁺/⁻-3.7.2C cell line. Resolution is at the 300-band level. The galactokinase (GK) gene is proposed to be distal to the thymidine kinase (TK) gene relative to the centromere, and to be associated with a gene, ℓ, whose loss is lethal to the cell. The TK gene is believed to be associated with a gene (or genes), σ, whose alteration by chromosomal damage results in decreased growth rate; under cloning conditions, this gives rise to the small-colony (σ) phenotype. Intra- and extragenic damage of various magnitudes are believed to be responsible for the λ (large-colony) and σTK⁻/⁻ mutants typically observed in this assay. Most clastogens produce σTK⁻/⁻ mutants. A few, such as the folate antagonists methotrexate and pyrimethamine, produce mostly λTK⁻/⁻ mutants, presumably by producing high proportions of minute chromosomal damage, the same damage as is known to occur in cells grown in the presence of methotrexate.

TABLE 1

DIFFERENTIAL MUTABILITIES OF TK, GK LOCI IN 3.7.2C AND 2 CHROMOSOME 11
TERMINAL DELETION MUTANTS

| Cell Line | Genotype | Locus | Mutation Frequency $(\times 10^6)$ | | |
			Spontaneous	EMS*	Hycanthone*
3.7.2C	TK^+GK^+	TK	$\sim$50†	758	$\sim$800†
C.692	$\overline{TK^-GK^-}$ $TK^{+/-}GK^{-/-}$	TK	2500‡	4100	11800
3.7.2C	TK^+GK^+	GK	7	156	60
C.540F	$\overline{TK^-GK^-}$ $TK^{-/-}GK^{+/-}$	GK	1.3	4.4	2.3

*EMS concentration = 500 μg/ml for 3.7.2C, 310 μg/ml for C.692 and C.540F; hycanthone concentration = 10 μg/ml.
†Typical historical frequency.
‡Confirmed with "cleansed" cells and by fluctuation assay.

significantly enhanced by the potent mutagens EMS or hycanthone (TABLE 1). On the other hand, the mutability of the TK^+ locus in mutant C.692 is greatly increased (TABLE 1). This is interpreted to be a consequence of dissociating the GK-linked lethal gene from the TK^+ gene. The resulting lethal-free TK^+ terminus in C.692 permits viability of a much higher proportion of chromosomal $TK^{-/-}$ mutants than does 3.7.2C, in which many potential $TK^{-/-}$ mutants would require a second "mutational" event such as a translocation in order to become viable. Altered kinetics of $TK^{-/-}$ mutant induction would be expected for mutant C.692; this is seen for the mutagens EMS and hycanthone (data not shown) in the form of a greater initial slope of the dose-response curve followed by a plateau, instead of the typical greater-than-linear kinetics seen with 3.7.2C.

CYTOGENETIC ANALYSIS OF SMALL-COLONY L5178Y TK$^{-/-}$ MUTANTS EARLY IN THEIR CLONAL HISTORY

John Hozier and Jeffrey Sawyer

Department of Biological Sciences
Florida Institute of Technology
Melbourne, Florida 32901

Donald Clive

Genetic Toxicology Laboratory
Burroughs Wellcome Co.
Research Triangle Park, North Carolina 27709

Martha M. Moore

Mutagenesis and Cellular Toxicology Branch
U.S. Environmental Protection Agency
Research Triangle Park, North Carolina 27709

The L5178Y TK$^{+/-}$ → TK$^{-/-}$ mouse lymphoma mutagen assay system, which allows quantitation of forward mutations at the autosomal thymidine kinase (TK) locus, uses a TK$^{+/-}$ heterozygous cell line, 3.7.2C. TK$^{-/-}$ colonies derived from the TK$^{+/-}$ 3.7.2C cell line display a bimodal frequency distribution of sizes for most mutagenic substances tested, with a small-colony (σTK$^{-/-}$) mutant class and a large-colony (λTK$^{-/-}$) mutant class.

Several lines of evidence based primarily on genetic features and relative growth patterns[1,2] have led to the hypothesis that σ and λ mutants result from different extents of genetic damage. In this hypothesis, λ mutants represent events possibly limited to the TK gene locus only, while σ mutants represent effects not only at the TK locus but at multiple loci, rendering the cell less genetically fit and therefore slower growing.

Direct evidence for this view has been obtained by cytogenetic analysis of TK$^{-/-}$ mutants. Banded metaphase chromosome analysis showed a high association (30 of 51 mutants) of the σ phenotype with the presence of abnormalities involving chromosome 11 after 20 to 30 cell generations,[2,3] while λ mutants did not have chromosome 11 rearrangements. The thymidine kinase gene has been mapped to chromosome 11 of the mouse by somatic cell genetic techniques.[4]

However, not all σTK$^{-/-}$ mutants show chromosome 11 abnormalities under these conditions. Because our studies indicated possible karyotype instability in some σTK$^{-/-}$ mutants 20–30 cell generations after mutant induction, we devised a technique to analyze karyotypes earlier in the history of TK$^{-/-}$ mutants, at 10 or fewer cell generations.

TABLE 1 is a summary of karyotype results to date at early cell generation times compared to later generation times for both σ and λ TK$^{-/-}$ mutants. Seven of the 26 σ mutants analyzed at 10 generations had cells with a dicentric chromosome 11 compared with only one of 51 at 20–30 cell generations. FIGURE 1 shows chromosomes 11 and 17 from a single σTK$^{-/-}$ mutant analyzed at the 10 cell generation stage. There are three cells (a, b, and c), two of which (a and b) show rearrangements involving chromosomes 11 and 17. In cell a, a dicentric between the chromosome 11 and a chromosome 17 is present. In cell b, a portion of the chromosome 17 is attached to the distal end of the chromosome 11. In cell c, an apparently normal set of chromosomes 11

423

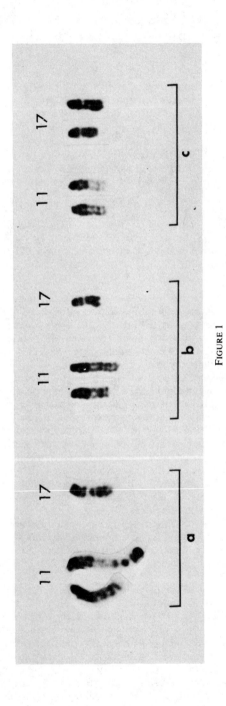

FIGURE 1

and 17 is present. That is, seen early in the history of the σTK$^{-/-}$ mutant, chromosome rearrangements may be rapidly evolving into near-normal chromosome complements from an initial unstable state (i.e., a dicentric chromosome).

There are two main conclusions from this study: (1) Analyzing σTK$^{-/-}$ mutant colonies earlier in their clonal history reveals a high proportion of mutants with chromosome 11 abnormalities (TABLE 1). All λTK$^{-/-}$ mutants thus far studied have normal chromosomes 11. (2) The character of the chromosome rearrangements found in early clonal history provides an explanation for karyotype evolution toward normal chromosome 11 morphology in later clonal history. That is, the presence of chromosome 11–containing dicentrics in a significant proportion of σTK$^{-/-}$ mutants at 10 generations provides a source of unstable chromosomes that may be resolved, by breakage between centromeres, into the entire range of abnormal, near-normal, and normal chromosomes 11 seen at the 20–30 cell generation stage. Karyotypically

TABLE 1

Stage Post–Mutant Induction	Mutant Phenotype	Number of TK$^{-/-}$ Mutants Analyzed	Number of TK$^{-/-}$ Mutants with Chromosome Rearrangements	Percent TK$^{-/-}$ Mutants with Chromosome Rearrangements
20–30 cell	σ	51	30	59
divisions	λ	14	0	0
10 cell	σ	26	23	88
divisions	λ	8	0	0

unstable σTK$^{-/-}$ mutants and at least some of the apparently normal karyotypes in σTK$^{-/-}$ mutants may originate in early unstable (dicentric) rearrangements of chromosome 11. (This does not rule out the possibility that some σTK$^{-/-}$ mutants contain minor chromosomal rearrangements not detected by standard metaphase chromosome banding techniques.) All of these results are consistent with the chromosomal nature of mutagenic events at the TK locus in σTK$^{-/-}$ mutants.

REFERENCES

1. CLIVE, D., *et al.* 1979. Mutat. Res. **59**: 61.
2. MOORE, M. M., *et al.* (In preparation.)
3. HOZIER, J., *et al.* 1981. Mutat. Res. **84**: 169.
4. KOZAK & RUDDLE. 1977. Somatic Cell Genet. **3**: 121.

AN *IN VITRO* HUMAN-ORGAN-SPECIFIC SYSTEM FOR TESTING GENOTOXIC AGENTS*

Steven M. D'Ambrosio, Carol T. Oravec, and
Ruth Gibson-D'Ambrosio

Department of Radiology
College of Medicine
The Ohio State University
Columbus, Ohio 43210

Most *in vitro* assays for genotoxicity utilize prokaryotic cells or cells derived from rodents or human fibroblasts. The main limitation of these assays is that they cannot provide information about the effects in other human organs. This is important since each organ may exhibit its own unique levels of metabolism, DNA repair, replication, and promotion and thus respond differently to genotoxins. We have developed cell-culturing conditions that allow establishment of cells from human brain, dermis, kidney, liver, lung, and intestine. Using this cell system we are able to characterize and compare the human-organ-specific responses to various chemical and physical agents.

Human cell cultures were established from fetal brain, dermis, kidney, intestine, liver, and lung using various modifications of modified minimal essential medium.[1-3] Fibroblasts were isolated from dermis and lung; epithelial cells from dermis, lung, intestine, and kidney; and hepatocytes from liver. Brain cell cultures consisted mainly of microglial, astrocytes, polygonal glial, and multipolar neuronal cells. Experiments were performed in the primary passage and, when noted, at later passages. Unscheduled DNA synthesis (UDS) was measured by determining the amount of ^{3}H-thymidine incorporated per microgram of DNA.[1] Hydroxyurea and serum-deprived medium were used to block over 99% of normal DNA synthesis. The effect of agents upon normal DNA synthesis (NDS) was determined by pulse labeling cells with ^{3}H-thymidine at various times posttreatment. The ability of cells to metabolize ^{3}H-7,12-dimethylbenz[*a*]anthracene (DMBA) to water- and organic-soluble metabolites and the extent of binding of DMBA to cellular DNA were determined as previously described.[3]

Representative data comparing the capacity of human cells, derived from a number of organs, to repair DNA damage induced by ultraviolet radiation (UV) and *N*-ethyl–*N*-nitrosourea (ENU) are shown in TABLE 1. Cells derived from human brain, intestine, and kidney exhibited the lowest levels of UDS, while dermal and lung cells

*Supported by U.S. Environmental Protection Agency Grant No. R807693.

exhibited the highest levels of UDS. These data are consistent with our other data showing the same differences in repair between dermis, brain, and kidney cells using the UV-endonuclease sensitive site assay.[1,2] Differences in the repair of ENU-induced DNA damage were also observed (TABLE 1). Both brain and kidney cells exhibited the lowest levels of UDS when compared to the other cell types. As expected both UV and ENU inhibited normal DNA synthesis. The greatest level of inhibition was observed in the brain, dermal, and kidney cells following UV irradiation. Intestine and kidney cells showed the greatest level of inhibition following ENU treatment.

TABLE 1

UNSCHEDULED AND SCHEDULED DNA SYNTHESIS FOLLOWING EXPOSURE
OF HUMAN CELLS TO UV AND ENU

| | UDS* (% Dermis UDS)† | | | | NDS‡ (% Control)§ | | | | |
| | UV¶ | | ENU‖ | | UV¶ | | ENU‖ | | |
Cell Origin	5 J/m²	10 J/m²	25 μM	3.5 mM	5 J/m²	10 J/m²	1 mM	2.5 mM	5 mM
Dermis	100	100	100	100	31	32	78	47	77
Brain	47	40	27	34	26	28	74	65	52
Lung	80	75	123	—	42	40	62	41	34
Intestine	37	43	71	—	62	53	40	22	45
Kidney	—	59	14	—	28	32	56	56	38

*UDS was measured by determining the amount of ³H-thymidine incorporated into cellular DNA over a two-hour period. Hydroxyurea (5×10^{-3} M) and serum-deficient (1% calf serum) medium were used to block normal DNA synthesis. Cells were in passages 0–3.

†Percent dermis UDS = (UDS exhibited by cell type) divided by (UDS exhibited by dermal cells) times 100.

‡NDS was measured two hours posttreatment by incubating cells for 15 minutes in medium containing ³H-thymidine.

§Percent control = (dpm/μg DNA in cells not treated) divided by (dpm/μg DNA in cells treated) times 100.

¶Cells were irradiated with UV radiation (254 nm) emitted from a germicidal lamp. Medium was removed and cells washed with phosphate-buffered saline immediately before irradiation.

‖Cells were incubated with ENU dissolved in Hanks buffered salts solution, pH 6.0, for 30 minutes at 37°C. After treatment, cells were incubated in prewarmed medium.

FIGURE 1 compares the rate of metabolism of DMBA by cells derived from human dermis, brain, liver, lung, kidney, and intestine. Liver, lung, dermis, and kidney appear to exhibit the highest rate of metabolism, 39–45% of DMBA. Brain on the other hand only metabolized 7% of DMBA. Qualitative and quantitative differences in the metabolites have been observed.[3] The level of binding after 24 hours of incubation with DMBA appeared to be highest in liver hepatocytes (94 μmoles DMBA/mole DNA-P) and lowest in brain cells (6 μmoles DMBA/mole DNA-P).

Our studies to date indicate that cells derived from different human organs respond selectively to repair DNA damage, replicate DNA following insult, and metabolize and bind DMBA. This system and the resulting data provide information about the genotoxic potential of DNA-damaging agents in various human organs.

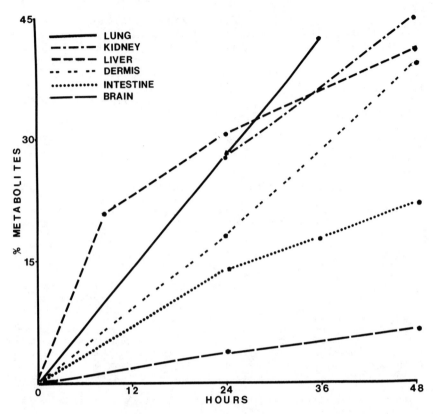

FIGURE 1. Metabolism of DMBA to aqueous- and organic-soluble metabolites. Cells were derived from organs of five different specimens. Cells from lung, kidney, and brain were in primary passage, while intestinal cells were in P1, dermis in P2, and liver in P5.

REFERENCES

1. GIBSON, R. E. & S. M. D'AMBROSIO. 1982. Photochem. Photobiol. 35: 181.
2. GIBSON-D'AMBROSIO, R. E., Y. LEONG & S. M. D'AMBROSIO. 1982. In Vitro 18: 299.
3. ORAVEC, C., R. E. GIBSON-D'AMBROSIO & S. M. D'AMBROSIO. 1982. In Polynuclear Aromatic Hydrocarbons. W. Cooke, A. Dennis & G. Fisher, Eds. 7. Battelle Press. Columbus, Ohio. (In press.)

SUPEROXIDE RESPONSE INDUCED BY INDOLE ALKALOID TUMOR PROMOTERS*

Jerry Formisano and Walter Troll

Department of Environmental Medicine
New York University Medical Center
New York, New York 10016

Takashi Sugimura

National Cancer Center Research Institute
Tokyo, Japan

The pleiotropic effects of the potent tumor promoter 12-O-tetradecanoyl-phorbol-13-acetate (TPA) upon cell growth and function are well known.[1] TPA is also a strong inflammatory agent that activates the reduced nicotinamide adenine dinucleotide phosphate (NADPH) oxidase of human neutrophils inducing superoxide anion (O_2^-) production.[2] Previous work in our laboratory, by Goldstein *et al.*, shows an apparent positive association between tumor promotion and O_2^- formation for phorbol esters and analogues of phorbol esters.[3] Slaga has shown that other oxygen radical–producing substances may be implicated in tumor promotion as well.[4]

Teleocidin (from *Streptomyces mediocidicus*) and lyngbyatoxin A, debromoaplysiatoxin, and aplysiatoxin (from the marine blue-green alga *Lyngbya majuscula*) have been shown to have an inflammatory effect on animal skin.[5] Lyngbyatoxin A and debromoaplysiatoxin cause dermatitis or "swimmer's itch" in humans.[6] All of these compounds have shown an ability to induce superoxide anion production in human polymorphonuclear leukocytes (PMNs) in comparable amounts to TPA. Superoxide anion production increased to a peak amount for various molar quantities of these materials, with debromoaplysiatoxin and aplysiatoxin apparently producing greater amounts of O_2^- than did TPA and teleocidin. Lyngbyatoxin A produced much less of a respiratory burst after a 10-minute incubation with the same number of cells, even at higher molar concentrations.

These substances were tested for their ability to produce O_2^- in PMNs by the reduction of ferricytochrome C, which was inhibited by superoxide dismutase.[3] Whole blood was obtained and sedimented in 1% dextran (M_r 240,000) at 25°C for 45 minutes. After osmotic shock to remove red cells, PMNs were resuspended in phosphate-buffered saline (PBS) (Ca^{2+} and Mg^{2+} free) at a cell concentration of 10^6–10^7/ml.

Dose-response curves for each compound were done from 1 ng/ml to several hundred ng/ml. The results of these tests, plotted as molar quantities $\times$ 10^{-7} on semilog scale, can be seen in FIGURE 1. All materials (teleocidin B, lyngbyatoxin A, aplysiatoxin, debromoaplysiatoxin, and TPA) showed a response in the cells, and a production of O_2^-. For all compounds, except lyngbyatoxin A, the range of 0.5–1.0 $\times$ 10^{-7} M showed a sharp rise in the stimulation of O_2^- production. Lyngbyatoxin seemed to show a similar rise over a range of 4–9 $\times$ 10^{-7} molar, or at least a fivefold greater amount.

*This investigation was supported by Public Health Service Grant No. CA 16060, awarded by the National Cancer Institute, Department of Health and Human Services.

The fact that all of these substances are able to stimulate the production of O_2^- is of interest, considering that nonpromoting substances (such as 4-O-mephorbolmyristate acetate and phorbol diacetate, analogues of TPA) do not show a response. The comparative abilities of these substances to produce O_2^- are also of interest. Fujiki *et al.* have looked at various indicators of tumor promotion for these compounds, including an assay for ornithine decarboxylase (ODCase), an assay of adhesion of HL-60 cells, and an inhibition of terminal differentiation of Friend erythroleukemia cells.[6] Teleocidin, TPA, and debromoaplysiatoxin seemed to result in a greater production of ODCase over time than did lyngbyatoxin. This relationship appears to be true for production of superoxide anion as well. Debromoaplysiatoxin showed far fewer cellular effects (e.g., tests of induction of cell adhesion of HL-60 cells and inhibition of terminal differentiation) than did teleocidin, lyngbyatoxin A, or TPA. Our prelimininary studies do not show such a relationship for O_2^- production.

Studies were also done to see the effects of various inhibitors on these materials, inhibitors such as antipain, retinoic acid, and the Bowman-Birk protease inhibitor isolated from soybeans (TABLE 1). Antipain at 60 mM concentration, retinoic acid at 0.1 mg/ml, and Bowman-Birk at 4 mg/ml showed distinct inhibition of the O_2^- response in cells over controls. For the inhibition studies, all compounds were incubated with 1.85×10^6 cells/ml for 10 minutes at 37°C. The inhibitory compounds were preincubated with the cells for 15 minutes at 37°C before the tumor promoters were added. The dose-response curves were carried out on the same batch of cells, incubating the tumor promoters with 1.72×10^6 cells/ml also for 10 minutes at 37°C.

All of these compounds caused production of O_2^- in human PMNs. A further analysis of tumor-promoter-related biochemical parameters will be necessary to determine if the relative rates of O_2^- production are correlated to the compound's potency as a promoter.

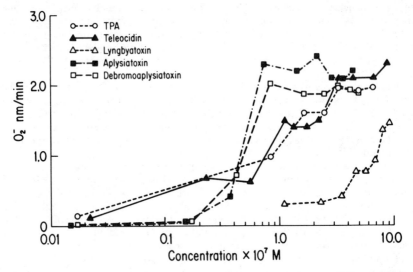

FIGURE 1. Superoxide anion production in polymorphonuclear leukocytes. Cell concentration, 1.7×10^6 cells/ml; cytochrome c, 33 μM. All compounds incubated with cells 10 minutes at 37°C.

Formisano *et al.*: Superoxide Response 431

TABLE 1

EFFECT OF BOWMAN-BIRK SOYBEAN INHIBITOR (BBSI), ANTIPAIN, AND RETINOIC ACID ON
SUPEROXIDE ANION IN POLYMORPHONUCLEAR LEUKOCYTES STIMULATED BY VARIOUS
TUMOR PROMOTERS (EXPRESSED AS PERCENT OF CONTROL)

Promoter	Concentration (M × 10⁻⁷)	BBSI (160 µM)	Antipain (60 mM)	*cis*-Retinoic Acid (0.1 ng/ml)
TPA	1.7	44.1*	18.6	37.7
Teleocidin	2.2	44.1*	22.5	44.3
Debromoaplysiatoxin	0.8	14.7*	19.0	52.0
Aplysiatoxin	0.7	30.8*	18.8	48.0
Lyngbyatoxin	4.6	24.4*	11.3*	58.1*

*Denotes a separate set of experiments, using 1.43 × 10⁶ cells/ml. Remainder of results obtained with 1.85 × 10⁶ cells/ml. All compounds incubated with cells for 10 minutes at 37°C. Inhibitors preincubated with cells for 15 minutes at 37°C.

REFERENCES

1. SLAGA, T. J., A. SIVAK & R. K. BOUTWELL, Eds. 1978. Carcinogenesis—A Comprehensive Survey. **2.** Raven Press. New York, N.Y.
2. SUZUKI, Y. & R. I. LEHRER. 1980. NAD(P)H oxidase activity in human neutrophils stimulated by phorbol myristate acetate. J. Clin. Invest. **66:** 1409–1418.
3. GOLDSTEIN, B. D., G. WITZ, M. AMORUSO, D. S. STONE & W. TROLL. 1981. Stimulation of human polymorphonuclear leukocyte superoxide anion radical production by tumor promoters. Cancer Lett. **11:** 257–262.
4. SLAGA, T. J., A. J. P. KLEIN-SZANTO, L. L. TRIPLETT, L. P. YOTTI & J. E. TROSKO. 1981. Skin tumor–promoting activity of benzoyl peroxide, a widely used free radical–generating compound. Science **213:** 1023–1025.
5. CARDELLINA, J. H., II, F. J. MARNER & R. E. MOORE. 1979. Seaweed dermatitis: structure of lyngbyatoxin A. Science **204:** 193–195.
6. FUJIKI, H., M. MORI, M. NAKAYASU, M. TERADA, T. SUGIMURA & R. E. MOORE. 1981. Indole alkaloids: dihydroteleocidin B, teleocidin, and lyngbyatoxin A as members of a new class of tumor promoters. Proc. Nat. Acad. Sci. USA **78:** 3872–3876.

DIFFERENTIAL IMMUNOMODULATORY EFFECTS OF ANTIBIOTICS ON CELLULAR IMMUNE REACTIVITY: AN INDICATOR OF POTENTIAL ANTIBIOTIC TOXICITY*

G. B. Wilson, J. F. Metcalf, E. Floyd, J. E. Smalls, C. J. Pickett, and H. H. Fudenberg

Department of Basic and Clinical Immunology and Microbiology
and Department of Medicine
Medical University of South Carolina
Charleston, South Carolina 29425

Clinical administration of aminoglycoside antibiotics (e.g., amikacin) and some other commonly used antibiotics can promote pronounced alterations in mitogen responsiveness of lymphocytes from some subjects but not from others (due perhaps to individual susceptibility, possibly genetic) as measured by in vitro tests for cell-mediated immunity.[1,2] The effects of clinical administration of amikacin on lymphocytes precede obvious ototoxicity or nephrotoxicity, suggesting that the effects of antibiotics on lymphocyte responsiveness might be used to predict the potential of a patient to express nephrotoxicity or ototoxicity as a consequence of receiving antibiotic therapy.[3] In the present study we evaluated three aminoglycosides (gentamicin, kanamycin, amikacin), tetracycline, and benzylpenicillin for their effects on basal lymphocyte metabolism and mitogen-induced lymphocyte activation and proliferation. Our goal was to define those culture conditions that would maximize one's ability to discriminate the potential immunomodulating effects of different classes of antibiotics and to develop an assay that could be used to predict which patients are prone to develop nephrotoxicity or ototoxicity when treated with an antibiotic at normal therapeutic amounts.

MATERIALS AND METHODS

Lymphocyte responsiveness to concanavalin A (con A), phytohemagglutinin (PHA), or pokeweed mitogen (PWM) with and without antibiotics added and basal lymphocyte metabolism (no mitogen added) with and without the addition of an antibiotic were determined using a microtiter system[1] (FIGURES 1 and 2). All antibiotics were tested over a wide range of concentrations, including concentrations equivalent to the mean blood level of each antibiotic obtained during therapy (FIGURES 1 and 2). The concentration of each mitogen was varied to determine antibiotic effects on lymphocyte responsiveness to mitogen when used at maximal, submaximal, and supramaximal stimulatory concentrations (FIGURE 1). Mitogens were added to mononuclear leukocytes (MNLs) either concurrently with each antibiotic or at appropriate time intervals after the addition of each antibiotic (results reported here are for 96 hours) to simulate the cells' environment in situ during treatment (FIGURES 1 and 2).

*Publication number 577 from the Department of Basic and Clinical Immunology and Microbiology.

0077-8923/83/0407-0432 $01.75/0 © 1983, NYAS

FIGURE 1. Effects of antibiotics on MNL responsiveness to con A. MNLs (10^5 per well) were cultured in wells of microtiter plates (flat bottom, Costar, Cambridge, Mass.) in 200 μl RPMI 1640 medium supplemented with 20% human AB serum with or without the addition of con A or each antibiotic at the final concentrations (per ml) shown on the abscissa. Con A was evaluated at submaximal (2.5 μg/ml), maximal (5.0 μg/ml), and supramaximal (10 μg/ml) stimulatory concentrations. The concentrations given for each antibiotic are in μg/ml except for benzylpenicillin (units/ml). The approximate mean blood level achieved for each antibiotic used during clinical treatment is noted in parentheses under its name. Antibiotics were added either simultaneously with mitogen (top) or preincubated with MNLs 96 hours prior to mitogen addition (bottom). Ninety-two hours after mitogen addition, the MNLs were pulsed with 2 μCi ^{3}H-(methyl)-thymidine and harvested using a Mash I semiautomated harvestor 4 hours later. (See FIGURE 2 for further details.)

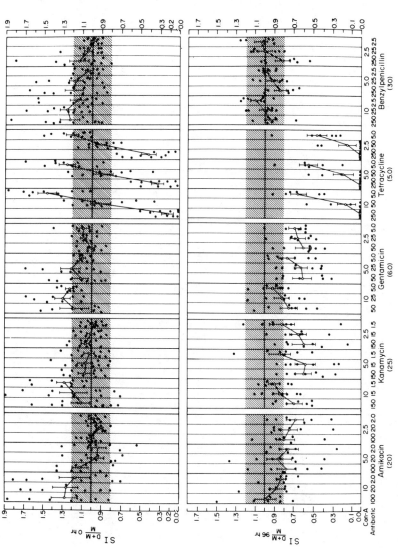

FIGURE 2. Effects of antibiotics on MNL responsiveness to PHA. Culture conditions and explanations of the abscissa are as noted in the legend to FIGURE 1. Results shown are mean stimulation indexes (SI) ± standard error of the mean for MNL cultures from 7 to 10 different normal healthy donors. Each solid point is the mean of four values for each subject's MNLs. The shaded area denotes values that fall in a "no-effect zone" (SI values of 0.80 to 1.20). The SI D+M/M equals the counts per minute (cpm) incorporated in the presence of the antibiotic plus the mitogen divided by the cpm incorporated in the presence of mitogen only. Controls were always run in which MNLs were cultured in medium only or in medium plus antibiotic (without mitogen). Top and bottom panels as in FIGURE 1.

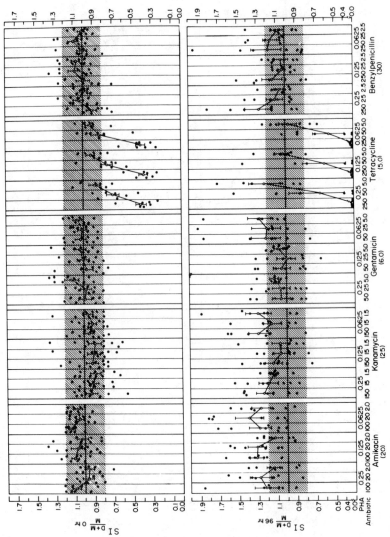

RESULTS AND DISCUSSION

Marked differences were observed for the effects of antibiotics on lymphocyte reactivity (FIGURES 1 and 2). MNLs from individual donors clearly expressed individual susceptibility patterns, suggesting that the assay system employed has prognostic value. If we take a 70% or greater change in immune reactivity to con A or PHA as suggestive of an individual being at risk for developing nephrotoxicity or ototoxicity, our results seem to correlate closely with the actual percentages observed historically for each antibiotic when used therapeutically. In addition, the following general conclusions can be made: (1) The most pronounced effects and best culture conditions for discriminating the effects of different classes of antibiotics were obtained with experimental protocols that simulated the cells' actual environment *in situ* during therapy (FIGURES 1 and 2; 96 hours). (2) The effects observed were dependent on the mitogen used and the concentration of each mitogen. Responsiveness to PWM was not significantly affected by any antibiotic (data not shown). Responses to PHA were potentiated by therapeutic amounts of aminoglycosides but not tetracycline, whereas benzylpenicillin was without effect. In contrast, lymphocyte responsiveness to con A was suppressed to some extent by all antibiotics tested except benzylpenicillin. (3) The differential effects of the various antibiotics tested on PWM-, PHA-, and con A–induced lymphocyte activation indicate that these antibiotics have their most profound effect on suppressor immunocytes. (4) The effects observed were not due to the death of a particular subpopulation of immunocytes, rather they seem to be primarily metabolic since no significant effect on lymphocyte viability (using trypan blue dye exclusion) was observed even when each antibiotic was tested at 5- to 10-fold its optimal therapeutic concentration. (5) Finally, the effect of each antibiotic on basal lymphocyte metabolism mimicked those effects seen for con A but not PWM or PHA. We feel that if the above test is to be used for monitoring during long-term antibiotic therapy or as a predictor of a patient's susceptibility to toxicity, then (1) the MNLs should be incubated with antibiotic prior to testing immune responsiveness and (2) the effects of an antibiotic on both con A and PHA responsiveness should be evaluated.

REFERENCES

1. METCALF, J. F., J. F. JOHN, G. B. WILSON, H. H. FUDENBERG & R. A. HARLEY. 1981. Am. J. Med. **71:** 485–492.
2. HAUSER, W. E., JR. & J. S. REMINGTON. 1982. Am. J. Med. **72:** 711–716.
3. WILSON, G. B., H. H. FUDENBERG, J. F. METCALF, C. J. PICKETT, J. E. SMALLS, E. FLOYD & M. DOPSON. 1980. Clin. Res. **28:** 734A.

A MODEL FOR SELECTING SHORT-TERM TESTS OF CARCINOGENICITY

Lester B. Lave, Gilbert S. Omenn, Kathleen D. Heffernan,
and Glen Dranoff

The Brookings Institution
Washington, D.C. 20036

One goal of toxicologists is the development of efficient and reliable means to identify, characterize, and control potentially harmful chemicals in the environment and workplace. There are two stages in achieving this goal: scientific identification and characterization of potential carcinogens, and private and public decisions to reduce or eliminate exposures.

In choosing which test or tests to use in identifying potential human carcinogens, we must consider the accuracy and cost of each test and combination of tests. Consider a simplified scheme, in which each chemical is assumed to be potentially carcinogenic or not. A good test provides data allowing correct classification, but inevitably there will be some cases in which carcinogens are incorrectly classified as noncarcinogens—false negatives (FN)—and some cases in which noncarcinogens are incorrectly classified as carcinogens—false positives (FP). Misclassification is costly to society.

In addition to the accuracy and cost of each test, we must consider the amount of additional information it provides to a given set of tests. Stepwise statistical techniques can indicate this. Cumulative cost and cumulative accuracy of a set of tests can then be graphed, as they are in FIGURE 1.

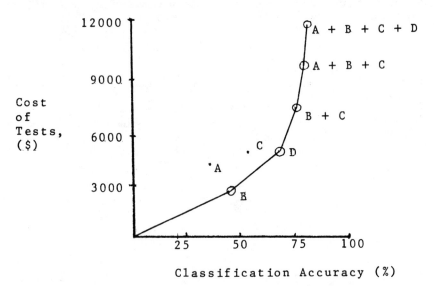

FIGURE 1. Accuracy-cost tradeoffs for short-term tests. (A, B, C, and D are short-term tests.)

436

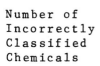
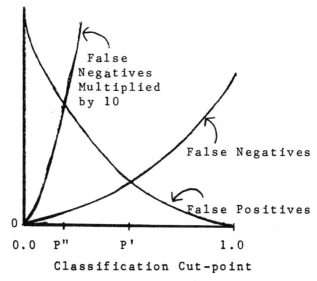

FIGURE 2. Choosing a cut point to minimize social cost of misclassification.

At this point, we have chosen a group of tests that are "best buys." In order to classify chemicals, given the undoubtedly inconsistent results of the various tests in the group, one must use any of several statistical techniques to transform test results into estimates of the probability that each chemical is a carcinogen. Some statistical procedures estimate weights to be applied to each test result in order to compute a probability that each chemical is a carcinogen, P.

A cut point, P', must be selected so that all chemicals with P greater than P' are classified as carcinogens, and chemicals with P lower than P' are classified as noncarcinogens. P' is the cut point that results in the lowest total social cost of misclassification when the procedure outlined above is used on an ideal data set. This data set would consist of test results on chemicals with known toxic properties that represent the entire set of chemicals in common use.

Since the goal is to minimize the social costs of false negatives and false positives, the best cut point will depend on the cost of a false negative relative to the cost of a false positive. If a false negative costs the same as a false positive, social cost will be minimized where FN + FP is smallest (P' in FIGURE 2). If the cost of a false negative were 10 times that of a false positive, P' would be smaller (P″ in FIGURE 2) and more chemicals would be classified as carcinogens.

BIOACTIVATION AND TOXICITY OF POLYCYCLIC AROMATIC HYDROCARBONS IN CULTURED HUMAN AND RAT EPIDERMAL AND ESOPHAGEAL KERATINOCYTES

R. Heimann* and R. H. Rice

*Charles A. Dana Laboratory of Toxicology
Harvard School of Public Health
Boston, Massachusetts 02115*

Two major problems confronting research and testing in toxicology are target-organ specificity and species variation in effects of toxic chemicals. These pose major obstacles in extrapolating studies done in animal models to human risk assessment. Research and testing strategies that incorporate cultured cells from target tissues of human and animal models appear advantageous both scientifically and economically in conjunction with other *in vitro* and *in vivo* systems.

Keratinocytes are the major cell type of the epidermis and the epithelial lining of the esophagus. This cell type provides a frontier of contact with the external environment and thus a major target for toxic chemicals. Recent advances in the culture of keratinocytes provide considerable opportunity for elucidating how these cells respond to environmental agents. Epidermal and esophageal epithelial cells cultured on a feeder layer of irradiated 3T3 cells form stratified colonies with characteristic epithelial morphology.[1,2] The cells retain structural and biochemical markers of differentiation of the epithelium *in vivo*. They synthesize abundant keratin proteins and undergo a coordinated program of terminal differentiation in which small cuboidal cells become enlarged superficial squames exhibiting cross-linked envelopes. The human cells have a finite lifetime upon serial culture (up to 150 generations),[3] whereas the rat cells turn into continuous lines after being passaged for 50–150 generations.

These keratinocytes exhibit markedly different sensitivities to the toxic effects of 3-methylcholanthrene. Colony growth of both epidermal and esophageal keratinocytes from humans and early-passage rat epidermal keratinocytes is greatly inhibited in the presence of 3-methylcholanthrene at concentrations of 0.4–40 μM in the medium. Rat esophageal and late-passage rat epidermal cells appear not to be affected. Consistent with these observations, sensitivity to the effects of 3-methylcholanthrene in reduction of colony-forming efficiency upon one day of exposure also serves as a measure of the toxic response. Except for late-passage rat epidermal cells, the keratinocytes metabolize polycyclic aromatic hydrocarbons at comparable rates and express aryl hydrocarbon hydroxylase activity, which is maximally inducible by similar concentrations of 3-methylcholanthrene. The lack of toxicity of 3-methylcholanthrene toward late-passage epidermal cells can be attributed to the low constitutive rate of biotransformation and lack of inducibility that these cells exhibit. The insensitivity of rat esophageal cells despite substantial metabolic activity appears to reflect intrinsic differences in keratinocytes among epithelia and species and could be due to different types of metabolites, detoxification and repair mechanisms, or intracellular targets. Qualitative differences in the resulting metabolites have not as yet been examined, but at least

---

*Present affiliation: Department of Pharmacology, New York University Medical Center, New York, N.Y. 10016.

438

measurements of total water-soluble metabolites and total metabolites retained intracellularly do not suggest much contrast in degrees of conjugation.

The present results indicate that intrinsic toxic responses are expressed in keratinocytes from different epithelia and in corresponding epithelia from different species. This approach may lead to a more fundamental knowledge of toxic mechanisms required for understanding target-organ specificity and interspecies variability. The usefulness of the cell-culture system presented is emphasized further by our observations that monkey esophageal as well as human and rat genital tract keratinocytes can be cultured and metabolize polycyclic aromatic hydrocarbons.

References

1. RHEINWALD, J. G. & H. GREEN. 1975. Cell **6:** 331–344.
2. HEIMANN, R. & R. H. RICE. J. Cell. Physiol. (Submitted.)
3. RHEINWALD, J. G. 1980. Methods Cell Biol. **21:** 229–254.

SURVIVAL, SISTER CHROMATID EXCHANGE, AND HOST CELL REACTIVATION IN MUTAGEN-HYPERSENSITIVE FIBROBLASTS

G. J. Hook and J. A. Heddle*

Department of Biology
York University
Downsview, Ontario, Canada M3J 1P3

and

Ludwig Institute for Cancer Research
Toronto, Ontario, Canada M4Y 1M4

The pattern of mutagen hypersensitivity in cancer-prone syndromes has been indicative of DNA-repair defects as in the case of Fanconi anemia.[1,2] Bloom syndrome (BS) is an autosomal recessive condition associated with a predisposition to cancer.[3] Recent results have demonstrated that BS fibroblasts are hypersensitive to N-ethyl-N-nitro-N-nitrosoguanidine,[4] and 254 nm ultraviolet light (UVC)[5,6] as measured by sister chromatid exchange (SCE) but not by cell survival. In addition three fibroblast strains from Japanese BS patients have been shown to be hypersensitive to mitomycin-C (MMC) as measured by cell survival.[7] However, we have found only normal sensitivity to MMC-induced SCEs using BS fibroblasts from non-Japanese donors.[8] It was of interest, therefore, to attempt to demonstrate clearly the nature of the hypersensitivity of BS fibroblasts to MMC. We have measured the sensitivity of three BS strains, including a hypersensitive strain from a Japanese donor (86NoKi)[7] and a BS strain of normal sensitivity to MMC-induced SCEs (1492),[8] using both SCEs and cell survival as end points. All BS strains tested showed a hypersensitivity to MMC cell killing (FIGURE 1), but only normal sensitivity as measured by induced SCEs (data not shown). We conclude that the reported difference in sensitivity of BS fibroblasts to MMC is not the result of procedural or strain differences, but is the result of differences in the end point being measured.

A second project involved the examination of an unusual mutagen-hypersensitive fibroblast strain (46BR), which has been isolated from an individual suffering from a disease with similarities to ataxia telangiectasia.[9] 46BR fibroblasts are unique in the variety of mutagens to which they are hypersensitive [e.g., UVC and N-methyl-N-nitro-N-nitrosoguanidine (MNNG)].[9,10] Arlett et al. using bichemical indicators found no evidence of defects in repair of UV- or MNNG-induced DNA damage.[10,11] As a further test of the DNA-repair capabilities of 46BR fibroblasts, we attempted to discover whether defective DNA repair of MNNG-induced damage in 46BR fibroblasts could be found using host cell reactivation (HCR) of MNNG-treated adenovirus 2 (Ad2) as a biological dosimeter. We have used a fluorescent-antibody procedure to measure the HCR of viral antigens.[12] We found that 46BR fibroblasts have a normal ability to support the production of viral proteins of MNNG-treated Ad2 (FIGURE 2). Therefore, we conclude that the hypersensitivity of 46BR fibroblasts to MNNG is not the result of a defect in a repair process involved in HCR of Ad2.

---

*Correspondence to both authors should be addressed to the Ludwig Institute.

0077–8923/83/0407–0440 $01.75/0 © 1983, NYAS

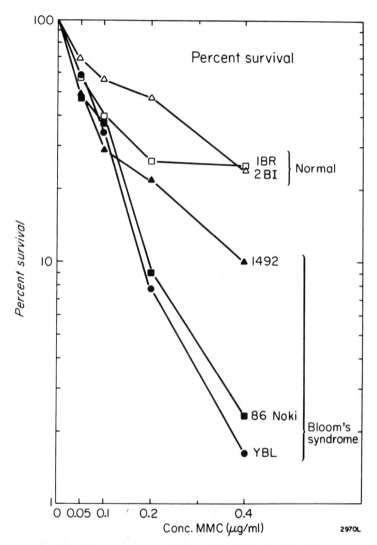

FIGURE 1. Survival of colony-forming ability for BS (bl/bl) (86NoKi, filled squares; YBL, filled circles; and 1492, filled triangles) and normal (+/+) (2BI, open triangles; 1BR, open squares) fibroblasts 14 days after MMC (suppled by Sigma) treatment. Each point represents the mean of at least three experiments. Strains 2BI and 1BR were supplied by Dr. C. Arlett (MRC, Cell Mutation Unit, Brighton, Sussex, United Kingdom), 1492 by the Human Genetic Cell Repository (Camden, N.J.), 86NoKi by Dr. M. Inoue (Kanazawa Medical University, Ishikawa-ken, Japan), and YBL was established in our laboratory.

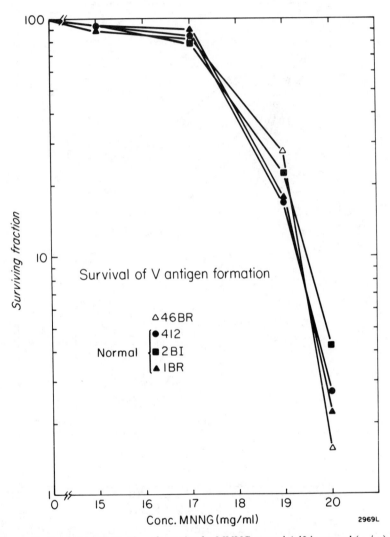

FIGURE 2. Survival of viral antigen formation for MNNG-treated Ad2 in normal (+/+) and 46BR fibroblasts (see text for background). Viral antigen formation was measured using rabbit Ad2 antiserum, then a fluorescein-conjugated antirabbit globulin. Fluorescing cells were counted at three serial dilutions. Data were analyzed as in Krepinsky et al.[6] Each point represents the mean of at least three experiments. 46BR fibroblasts were supplied by Dr. C. Arlett (see FIGURE 1), and 412 were supplied by Dr. M. Buchwald (Department of Genetics, Hospital for Sick Children, Toronto, Ontario, Canada).

REFERENCES

1. FUJIWARA, Y. & M. TATSUMI. 1975. Repair of mitomycin C damage to DNA in mammalian cells and its impairment in Fanconi's anemia cells. Biochem. Biophys. Res. Commun. 66: 592–598.
2. SASAKI, M. S. & A. TONOMURA. 1973. A high susceptibility of Fanconi's anemia to chromosome breakage by DNA crosslinking agents. Cancer Res. 33: 1829–1836.
3. GERMAN, J. 1974. Bloom's syndrome. II. The prototype of human genetic disorders predisposing to chromosome instability and cancer. In Chromosomes and Cancer. J. German, Ed.: 601–617. John Wiley and Sons, Inc. New York, London, Sydney & Toronto.
4. KREPINSKY, A. B., J. GINGERICH & J. A. HEDDLE. 1982. Further evidence for the hypersensitivity of Bloom's syndrome cells to ethylating agents. In Progress in Mutation Research. K. C. Bora, G. R. Douglas & E. R. Nestmann, Eds. 3: 175–178. Elsevier/North-Holland Biomedical Press. Amsterdam, the Netherlands.
5. KURIHARA, T., I. MASAO, H. KAWASHIMA, I. YAGI & H. TAKEBE. 1981. DNA repair in Bloom's syndrome fibroblasts after ultraviolet light irradiation. J. Kanazawa Med. Univ. 6: 40–44.
6. KREPINSKY, A. B., A. J. RAINBOW & J. A. HEDDLE. 1980. Studies on the ultraviolet light sensitivity of Bloom's syndrome fibroblasts. Mutat. Res. 69: 357–368.
7. ISHIZAKI, K., T. YOGI, M. INOUE, O. NIKIEDO & H. TAKEBE. 1981. DNA repair in Bloom's syndrome fibroblasts after UV irradiation or treatment with mitomycin C. Mutat. Res. 80: 213–219.
8. KREPINSKY, A. B., J. A. HEDDLE, A. J. RAINBOW & A. KWOK. 1979. Sensitivity of Bloom's syndrome cells to specific mutagens. Environ. Mutagen. 1: 188–189.
9. WEBSTER, D., C. F. ARLETT, S. A. HARCOURT, I. A. TEO & L. HENDERSON. 1981. A new syndrome of immunodeficiency and increased cellular sensitivity to DNA damaging agents. In Ataxia Telangiectasia—A Cellular and Molecular Link between Cancer, Neuropathology and Immune Deficiency. B. A. Bridges & D. G. Harnden, Eds.: 379–386. John Wiley and Sons Ltd. London, England.
10. TEO, I. A., C. F. ARLETT, S. A. HARCOURT, A. PRIESTLY & B. C. BROUGHTON. 1983. Multiple hypersensitivity to mutagens in a cell strain (46BR) derived from a patient with immuno-deficiencies. Mutat. Res. 107: 371–386.
11. ARLETT, C. F. Personal communication.
12. RAINBOW, A. J. & M. HOWES. 1977. Reduced host cell reactivation of UV irradiated adenovirus in Fanconi's anemia fibroblasts. Biochem. Biophys. Res. Commun. 74: 714–719.

A CRITICAL ANALYSIS OF THE AUTORADIOGRAPHIC
DETECTION OF DNA REPAIR IN PRIMARY CULTURES
OF RAT HEPATOCYTES

M. Lonati-Galligani, P. H. M. Lohman, and F. Berends

Medical Biological Laboratory TNO
2280 AA Rijswijk, the Netherlands

The induction of DNA repair in hepatocytes has been used for the *in vitro* detection of genotoxic agents. The hepatocyte system provides a very suitable model for studies of DNA-repair synthesis because replicative synthesis is minimal. Furthermore, hepatocytes possess a broad capability for metabolizing precarcinogens to reactive intermediates. The study of DNA repair in hepatocytes has the great advantage that the DNA target is present within the same cell where bioactivation takes place.

The most widely used method for the detection of DNA repair is the autoradio-graphic measurement of ^{3}H-thymidine (TdR) incorporation into nuclei of non-S-phase cells (unscheduled DNA synthesis, or UDS). A serious technical problem in the measurement of UDS in hepatocytes is the high and variable cytoplasmic background labeling, which is generally attributed to nonspecific binding of ^{3}H-TdR[1,2] or to incorporation into proteins of its catabolic products. Therefore, correction of nuclear counts for cytoplasmic counts is common practice.[2,3]

In our opinion, the cytoplasmic labeling mainly results from the incorporation of ^{3}H-TdR into mitochondrial DNA (mtDNA). Liver cells are rich in mitochondria, mtDNA has a high turnover, and strong incorporation of ^{3}H-TdR into mtDNA of cells is known to occur, both *in vivo* and *in vitro*. Our results support this conception: (a) in hepatocytes exposed to ^{3}H-TdR, low doses of ethidium bromide (EB)—a specific inhibitor of mtDNA synthesis—caused a strong inhibition of the cytoplasmic labeling (FIGURE 1); (b) benzo[a]pyrene (BaP), too, reduced cytoplasmic labeling, in agreement with the inhibition of mtDNA synthesis reported for its ultimate carcinogenic metabolite in mouse embryo cells (FIGURE 2). Then, two independent DNA targets are present in hepatocytes and nuclear DNA is not necessarily the more sensitive one; certain carcinogens are known to have a preference for mtDNA.

Our results demonstrate that the cytoplasmic labeling may be influenced by the compound tested. Reduction (FIGURE 1, EB; FIGURE 2, BaP) as well as enhancement (FIGURE 2, thiourea, or THU) was observed. With strong UDS inducers such as EB (FIGURE 1), ultraviolet light (UV), methyl methanesulfonate (MMS), and 2-aminofluorene (2AF) (FIGURE 2), the induction of repair synthesis in nuclear DNA is obvious. With THU (FIGURE 2) and BaP (FIGURE 2, BaP I) the effect on cytoplasmic labeling strongly influences the nuclear counts and makes the interpretation of the results debatable.

The reproducibility of this short-term test is affected by the large variations in the functional state of the hepatocytes. When different cell isolates are subjected to an identical treatment with bioactivation-requiring carcinogens, *quantitative* and *qualitative* differences in the induction of DNA repair can be found. The *quantitative* variability is illustrated in FIGURE 1, which shows the response to EB by three different cell preparations. Greater problems arise with genotoxic agents that induce a low amount of DNA-repair synthesis. In this case, *qualitatively* different responses can be obtained with different isolates. As an example, in FIGURE 2 two dose-response curves for BaP are reported.

444

Also, among replicate samples of the same cell isolate, variations were found. Statistical analysis showed the presence of different cell populations within the same isolate.

Furthermore, differences in the amount of grains scored over nuclei of different size were observed. These variations could be eliminated by reporting the grains counted above the nucleus in proportion to the size of the nuclear area scored.

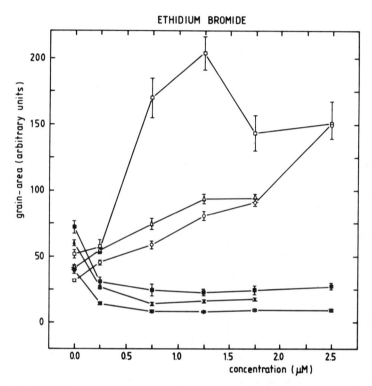

FIGURE 1. UDS induction in hepatocyte cultures treated with EB. Cells were treated with EB and ^{3}H-TdR for 18 hours and then processed for autoradiography. The results obtained with different isolates are reported. For each cell, the silver grains above the nucleus and an area of equal size of the cytoplasm immediately adjacent to the nucleus were scored; and for each experiment, the nuclear (open symbols) and the cytoplasmic (filled symbols) counts were reported. Data points represent the mean grain counts of 100 cells ± standard error of the mean (SEM).

CONCLUSIONS

Assay of repair synthesis of nuclear DNA in hepatocytes is hampered by a high cytoplasmic labeling, which is attributed to incorporation of ^{3}H-TdR into mtDNA. Since mtDNA constitutes a second DNA target for genotoxic agents, cytoplasmic counts should not be subtracted from nuclear counts. For each compound a dose-response curve should be given for both nuclear and cytoplasmic incorporation.

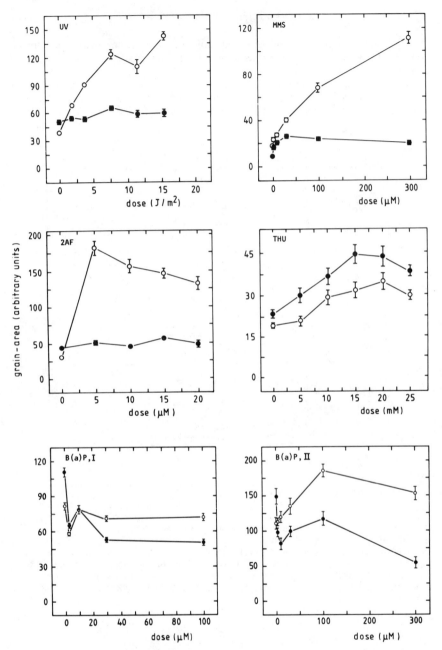

FIGURE 2. UDS in primary hepatocyte cultures treated with different agents. Cells were incubated with the test chemical and ^{3}H-TdR for 18 hours and then processed for autoradiography. In the case of UV, after irradiation the cells were incubated for 18 hours with ^{3}H-TdR. Analysis of tritium incorporation was performed as described in the legend to FIGURE 1. Grain counts are given as mean ± SEM; open symbols, nuclear counts; filled symbols, cytoplasmic counts.

The reproducibility of the system is affected by large variations in the functional state of the isolated cells. If false negatives with weak repair inducers are to be avoided, all test compounds should be studied in conjunction with a potent DNA-repair-inducing analogue.

REFERENCES

1. WILLIAMS, G. M. 1978. Cancer Lett. **4:** 69.
2. PROBST, G. S., R. E. MCMAHON, L. E. HILL, C. Z. THOMPSON, J. K. EPP & S. B. NEAL. 1981. Environ. Mutagen. **3:** 11.
3. WILLIAMS, G. M. 1982. Mutat. Res. **97:** 359.

INHIBITION OF METABOLIC COOPERATION BETWEEN CHINESE HAMSTER V79 CELLS BY TUMOR PROMOTERS AND OTHER CHEMICALS

A. R. Malcolm and L. J. Mills

Biological Effects Division
Environmental Research Laboratory
U.S. Environmental Protection Agency
Narragansett, Rhode Island 02882

E. J. McKenna

Department of Pharmacology and Toxicology
College of Pharmacy
University of Rhode Island
Kingston, Rhode Island 02881

Inhibition of metabolic cooperation (contact feeding) between cells may result in tumor promotion.[1,2] Chemical inhibition of this type of intercellular communication may be measured in selected cell-culture systems by adding test chemical and selective agent (6-thioguanine) simultaneously to cocultivated mutant (HGPRT⁻) and wild-type (HGPRT⁺) cells in reconstructed selection experiments. Experiments employ wild-type cell densities high enough so that most mutants establish contact with wild-type cells and are killed when the latter cells metabolize the selective agent to a toxic form, which is then passed to mutants through gap junctions. The ability of test chemicals to inhibit contact feeding is measured as an increase in mutant cell recovery over background. Using an assay developed with V79 cells,[1,3] we tested several chemicals for their ability to block metabolic cooperation. The potent promoter phorbol myristate acetate[4,5] induced almost complete mutant recovery at 1 ng/ml. Mutant recovery then declined without increased cytotoxicity at 10 and 100 ng/ml (FIGURE 1). The weaker promoter butylated hydroxytoluene[6] induced as much as a twofold recovery over background at 20–25 μg/ml. Phenol, a promoter for mouse skin,[7,8] failed to inhibit metabolic cooperation at concentrations of 5–50 μg/ml, but activity was observed in preliminary tests with three of its metabolites (catechol, quinol, and hydroxyquinol) at low (1–2 μg/ml) doses. Catechol and quinol have been reported to be inactive as promoters for mouse skin.[7,8] The suspected promoter sodium cyclamate[9] inhibited metabolic cooperation in a dose-dependent fashion at high (1–4 mg/ml) doses. The complete carcinogen 3,4-benzopyrene[10] was inactive when tested directly. The trisodium salt of nitrilotriacetic acid (NTA), a carcinogen at high doses in rats,[11,12] and di-(2-ethylhexyl) phthalate (DEHP), similarly carcinogenic in rats and mice,[13] inhibited metabolic cooperation in a dose-related manner. DEHP-induced mutant recovery reached a maximum level at 10 μg/ml, and remained constant when the concentration was increased to 15 and 30 μg/ml (FIGURE 2). Ethanol was inactive at doses as high as 5 mg/ml. Dimethylsulfoxide was inactive at concentrations of 2.5 mg/ml or less, but inhibited metabolic cooperation at 5 mg/ml. The monochlorinated biphenyl 4-chlorobiphenyl also inhibited metabolic cooperation at noncytotoxic doses (10–15 μg/ml). The assay appears promising as a screening test for some tumor promoters, but additional studies are required before any conclusions concerning this use of the assay are justified.

448

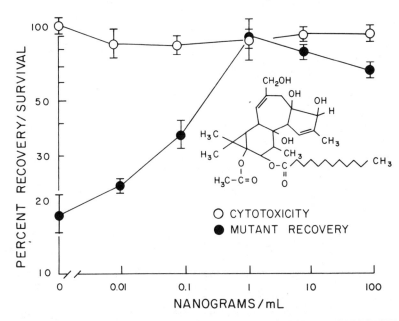

FIGURE 1. Inhibition of metabolic cooperation between V79/HGPRT⁻ and V79/HGPRT⁺ cells by phorbol myristate acetate.

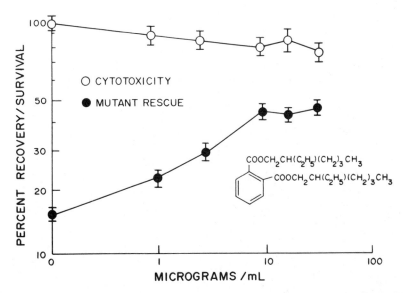

FIGURE 2. Inhibition of metabolic cooperation between V79/HGPRT⁻ and V79/HGPRT⁺ cells by di-(2-ethylhexyl) phthalate.

REFERENCES

1. YOTTI, L. P., C.-C. CHANG & J. E. TROSKO. 1979. Science 206: 1089–1091.
2. TROSKO, J. E., L. P. YOTTI, S. T. WARREN, G. TSUSHIMOTO & C.-C. CHANG. 1982. *In* Carcinogenesis. E. Hecker *et al.*, Eds. 7: 565–585. Raven Press. New York, N.Y.
3. TROSKO, J. E., L. P. YOTTI & C.-C. CHANG. 1981. *In* Short-Term Tests for Chemical Carcinogens. H. F. Stich & R. H. C. San, Eds.: 420–426. Springer-Verlag. New York, N.Y.
4. VAN DUUREN, B. L., L. ORRIS & E. ARROYO. 1963. Nature 200: 1115–1116.
5. HECKER, E., H. KUBINYI & H. BRESCH. 1964. Angew. Chem. Int. Ed. 3: 747–748.
6. PERAINO, C., R. J. M. FRY, E. STAFFELDT & J. P. CHRISTOPHER. 1977. Food Cosmet. Toxicol. 15: 93–96.
7. BOUTWELL, R. K. & D. K. BOSCH. 1959. Cancer Res. 19: 413–424.
8. VAN DUUREN, B. L. 1976. ACS Monogr. 173: 24–51.
9. IARC-WHO. 1980. IARC Monographs on the Evaluation of the Carcinogenic Risk of Chemicals to Humans 22: 95. International Agency for Research on Cancer. Lyon, France.
10. IARC-WHO. 1973. IARC Monographs on the Evaluation of the Carcinogenic Risk of Chemicals to Man 3: 115. International Agency for Research on Cancer. Lyon, France.
11. GOYER, R. A., H. L. FALK, M. HOGAN, D. D. FELDMAN & W. RICHTER. 1981. J. Nat. Cancer Inst. 66: 869–880.
12. NCI. 1977. Carcinogenesis. National Cancer Institute Technical Report Series. Publication No. 6. National Cancer Institute. Bethesda, Md.
13. NTP. 1982. National Toxicology Program Technical Report Series No. 217. National Institute of Environmental Health Sciences. U.S. Department of Health and Human Services. Research Triangle Park, N.C.

DNA REPAIR IN HEPATOCYTE PRIMARY CULTURES DERIVED FROM THE RAT, MOUSE, HAMSTER, GUINEA PIG, AND RABBIT

Carol J. Maslansky* and Gary M. Williams*†

*Department of Pathology
New York Medical College
Valhalla, New York 10595

†The Naylor Dana Institute for Disease Prevention
American Health Foundation
Valhalla, New York 10595

INTRODUCTION

An important application of hepatocyte primary cultures (HPCs) is the study of the biotransformation, cytotoxicity, and genotoxicity of xenobiotics. The liver is the principal organ for biotransformation of xenobiotics, and has been recognized as having the capacity to metabolize all carcinogens that undergo biotransformation, except those that require a prior step of bacterial modification. Variations between species have been demonstrated for a variety of hepatocellular functions as well as in hepatic response to xenobiotics. Species differences in liver metabolism, toxicity, and carcinogenicity of numerous compounds also exist. Because cultured hepatocytes maintain many *in vivo* functions, differences in the biological activity of many exogenous compounds can be demonstrated *in vitro*. To facilitate the study of species variability in biotransformation of and susceptibility to toxicity of xenobiotics, we have developed hepatocyte culture systems derived from mouse, hamster, guinea pig, and rabbit that are comparable to rat hepatocyte cultures. Conditions that optimize recovery, attachment, and survival of hepatocytes from each species have been described, and the extent of preservation of cytochrome P_{450} has also been examined.[1]

Isolated hepatocytes from these additional species have also been utilized in the HPC/DNA-repair assay. Differences in the amount of DNA repair induced in different species have been noted. These differences have been interpreted to reflect variations in the metabolic activation of the compounds involved. It has been suggested, however, that there are major differences in the repair capacity of species with differing longevities. The repair capabilities of HPCs have never been compared. Such a comparison is necessary, however, in order to interpret differences in the induction of DNA repair in response to activation-dependent genotoxins. In this study, we measured the amount of DNA damage in HPCs from the mouse, rat, hamster, guinea pig, and rabbit. In addition, the amount of repair induced in response to 2-acetylaminofluorene (AAF) in HPCs derived from the guinea pig, a species refractory of AAF-induced carcinogenesis *in vivo,* was also measured.

MEASUREMENT OF DNA REPAIR

Freshly isolated hepatocytes (5×10^5) were seeded onto coverslips; after attachment, the cells were washed and exposed to ultraviolet light (UV) or N-acetoxy-

451

acetylaminofluorene (NAcAAF) or AAF, and ^{3}H-thymidine (10 μCi/ml; specific activity, 60–80 Ci/mmol). UV irradiation (254 nm) was emitted by a germicidal lamp with an incident dose rate of 0.5 J/m^2 per second and 5 J/m^2 per second. NAcAAF and AAF were dissolved in dimethylsulfoxide (DMSO) and immediately added to the cultures. NAcAAF was removed from the cultures after 30 minutes, and fresh medium containing ^{3}H-thymidine was added. After 18 hours, the cells were fixed and processed for autoradiography. Nuclear grain counts were determined by subtracting the highest of three adjacent cytoplasmic counts from each nuclear count.

RESULTS

Except at very low fluences, there were no significant differences in the amount of DNA repair induced in response to UV light in HPCs derived from the F-344 rat, B6C3F1 mouse, Syrian hamster, Hartley guinea pig, and New Zealand white rabbit (FIGURE 1). At low fluences, there appears to be a correlation between longevity and the capacity to repair UV-induced DNA damage in hepatocytes isolated from species with differing life spans: rat, 3–4 years; mouse, 2–3 years; hamster, 1–3 years; guinea pig, 2–5 years; rabbit, 5–15 years. There were considerable species differences in the induction of DNA repair in HPCs exposed to the direct-acting genotoxin NAcAAF (FIGURE 2). Variations in the extent of repair as well as in toxicity were noted. These data suggest that there are significant differences between species in the rate and/or extent of detoxication of NAcAAF. Variations in the extent of reutilization of the products of degradation of NAcAAF may also be involved.

Guinea pig hepatocytes were able to metabolize AAF to a reactive species. The mean net nuclear grain count induced in response to 10^{-4} M AAF was 47.4 ± 3.8.

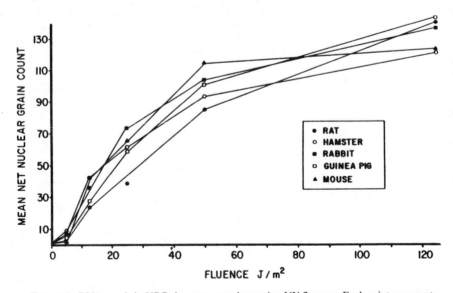

FIGURE 1. DNA repair in HPCs in response to increasing UV fluences. Each point represents the mean of at least three experiments.

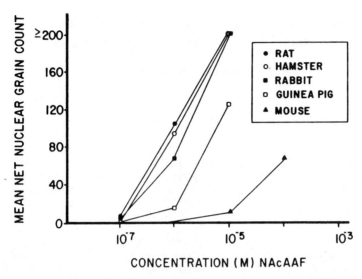

FIGURE 2. DNA repair in HPCs in response to increasing concentrations of NAcAAF. Each point represents the average of at least two experiments. In each species, maximal repair was elicited at the highest nontoxic concentration.

Subsequent studies with the N-hydroxylated derivative of AAF produced net nuclear grain counts of 141.6 ± 5.2 at a concentration of 10^{-5} M. These data demonstrate that guinea pig hepatocytes are capable of activating aromatic amines, and further demonstrate the effectiveness of the HPC/DNA-repair assay in detecting genotoxins even in species in which *in vivo* experiments fail to elicit a tumorigenic response.

REFERENCE

1. MASLANSKY, C. J. & G. M. WILLIAMS. 1982. In Vitro **18**: 683–693.

DNA REPAIR IN PRIMARY HEPATOCYTES AND FIBROBLASTS AFTER HYDRALAZINE TREATMENT

Dieter Müller and Emilio C. Puri

Research Laboratories
CIBA-GEIGY Limited
CH-4002 Basel, Switzerland

The antihypertensive agent hydralazine (HZ) proved weakly mutagenic in the Ames test[1-3] and has been reported to increase the incidence of lung tumors in the mouse.[4] Williams *et al.* and McQueen *et al.* investigated the drug using the DNA-repair test (RT) on hepatocytes (HPC) and found that it induced repair processes.[3,5] In view of the usefulness of HZ, possible risks attendant on its use have to be assessed with care.

Hepatocytes were isolated from RAI and Fischer rats[6,7] and cultured in Williams' medium E containing 10% fetal bovine serum (FBS) on coverslips (10^5 cells/ml; 2 ml/compartment). After an attachment period of 1.5–2 hours followed by washing in Hanks balanced salt solution (BSS), the cells were incubated for 18 hours in medium with and without FBS containing HZ (for concentrations see figure), dimethyl-nitrosamine (DMN, 100 mM), or—as a negative control—no substances. At the same time, 6-^{3}H-thymidine (^{3}H-TdR, The Radiochemical Centre, Amersham, England; specific activity 22 Ci/mmol) at 4 or 10 μCi/ml was added. Five-hour incubations were performed after overnight culture in the same medium containing FBS. After fixation (ethanol/acetic acid, 3:1), autoradiograms were prepared on stripping film AR 10 (Kodak) and exposed for six days. Silver-grain counts were made under oil immersion at 100× over the nuclei and in correspondingly large areas of the cytoplasm. The results (silver-grain counts over the nuclei = GPN) are shown in FIGURE 1. After 18 hours incubation with and without FBS, GPN decreased up to the highest concentration of HZ (4 mM) by comparison with the negative control. The same result was observed after 5 hours incubation without FBS. In 5-hour incubations with FBS, HZ could be added in concentrations up to 10 mM. In the positive controls GPN was invariably increased, on the average by a factor of 7.2 in relation to the negative control.

RT was also performed on human fibroblasts[8] CRL 1121 (HF). After culture for 1 day in Dulbecco's minimum essential medium (MEM) with 10% FBS on coverslips (initial density: 3 × 10^4 cells/ml, 1 ml/compartment), the cells were incubated for 5 hours in medium containing FBS [0.082, 0.41, 2.04, and 10.2 mM HZ; 5 μM 4-nitroquinoline 1-oxide (4-NQO); and negative control]. 4-NQO increased GPN by a factor of 49.5 in relation to the negative control. HZ reduced GPN by up to one-third in a concentration-dependent manner.

The possibility that HZ might exert an inhibitory effect on DNA synthesis in HF was tested. Replicative DNA synthesis was examined after 5 hours incubation with 0 and 12 concentrations (0.0005–10.0 mM) HZ and 2 μCi ^{3}H-TdR/ml. After washing (trichloroacetic acid 10%, 3 × 30 minutes, 4°C), radioactivity was measured with a liquid scintillation counter and the protein content determined.[9] The quotient dpm/μg protein was reduced in relation to the control from 0.05 upwards (0.05 mM = 77%; 0.5 mM = 14%; 5.0 mM = 0.014% of control). The effect of HZ on reparative DNA synthesis in HF was determined autoradiographically after ultraviolet (UV) irradia-

0077–8923/83/0407–0454 $01.75/0 © 1983, NYAS

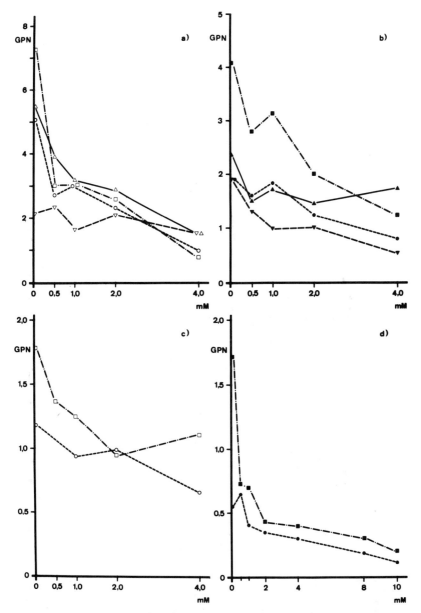

FIGURE 1. Dose response in ³H-TdR incorporation into DNA in hepatocytes exposed to hydralazine. Grains per nucleus. Mean values from 50 nuclear counts. Hepatocyte cultures were exposed continuously to the indicated concentrations of hydralazine and to ³H-TdR. (a) Exposure time 18 hours, medium without serum. (▽) RAI rat, 4 μCi ³H-TdR/ml; (△) RAI rat, 10 μCi/ml; (○) Fischer rat, 4 μCi/ml; (□) Fischer rat, 10 μCi/ml. (b) Exposure time 18 hours, medium with serum. (▼) RAI rat, 4 μCi/ml; (▲) RAI rat, 10 μCi/ml; (●) Fischer rat, 4 μCi/ml; (■) Fischer rat, 10 μCi/ml. (c) Exposure time 5 hours, medium without serum. (○) Fischer rat, 4 μCi/ml; (□) Fischer rat, 10 μCi/ml. (d) Exposure time 5 hours, medium with serum. (●) Fischer rat, 4 μCi/ml; (■) Fischer rat, 10 μCi/ml. Silver-grain counts calculated for an exposure time of 1 day. Positive controls: DMN, 100 mM. Factor in relation to negative controls 2.5 to 12.2, average 7.22.

tion (254 nm, 177 J/m²) and incubation for 5 hours (0.049–6.10 mM HZ; 2 µCi ³H-TdR); 0.049 mM already exerted an inhibitory effect on reparative synthesis.

CONCLUSIONS

1. The concentration-dependent reduction in GPN can be attributed to the inhibitory effect of HZ on DNA synthesis.

2. The discrepancy between the above-mentioned results[3,5] and the present findings is most likely due to differences in the autoradiographic evaluation. The use of a net parameter GPN (i.e., grains per nucleus minus silver-grain counts/equivalent area cytoplasm)[10,11] seems inadmissible for spatial geometric reasons, as the amount of

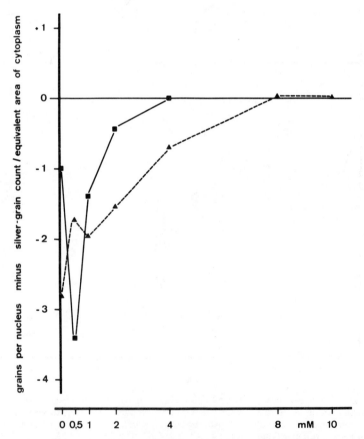

FIGURE 2. Example of the results of calculation of a "net value grains per nucleus" (grains/nucleus minus silver-grain count per equivalent area of cytoplasm). Silver-grain counts calculated for an exposure time of 1 day. (▲) Fischer rat, medium with serum, 4 µCi/ml, exposure time 5 hours. (■) Fischer rat, medium without serum, 10 µCi/ml, exposure time 18 hours.

cytoplasm overlying the nucleus, or in its vicinity, is not the same as in the surrounding area. Moreover, substances may affect the incorporation of ^{3}H-TdR and especially its metabolites into the cytoplasm. Negative values resulting from the net calculation bear out this view (FIGURE 2).

3. It appears desirable to limit the incubation time in RT to about 5 hours, because (a) test substances can affect the breakdown of thymidine[12,13] and hence the amount available; and (b) ^{3}H-TdR itself causes DNA damage, and it is possible that a potentiating effect could occur during longer incubation periods.

REFERENCES

1. TOSK, J., I. SCHMELTZ & D. HOFFMANN. 1979. Mutat. Res. **66:** 247–252.
2. SHAW, C. R., M. A. BUTLER, J.-P. THENOT, K. D. HAEGLE & T. S. MATNEY. 1979. Mutat. Res. **68:** 79–84.
3. WILLIAMS, G. M., G. MAZUE, C. A. McQUEEN & T. SHIMADA. 1980. Science **210:** 329–330.
4. TOTH, B. 1978. J. Nat. Cancer Inst. **61:** 1363–1365.
5. McQUEEN, C. A., C. J. MASLANSKY, I. B. GLOWINSKI, S. B. CRESCENZI, W. W. WEBER & G. M. WILLIAMS. 1982. Proc. Nat. Acad. Sci. USA **79:** 1269–1272.
6. BERRY, M. N. & D. S. FRIEND. 1969. J. Cell Biol. **43:** 506–520.
7. SCHWARZ, L. R., K.-H. SUMMER & M. SCHWENK. 1979. Eur. J. Biochem. **94:** 617–622.
8. SAN, R. H. C. & H. F. STICH. 1975. Int. J. Cancer **16:** 284–291.
9. LOWRY, O. H., N. J. ROSEBROUGH, A. L. FARR & R. J. RANDALL. 1951. J. Biol. Chem. **193:** 265–275.
10. WILLIAMS, G. M. 1976. Cancer Lett. **1:** 231–236.
11. WILLIAMS, G. M. 1978. Cancer Lett. **4:** 69–75.
12. LANG, W., D. MÜLLER & W. MAURER. 1966. Exp. Cell Res. **44:** 645–648.
13. YAGER, J. D., JR. & J. A. MILLER, JR. 1978. Cancer Res. **38:** 4385–4394.

NEUROTOXICITY TESTING: AN *IN VITRO* STRATEGY*

Roland M. Nardone

Department of Biology
The Catholic University of America
Washington, D.C. 20064

The expanding use of *in vitro* systems for carcinogenesis, mutagenesis, and general toxicity testing attests to their potential and real value and ever-increasing role in monitoring the quality of our environment. Expectations associated with the use of *in vitro* systems include benefits in cost, time, and understanding.[1]

Despite the fact that mammalian cell cultures have been used for toxicity studies for about two decades, the field is still plagued by a number of inadequacies. These include, until 1975, a lack of any systematic studies aimed at the standardization and validation of *in vitro* test systems, and a failure to exploit the opportunity presented by the dramatic advances in tissue-culture techniques and somatic cell genetics made during the past decade. These advances include some that offer great promise for the attainment of an expanded repertoire of cell types suitable for toxicity testing in general and for the testing of target-specific substances, such as neurotoxic agents, that may be of special interest to those responsible for monitoring the environment.

Among the significant advances in neuronal cell cultures have been those that have involved the culture of continuously proliferating neuroblastoma and glioma cells, which, *in vitro*, continue to manifest their differentiated properties. These include for neuroblastoma cells neurite formation, action-potential generation, synapse formation, and specialized neurotransmission-related biochemical properties. Glioma cells in culture also show differentiated properties including the production of nervous-system-specific proteins and inducibility to catecholamine and cortisol. Primary cell cultures, aggregation cultures of brain cells, and organotypic cultures have been used with some success.

Research in the area of *in vitro* mutagenesis and carcinogenesis has emphasized the need for testing to involve batteries of tests with overlapping and complementary end points.[2] Indeed, that view is a basic underpinning of this research program. Hence, in addition to a variety of cellular targets (embryonic brain cells and organ culture, established cell lines of neuronal and nonneuronal origin, and primary spinal cord cultures, see TABLE 1), a matrix of biochemical, cell-viability, and cell-behavior tests relating to neuronal and nonneuronal function are used to test the xenobiotics for their toxicity (TABLE 2).

The development of a model system for testing xenobiotics requires extensive validation. For such purposes, related and unrelated chemicals whose *in vivo* toxicity ranking is known are employed. Ideally, the specific mode of action of the test chemicals should also be known. Acrylamides and organophosphates satisfy these criteria (at least in part) and have been selected as the test chemicals used in the initial validation. The results obtained to date are very encouraging. The *in vitro* toxicity ranking of acrylamide, *n*-methylacrylamide, and crotonamide (listed in descending order of toxicity) parallels the *in vivo* ranking of these chemicals.[3] Furthermore, there was similar inhibition of neuron-specific enolase (a putative target for acrylamides)

---

*Supported by Grant 81-0219 from the U.S. Air Force Office of Scientific Research.

TABLE 1

THE COMPONENTS OF THE TEST BATTERY

1. Mouse neuroblastoma NIE-115
2. Neuroblastoma × glioma hybrid cell line NG108-15
3. Chick brain cortex organ culture
4. Chick brain cell aggregate
5. Mouse primary spinal cord culture

but not of total enolase. Studies with organophosphates and the neuroblastoma cell line NIE-115 have resulted in the development of an *in vitro* alternative to the hen brain assay for neurotoxic esterase.[4]

REFERENCES

1. NARDONE, R. M. 1980. The interface of toxicology and tissue culture and reflections on the carnation test. Toxicology **17**: 106-111.
2. NARDONE, R. M. 1981. Ocular toxicity testing; an *in vitro* strategy. *In* Proceedings of the Symposium on Trends in Bioassay Methodology: In Vivo, In Vitro and Mathematical Approaches: 163–175. NIH Publication No. 82-2382. National Institutes of Health. Bethesda, Md.
3. KRAUSE, D. & R. M. NARDONE. 1982. Effect of acrylamide on acetylcholinesterase and neuron specific enolase activity in mouse neuroblastoma cells. In Vitro **18**: 325.
4. FEDALEI, A. & R. M. NARDONE. An *in vitro* alternative for testing the effect of organophosphates on neurotoxic esterase activity. *In* Product Safety Evaluation, Development of New Methodological Approaches. A. M. Goldberg, Ed. Mary Ann Liebert Publ., Inc. New York, N.Y. (In press.)

TABLE 2

NEURONAL AND NONNEURONAL END POINTS USED IN THE TEST BATTERY

The Neuronal End Points
1. Neuron-specific enolase activity
2. Neurotoxic esterase activity
3. Acetylcholine esterase activity and localization
4. Choline acetyltransferase
5. Tyrosine hydroxylase
6. Microtubular organization in neurites
7. Vesicles in neurites
8. Brain aggregate topography and neuronal and glial cell distribution

The Nonneuronal End Points
1. Cell viability
2. Plating efficiency
3. Cell proliferation
4. Macromolecular content and synthesis rates (DNA, RNA, and protein)
5. ATP levels
6. Total enolase activity
7. Total esterase activity

INDUCTION OF DNA-STRAND BREAKS IN CULTURED HUMAN FIBROBLASTS BY REACTIVE METABOLITES

Magnus Nordenskjöld* and Peter Moldéus†

*Department of Clinical Genetics
†Department of Forensic Medicine
Karolinska Institute
S-104 01 Stockholm, Sweden

Chemically induced DNA damage may give rise to strand breaks in cellular DNA, either through facilitated hydrolysis or through enzymatic strand scission involved in DNA repair. DNA breaks are usually formed rapidly after the introduction of DNA damage. Analysis of DNA-strand breaks by DNA unwinding in alkali followed by hydroxyl apatite chromatography is therefore a useful method to elucidate the mechanisms involved in the metabolic activation of genotoxic agents. This is shown in studies with phenacetin, phenacetin metabolites, and dopamine.

Penacetin is a component in several analgesic and antipyretic formulations and has been reported to cause severe kidney damage and an increased incidence of urinary tract cancer after long-term treatment. In contrast, phenacetin and its major metabolites have been found negative in the Ames test in the presence of liver microsomes. We therefore considered the possibility that phenacetin or one of its metabolites may undergo organ-specific metabolic activation in the urinary tract. One such mechanism may be cooxygenation catalyzed by prostaglandin synthetase, an enzyme activity that is present in high levels in the kidney but not in the liver.

Cultured human skin fibroblasts were incubated with ram seminal vesicle microsomes, containing high concentrations of prostaglandin synthetase (PGS), and the induction of DNA-strand breaks in the fibroblasts was detected. In these experiments p-phenetidine, a primary phenacetin metabolite, caused DNA-strand breaks (TABLE

TABLE 1

EFFECT OF PHENACETIN AND PHENACETIN METABOLITES ON THE INDUCTION OF DNA BREAKS IN CULTURED HUMAN FIBROBLASTS*

	+AA	−AA	+AA +IM (0.1 mM)	+AA +AC (1.0 mM)
Solvent control (dimethyl sulfoxide)	0 ± 0.6	0 ± 1.8	0 ± 1.2	0 ± 1.0
Phenacetin (200 μM)	−4.3 ± 1.6	—	—	—
Paracetamol (50 μM)	−2.7 ± 1.8	—	—	—
N-Hydroxyphenacetin (50 μM)	−3.0 ± 1.8	—	—	—
2-Hydroxyphenacetin (50 μM)	0.0 ± 1.3	—	—	—
p-Phenetidine (50 μM)	21.7 ± 2.4†	−4.4 ± 2.3	2.1 ± 1.4	3.7 ± 1.6
p-Aminophenol (50 μM)	19.4 ± 4.4†	1.4 ± 2.8	−0.6 ± 0.5	0.1 ± 1.5

*Means ± standard errors are given. Background values of 18.7 ± 1.2 to 24.3 ± 0.6 were subtracted. The concentration of ram seminal vesicle microsomes was 1.0 mg/ml and of arachidonic acid, 0.1 mM. (Data from Reference 1.) Abbreviations: AA, arachidonic acid; IM, indomethacin; AC, acetylsalicylic acid.
†p < 0.001.

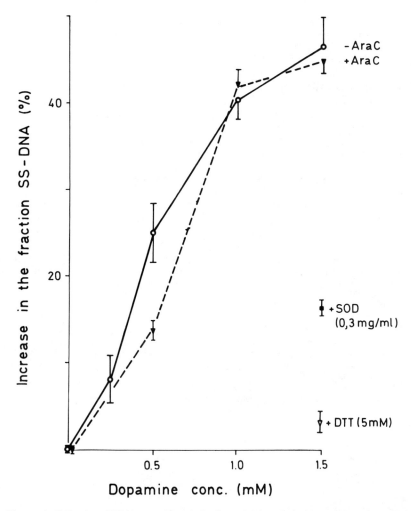

FIGURE 1. Induction of DNA-strand breaks by dopamine in cultured human fibroblasts. Cells were incubated with dopamine for 30 minutes at 37°C in the absence or presence of 100 μM AraC, as indicated. DNA-strand breaks are given as the increase in the fraction of DNA eluted from hydroxyl apatite in single-strand form. Mean values and standard error of four different determinations are given. (Data from Reference 2.)

1). A similar effect was also found in cells exposed to the possible secondary phenacetin metabolite p-aminophenol (TABLE 1). No strand breaks were detected when the prostaglandin synthetase activity was inhibited. Phenacetin, paracetamol, and other primary phenacetin metabolites caused no detectable DNA-strand breakage in these experiments (TABLE 1). These results thus indicate that PGS-catalyzed activation of p-phenetidine may be of relevance for the adverse effects of phenacetin *in vivo*.

Dopamine, an endogenous catecholamine, was found to cause DNA-strand breaks in cultured human fibroblasts (FIGURE 1). However, when the reducing agent dithiothreitol was added to the incubations, no strand breaks were detected. In addition, when superoxide dismutase was added to the incubation, a decrease in the number of strand breaks was observed (FIGURE 1). Furthermore, no increase in the number of dopamine-induced DNA-strand breaks was found when fibroblasts were incubated with the nucleoside analogue AraC (FIGURE 1). This inhibitor of long-patch DNA-repair replication causes a drastic increase in ultraviolet-light- and benzo[a]pyrene-induced DNA-strand breaks.

In cell-free experiments, radioactive dopamine bound to calf thymus DNA *in vitro*, as determined by ethanol extractions and CsCl gradient centrifugations. It was also shown, using CsCl-ethidiumbromide gradient centrifugation, that dopamine caused nonenzymatic DNA-strand scissions at neutral pH of circular double-stranded plasmid DNA.

These results suggest that dopamine may undergo spontaneous oxidation to yield semiquinone, quinone, or reactive oxygen species causing nonenzymatic strand breakage of cellular DNA. Dopamine was also found to be positive in a mouse lymphoma forward-mutation assay, but negative in the Ames test and micronucleus tests in mouse and rat, and did not induce sex-linked recessive lethals in *Drosophila melanogaster* or sister chromatid exchanges in human lymphocytes.

In conclusion, the analysis of DNA-strand breaks in human fibroblasts is a sensitive and useful technique for the study of mechanisms involved in the induction of DNA damage and metabolic activation of certain genotoxic agents.

REFERENCES

1. ANDERSSON, B., M. NORDENSKJÖLD, A. RAHIMTULA & P. MOLDÉUS. 1982. Prostaglandin synthetase catalyzed activation of phenacetin metabolites to genotoxic products. Mol. Pharmacol. **22:** 479–485.
2. MOLDÉUS, P., M. NORDENSKJÖLD, G. BOLCSFOLDI, E. EICHE, U. HAGLUND & B. LAMBERT. Genetic toxicity of dopamine. Mutat. Res. (Submitted.)

A RAPID TECHNIQUE FOR THE QUANTITATION OF DNA-REPAIR SYNTHESIS IN THE HEPATOCYTE/DNA-REPAIR TEST FOR CHEMICAL CARCINOGENS

Felix R. Althaus* and Henry C. Pitot

McArdle Laboratory for Cancer Research
University of Wisconsin
Madison, Wisconsin 53706

This report demonstrates the application of a new procedure ("nuclei procedure") to the quantitation of DNA-repair synthesis in the hepatocyte/DNA-repair test for carcinogen screening. The nuclei procedure involves the biochemical quantitation of methyl-^{3}H-thymidine incorporated into DNA following isolation of nuclei from hepatocytes treated with the agent under study.[1-3]

METHODOLOGICAL CONSIDERATIONS

Three different methods were evaluated to quantify methyl-^{3}H-thymidine incorporated into hepatocellular DNA following treatment of cells with the same dose of ultraviolet light. These data are summarized in TABLE 1.

The highest stimulation of DNA-repair synthesis (19-fold) was measured when nuclei were isolated prior to DNA extraction ("nuclei procedure"). This procedure yielded relatively low control values, i.e., 93 dpm/μg DNA. In contrast, direct extraction of DNA by the same procedure from whole hepatocytes resulted in a much higher "background" specific activity, i.e., 1,700 dpm/μg DNA. Purification of DNA on cesium chloride gradients reduced this activity more than 3-fold. These data demonstrate that most of the methyl-^{3}H-thymidine radioactivity contained in the acid-insoluble fraction from hepatocytes is not incorporated into nuclear DNA but likely represents contamination from nonnuclear sources.[1,4] This contamination can completely mask the repair response of these cells.[5] Therefore, isolation of nuclei prior to DNA extraction yields the lowest background radioactivity and the highest repair response. It should also be noted that in this system there is little thymidine incorporation into hepatocellular DNA due to replicative DNA synthesis, which is very low in these cultures.[6] This low but measurable activity can be further inhibited by hydroxyurea (10 mM), which does not affect DNA-repair synthesis.[4-6]

By combining the nuclei procedure with a double-labeling technique,[1] test results can be obtained within a few hours after exposure of hepatocytes to the test agents.

RESULTS

A summary of results obtained with 41 agents tested in the hepatocyte/DNA-repair assay using the nuclei procedure for determination of DNA-repair synthesis is

*Present affiliation: University of Zurich, Institute of Pharmacology and Biochemistry, Winterthurerstrasse 260, 8057 Zurich, Switzerland.

0077–8923/83/0407–0463 $01.75/0 © 1983, NYAS

seen in TABLE 2. Hepatocyte cultures were treated with increasing concentrations of the agent under study for 18 hours prior to harvesting, which was carried out between 45 and 48 hours after plating the cells. Hydroxyurea (10 mM) was added to the culture medium 1 hour prior to treatment with the test compound, and methyl-^{3}H-thymidine (10 μCi/plate, 42–58 Ci/mmol) was added to the medium immediately after the agent under study. (Further experimental details are described in Reference 1.)

DISCUSSION

The 41 agents we tested using the nuclei procedure comprise a broad spectrum of chemical classes of procarcinogens as well as direct-acting carcinogens, liver-tumor promoters, and noncarcinogenic compounds. All the compounds that have been recognized as carcinogens were positive in the test with the exception of diethylstilbestrol, ethylenethiourea, and thioacetamide. Diethylstilbestrol was also reported nega-

TABLE 1

A COMPARISON OF ANALYTICAL TECHNIQUES FOR THE QUANTIFICATION
OF DNA-REPAIR SYNTHESIS IN CULTURED HEPATOCYTES*

Analytical Procedure	Percent Recovery of DNA	dpm/μg DNA		Stimulation of DNA-Repair Synthesis
		Control	UV 60 J/m^2	
Direct extraction of DNA	100	1700 ± 81	6400 ± 63	3.8
Solubilization of cells— purification of DNA on cesium chloride gradients	58	519 ± 21	2023 ± 25	3.9
Nuclei procedure	84	93 ± 4	1788 ± 19	19.2

*Hepatocytes were irradiated with ultraviolet light (60 J/m^2) and were subsequently maintained in the presence of methyl-^{3}H-thymidine in the culture medium. Eighteen hours after UV treatment, the hepatocytes were harvested and DNA-repair synthesis was measured by three different methods. (For detailed procedures see Reference 1, from which the data were taken.)

tive in the autoradiographic hepatocyte/DNA-repair test.[7] Although there is evidence for the carcinogenicity of ethylenethiourea, the absence of a response in our hepatocyte/DNA-repair test is paralleled by the absence of detectable DNA binding in rats.[8] The lack of a response to thioacetamide could be related to the fact that this agent is a nonmutagenic carcinogen.[9]

The positive results obtained with methapyrilene hydrochloride, 4-acetylaminofluorene, benzo[e]pyrene, pyrene, and α-naphthylamine were somewhat surprising. Methapyrilene has been reported negative in the autoradiographic hepatocyte/DNA-repair assay,[7] whereas we found a very strong dose-dependent stimulation of DNA-repair synthesis. We have recently discussed this discrepancy of results elsewhere.[3] A similar difference of results associated with the use of different methods to quantify DNA-repair synthesis was observed for benzo[e]pyrene as well as pyrene.[1] Our test would indicate that α-naphthylamine has a carcinogenic potential although it has been considered a noncarcinogenic agent. However, the evidence for the noncarcinogenicity of α-naphthylamine is not convincing.[10]

Table 2

SUMMARY OF RESULTS WITH THE HEPATOCYTE DNA-REPAIR ASSAY USING THE
NUCLEI PROCEDURE AND QUANTITATING ^{3}H-THYMIDINE INCORPORATION
BY SCINTILLATION SPECTROMETRY*

Compound	DNA-Repair Synthesis at ED_{max}† (% of control)‡	Interpretation§
N-Acetoxy-2-acetylaminofluorene	1174 ± 290	P
2-Acetylaminofluorene¶	533 ± 47	P
4-Acetylaminofluorene¶	301 ± 86	P
Aflatoxin B_1¶	432 ± 81	P
4-Aminobiphenyl	270 ± 20	P
Aroclor 1254	217 ± 4	P
L-Ascorbic acid, Na salt¶	109 ± 10	N
Benzidine	142 ± 5	P
Benzo[a]pyrene¶	271 ± 40	P
Benzo[e]pyrene	133 ± 3	W
Chloroform¶	106 ± 13	N
Cyclophosphamide¶	267 ± 49	P
Diethylnitrosamine	149 ± 8	P
Diethylstilbestrol¶	96 ± 6	N
7,12-Dimethylbenz[a]anthracene	184 ± 6	P
17β-Estradiol	153 ± 19	P
Ethionine¶	127 ± 8	W
Ethylenethiourea¶	107 ± 14	N
α-Hexachlorocyclohexane	115 ± 6	W?
Methapyrilene hydrochloride	685 ± 49	P
Methionine¶	104 ± 18	N
3-Methylcholanthrene	216 ± 10	P
2-Methyl-14-dimethylaminoazobenzene	315 ± 10	P
Methyl methanesulfonate	521 ± 54	P
α-Naphthylamine¶	211 ± 38	P
β-Naphthylamine¶	252 ± 18	P
4-Nitroquinoline-1-oxide	1384 ± 156	P
Nitrosodiphenylamine¶	193 ± 34	P
N-Nitrosomethylurea¶	827 ± 77	P
β-Pentachlorocyclohexane	150 ± 7	P
Phenobarbital	103 ± 7	N
Proflavine	193 ± 8	P
Pyrene¶	166 ± 45	P
Retinylacetate¶	112 ± 11	N
Safrole	135 ± 12	P
2,3,7,8-Tetrachlorodibenzo-p-dioxin	101 ± 4	N
Thioacetamide	104 ± 30	N
Thiourea	124 ± 12	W
1,1,1-Trichloroethane¶	114 ± 12	N
Urethan (ethyl carbamate)	124 ± 12	W
UV light	1707 ± 109	P

*See text for details.

†Where no stimulation of DNA-repair synthesis was found, the compound was tested up to concentrations that caused cytotoxicity and detachment of hepatocytes.

‡Control (solvent) = 100%.

§The interpretations were based on a paired t-test of results obtained following duplicate testing of each compound in 2 to 19 separate cell preparations. P = positive (p < 0.01), W = weakly positive (p < 0.05), N = negative. (For more statistical information, see Reference 1.)

¶Tested in double-blind manner.

REFERENCES

1. ALTHAUS, F. R., S. D. LAWRENCE, G. L. SATTLER, D. G. LONGFELLOW & H. C. PITOT. 1982. Cancer Res. **42:** 3010–3015.
2. ALTHAUS, F. R., S. D. LAWRENCE, G. L. SATTLER & H. C. PITOT. 1980. Biochem. Biophys. Res. Commun. **95:** 1063–1070.
3. ALTHAUS, F. R., S. D. LAWRENCE, G. L. SATTLER & H. C. PITOT. 1982. Mutat. Res. **103:** 213–218.
4. YAGER, J. D., JR. & J. A. MILLER, JR. 1978. Cancer Res. **38:** 4385–4394.
5. MICHALOPOULOS, G., G. L. SATTLER, L. O'CONNOR & H. C. PITOT. 1978. Cancer Res. **38:** 1866–1871.
6. SIRICA, A. E., G. C. HWANG, G. L. SATTLER & H. C. PITOT. 1980. Cancer Res. **40:** 3259–3267.
7. PROBST, G. S., R. E. MCMAHON, L. E. HILL, C. Z. THOMPSON, J. K. EPP & S. B. NEAL. 1981. Environ. Mutagen. **3:** 11–32.
8. RUDDICK, J. A., D. T. WILLIAMS, L. HIERLIHY & K. S. KHERA. 1976. Teratology **13:** 35–40.
9. MCCANN, J. & B. N. AMES. 1976. Proc. Nat. Acad. Sci. USA **73:** 950–954.
10. RINKUS, S. J. & M. S. LEGATOR. 1979. Cancer Res. **39:** 3289–3318.

LUMINESCENCE METHODS FOR IDENTIFYING DNA-DAMAGING AGENTS

Stanley Scher

Environmental Health Research Laboratory
Richmond Field Station
School of Public Health
University of California
Berkeley, California 94720

To detect agents that damage DNA, most conventional short-term *in vitro* microbial test systems are designed to measure the frequency of mutations from dependence to independence, auxotrophy to prototrophy, sensitivity to resistance, or phage induction in appropriate microorganisms. Alternatively, differential growth or oxygen uptake can be measured in DNA-deficient and DNA-competent cells or revertants. In principle, it should be possible to determine more directly the quantity or rate of production of a newly expressed gene product as a measure of damage to DNA. We describe here recent experience in developing a luminescence method based on the induction of emitted light in luminescence-deficient strains of *Photobacterium phosphoreum*. In this test system, the kinetics of photon emission serve as a convenient measure of genotoxicity.

Emission at 500 nm was monitored with a liquid scintillation counter using ambient temperature photomultiplier tubes. Light emitted from 10^5 cells was integrated over 1.0-minute periods and recorded as counts per minute (cpm). Counting was recorded in the $^{14}C + ^3H$ mode. Results are shown in TABLE 1.

When exposed to sublethal concentrations of chemicals known to damage DNA by base modification, interstrand cross-linking, or intercalation, luminescence-deficient cells displayed dose-dependent increases in photon-emission rates over at least four

TABLE 1

CLASSES OF DNA-DAMAGING AGENTS IDENTIFIED BY LUMINESCENCE METHODS

Chemical Class	Example	Concentration/Dose	Reference
Base modifying			
Alkylation	ethylmethane sulfonate	$10\ \mu g/ml$	1
Deamination	hydroxylamine	$1\ \mu g/ml$	1, 2
Dimerization	ultraviolet light	$400\ ergs/mm^2$	3
Interstrand cross-linking			
Antibiotics	mitomycin C	$5\ \mu g/ml$	3
DNA synthesis inhibitors	naladixic acid	$2.5\ \mu g/ml$	3
Intercalating agents			
Acridine dyes	proflavin	$1\ \mu g/ml$	1, 2, 3, 4
Methylated xanthines	caffeine	$2\ mg/ml$	2, 4, 5
Carcinogens & suspected carcinogens			
Organic solvents	benzene	$0.1\ ml/disc$	4
Other	ethyl carbamate	$0.1\ ml/disc$	4

orders of magnitude. These results are consistent with the hypothesis that emitted light serves as a direct measure of gene expression.

Automated monitoring of photon-emission kinetics in luminescence-deficient strains provides a model system for studying cellular responses to DNA damage and a rationale for developing alternative approaches for identifying agents that pose a genetic risk to human health and the environment. Substitution of a physical method of amplifying newly expressed gene products for biological amplification in conventional short-term microbial assays makes it possible to shorten the expression time from days to hours or minutes, reduce the cost of the procedure by an order of magnitude, and increase the sensitivity of the assay.

REFERENCES

1. ULITZUR, S. *et al.* 1980. Mutat. Res. **74:** 113–124.
2. SCHER, S. & R. A. WECHER. 1982. Environ. Sci. Res. **25:** 607–617.
3. WEISER, I., *et al.* 1981. Mutat. Res. **91:** 443–450.
4. WECHER, R. A. & S. SCHER. 1982. *In* Luminescence Assays. M. Serio & M. Pazzagli, Eds.: 109–113. Raven Press. New York, N.Y.
5. ULITZUR, S. & I. WEISER. 1981. Proc. Nat. Acad. Sci. USA **78:** 3338–3342.

COMPARISON OF *IN VITRO* AND *IN VIVO* IMMUNOTOXICOLOGY ASSAYS*

Melinda J. Tarr, Richard G. Olsen, and Debra L. Jacobs

Department of Veterinary Pathology
The Ohio State University
Columbus, Ohio 43210

Routine toxicology screening programs are rapidly incorporating evaluation of immunotoxic properties of compounds in addition to carcinogenic and general toxic potential. Described here are examples of immunotoxic evaluation of two chemicals for which *in vitro* results correlated with *in vivo* testing. The advantages and limitations of *in vitro* immunoassays are briefly discussed.

In the first set of experiments (see TABLE 1), methylnitrosourea (MNU), a potent resorptive carcinogen, caused a dose-related suppression of the lymphocyte blast transformation (LBT) response of splenocytes from treated Balb/c mice, using the mitogens concanavalin A (con A) and lipopolysaccharide (LPS). Suppression occurred at 50 mg/kg. Similarly, isolated splenocytes from normal mice showed a concentration-related decreased LBT response when preincubated for 30 minutes with 25 µg/ml MNU. This suppression was not due to cytotoxicity. Both *in vivo* and *in vitro* exposure of immunocytes to MNU resulted in a suppressed LBT response.

In a second set of experiments, 1,1-dimethylhydrazine (UDMH) was evaluated in a variety of *in vivo* and *in vitro* immunoassays. UDMH-treated mice (25 mg/kg) showed enhancement of a number of Jerne plaque-forming cells (PFC), decreased con A–induced suppressor activity, no change in LBT response, and a decreased mixed lymphocyte reaction (MLR). *In vitro* UDMH treatment of splenocytes resulted in a low dose (≤ 25 µg/ml) enhancement and high dose (≥ 50 µg/ml) suppression of the LBT responses to LPS, decreased con A–induced suppressor activity, and a decreased MLR. In addition, pretreatment of the adherent responder cell population (which includes suppressor activity) with UDMH resulted in enhancement of the MLR. Both *in vitro* and *in vivo* exposure of immunocytes to UDMH thus resulted in a low dose-related decrease in suppressor activity (increased PFC, decreased con A–induced suppressor activity, decreased adherent cell suppressor activity in the MLR) as well as a higher dose-related suppression of proliferative response (MLR). However, the effects on the LBT response differed between *in vivo* UDMH exposure (no change) and *in vitro* UDMH exposure (decreased response) (see TABLE 2).

In vitro immunotoxicology screening assays would obviously be preferred to *in vivo* assays because of rapidity, low cost, and ease of performance. However, results may vary depending upon which assays are used, as the above experiments have shown. The complex nature and interaction of the immune system within the body may preclude rapid mass screening of chemicals for immunotoxic properties. Other important factors to consider for *in vitro* testing are compound solubility and metabolism. While we have demonstrated two chemicals for which *in vivo* and *in vitro* exposure experiments yielded similar results, we conclude that more understanding of the mechanisms of immunocyte interaction is necessary before one or two immunoassays can be selected for routine chemical screening.

*Supported in part by Contract F49620-79-C-0163 from the Air Force Office of Scientific Research.

469

TABLE 1

CORRELATION OF *In Vivo* AND *In Vitro* LYMPHOCYTE BLAST TRANSFORMATION ASSAY FOR IMMUNOTOXICITY EVALUATION OF METHYLNITROSOUREA

Dose MNU (mg/kg)	LBT Response to LPS: In Vivo Mouse Treatment*	
	Counts per Minute‡	Percent Control Response
Diluent (control)	38594 ± 13920	—
10	34033 ± 11444	88.2
25	42611 ± 16198	110.4
50	22545 ± 7149	58.4 (p < 0.005)§
75	4342 ± 3446	11.3 (p < 0.001)§

Concentration MNU (µg/ml)	LBT Response to LPS: In Vitro Splenocyte Exposure†	
	Counts per Minute‡	Percent Control Response
0 (control)	26325 ± 9862	—
1	25007 ± 8942	95.0
5	22874 ± 7481	86.9
10	23944 ± 8738	91.0
25	17015 ± 6304	64.6 (p < 0.05)¶
50	7244 ± 2640	27.5 (p < 0.001)¶
75	1495 ± 585	5.7 (p < 0.001)¶
100	419 ± 236	1.6 (p < 0.001)¶

*Swiss outbred mice were injected intraperitoneally one time with MNU or diluent, and sacrificed seven days later. Splenocytes were cultured in the lymphocyte blast transformation assay.

†Splenocytes from Swiss outbred mice were incubated with MNU for 30 minutes, then washed and cultured in the lymphocyte blast transformation assay.

‡Means ± standard error of the mean.

§p values determined by Student's t-test.

¶p values determined by paired t-test.

TABLE 2

LACK OF CORRELATION OF *In Vivo* AND *In Vitro* LYMPHOCYTE BLAST TRANSFORMATION ASSAYS
FOR IMMUNOTOXICITY EVALUATION OF 1,1-DIMETHYLHYDRAZINE

	LBT Response to LPS: *In Vivo* Mouse Treatment*			LBT Response to LPS: *In Vitro* Splenocyte Exposure†	
Dose UDMH (mg/kg)	Counts per Minute‡	Percent Control Response	Concentration UDMH (µg/ml)	Counts per Minute‡	Percent Control Response
Diluent (control)	74391 ± 10301	—	0 (control)	81968 ± 10051	
10	70118 ± 10880	94.3 NS§	10	87597 ± 10454	106.9
25	57876 ± 16147	77.8 NS§	25	89260 ± 12622	108.9
50	63005 ± 7031	84.7 NS§	50	76089 ± 15252	92.8
100	67329 ± 9207	90.5 NS§	100	24688 ± 7647	30.1 ($p < 0.001$)¶
150	67999 ± 11735	91.4 NS§	150	11095 ± 6758	13.5 ($p < 0.001$)¶

*Swiss outbred mice were injected intraperitoneally once daily for seven days, then sacrificed. Splenocytes were cultured in the lymphocyte blast transformation assay.

†Splenocytes from Swiss outbred mice were cultured in the lymphocyte blast transformation assay in the presence of UDMH.

‡Means ± standard error of the mean.

§Not significant (Student's t-test).

¶p value determined by paired t-test.

QUANTITATIVE, CLONAL STUDIES OF NEOPLASTIC PROGRESSION OF RAT TRACHEAL EPITHELIAL CELLS

David G. Thomassen, Thomas E. Gray, S. Balakrishna Pai,
Marc J. Mass, Paul Nettesheim, and J. Carl Barrett

Department of Health and Human Services
National Institute of Environmental Health Sciences
Research Triangle Park, North Carolina 27709

We have developed a system for quantitating early carcinogen-induced changes of rat tracheal epithelial (RTE) cells and their subsequent progression to neoplasia.[1-3] Methods are described for growing normal RTE cells in culture and for selecting and quantitating the appearance of enhanced growth (EG) variants. Normal RTE cells were isolated by pronase treatment of Fischer 344 rat tracheas and grown on lethally irradiated 3T3 cells in Ham's F12 medium with 5% fetal bovine serum, insulin (1 μg/ml), and hydrocortisone (0.1 μg/ml). Colony-forming efficiencies were 5–10% for at least 5 passages (>20 population doublings). N-Methyl-N'-nitro-N-nitrosoguanidine (MNNG) treatment resulted in a dose-dependent reduction in colony-forming efficiency. One to seven days after MNNG treatment, EG variants were selected by removing feeder cells and refeeding cultures with complete medium or by replating RTE cells onto dishes without feeders. After four weeks, large colonies of small hyperchromatic cells with increased nuclear-cytoplasmic ratios were observed.

Treatment of RTE cells with MNNG also resulted in a dose-dependent increase in the frequency of EG variants. The maximum transformation frequency in the treated cultures (3.7% per colony-forming cell) was 100-fold greater than in control cultures when selection for EG variants was applied seven days after treatment. The dose response of MNNG-induced transformation was linear with a slope of 1.09 when plotted as the logarithm of cell-transformation frequency versus the logarithm of dose, which is consistent with a one-hit phenomenon. Similar results were obtained with a wide variety of carcinogens including γ radiation, benzo[a]pyrene, benzo[a]pyrene diol epoxide, and nickel salts.

The frequency of carcinogen-induced preneoplastic RTE cells was calculated based on the Poisson distribution (P_0) method from the fraction of dishes without transformed colonies. The distribution of EG variants statistically followed a Poisson distribution and allowed the use of this method for calculation of the transformation frequency. In addition, comparable results were obtained when the transformation frequencies were calculated based on the actual number of variants obtained.

EG variants could be subcultured, and about 50% formed permanent cell lines. Continued propagation of cultures of EG variants resulted in the accumulation of cells capable of growing in semisolid agarose medium. These colonies when isolated had high colony-forming efficiencies in agarose medium (5–40%) and were nontumorigenic in nude mice. After extensive propagation *in vitro*, EG variants developed neoplastic potential and formed squamous cell carcinomas when injected into nude mice.

This work provides a system for characterizing and quantitating the frequency of early carcinogen-induced changes and the development of neoplasia in respiratory epithelial cells.

472

REFERENCES

1. BARRETT, J. C., T. E. GRAY, M. J. MASS & D. G. THOMASSEN. 1982. *In* Short-Term Bioassays in the Analysis of Complex Environmental Mixtures. M. Waters *et al.,* Eds. **3:** 325–340. Plenum Press. New York, N.Y.
2. PAI, S. B., V. E. STEELE & P. NETTESHEIM. 1983. Carcinogenesis. (In press.)
3. MARCHOK, A. C., J. C. RHOTON & P. NETTESHEIM. 1978. Cancer Res. **38:** 2030–2037.

COMPARISON OF SISTER CHROMATID EXCHANGE AND MAMMALIAN CELL MUTAGENESIS AT THE HYPOXANTHINE GUANINE PHOSPHORIBOSYL TRANSFERASE LOCUS IN ADULT RAT LIVER EPITHELIAL CELLS

S. Ved Brat, C. Tong, S. Telang, and G. M. Williams

Naylor Dana Institute for Disease Prevention
American Health Foundation
Valhalla, New York 10595

Adult rat liver epithelial cells offer diverse xenobiotic biotransformation capability and a variety of reliable and relevant genetic end points of biological significance for use in identification of genotoxins.[1] Liver systems are particularly suitable for a comparative study of more than one end point for a variety of activation-dependent chemical compounds that are metabolized by intact cells. Thus, we have studied sister chromatid exchange (SCE) and mutagenesis at the hypoxanthine guanine phosphoribosyl transferase (HGPRT) locus in the adult rat liver epithelial cell line ARL18 in response to a number of polycyclic aromatic hydrocarbons, mycotoxins, and aromatic amines (TABLE 1). Sister chromatid differentiation was obtained by the fluorescence

TABLE 1

INCIDENCE OF 6 THIOGUANINE–RESISTANT MUTANTS (TG^r) AND SISTER CHROMATID EXCHANGES IN ADULT RAT LIVER EPITHELIAL CELLS (ARL18) AFTER EXPOSURE TO A NUMBER OF ACTIVATION-DEPENDENT COMPOUNDS

Chemical	Dose	TG^r/10^6 CFC*	SCEs/Cell
Benzo[a]pyrene	10^{-4}M	610 ± 37	103.3 ± 5.1
	10^{-5}M	335 ± 11	59.8 ± 3.8
	10^{-6}M		16.2 ± 0.9
Benzo[e]pyrene	10^{-3}M		15.8 ± 1.3
	10^{-4}M	9 ± 6	
Pyrene	10^{-3}M		13.2 ± 1.1
	10^{-4}M	5 ± 3	
Dimethyl sulfoxide (DMSO)		6 ± 3	11.2 ± 0.9
			14.2 ± 1.3
7,12 Dimethylbenz[a]anthracene	10^{-4}M	597 ± 23	56.2 ± 3.0
	10^{-5}M	513 ± 11	40.4 ± 3.9
	10^{-6}M	129 ± 16	29.5 ± 1.9
Benz[a]anthracene	10^{-3}M		26.2 ± 1.6
	10^{-4}M	11 ± 5	29.2 ± 1.5
	10^{-5}M	6 ± 4	21.2 ± 2.1
Anthracene	10^{-3}M		15.3 ± 1.1
	10^{-4}M		16.8 ± 0.9
DMSO		4 ± 5	14.9 ± 1.1
			15.8 ± 1.2
N-2-Fluorenylacetamide	10^{-3}M		20.5 ± 1.0
	5 × 10^{-4}M	53 ± 11	
	10^{-4}M	38 ± 8	14.7 ± 1.5
	10^{-5}M	0 ± 0	17.6 ± 1.1

TABLE 1 (*continued*)

Chemical	Dose	TG^r/10^6 CFC*	SCEs/Cell
N-4-Fluorenylacetamide	10^{-3}M		17.8 ± 1.1
	10^{-4}M	15 ± 4	14.6 ± 0.8
DMSO		8 ± 3	15.8 ± 1.1
Aflatoxin B$_1$	5 × 10^{-5}M	130 ± 21	51.1 ± 2.9
	5 × 10^{-6}M	108 ± 31	42.6 ± 2.5
	10^{-6}M	46 ± 8	
	5 × 10^{-7}M		28.5 ± 1.9
Aflatoxin G$_2$	10^{-5}M	8 ± 2	20.3 ± 1.0
	10^{-6}M		15.7 ± 0.8
	10^{-7}M		15.8 ± 1.1
DMSO		7 ± 2	15.9 ± 1.3

*CFC = colony-forming cell.

plus giemsa (FPG) technique, and mutagenesis was studied by selecting 6 thioguanine–resistant mutants.[2,3]

The system appears to be ideally suited for the activation-dependent polycyclic aromatic hydrocarbons (PAH) benzo[*a*]pyrene, 7,12,-dimethylbenz[*a*]anthracene, and aflatoxin B$_1$, which elicit a parallelism between the log dose response in mutagenesis and SCE. The noncarcinogenic PAH benzo[*e*]pyrene, pyrene, and anthracene were all negative in both assays. The weak carcinogen benz[*a*]anthracene elicited a positive response in SCE and negative in mutagenesis. The mycotoxin aflatoxin G$_2$, which has been found to elicit a positive response in the hepatocyte primary culture/DNA-repair test in hamsters,[4] was weakly positive in the SCE assay and negative in mutagenesis. The aromatic amine *N*-2-fluorenylacetamide, however, elicited a weak positive response in mutagenesis as well as SCE, in contrast to *N*-4-fluorenylacetamide, which produced a negative response in both assays.

Thus, the ARL18 cells are very well suited for a comparative study of a variety of activation-dependent genotoxins in SCE and mutagenesis. Also, SCE appears to be a more sensitive assay for studying weak carcinogens.

REFERENCES

1. WILLIAMS, G. M. 1980. The detection of chemical mutagens/carcinogens by DNA repair and mutagenesis in liver cultures. *In* Chemical Mutagens. A. Hollaender, Ed. **6**: 61–79. Plenum Press. New York, N.Y.
2. TONG, C. T., S. VED BRAT & G. M. WILLIAMS. 1982. Sister chromatid exchange induction by polycyclic aromatic hydrocarbons in an intact cell system using adult rat liver epithelial cells. Mutat. Res. **91**: 467–473.
3. TONG, C. T., M. F. LASPIA, S. TELANG & G. M. WILLIAMS. 1981. The use of adult rat liver cultures in the detection of the genotoxicity of various polycyclic aromatic hydrocarbons. Environ. Mutagen. **3**: 477–487.
4. MCQUEEN, C. A., D. KAISER & G. M. WILLIAMS. 1983. The hepatocyte primary culture/ DNA repair assay using mouse or hamster hepatocytes. Environ. Mutagen. **5**: 1–8.

THE GRANULOMA POUCH ASSAY

P. Maier and G. Zbinden

Institute of Toxicology
Swiss Federal Institute of Technology
and University of Zurich
CH-8603 Schwerzenbach, Switzerland

There is a need for tests in mammals in which the mutagenic activity of chemicals can be verified and which provide a basis for risk assessment. In the granuloma pouch assay (GPA), cells are subjected to test compounds in intact adult rats, and genetic alterations are recovered *in vitro* in individual cells at different organization levels of the genome (DNA/gene/chromosome/genome).[6] This concept involves the complexity of xenobiotic metabolism, covers the important relationship between mutagenicity and toxicity *in vivo,* and amplifies or visualizes genetic damage otherwise not detectable.

METHODS

On the back of adult male rats, a subcutaneous 25 ml air pouch is established. The mechanical stretching of the loose connective tissue stimulates growth of mesenchymal cells at the inside wall of the pouch, mainly fibroblastlike cells and endothelial cells. After two days, when the highest number of DNA-synthesizing cells is reached, the test compound is either injected into the pouch, which assures a direct interaction with the dividing target cells, or is administered by systemic routes. Furthermore target cells can be subjected *in vivo* to various experimental procedures before and after treatment (e.g., repair inhibitors, tumor promoters).

Depending on the genotoxic events to be analyzed, the granulation tissue is dissected out between 4 and 48 hours after treatment (FIGURE 1). For single-cell analysis the tissue is dissociated enzymically. The 30 to 70×10^6 recoverable living cells can be subjected to tests similar to those with established cell lines.

So far initial DNA damage (alkali-labile lesions),[2] gene mutations (6-TG^r and Oua^R),[4,5] and sister chromatid exchanges (SCEs)[7] have been analyzed. Furthermore, the GPA can be used for the assessment of carcinogenic effects. Rats are kept alive after treatment, and the development of tumors, including fibrosarcomas in the area of the granuloma pouch, is observed.[8]

RESULTS

Among 12 model mutagens tested in a gene-mutation assay or SCE test (TABLE 1), mainly three pharmacokinetic patterns of the genotoxic metabolites were distinguishable. (1) With MNNG, the genotoxic activity was restricted to the injection site. When the compound was injected directly into the pouch, a dose-dependent increase of 6-TG^r and Oua^R was obtained.[4] However, after intraperitoneal application, we could not induce detectable mutation frequencies in the subcutaneous target cells. (2) A rapid distribution of genotoxic metabolites in the organism was found with MMC, ENU, DR, and CP. These chemicals induced mutations in a dose/tissue-dependent relationship after local as well as systemic application routes. (3) With procarbazine, identical doses caused nearly similar mutation frequencies independent of their application route.[3] This identical response, although the drug could interact directly with the target tissue when applied locally, indicates that the chemical was first absorbed and, after passage through the liver, stable metabolites were formed which were distributed again in the body. The final genotoxic species was then produced in a second intracellular activation/degradation step in extrahepatic tissues.

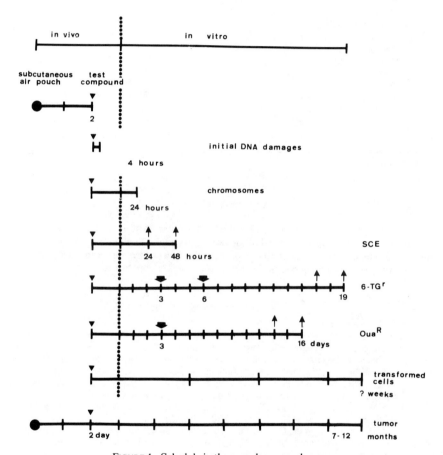

FIGURE 1. Schedule in the granuloma pouch assay.

The metabolic competence of the granulation tissue cells was investigated. Cytochrome P_{448} (7-ethoxy-resorufin de-ethylase) and P_{450}-dependent monooxygenase (aldrin epoxidase) activities were found.[1] Although the activities were much lower than in liver, BP and DMN were transformed to mutagenic metabolites after local injection into the air pouch.

TABLE 1

MODEL MUTAGENS TESTED IN THE GRANULOMA POUCH ASSAY

Test Compound	Recovery of Specific Locus Mutations (6-TG^r) or SCEs	
	Systemic Application	Local Application
MNNG (N-methyl-N'-nitro-N-nitrosoguanidine)	−	+
EMS (ethylmethanesulfonate)	+	+
MMS (methylmethanesulfonate)	+	+
ENU (1-ethyl-1-nitrosourea)	+	+
DR (doxorubicin)	+	+
MIH (procarbazine)	+	+
AF-2 (furylfuramide)	+/−	+
MMC (mitomycin-C)	+	+
BP (Benzo[a]pyrene)	NT*	+
CP (cyclophosphamide)	+	NT*
DMN (dimethylnitrosamine)	+/−	+
AA (9-aminoacridine)	−	−
X-Rays		+

*NT = not tested.

CONCLUSION

The GPA is an vivo mutagenicity and carcinogenicity test in an extrahepatic tissue of rats. Several genetic events including gene mutations are efficiently recoverable in individual cells. Target cells can be exposed directly to test compounds. This concept allows study of the pharmacokinetic profile of genotoxic metabolites.

REFERENCES

1. FREI, K., P. MAIER & G. ZBINDEN. 1982. Cytochrome P-450 dependent monooxygenase activities in extrahepatic cells. (abst.) Experientia 38/6: 754.
2. LEE, I. P. & G. ZBINDEN. 1979. Differential DNA damage induced by chemical mutagens in cells growing in a modified Selye's granuloma pouch. Exp. Cell Biol. 47: 92–106.
3. MAIER, P. & G. ZBINDEN. 1980. Specific locus mutations induced in somatic cells of rats by orally and parenterally administered procarbazine. Science 209: 299–300.
4. MAIER, P., P. MANSER & G. ZBINDEN. 1978. Granuloma pouch assay. I. Induction of ouabain resistance by MNNG in vivo. Mutat. Res. 57: 157–165.
5. MAIER, P., P. MANSER & G. ZBINDEN. 1980. Granuloma pouch assay. II. Induction of 6-thioguanine resistance by MNNG and Benzo[a]pyrene in vivo. Mutat. Res. 77: 165–175.
6. MAIER, P. 1980. The granuloma pouch assay for mutagenicity testing. Arch. Toxicol. 46: 151–157.
7. MAIER, P., K. FREI, B. WEIBEL & G. ZBINDEN. 1982. Granuloma pouch assay. IV. Induction of sister chromatid exchanges in vivo. Mutat. Res. 97: 349–357.
8. ZBINDEN, G., P. MAIER & S. ALDER. 1979. Granuloma pouch assay. III. Enhancement of the carcinogenic effect of MNNG. Arch. Toxicol. 45: 227–232.

SITE SPECIFICITY IN THE INDUCTION OF NUCLEAR ANOMALIES BY CARCINOGENS

Amiram Ronen,* John A. Heddle, and Alessandra M. V. Duncan

Ludwig Institute for Cancer Research
Toronto, Ontario, Canada M4Y 1M4

INTRODUCTION

Nuclear anomalies are induced in the mucosal cells of the mouse colon by various carcinogens; thus, although cells containing them will not reproduce, nuclear anomalies represent a potential assay for colon carcinogens.[1,2] The finding that six of six known colon carcinogens and only one of four carcinogens active at other sites (but not in the colon) could induce nuclear aberrations in the colonic mucosa[2] suggests that the assay might be specific for colon carcinogens. We investigated (a) the possibility of extending this assay to other tissues and (b) the site specificity of the assay for nuclear anomalies, by studying the induction of these anomalies by various carcinogens at five sites along the gastrointestinal tract.

MATERIALS AND METHODS

Female C57BL/6 mice were treated with a single 100-rad dose of gamma rays or with single intraperitoneal (ip) injections of 1,2-dimethylhydrazine (DMH, 20 mg/kg), N-methyl-N-nitrosourea (MNU, 50 mg/kg), or benzo[a]pyrene (BaP, 250 mg/kg). The appearance of anomalous nuclei (karyorrhectic, apoptotic, or micronucleated) in the mucosal cells at five sites—the rectal and cecal ends of the colon, the duodenum, the forestomach, and the esophagus—was measured in histological sections made from tissue samples taken at various times after treatment. Sampling times up to 96 hours for gamma rays, DMH, and MNU were as indicated in FIGURE 1; since the experiment with BaP was preliminary, samples were taken only at 30 and 48 hours.

RESULTS AND DISCUSSION

The frequencies of mucosal cells with nuclear anomalies are summarized in TABLE 1. Gamma rays, as expected, induce nuclear anomalies at all sites, albeit with different efficiencies. The duodenum appears to be twice as sensitive to the induction of nuclear anomalies by gamma rays as is the colon. In contrast, DMH induces nuclear anomalies preferentially in the anal end of the colon, but also induces some in the duodenum and other sites. This agent is known to induce tumors predominantly in the anal end of the colon in mice given ip injections.[3] Similarly, MNU, which is highly specific for the forestomach when administered ip to mice,[4] induces many nuclear anomalies at that site. However, other sites (notably the esophagus) are also highly sensitive to the induction of nuclear anomalies by this agent. This indicates that MNU's site specificity for carcinogenicity does not result solely from a failure of the active form to

*Permanent address: Department of Genetics, The Hebrew University of Jerusalem, Jerusalem, Israel.

479

reach sites at which little or no cancer is induced. With BaP, a potent carcinogen, there is also a discrepancy between the induction of nuclear anomalies and cancer: BaP induces nuclear anomalies in the duodenum but is not known to induce tumors in the gastrointestinal tract when injected intraperitoneally.[5]

These results show that the assay for nuclear anomalies works for several sites in

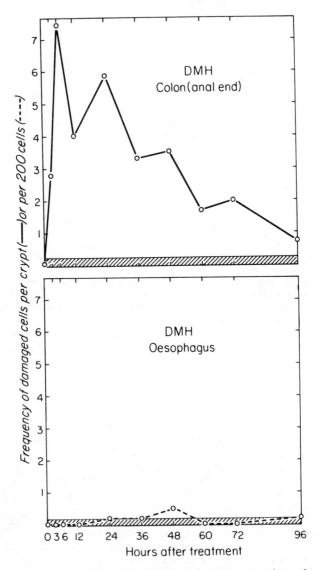

FIGURE 1. Nuclear damage induced by DMH at the most responsive and at the least responsive sites studied. In the anal end of the colon, damage is expressed as the frequency of cells with nuclear anomalies per crypt. In the esophagus, it is expressed as the frequency of cells with nuclear anomalies per 200 mucosal cells. Shaded area: control frequency ± 1 standard error.

TABLE 1

INDEX OF INDUCED NUCLEAR ANOMALIES*

		Index*	Relative Index†
Gamma rays	duodenum	143	1.0
(100 rad)	colon (anal end)	89	1.0
	colon (cecal end)	68	1.0
	forestomach	53	1.0
	esophagus	31	1.0
DMH	colon (anal end)	272	3.1
(20 mg/kg)	duodenum	118	0.8
	colon (cecal end)	65	1.0
	forestomach	27	0.5
	esophagus	(−6)	—
MNU	forestomach	344	6.5
(50 mg/kg)	duodenum	300	2.1
	colon (anal end)	185	2.1
	esophagus	144	4.6
	colon (cecal end)	107	1.6
BaP*	duodenum	266	1.9
(250 mg/kg)	colon (cecal end)	81	1.2
	colon (anal end)	54	0.6
	forestomach	14	0.3
	esophagus	7	0.2

*This index for gamma rays, DMH, and MNU is the area under the corresponding time-course curve (such as in FIGURE 1); it is expressed in terms of cells with anomalous nuclei per crypt (in the colon and duodenum) or per 200 cells (in the forestomach and esophagus) × hours. For BaP, the index is derived from the frequency of nuclear anomalies found at 30 hours and 48 hours after treatment.
†The relative index is the index divided by the gamma ray value for the same tissue.

the gastrointestinal tract and that each site has its own characteristic time response following treatment by any agent (e.g., FIGURE 1).

From the ability of carcinogens to induce nuclear anomalies at sites that fail to show measurable response in carcinogenicity tests, it seems one cannot conclude that an agent will necessarily be a strong carcinogen at a site where it induces nuclear anomalies. This finding is puzzling in light of the successful validation of the assay in the colon.[2] Apart from trivial explanations such as differences between strains of mice used in the cancer studies and our own, this difference may arise because nuclear anomalies include several types of nuclear damage, some of which may be more closely related to the carcinogenic lesions than others. The difference may also reflect the ability of some tissues (e.g., the duodenal mucosa) to repair lesions that would otherwise be carcinogenic. It should be borne in mind that the nuclear anomalies we observe are taken to be a measure of the carcinogenic dose to the tissue but, being cell-lethal events, are not themselves precursors to cancer.

SUMMARY AND CONCLUSIONS

Nuclear anomalies can be measured in the esophagus, forestomach, and duodenum after treatment with carcinogens. MNU and DMH induce nuclear anomalies preferentially at sites where these agents are known to be carcinogens, but their site specificity is not completely reflected by this assay. Nuclear anomalies also occur at sites at which there are few, if any, tumors. Notable among these sites is the duodenum, which shows a strong response to all four carcinogens tested. This suggests that the site specificity of carcinogenesis is not solely the result of the distribution of reactive materials among tissues.

REFERENCES

1. HEDDLE, J. A., D. BLAKEY, A. M. V. DUNCAN, M. T. GOLDBERG, H. NEWMARK, M. J. WARGOVICH & W. R. BRUCE. 1982. Micronuclei and related anomalies as a short-term assay for colon carcinogens. *In* Indicators of Genotoxic Exposure. Banbury Reports. B. A. Bridges, B. E. Butterworth & I. B. Weinstein, Eds. **13:** 367–375. Cold Spring Harbor Laboratory. Cold Spring Harbor, N.Y.
2. WARGOVICH, M. J., M. T. GOLDBERG, H. NEWMARK & W. R. BRUCE. Nuclear aberrations as a short term test for genotoxicity to the colon: evaluation of 19 agents. J. Nat. Cancer Inst. (In press.)
3. IARC. 1974. IARC Monographs on the Evaluation of Carcinogenic Risk to Man **4:** 145–152. International Agency for Research on Cancer. Lyon, France.
4. IARC. 1978. IARC Monographs on the Evaluation of Carcinogenic Risk to Man **17:** 227–255. International Agency for Research on Cancer. Lyon, France.
5. IARC. 1973. IARC Monographs on the Evaluation of Carcinogenic Risk to Man **3:** 91–136. International Agency for Research on Cancer. Lyon, France.

Index of Contributors

483

DATE DUE

8/19/83	D. Thompson — Top		
GAYLORD			PRINTED IN U.S.A.